BASIC ELECTRONICS

BASIC ELECTRONICS

By

Rakesh Kumar Garg
B.Tech. (Electronics), M.Tech. (EC)
Sr. Lecturer & HOD
JMIT, Radaur
Haryana

Ashish Dixit
B.Tech. (Hons.), E & T
M.Tech. (V.L.S.I. Design)
Lecturer, E.C. Deptt.
Amity University
Lucknow Campus, Lucknow
Uttar Pradesh

Pavan Yadav
B.Tech. (EC), M.Tech. (V.L.S.I. Design)
Lecturer in ECE Deptt.
St. Margaret Engg. College, Nimrana
Rajasthan

UNIVERSITY SCIENCE PRESS
(An Imprint of Laxmi Publications Pvt. Ltd.)
An ISO 9001:2015 Company

BENGALURU • CHENNAI • GUWAHATI • HYDERABAD
KOCHI • KOLKATA • LUCKNOW • RANCHI
NEW DELHI

Printed and bound in India

First Edition : 2008, *Reprint* : 2009; *Edition* : 2025

ISBN : 978-81-318-0640-1

Branches

Bengaluru	080-26 75 69 30
Chennai	044-24 34 47 26
Guwahati	0361-254 36 69
Hyderabad	040-27 55 53 83
Kochi	0484-405 13 03
Kolkata	033-40 04 77 79
Lucknow	0522-430 36 13
Ranchi	0651-224 24 64

PUBLISHED IN INDIA BY

UNIVERSITY SCIENCE PRESS

(An Imprint of Laxmi Publications Pvt. Ltd.)

An ISO 9001:2015 Company
113, GOLDEN HOUSE, GURUDWARA ROAD, DARYAGANJ,
NEW DELHI - 110002, INDIA

Corporate Office : B-33, Sector-5, Noida,
Gautambuddha Nagar (U.P.) - 201 301, INDIA
Telephone : +91-120-44 46 900
www.laxmipublications.com
info@laxmipublications.com

To Our Parents With
Immense Love and Regard

CONTENTS

Preface .... (*xi*)

Syllabus .... (*xiii*)

INTRODUCTION .... **1—4**

UNIT—I
(Pages 5—178)

CHAPTER 1 : SEMICONDUCTOR MATERIALS AND JUNCTION DIODE .... **7—78**

1.1. Brief Introduction 7
1.2. Important Parameters Regarding to Semiconductor Materials 8
1.3. Covalent Bond 9
1.4. Electron-Hole Concept 10
1.5. Energy Bands 12
1.6. Doping 16
1.7. Fermi Level 21
1.8. Donor and Acceptor Impurities 23
1.9. The Mass-Action Law 25
1.10. Mobility 27
1.11. Diffusion 33
1.12. Hall Effect 34
1.13. Introduction of Junction Diode 37
1.14. P-N Junction 37
1.15. V-I Characteristics of a PN-Junction Diode 43
1.16. Static and Dynamic Resistance of a PN-Junction Diode 47
1.17. Diode Capacitance 48
1.18. Different Ratings of Diode 50

CHAPTER 2 : DIODE APPLICATIONS .... **79—147**

2.1. Introduction 79
2.2. Rectifiers 79
2.3. Half-Wave Rectifier (HWR) 80

2.4. Full Wave Rectifier (FWR) 89
2.5. FWR with Center Tapped Transformer 89
2.6. Full Wave Bridge Rectifier 95
2.7. Comparison of Rectifier Circuits 98
2.8. Filters 98
2.9. Clipping Circuits 104
2.10. Clamping Circuits 108

CHAPTER 3 : ZENER DIODE AND LED .... 148—178

3.1. Introduction 148
3.2. Breakdown Mechanism 148
3.3. Breakdown Characteristic of a Zener Diode 150
3.4. Zener Resistance 151
3.5. Zener Diode Specification 152
3.6. Zener Diode Application as a Shunt Regulator 152
3.7. Light-Emitting Diode : (Construction and Working Principle) 154

UNIT—II
(Pages 179—328)

CHAPTER 4 : BIPOLAR JUNCTION TRANSISTOR .... 181—244

4.1. Introduction 181
4.2. Transistor 181
4.3. Basic Construction of a Junction Transistor 182
4.4. Transistor Action (Operation of Transistor) 183
4.5. Different Types of Transistor Configuration 185
4.6. Regions of a Transistor 195
4.7. Stabilization 196
4.8. Operating Point 198
4.9. Transistor Biasing 201

CHAPTER 5 : TRANSISTOR AMPLIFIER .... 245—328

5.1. Introduction 245
5.2. Classification of Amplifiers 245
5.3. Graphical Analysis of CE Amplifier 246
5.4. *h*-Parameter Model for Low Frequency 251

5.5. Computation of A_p, A_V, Z_i, Z_o of Single Stage Transistor CE and CC Amplifier Configuration 262
5.6. Introduction to Feedback in Amplifier 265
5.7. Basic Concept of Feedback 266
5.8. Types of Feedback 268
5.9. Principle of Feedback in Amplifiers 275
5.10. Single Loop Feedback Amplifier (Principle) 277
5.11. Introduction to Oscillators 279
5.12. Oscillatory Circuit 282
5.13. Conditions for Oscillations (Bark Hausen Criterion) 284
5.14. RC Oscillators 286
5.15. Hartley and Colpitts Oscillator 294
5.16. Crystal Oscillator 296
5.17. Frequency Stability of Oscillator 299

UNIT—III
(Pages 329—390)

CHAPTER 6 : OPERATIONAL AMPLIFIERS ..... 331—390

6.1. Introduction 331
6.2. Differential Amplifier 331
6.3. Concept of op-amp 333
6.4. Ideal Voltage Transfer Curve 335
6.5. Op-amp IC 335
6.6. Ideal op-amp 336
6.7. Op-amp Applications 338
6.8. Need of Feedback 338
6.9. Inverting Amplifier 339
6.10. Non-Inverting Amplifier 343
6.11. Op-amp Applications 345

UNIT—IV
(Pages 391—451)

CHAPTER 7 : JUNCTION FIELD EFFECT TRANSISTOR ..... 393—432

7.1. Introduction 393
7.2. Classification of Field-Effect Devices 393

7.3. Comparison between BJT and FET 394
7.4. Structure and Operation of JFET or FET 395
7.5. Pinch Off Voltage 403
7.6. Small Signal Model for JFET or FET 405
7.7. JFET Parameters 406
7.8. Biasing of JFET 409
7.9. FET as an Amplifier 411
7.10. FET Applications 416

CHAPTER 8 : METAL OXIDE SEMICONDUCTOR FIELD EFFECT TRANSISTOR (MOSFET) .... 433—443

8.1. Introduction 433
8.2. Depletion-Type MOSFET 433
8.3. Enhancement-Type MOSFET 436
8.4. Circuit Symbols for FET and MOSFET 440
8.5. Comparison of JFET and MOSFET 440

CHAPTER 9 : POWER ELECTRONIC DEVICES .... 444—451

9.1. Introduction 444
9.2. Silicon Controlled Rectifier (SCR) 444
9.3. The Triac 447
9.4. Unijunction Transistor (UJT) 449

EXAMINATION PAPERS .... 452—466

INDEX .... 467—470

PREFACE

We feel immense pleasure in bringing the book **Basic Electronics** to you. The purpose of writing this book is to present the subject matter in a most concise, compact and lucid manner. This book is strictly based on the revised syllabus of B.Tech 1st year, Kurukshetra University, Kurukshetra. This book is also useful for the students studying in other universities.

The book comprises of four chapters. Each chapter contains adequate text supported by self-explanatory figures, and large numbers of solved examples, proper reasons, basic concepts and summary at the end of the chapter.

While preparing manuscript, we constantly kept in mind the requirements of all the students, regarding the latest as well as changing trend of examination. The contents of this book are adequate with simple and easy to understand language. In order to explain the topics in very simple way, analytical and mathematical approach has been provided wherever necessary. Neat and clean diagrams have been used for explanation. We hope the book will meet the requirement of the students for whom it has been designed.

Although every care has been taken in printing the book, yet the very factor of human error can't be ruled out. Any errors, omissions and suggestions for the improvement of this book, brought to our notice, will be thankfully acknowledged and incorporated in next edition.

—AUTHORS

ACKNOWLEDGEMENT

We greatly acknowledge our indebtedness to innumerable students who requested to write a book on "Basic Electronics". They have been a great source of inspiration for us.

We are thankful to our friends Mr. Dinesh Gupta, Mr. Manoj Arora, Mr. Ranjeet Chauhan, Mr. Sandeep Sachdeva, Mr. Vikram Verma, Mr. Ashish Jasuja for the encouragement during the preparation of the book.

We are also thankful to Dr. S.N. Chaturvedi, Ex. Director USIC, Kurukshetra University and presently working as Professor and Head of ECE department, Technical Education & Research Institute, Kurukshetra for his invaluable guidance and encouragement.

We greatly acknowledge our indebtedness to Director Naveen Goswami, Professor H.S. Yadav (Dean), Vinith Chauhan (HOD), F.B. Sharma (Procter), Abhishek Ramniwas, K.M. Rai, Hava Singh, Gajanand Sharma, Shailendra Soni, Bram Dev, Shashank of St. Margaret Engineering College, Additional Director Deepak Kumar Syal, Design Engineers Gurmohan and Shivani of CDAC Mohali, Vice-Chairman Gyan Sharma of JMIT Radaur for their guidance and encouragement.

We duly acknowledge the efforts of our students who inspite of their difficulties and problems, provided the main incentive to write this book.

Any constructive comments, suggestions, criticisms from the students and colleagues will be highly appreciated and greatly acknowledged.

—AUTHORS

SYLLABUS

ELEMENTS OF ELECTRONICS ENGINEERING (EL-101E)

L T P		Sessional :	50 Marks
3 1 –		Exam :	100 Marks
		Total :	150 Marks
		Time :	3 Hrs.

UNIT–I

Semiconductors *p*-type, *n*-type, *pn* junction diodes, *pn* junction as a circuit element, its characteristics, half wave and full wave and bridge type rectifier circuits basic filter circuits. Diode as voltage multiplier, clipper and clamper circuit. Zener diode as a voltage regulator. LED its characteristics construction and applications.

UNIT–II

Characteristics of transistors in different configuration. Concept of d.c. and a.c. load line and operating point selection. Various amplifiers configurations their h-parameter equivalent circuits determination of voltage gain, current gain, input resistance and output resistance and power gain. Concept of feedback in amplifiers, different oscillators circuits (without analysis).

UNIT–III

Differential amplifier and its transfer characteristics. IC Op-Amps, its ideal and practical specifications and measurement of parameters. Op-Amp in different modes as inverting amplifier non-inverting amplifier scale changer, differentiator and integrator.

UNIT–IV

Characteristics of JFET, MOSFET, Various amplifier configurations using FET. Characteristics and Construction of SCR, TRIAC, UJT. Their basic areas of applications.

Note : The Examiners will set eight questions, taking two from each unit. The students are required to attempt five questions in all selecting atleast one from each unit. All questions will carry equal marks.

INTRODUCTION

1. MEANING OF ELECTRONICS

The word "Electronics" has been derived from the greek word 'Electron' plus 'mechanism'. The meaning of this term signifies the study of electron behaviour under different field conditions. Thus electronics is that Branch of Science and Technology which deals with study of electron motion in various vacuum tubes, glass tubes and semiconductor devices. Generally, whole electronics deals with two basic parts:

(1) Physical Electronics and

(2) Electronics Engineering.

Physical Electronics deals with the actual phenomenon occurring within the device during its operation. However, in the part of Electronics, we study how an electronic device actually works. On the other hand, electronics engineering is the application part of core electronics and under this section we deal with successive development. Finally, electronics can be defined as "the field of science and engineering which deals with Electron devices and their utilization".

2. BRIEF REVIEW OF HISTORICAL SIGNIFICANCE AND DEVELOPMENT OF ELECTRONICS

Probably, the era of electronics started when a British Physicist "Thomson" invented the fundamental particle of all the atoms *i.e.*, electron (in 1890). By the very advent of this particle many electron devices like vacuum tubes—diode, triode etc. came into existence. Their operating principle was solely based upon the controlled movement of these electrons. Inventions of fleming (Invented Diode valve in 1904, having capability of unidirectional current flow), Leedeforest (invented triode valve in 1907, having capability of amplifying A.C. signals of relatively lower magnitude) and many more revolutionized the development of Electronics. After that, in early 1920's devices having more than three terminals like tetrode, pentode came into existence and this was the time when first picture tube (Kinescope developed by American's in 1920's) came into existence.

During the second world war the major requirement was to establish long distance communication links, for this very purpose microwave devices like magnetron, klystron, travelling wave tube (TWT) and many more came into existence and as such they facilitated the researchers to invent Radar and other communication systems. Clearly it was the end of tube devices when in 1947, Walter Britain. John Bardeen and Shockley invented the first transistor at Bell Laboratories. General Electric (GE), RCA and Western Electric was producing the transistor commercially in 1951.

The Discrete transistor, for Engineers and Researchers was like a panacia which solved their numerous problems. They started using this transistor as basic building block of all the Electronic circuits. In case of analog electronics, it was used as an amplifier. On the other hand, for digital circuits it served as an electronic switch. After further advancement in Electronic Industry several discrete components were fabricated on a single wafer and very first monolithic Circuits (ICs) came into existence after early 1960's. The invention of IC fabrication technology reduced the size of devices through further miniaturization the packing density of the IC increasing many folds and the speed and accuracy were found to be up to the desired level. Several levels of IC series are given under here:

Series	No. of Components	Years
SSI (Small Scale Integration)	3 to 30	1960–1965
MSI (Medium Scale Integration)	30 to 300	1965–1970
LSI (Large Scale Integration)	300 to 3000	1970–1975
VLSI (Very Large Scale Integration)	more than 3000	late 1975
First Commercial VLSI chip :	64 K RAM	late 1970s
	256 K RAM	early 1980s.

3. THE ADVENT OF TRANSISTORS AS BASIC BUILDING BLOCKS OF SOLID STATE DEVICES AND CIRCUITS

In 1947, the First Germanium transistor was introduced. The name transistor suggests that this device employs transfer of resistance. It means that whatever be the signal at input terminal, it will travel from low resistance side towards high resistance side and all this phenomenon depends upon the concentration profile of the semiconductor material used to form a three layer three terminal device *i.e.,* so called transistor. This three terminal (Emitter, Base and Collector) device can be configured in many ways depending upon the use. At some instant it works as an amplifier (or oscillator) and at other instant it can be made to work as switch. The evolution of transistor as a discrete circuit component reduced size, cost and circuit complexity. At the same time the use of transistor increased reliability, portability and accuracy of the electronic circuits. This is the reason that's why the transistors found tremendous and enormous applications in almost all the fields concerned with electronic industry.

4. COMPARISON BETWEEN ELECTRON TUBES AND SOLID STATE DEVICES

For a beginner it has always been a confusion that why and how the electron tubes got abundant and their place substituted by the solid state (semiconductor) devices. The following table better describes the inherent characteristics and comparative study of the both class of the devices.

S. No.	Particular	Electron Tubes	Solid State Devices
1.	Operating principle	They were based upon the conduction of electron beam from cathode to anode *i.e.,* they were based upon the principle of thermionic emission.	These are the Semiconductor (Si and Ge) based devices their operation solely depend upon the impurity concentration of dopants.
2.	Size	These devices employed bigger size and delicate glass Jackets (valve).	Their size is comparatively very small.
3.	Portability	Because of their size and glass tube they were not easily transportable.	These are very portable and handy.
4.	Temperature dependency	Because of thermionic emission they suffer the problem of excessive heating and require extra circuitary for heat absorption. They require initial heating of cathode by other means.	Although their operation could be interrupted. Because of function temperature but they can sustain a wide range of external temperature.

5.	Power consumption	They require high external field for thermionic emission and plate potential.	They require comparatively few volts for their biasing purpose.
6.	Speed	Their turn-on and turn-off is very slow.	Their speed of operation is critical *i.e.*, they can be used as excellent switches.
7.	Reliability	By the passage of time their response get diminished.	For a particular-span of time (Several years) they remain equally reliable.
8.	Durability	They prove to be destructed easily because of delicate glass tubes.	They are highly durable.
9.	Cost	Their manufacturing cost is very high.	They are relatively cheaper and so can be replaced easily if damaged.

5. VARIOUS SOLID STATE DEVICES AND APPLICATIONS

Now-a-days there is not a single corner in which we can observe the absence of electronic devices and appliances. Life in today's scenario offers many conveniences which utilize the foremost use of Electronics. The progress of a society and a country as a whole depends upon several factors. Among them, the development of Electronics and communication industry has got an indespensable and significant role. Various solid state devices like diode, transistor, oscillator, FET's (Field Effect Transistor). MOSFET's (Metal Oxide Field Effect Transistor) and OP-amp (operational amplifier) find particular uses in respective fields of developments.

The tree representation given in next page shows a brief overview of various applications of electronics.

6. ELECTRONIC COMPONENTS

Usually electronic components may be broadly classified as active and passive components. Passive components are those which are to be connected with devices to perform particular mathematical and logical functions. These components are named as passive because they cannot generate their own signal and require external power to get energized. On the other hand, active components are the discrete electronic devices and they can generate internal signals for the circuit operation in which they are attached. One more difference between these components is that passive components have linear relationship between excitation variable and response (Output) variable, but the graph of active components is not linear.

Any complex electronic circuit may contain thousands of components yet each component will necessarily be one of the following five types.

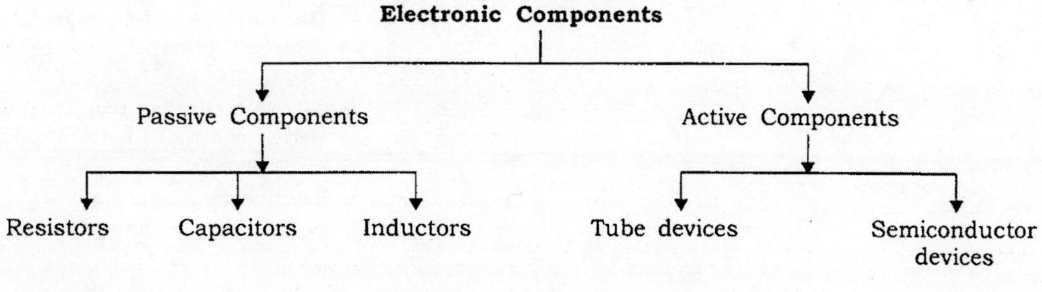

Table 1. Various Applications of Electronics

ELECTRONICS

- Communications and Entertainment
  - Line Communication
    - Telegraphy, Telephony, Telex, Teleprinters
  - Wireless
    - Radio Broadcasting
      - Radio Broadcasting, TV Broadcasting, Fascsimile (Radio Photo), Satellite Communication
  - Audio Systems
    - PA Systems, Stereo Amplifiers, Record Players, Tape Recorders
- Defence
  - Radar, Guided missiles, Coded-Communication
- Industry
  - Automatic Control Systems, Computers
- Medical Sciences
  - X-ray Electron Microscope, Electro Therapy, Magnetic Resonance Imaging (MRI), Electrical Encep-lograph (EEG)
- Instrumentations
  - Precision Measuring Instruments *e.g.* VTVM, CRO, Frequency Counters, pH-meter, Strain Guage, etc.

In order to further classify various components of active class we have to go for following description.

ACTIVE COMPONENTS

- TUBE DEVICES
  - 1. Vacuum tubes (rectifier, detector)
    2. Vacuum triode (amplifier, oscillator)
    3. Vacuum pentode (amplifier, oscillator)
  - 1. Gas tubes (Voltage regulator, non-signs)
    2. Thyratron (controlled rectifier)
- SEMICONDUCTOR DEVICES
  1. Function diode (rectifier, detector switching circuits)
  2. Transistor (BJT amplifier, oscillator)
  3. Field effect transistor (FET) (amplifier oscillator)
  4. Unijunction transistor (UJT) (Power control switching circuits)
  5. Silicon controlled rectifier (SCR) (speed control of motors, power electronics)
  6. Tunnel diode (oscillators)
  7. Zener diode (voltage regulator).

UNIT—I

(Chapters 1, 2 and 3)

- At the end of this unit you will able to learn about the :
- Semiconductors (P-type, N-type)
- PN Junction diodes and its characteristics
- Rectifiers (Half wave, full wave, bridge type)
- Filters
- Clipping, clamping and voltage multipliers
- Breakdown mechanism
- LED and its characteristics.

1

SEMICONDUCTOR MATERIALS AND JUNCTION DIODE

1.1. BRIEF INTRODUCTION

Semiconductors are those materials which have electrical conductivities lying between those of good conductors and insulators. The resistivity of semiconductors varies from 10^{-5} to 10^{+4} ohm-meter as compared to the values ranging from 10^{-8} to 10^{-6} ohm-meter for conductors and from 10^{7} to 10^{8} ohm-meter for insulators. The most important elemental semiconductors such as germanium (Ge) and silicon (Si) which belongs to Group IV of the periodic table and have resistivity of about 0.6 and 1.5×10^{3} ohm-meter respectively. Besides these, there are certain compound semiconductors such as gallium arsenide (GaAs), indium phosphide (InP), cadmium sulphide (Cds) etc. which are formed from the combinations of the elements of group III and V. Another important characteristics of semiconductors is that they have small band gap. One thing that should be always kept in mind that all the semiconductors have negative temperature coefficient of resistance. This is due to the fact that the number of carrier in a semiconductor will increase significantly with temperature, resulting reduces the resistance of the semiconductor.

The study of semiconductors resulted in development of discrete semiconductor devices such as diodes, transistors, field effect transistor, metal-oxide semiconductor field-effect transistor and latter on integrated circuits, microprocessors etc.

Group-IV Materials

The conductivity of semiconductor is considerably greater than that of insulator but considerably less than that of good conductor.

The elemental semiconductors such as germanium (Ge) and silicon (Si) which belongs to Group-IV of the periodic table and have resistivity of about 0.6 and 1.5×10^{3} ohm-meter respectively. The band gap energies for the elements of group IV at OK are as follows :

C (diamond) $\longrightarrow$ 5.51 eV

Ge $\longrightarrow$ 0.75 eV

Si $\longrightarrow$ 1.16 eV

Sn (grey) $\longrightarrow$ 0.08 eV

Pd $\longrightarrow$ $\cong 0$ eV

From the following figure it is clear that, at room temperature (*i.e.*, 0 K) diamond behaves as an insulator, Ge and Si as semiconductors and Sn and Pd as conductors.

The importance of semiconductors is further increased due to the fact that the conductivity and the effective band gaps of these materials can be modified by the introduction of impurities which strongly affect their electronic and optical properties. The process of introduction of impurities in semiconductors in a precisely controlled manner is called *doping.*

1.2. IMPORTANT PARAMETERS REGARDING TO SEMICONDUCTOR MATERIALS

The term semi is normally applied to a range of levels midway between two limits.

- The term conductor is applied to any material that will support a continuous flow of charge when a voltage source of limited magnitude is applied across its terminals.
- An insulator is a material that offers a very low level of conductivity under presence of an applied voltage source.
- However, a semiconductor, therefore, is a material that has a conductivity level somewhere between the extremes of an insulator and a conductor.

Resistivity (ρ)

Resistivity is the property of the material. It is inversely related to the conductivity of the material (*i.e.*, $\rho \propto \frac{1}{\sigma}$). That is, the higher conductivity level, the lower the resistance level. The term resistivity is often used in order to compare the resistance levels of materials. It is measured in ohm-meter. Mathematically, resistivity (ρ) may be defined as :

$$\rho = \frac{RA}{l} \ (\Omega \text{ cm}) \qquad \text{...(1.1)}$$

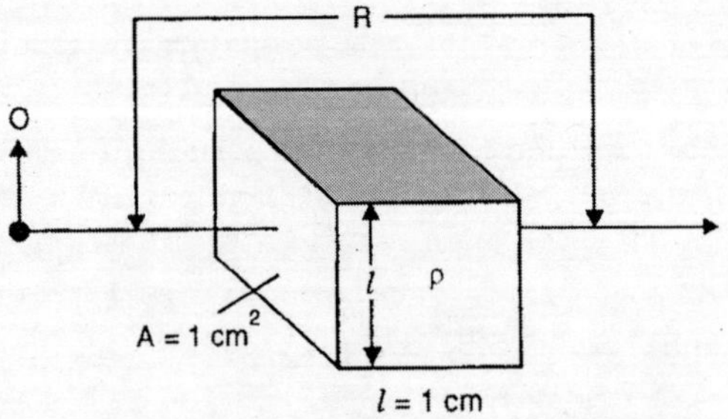

Fig. 1.1. *Defining the units of resistivity.*

From eqn. (1.1),

$$R = \frac{\rho l}{A} \qquad \text{...(1.2)}$$

where, R = Resistance

l = Length of the conductor

A = Area of the cross-section

ρ = Resistivity of the material.

Some resistivity values for conductor, semiconductor and insulator are given below :

• Conductor	$\rho \cong 10^{-6}\ \Omega\text{-cm}$	(copper)
• Semiconductor	$\rho \cong 50\ \Omega\text{-cm}$	(germanium)
	$\rho \cong 50 \times 10^3\ \Omega\text{-cm}$	(silicon)
• Insulator	$\rho \cong 10^{12}\ \Omega\text{-cm}$	(mica)

1.3. COVALENT BOND

Germanium (Ge) and silicon (Si) are the two most important semiconductors used in electronic devices. Some of the unique qualities of Ge and Si are due to their atomic structure. The atoms of both materials form a very definite pattern that is periodic in nature. One complete pattern is called *crystal* and the periodic arrangement of atoms a *lattice*. For Ge and Si the crystal has the three-dimensional diamond structure as shown in Fig. 1.2.

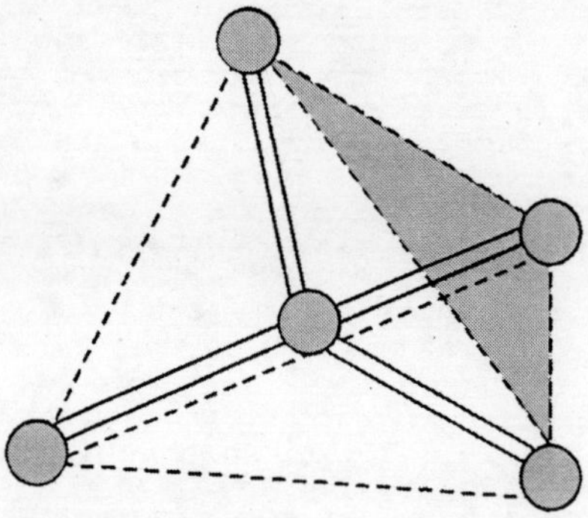

Fig. 1.2. *Ge and Si single-crystal structure.*

Any material composed only of repeating crystal structures of same kind is called a single-crystal structure.

Let us now examine the structure of the atom itself and note how it might affect the electrical characteristics of the material. As we know that any atom is composed of three basic particles : the electron, the proton and the neutron. In the atomic lattice, the neutrons and protons form the nucleus, while the electrons revolve around the nucleus in a fixed orbit. The Bohr models of the two most commonly used semiconductors, germanium and silicon are shown in Fig. 1.3.

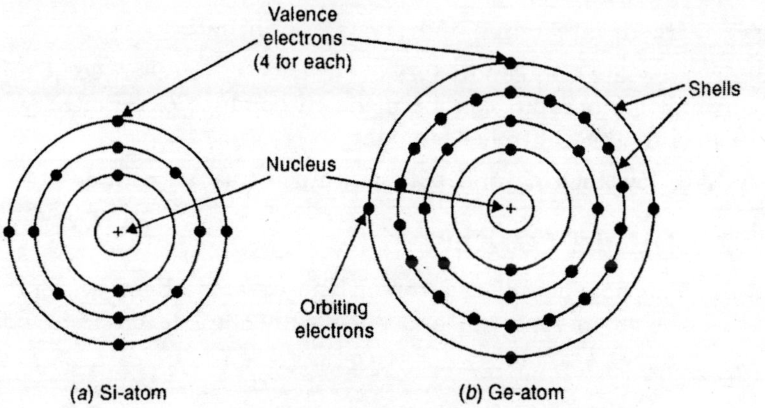

Fig. 1.3. *Atomic structure.*

As indicated by Fig. 1.3 (*a*) silicon atom has 14 orbiting electrons, while germanium atom has 32 orbiting electrons. In each case there are 4 electrons in the outermost (valence) shell. In pure germanium or silicon crystal these 4 valence electrons are bonded to 4 adjoining atoms as shown in Fig. 1.4. Both Ge and Si are referred to as *tetravalent atoms* because they each have four valence electrons.

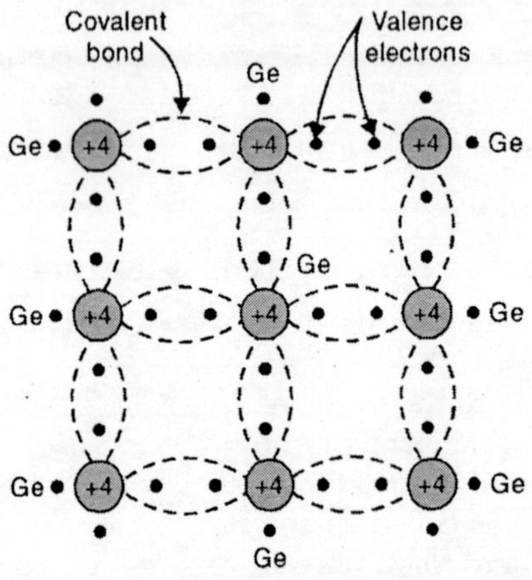

Fig. 1.4. *Crystal structure of germanium, illustrated symbolically in two dimensions.*

A bonding of atoms, strengthened by the sharing of electrons is called covalent bonding.

The binding forces between neighbouring atoms result from the fact that each of valence electrons of a germanium atom is shared by one of its four nearest neighbours. This *electron-pair or covalent bond* is represented in Fig. 1.4 by the two dashed lines which join each atom to each of its neighbours.

A covalent bond is formed by sharing of one pair of electrons between the atoms is called a *single covalent bond.* The covalent bonds are not as strong as ionic bonds. Some of the interesting properties of covalent solids are as follows :

1. They are rigid, directional and crystalline in nature.
2. Usually, they are bad conductors of electricity. However, some of solids such as silicon and germanium can be made to conduct electricity by adding certain impurities in the controlled amount.
3. They have low melting and boiling temperature as compared to ionic solids.

1.4. ELECTRON-HOLE CONCEPT

The Hole. The absence of electron in the covalent bond is represented by the small circle as shown in Fig. 1.5, and such a incomplete covalent bond is called a *hole.*

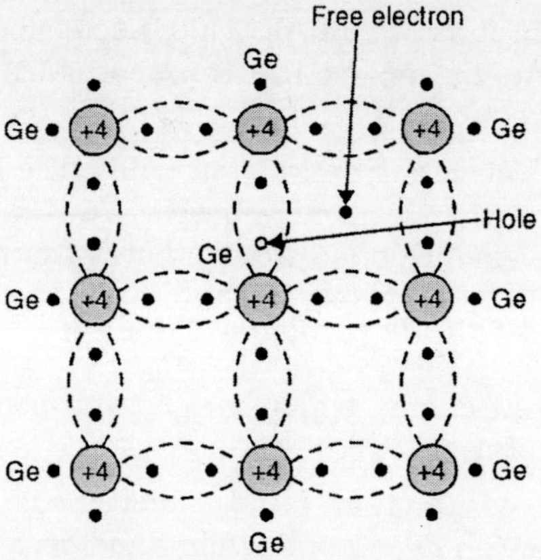

Fig. 1.5. *Germanium crystal with a broken covalent bond.*

The importance of the hole is that it may serve as a carrier of electricity comparable in effectiveness with the free electron.

As we know that at a very low temperature (say 0° K) the ideal structure of Fig. 1.4 is approached, and the crystal behaves as an insulator. Since, no free carriers are available. However, at room temperature, some of the covalent bonds will be broken because of the thermal energy supplied to the crystal and conduction is made possible. This situation is illustrated in Fig. 1.5. The energy E_G (will described latter) required to break such a covalent bond is about 0.72 eV for germanium and 1.1 eV for silicon at room temperature.

Hole Mechanism. The mechanism by which a hole contributes to the conductivity is as follows :

When a bond is incomplete so that a hole exists, it is relatively easy for a valence electron in a neighbouring atom to leave its covalent bond to fill this hole. An electron moving from a bond to fill a hole leaves a hole in its initial position. Hence, the hole effectively moves in the direction opposite to that of electron. This hole, in its new position, may now be filled by an electron from another covalent bond and the hole will correspondingly move one more step in the direction opposite to the motion of the electron. Here we have a mechanism for the conduction of electricity which does not involve free electrons. This phenomenon is illustrated in Fig. 1.6, where a circle with a dot in it represents a completed bond and an empty circle represents a hole. Fig. 1.6 (*a*) shows a row of 12 ions, with a broken bond or hole at ion 5. Now imagine that an electron from ion 6 moves into the hole at ion 5, so that the configuration of Fig. 1.6 (*b*) results. If we compare this figure with Fig. 1.6 (*a*), it looks as if the hole in (*a*) has moved

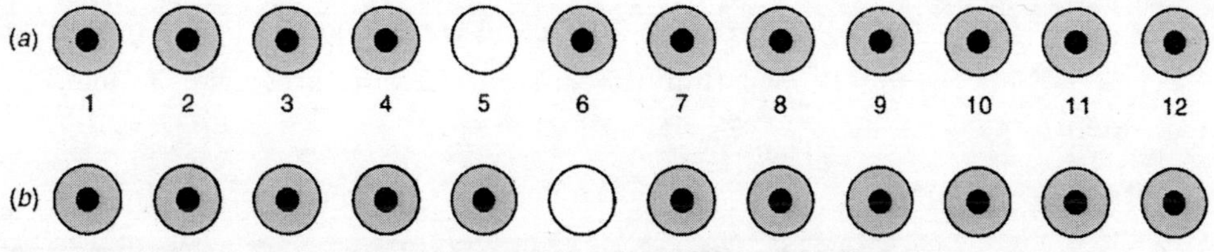

Fig. 1.6. *The mechanism by which a hole contributes to the conductivity.*

towards the right in (*b*) (from ion 5 to ion 6). This discussion indicates that the motion of hole in one direction actually means the transport of a negative charge an equal distance in the opposite direction.

So, far as the flow of electric current is concerned, the hole behaves like a positive charge equal in magnitude to the electronic charge. We can consider that the holes are physical entities whose movement constitutes a flow of current.

Some Conclusions related to Electron-Hole Concept

- 1 eV is defined as that energy which an electron occupies in moving through a potential difference of 1 V.
- An electron in the conduction band experiences almost negligible nuclear attraction. In fact electron in an conduction band does not belong to any particular atom. But it moves randomly throughout the solid. This is the reason why an electron in the conduction band are called **free electron.**
- When an electron breaks a covalent bond and moves away, a vacancy is created in the broken covalent bond. This vacancy is **called a hole.** Whenever a free electron is generated, a hole is created simultaneously *i.e.*, free electrons and holes are generated in a pairs. Therefore, the concentration of hole and free electrons will always equal in an intrinsic semiconductor. This type of generation of free-electron hole pairs is referred to as thermal generation.
- The absence of electron in a covalent bond is called a **hole.** Hole acts as a virtual charge, although there is no physical charge on it.
- It may be noted that hole current is due to movement of (unlike the normal current) *valence electron* from one covalent bond to another.

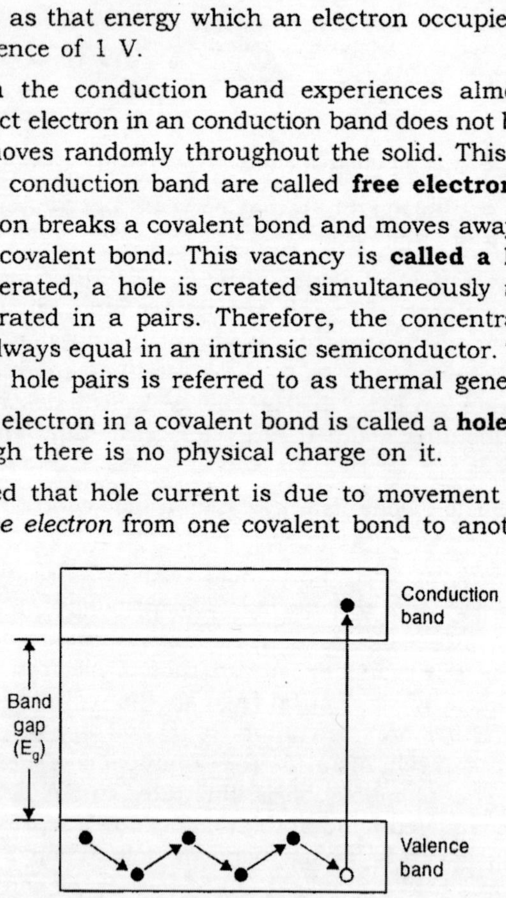

Fig. 1.6 (*a*)

- Why it is called a hole current ? When the conduction is again by electron ?
- The reason behind this is that the current flow in presence of holes in the *Valence band.*

1.5. ENERGY BANDS

Modern electronics is based on electrical conduction in solids. Electrical conductivity allows us to classify solids into three categories : Conductors, Semiconductors and

Insulators. All three classes of materials are used in microelectronics. However, the key component is the semiconductor. The use of metals and insulators to design integrated circuits does not require specific knowledge. Yet, semiconductor devices does require a good knowledge in solid-state physics. The first fundamental notion is the band diagram.

The band diagram defines the relationship between the energy of electrons and the momentum (P = mv). For electrons in the free space the E versus P relationship is quite simple; it is given by

$$E = \frac{P^2}{2m}$$

derived directly from the basic laws of physics. Fig. 1.7 (*a*) shows a typical E versus P diagram. The diagram is drawn together with the parabola of free electrons and reveals that only given energy intervals (allowed bands) have a corresponding wave propagation vector. We can see that for large energies, the allowed bands become broad and the forbidden regions become a hit more larger.

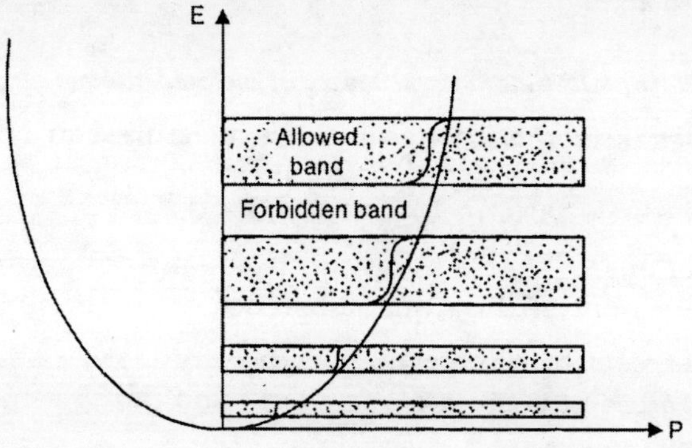

Fig. 1.7 (*a*). *Simplified E versus P diagram for electrons in solids.*

X-ray and other studies reveal that most metals and semiconductors are crystalline in structure. A crystal consists of a space array of atoms or molecules (strictly speaking, ions) built up by regular repetition in three dimensions of some fundamental structural unit.

The large number of discrete but closely spaced energy levels is called an *energy band.* Because it has been found that when atoms come together to form a molecule, they exert an influence on one another. As a result of this, the allowed energy levels 'split' into closely spaced levels. A similar situation exists, when many atoms come together to form a solid.

The splitted energy levels form continuous bands of allowed energy, which the electrons may occupy. Since the motion of electrons in a solid is controlled by the energy levels. Each material will, in fact, have its own set of permissible energy levels for the electrons in its atomic structure. The nature of energy bands is of primary interest in determining the electrical properties of the solid.

1.5.1. Energy Levels

In the isolated atomic structure there are discrete (individual) energy levels associated with each orbiting electron as shown in Fig. 1.7 (*b*). Between the discrete

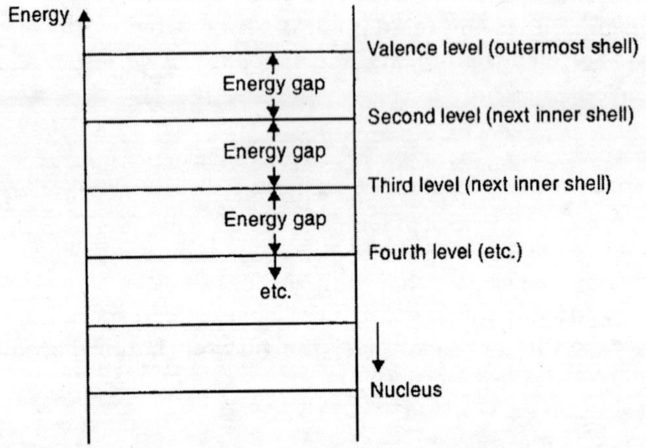

Fig. 1.7 (*b*). *Discrete energy levels in isolated atomic structure.*

energy levels no electrons in the isolated atomic structure can appear.

Some Important Points Regarding to Energy Bands

1. The number of energy bands in a solid is equal to the number of energy levels in any isolated atom of the same substance.
2. Each band consists of a large number of very closely spaced discrete energy levels. The number of such levels in each band is the same as that of number of atoms in the solids.
3. In considering the electrical properties, we are only interested in the two upper bands. These two bands are the conduction band and the valence band. The electrons, having energies in the conduction band, are those which take part in the conduction of current.

1.5.2. Simplified Representation of Energy Level Diagram of a Solid

Fig. 1.8 shows a simplified representation of an energy level diagram of a solid. In this diagram only the conduction and the valence bands are shown. The inner (or lower) bands, which are completely filled, are not shown in the diagram. This has been done due to the fact that the inner bands are not important in determining the electrical as well as electronic properties of a solid. It may be noted that at absolute zero (0 K) temperature, the electrons will occupy the lowest energy levels available to them. Thus at 0 K, the valence band is filled, while the conduction band is completely empty.

This arrangement of filled and empty energy bands has an important effect on the electrical as well as electronic properties of a solid.

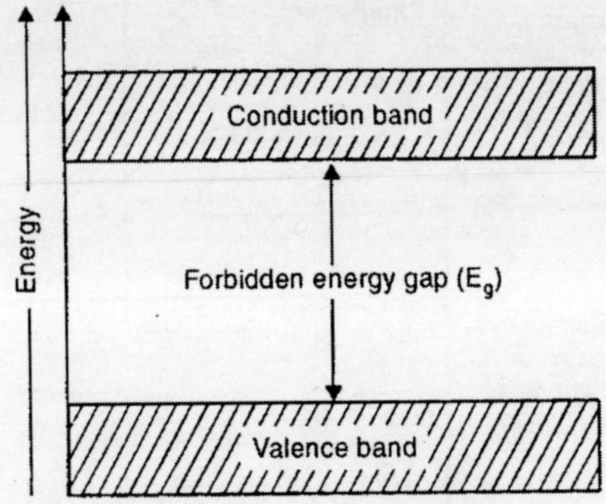

Fig. 1.8. *Simplified energy level diagram of a solid.*

The energy gap between the conduction band and the valence band is called the forbidden energy gap. It is represented by E_g and measured in electron volt (*i.e.*, eV) mathematically :

$$1 \text{ eV} = 1.6 \times 10^{-19} \text{ J}$$

If an electron in the valence band is to move throughout the crystal, it must be elevated first from valence band into the conduction band. This can be done only, if a certain amount of energy equal to energy gap is supplied or added to the material. When a electron in the valence band acquires enough energy to move into the conduction band, it is called a *free electron.* The free electron can move throughout the crystal. This provides an electric current, when a voltage source is connected across the material.

1.5.3. Insulators, Conductors and Semiconductors

As a matter of fact, every solid has its own characteristics energy band structure. This variation in band structure is responsible for the wide range of electrical properties observed in various materials. Now, we shall discuss the electrical properties of insulators, conductors and semiconductors.

Insulators. The materials, in which the conduction and valence bands are separated by a wide energy gap (E_g > 3 eV) as shown in Fig. 1.9 (*a*) are called insulators (also referred to as dielectric materials). A wide energy gap means that a large amount of energy is required, to free the electrons, by moving them from valence band into the conduction band. Since, at room temperature, the valence electrons of an insulator do not have enough energy to jump into the conduction band, therefore insulators do not have an ability to conduct current. Thus, insulators have very high resistivity (or extremely low conductivity) at room temperatures. However, if the temperature is raised, some of the valence electrons may acquire energy and jump into the conduction band. It causes the resistivity of insulators to decrease. Therefore, the insulators have negative temperature coefficient of resistance.

Conductors. The materials in which conduction and valence bands overlap as shown in Fig. 1.9 (*c*) are called conductors. The overlapping indicates a large number of electrons available for conduction. Hence, the application of small amount of voltage results in a large amount of currents.

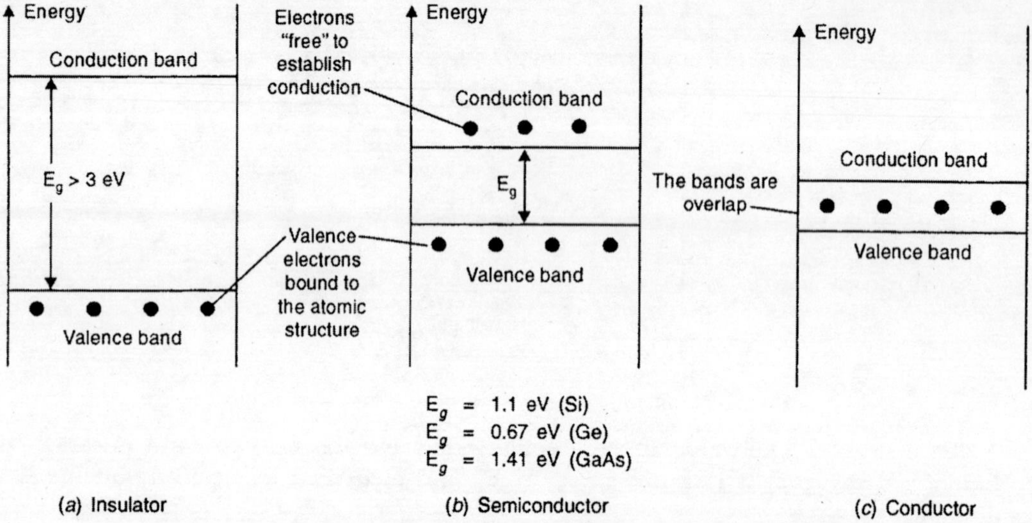

Fig. 1.9. *Energy levels in insulator, semiconductor and conductor.*

Semiconductors. The materials, in which the conduction and valence bands are separated by a small energy gap are called semiconductors shown in Fig. 1.9 (*b*). Silicon and Germanium are commonly used semiconductors. A small energy gap means that a small amount of energy is required to free the electrons by moving them from the valence band into the conduction band. The semiconductor behaves like insulators at 0 K, because no electrons are available in the conduction band. However, at room temperature, a significant number of electrons are available in the conduction band.

If the temperature is further increased, more valence electrons will acquire energy to jump into the conduction band. Thus, like insulators, semiconductors also have negative temperature coefficient of resistance. It means that conductivity of semiconductors increases with the increase in temperature.

Table 1.1 shows the Energy gap of solid materials.

Table 1.1

Material	***Energy gap***
Metal	None
Semiconductor	0.5 – 3.0 eV
Insulator	> 3 eV

1.6. DOPING

The process of introduction of impurities in semiconductors in a precisely controlled manner is called *doping.* Depending on the nature of impurities added, the semiconductors are classified as follows :

(*a*) Pure or intrinsic semiconductors.

(*b*) Impurity or Extrinsic semiconductors.

The semiconductors are in their pure form is referred to as pure or intrinsic semiconductors, whereas the extrinsic semiconductors are doped semiconductors in which suitable impurity atoms are added to modify the properties.

1.6.1. Intrinsic Semiconductors

As stated above, a semiconductor, which is in its extremely pure form, is known as an intrinsic (or pure) semiconductor. The nature of semiconductors is such that even a small amount of certain impurities can change their electrical properties drastically. For Germanium, the impurity level is less than 1 part in 10^8 parts. And for silicon, it is less than 1 part in 10^{12} parts. In actual practice, however, a semiconductor material, with somewhat larger impurity concentrations, than those mentioned above is still called as an intrinsic semiconductor.

The silicon and germanium are the two most widely used intrinsic semiconductors. The crystal structure of these materials consists of a regular repetition in three dimensions of a unit cell having the form of tetrahedron, with one atom at each vertex. The two dimensional representation of such a structure is shown in Fig. 1.10. For example, the silicon has a total 14 electrons in its atomic structure. Each atom in a silicon crystal contributes 4 valence electrons, so that the atom is tetravalent. These atoms are bonded together by a force, which results from the sharing of the neighbouring atoms. This bonding force is called a covalent bond and is shown by a pair of curved lines in the figure.

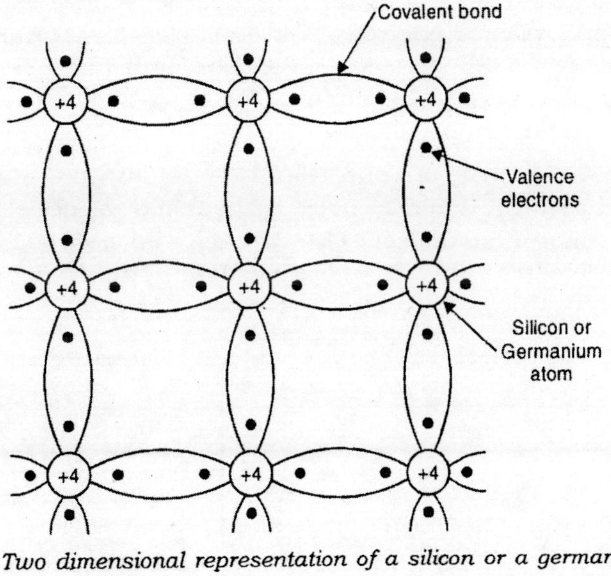

Fig. 1.10. *Two dimensional representation of a silicon or a germanium crystal.*

1.6.2. Extrinsic Semiconductors

In order to make the intrinsic semiconductor useful, their characteristics have to be changed by adding a certain amount of desired impurity atoms. The resulting semiconductors are called impure or extrinsic semiconductors.

As we have already discussed that the process of introduction of impurities in semiconductors in a precisely controlled manner is called *doping*. Generally, the impurities are added at the rate of only one atom of impurity per 10^6 to 10^{10} semiconductor atoms. As a result of this, the impurity atoms are separated in the crystal by hundreds or thousand of semiconductor atoms in all directions. It may be noted that an impurity atom enters the crystal by substituting for one germanium or silicon atoms.

1.6.3. Purpose of Doping

The purpose of doping *i.e.,* process of adding impurity atoms, is to increase either the number of free electrons or holes in a semiconductor. Generally, two types of impurity atoms are added to the semiconductor namely the impurity atoms containing 5 valence electrons (called pentavalent impurity atoms) and the impurity atoms containing 3 valence electrons (called trivalent impurity atoms). Depending upon the type of impurity atoms added to the semiconductor, the resulting semiconductor (*i.e.,* extrinsic semiconductor) may be of the following two types :

(*i*) N-type Semiconductors

(*ii*) P-type Semiconductors.

1.6.4. N-type Semiconductors

The semiconductors, which are obtained by introducing pentavalent impurity atoms, *i.e.,* atoms containing 5 valence electrons are known as N-type semiconductors. The pentavalent impurity is an element from group V of the periodic table. The elements in this group contain 5 valence electrons. The example of pentavalent impurities are phosphorus (P), antimony (Sb), arsenic (As) and bismuth (Bi). These elements donate excess **negative charge carriers** (electrons). Therefore, such elements are known as donor or N-type impurities.

Fig. 1.11 shows the structure of a silicon crystal lattice containing as antimony atom at the central position. It may be noted that out of 5 valence electrons, 4 electrons will form covalent bonds by sharing one electron each with electrons of the neighbouring atoms. The 5th electron is an extra electron and is loosely bound with

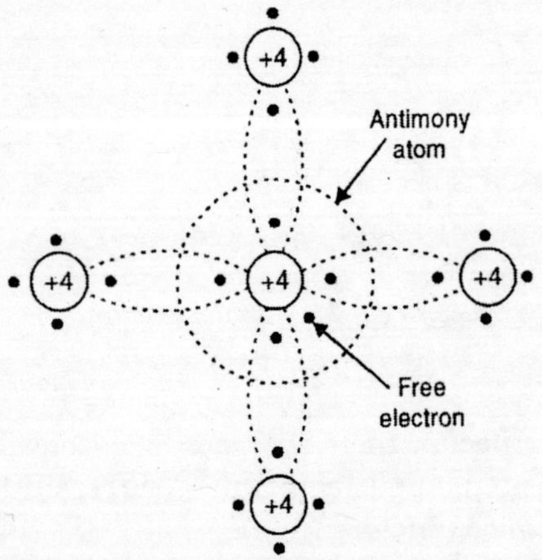

Fig. 1.11. *A silicon atom replaced by stationary atom in a crystal lattice.*

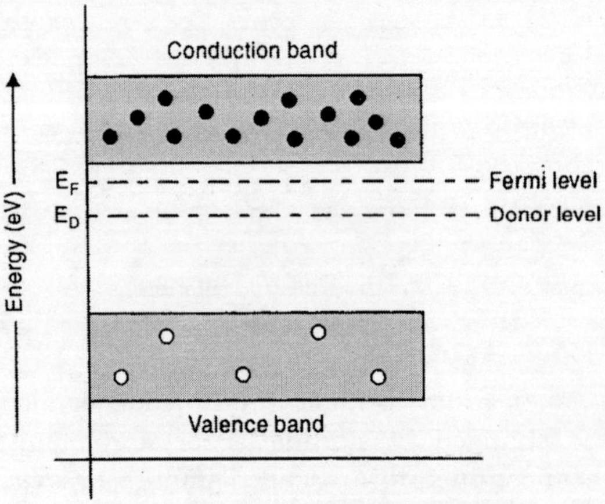

Fig. 1.12. *Energy band diagram of an N-type semiconductor.*

the antimony atom. This extra electron, if detached from the impurity atom will be available as a carrier of the current. The energy required to detach this electron is of the order of 0.05 eV for Silicon and 0.01 eV for Germanium.

It will be interesting to know that if a pure semiconductor is doped with N-type impurities, the number of electrons in the conduction band increases above a level that would be available in a pure semiconductor. However, the number of holes in the valence band decreases below a level, that would be available in a pure semiconductor.

It is due to the fact that a large number of electrons present in the semiconductor increases the rate of recombination of electrons with holes. Because of the greater number of electrons in a conduction band than that of holes in the valence band. The Fermi level shifts upward towards the bottom of the conduction band as shown in Fig. 1.12.

One thing should always kept in mind that in an N-type semiconductor, the current flows due to the movement of electrons and holes. But a major part of the current flows due to the movement of electrons. Therefore, the *electrons* in an N-type semiconductor are known as *majority carriers* and *holes* as *minority carriers*.

1.6.5. P-type Semiconductors

The semiconductors which are obtained by introducing a trivalent impurity atoms (*i.e.*, atoms containing 3 valence electrons) are known as P-type semiconductors. The trivalent impurity is an element of group III of the periodic table. The elements in this group contain 3 valence electrons. Typical examples of such elements are gallium (Ga), indium (In), aluminium (Al), boron (B) etc. These elements make available positive charge carriers because they create holes, which can accept electrons. Therefore, such elements are known as acceptor or P-type impurities.

It has been observed that when a trivalent impurity is added to a pure semiconductor, it displaces some of the atoms. Fig. 1.12 shows the structure of silicon crystal containing an indium atom at the central position. The 3 valence electrons, of an indium atoms, forms 3 covalent bonds by sharing one electron with the electrons of neighbouring atoms. However, the fourth covalent bond is incomplete. A vacancy, which exists in the incomplete covalent bond constitutes a hole. Now, the indium atom seeks its surrounding atoms so as to acquire the 4th electron, to complete the covalent bond. Thus, an electron which is in a favourable position, is captured by an indium atom. After doing this, the indium atom becomes an immobile ion. The energy, involved in capturing the electron to 0.05 eV for silicon and 0.01 eV for Germanium.

It will be interesting to know that if a pure semiconductor is doped with P-type impurities, the number of holes in the valence band increases above a level, which would be available in a pure semiconductor. However, the number of electrons in the conduction band decreases below a level, which would be available in a pure semiconductor. It is due to the fact, that a large number of holes present in the semiconductor increases the rate of recombination of holes with electrons. Because of the greater number of holes in the valence band than that of electrons in conduction band, the fermi-level shifts downwards towards the top of the valence band as shown in Fig. 1.13.

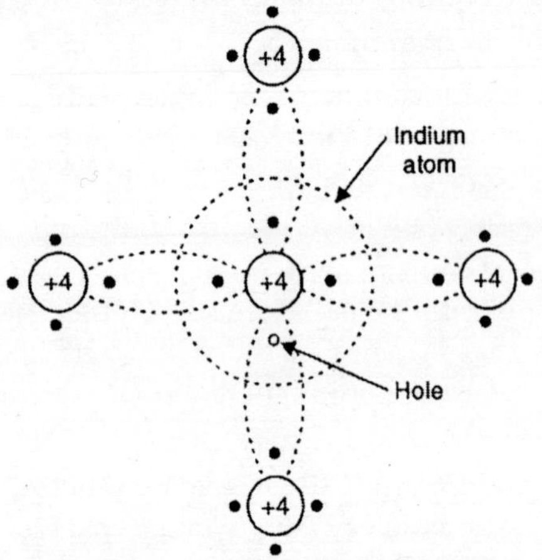

Fig. 1.13. *A silicon atom replaced by indium atom in a crystal lattice.*

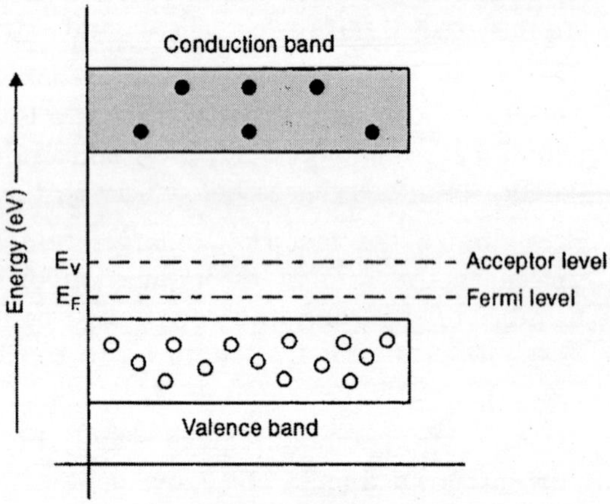

Fig. 1.14. *Energy band diagram of a P-type semiconductor.*

One thing should always kept in mind that in an P-type semiconductor, the current flows due to the movement of electrons and holes. But a major part of the current flows due to movements of holes.

Therefore, the *holes* in an P-type semiconductor are known as *majority carriers* and *electrons* as *minority carriers.*

1.6.6. Majority and Minority Charge Carriers

We have already discussed that a pure (or intrinsic) semiconductor has no free charge carriers at absolute zero temperature. However, as the temperature of the semiconductor is raised to room temperature, some of the covalent bonds are broken due to which few electron-hole pairs are created. These electron-hole pairs are called thermally generated charge carriers.

For instance, an N-type semiconductor has a large number of electrons. Besides this N-type semiconductor also contains fewer number of thermally generated holes. Since, the number of electrons, are much larger than the number of holes in a N-type semiconductor, therefore electrons are called majority carriers and holes a minority carriers. This situation is as shown in Fig. 1.15 (*a*).

Majority carriers : Electrons
Minority carriers : Holes

(*a*) N-type

Majority carriers : Holes
Minority carriers : Electrons

(*b*) P-type

Fig. 1.15.

Similarly, in a P-type semiconductor, the number of holes are much larger than the number of electrons. Therefore, in a P-type semiconductor, holes are called majority carriers and electrons as minority carriers. This situation is as shown in Fig. 1.15 (*b*).

1.7. FERMI LEVEL

Fermi level is a measure of the probability of occupancy of electrons or holes in the allowed energy states. We have already discussed that in N-type semiconductor the fermi level shifts upward towards the bottom of the conduction band. However, in the case of P-type semiconductor it shifts downwards towards the top of the valence band. One thing always should be kept in mind that in pure semiconductor having number of electrons in the conduction band is equal to the number of holes in the valence band, it lies between the middle of the conduction and valence band.

From the equation of current density [*i.e.*, $J = (n.\mu_n + p.\mu_p)\, q.E.$] and conductivity of the material [$\sigma = (n.\mu_n + p.\mu_p)\, q$] it is clear that the electrical characteristics of a semiconductor material depend on the concentration of free electrons and holes. Also the change in the concentration of electrons and holes may shifts the Fermi level of the energy band. Thus, we conclude that the only parameter which changes with impurities is the *Fermi level* E_F. Now we will discuss Fermi level in detail.

1.7.1. The Fermi-Dirac Function

The equation for $f(E)$ is called the Fermi-Dirac probability function and specifies the fraction of all states at energy E (in eV) occupied under the condition of thermal equilibrium. $f(E)$ is given by relation :

$$\boxed{f(E) = \frac{1}{1 + e^{\frac{(E - E_F)}{kT}}}} \qquad \text{...(1.3)}$$

where, k = Boltzmann constant, eV/°K

T = Temperature

E_F = Fermi level or characteristic energy, for the crystal, eV.

The Fermi level represents the energy state with 50% probability of being filled if no forbidden band exists. The reason for the last statement is that, for an energy E equal to the Fermi level energy E_F, the occupation probability is :

$$f(E) = \frac{1}{1 + e^{\frac{0}{kT}}} \qquad (\because E = E_F)$$

or

$$f(E) = \frac{1}{1+1} = \frac{1}{2}$$

Thus, an energy state at the fermi level has a probability of $\frac{1}{2}$ of being occupied by an electron.

A closer examination of $f(E)$ indicates that 0 K the distribution takes the simple rectangular form shown in Fig. 1.16. With $T = 0$, in the denominator of exponent, $f(E)$ is $\frac{1}{1+0} = 1$, when the exponent is negative (*i.e.*, $E_F > E$) and $f(E)$ is $\frac{1}{1+\infty} = 0$ when the exponent is positive (*i.e.*, $E > E_F$). This rectangular distribution implies that at 0 K every available energy state up to E_F is filled with electrons and all states above E_F are empty, since there are no electrons at 0 K which have energies in excess of E_F.

For better understanding at absolute zero temperature there exist two cases that are as follows:

Case I. When $E >> E_F$ then

$$f(E) = \frac{1}{1+e^{\infty}} = 0$$

This shows that the probability of finding an occupied quantum state of energy greater than E_F is zero at absolute zero temperature.

Case II. When $E << E_F$ then

$$f(E) = \frac{1}{1+e^{-\infty}} = 1$$

This shows that all quantum levels with energies less than E_F is occupied at absolute zero temperature.

At absolute temperature,

$$P_E = \begin{cases} \gamma E^{1/2} \text{ for } E < E_F \\ 0 \quad \text{ for } E > E_F \end{cases}$$

where,

P_E = distribution density of electrons in number of electrons per electron volt per cubic metre.

γ = constant = $6.82 \times 10^{27}/\text{m}^3\text{ev}^{3/2}$

It may be noted that *Fermi energy is the maximum energy that any electron may possess at absolute zero temperature.*

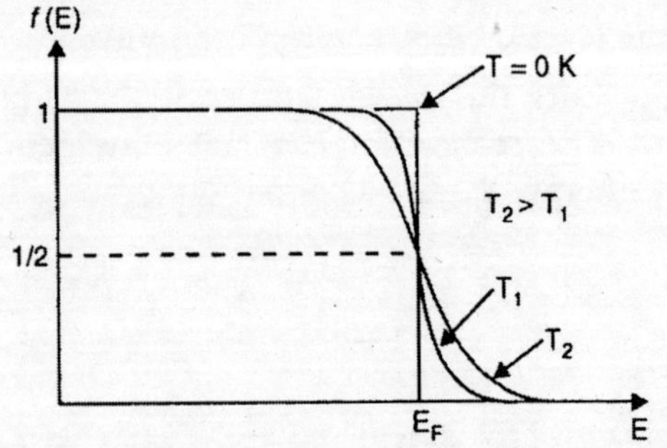

Fig. 1.16. *The Fermi-Dirac distribution function.*

NOTE : (*i*) Shift in energy level is given by equation

$$E_0 = kT \ln \frac{N_D N_A}{n_i^2}$$

where, kT = 26 mV

N_D = Donor concentration

N_A = Acceptor concentration

n_i = Intrinsic concentration

(*ii*) Other expression for E_0 is

$$E_0 = kT \ln \frac{P_p}{P_n} = kT \ln \frac{n_n}{n_p}$$

(*iii*) Einstein Relation :

Since both diffusion and mobility are statistical phenomenon, D and μ are not independent.

$$\frac{D_p}{\mu_p} = \frac{D_n}{\mu_n} = \frac{kT}{q} = V_T = 26 \text{ mV.}$$

This ratio depends upon :

(*a*) Temperature of the semiconductor.

(*b*) The type of semiconductor.

1.8. DONOR AND ACCEPTOR IMPURITIES

If in intrinsic silicon or germanium, there is added a small percentage of trivalent or pentavalent atoms, a doped, impure or extrinsic semiconductor is formed.

1.8.1. Donors

If the dopant has five valence electrons (example, Antimony), the crystal structure of Fig. 1.17 (*a*) is obtained. The impurity atoms will displace some of the germanium atoms in crystal lattice. Four of the five valence electrons will occupy covalent bonds and the fifth will be nominally unbound and will be available as a carrier of current.

The energy required to detach the fifth electron from the atom is of the order of only 0.01 eV for Ge or 0.05 eV for Si. Such impurities donate excess (negative) electron carriers and are therefore referred to as donor or *n*-type impurities.

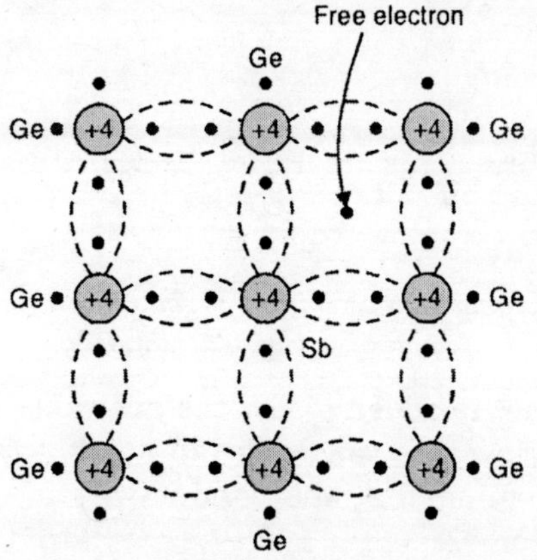

Fig. 1.17 (*a*). *Crystal lattice with a germanium atom displaced by a pentavalent impurity atom.*

When donor impurity is added to Ge, the allowable energy levels are introduced at a very small distance below conduction band as shown in Fig. 1.17 (*b*). These new allowed energy levels are discrete levels. The energy of new levels is only 0.01 eV in Ge and 0.05 eV in Si below conduction band. Fifth electrons of donor material are raised into the conduction band.

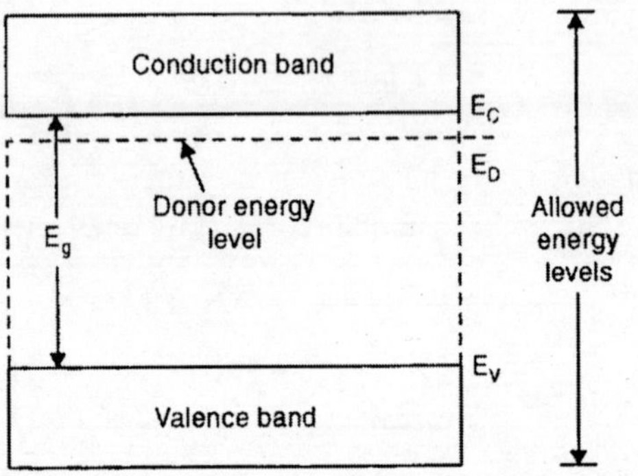

Fig. 1.17 (*b*)

1.8.2. Acceptors

If a trivalent impurity (boron, gallium or indium) is added to an intrinsic semiconductor, only three of the covalent bonds can be filled and the vacancy that exists in the fourth bond constitutes a hole as shown in Fig. 1.18 (*a*). Such impurities make available positive carriers because they create whole which can accept electrons. These impurities are consequently known as acceptor or *p*-type impurities.

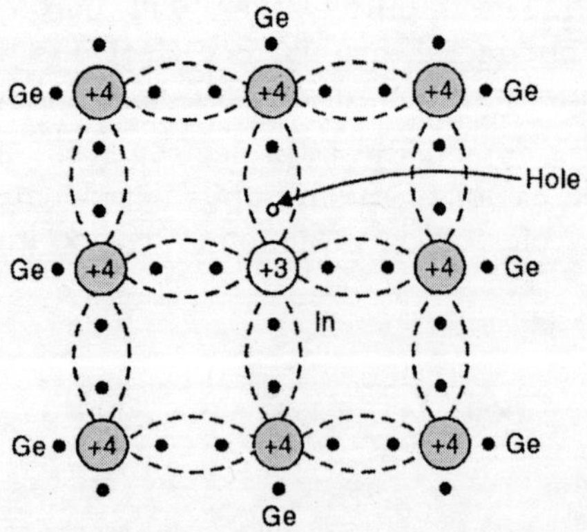

Fig. 1.18 (*a*). *Crystal lattice with a germanium atom displaced by an atom of a trivalent impurity.*

When Acceptor impurity is added to Ge, the allowable energy levels are introduced at a very small distance above the valence band as shown in Fig. 1.18 (*b*). One thing should be always kept in mind that the conductivity of semiconductor increases by increasing the impurity atoms.

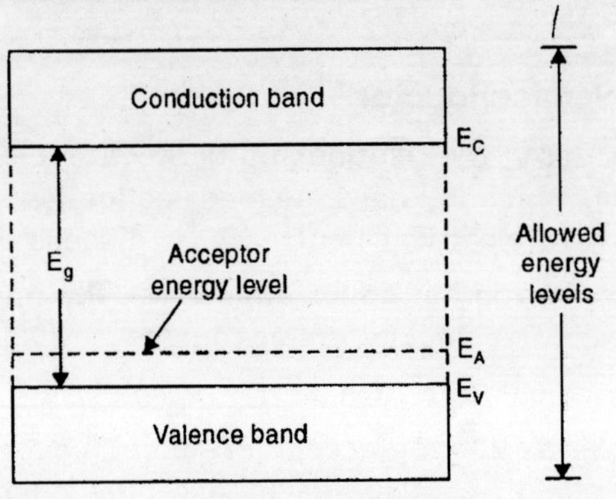

Fig. 1.18 (*b*)

As we have already described that in a pure (intrinsic) semiconductor the number of holes is equal to the number of free electrons. Thermal agitation continues to produce new hole-electron pairs, whereas other hole-electron pairs disappear as a result of recombination. The hole concentration p must equal to the electron concentration n, so that,

$$\boxed{n = p = n_i} \quad \text{...(1.4)}$$

where n_i is called the intrinsic concentration.

It may be noted that with increasing temperature, the density of hole-electron pairs increases and correspondingly, the conductivity increases. It is found that the intrinsic concentration n_i varies with T as,

$$\boxed{n_i^2 = A_0 T^3 e^{\frac{-E_{Go}}{kT}}} \quad \text{...(1.5)}$$

where, E_{G0} = Energy gap at 0° K in electron volts

k = Boltzmann constant *i.e.*, eV/°K

A_0 = Constant independent of T.

1.9. THE MASS-ACTION LAW

As we have already discussed that adding n-type impurities decreases the number of holes. Similarly, doping with p-type impurities decreases the number of electrons below a level which would have been available in pure semiconductor. It has been experimentally found that under thermal equilibrium, the product of number of holes and the number of electrons is constant and is independent of the amount of donor and acceptor impurity atoms. This relation is known as mass-action law and is given by the relation

$$\boxed{n \cdot p = n_i^2} \quad \text{...(1.6)}$$

where, n = Free electron concentration (*i.e.*, no. of free electrons per m^3)

p = Hole concentration (*i.e.*, number of holes per m^3)

n_i = Intrinsic concentration.

Charge Densities in a Semiconductor

We have already discussed the relationship between free electron concentration (n) and hole concentration (p), which is known as mass-action law. These concentration are further related by the law of electrical neutrality as discussed below.

Let, N_D = Concentration of donor atoms (*i.e.*, the number of positive charges /m^3 contributed by donor ions)

p = Hole concentration (*i.e.*, no. of holes/m^3)

N_A = Concentration of acceptor atoms (*i.e.*, the number of negative charges/m^3 contributed by acceptor ions)

n = Free electron concentration (*i.e.*, number of free electrons/m^3)

$\therefore$ Total number of positive charges/m^3 $= N_D + p$

and total number of negative charges/m^3 $= N_A + n$

Since the semiconductor, as a whole is electrically neutral, therefore magnitude of total positive charge density is equal to the total negative charge density, *i.e.*,

$$\boxed{N_A + n = N_D + p} \qquad ...(1.7)$$

Consider an n-type material having $N_A = 0$. Since, the number of electrons is much greater than the number of holes in an n-type semiconductor ($n >> p$), then eqn. (1.7) reduces to :

$$\boxed{n \approx N_D} \qquad ...(1.8)$$

Thus, we can also say that in an n-type material the free electron concentration is approximately equal to the density of donor atoms.

The concentration of holes (*i.e.*, here minority carriers) in an n-type semiconductor, may be obtained by using mass-action law (*i.e.*, $n.p = n_i^2$) and its value is given by the relation

$$\boxed{P_n = \frac{n_i^2}{n} = \frac{n_i^2}{N_D}} \qquad ...(1.9)$$

where, P_n denotes the concentration of holes in n-type semiconductor at room temperature (300°K), $n_i \approx 10^{13}/cm^3$ for Ge and $n_i \approx 10^{12}/cm^3$ for Si.

Similarly, in a p-type semiconductor, the concentration of donor atoms is zero (*i.e.*, $N_D = 0$) and the number of holes is much larger than the number of free electrons (*i.e.*, $p >> n$). Therefore, for p-type semiconductor, eqn. (1.7) may be written as

$$\boxed{p = N_A} \qquad ...(1.10)$$

The above relation shows that in a p-type semiconductor, the hole concentration (p) is equal to the number of acceptor atoms/m^3. The concentration of free electron is obtained by using mass action law and its value is given by the relation,

$$\boxed{n_p = \frac{n_i^2}{p} = \frac{n_i^2}{N_A}} \qquad ...(1.11)$$

where, n_p denotes the concentration of electrons in p-type semiconductor. It will be interesting to know that,

It is possible to add donor atoms into a p-type semiconductor. Similarly, it is also possible to add acceptor atoms into an n-type semiconductor. If an equal concentrations of donors and acceptors is added into an intrinsic semiconductor, it remains intrinsic semiconductor.

An extension of the above argument indicates that if the concentration of donor atoms added to a p-type semiconductor exceeds the acceptor concentration ($N_D > N_A$), the specimen is changed from p-type to an n-type semiconductor.

NOTE : One thing should be always kept in mind that the current in a metal is due to the flow of negative charges (electrons) whereas the current in a semiconductor results from the movement of both electrons and positive atoms so that the current is due to predominantly either to electrons or to holes. *The transport of the charges in a crystal under the influence of an electric field* (a drift current) and also as a result of non-uniform concentration gradient (a diffusion current).

1.10. MOBILITY

If a steady Electric field (E = v/m) is applied to the metal, free electrons get accelerated. The velocity of free electrons increases considerably with time. Since electrons lose energy at every inelastic collision, a steady state is reached and resulted a finite value of drift velocity, v_d in direction opposite to electric field.

The term mobility may be defined as the drift velocity per unit electric field. Mathematically,

$$\boxed{\mu = \frac{v_d}{E}} \qquad \text{...(1.12)}$$

where, μ = Mobility of electrons (measured in square meters per-volt second)

v_d = Drift velocity

E = Applied electric field.

Mobility means how freely charge carriers move in a crystal without collision. Mobility is a result of both lattice and impurity scattering.

Mobility can also be represented as,

$$\mu = \frac{qt}{m}$$

Conductivity in terms of mobility is given by $\sigma = nq\mu$

Therefore, current density J is expressed as

$$J = \sigma E = nq\mu E$$

The drift velocity is related to Electric field as

$$v_d \propto E$$

or $$v_d = \mu E$$

The steady flow of electrons in one direction caused by the applied electric field constitutes currents and is called the **drift-current.**

It can be easily noted that the current in the conductor is proportional to the applied electric field, which is also true from the ohm's law. The electrons that constitute the current acquire energy from the applied field and transfer it to lattice ions through inelastic collisions.

1.10.1. Temperature Dependency of Mobility

The parameter mobility (μ) of an intrinsic semiconductor varies as T^{-m} over a temperature range (T) of 100 to 400 K. For silicon, the value of m = 2.5 for electrons and 2.7 for holes. Similarly, for Germanium, the value of m = 1.66 for electrons and 2.33 for holes. It may be noted that drift velocity of a carrier depends upon the free path available to it for acceleration between two collisions (mean free path) as the mobility of an intrinsic semiconductor decreases with the increase in temperature because at higher temperatures, the number of carriers are more and they are more energetic also. This causes an increased number of collisions with the atoms causing scattering and thus the mobility decreases.

1.10.2. Effect of Strong Field

Fig. 1.19 shows a graph of drift velocity verses field intensity. It is linear for the initial portion, where field intensity is less than 10^3 V/cm (*i.e.*, $E < 10^3$ V/cm). Beyond that the graph tapers off and velocity reaches an upper limit of 10^7 cm/sec when the field is increased to 10^5 V/cm.

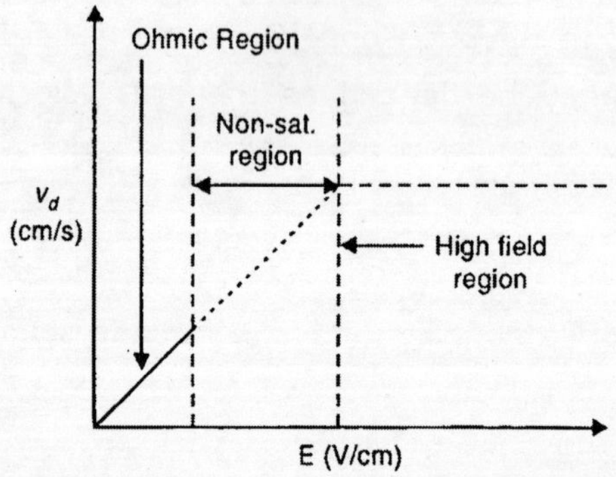

Fig. 1.19. *Variation of mobility with electric field.*

For better understanding the variation in mobility with electric field under different condition given below:

- μ = constant when $E < 10^3$ V/cm.
- $\mu \propto \frac{1}{\sqrt{E}}$ when $10^3 < E < 10^4$ V/cm.
- $\mu \propto \frac{1}{E}$ when $E > 10^4$ V/cm.

Mobility also depends upon the concentration of the charge carriers. The electrons ih germanium more about 3 times faster than silicon and hence is suitable for high frequency applications.

Type	μ_n (cm^2/V-S)	μ_p (cm^2/V-S)
Silicon	1350	480
Germanium	3900	1900
Gallium Arsenide	8500	4000

1.10.3. Conduction in Metals

As a matter of fact, the fundamental difference between semiconductor and metal is that the former is bipolar and latter is unipolar *i.e.,* it conducts current by means of charge (electrons) of one sign only. If an electric field is applied to a metal, then due to the electrostatic force, the electrons would be accelerated and the velocity would increases indefinitely, if there would have been no collision with the ions. However, at each collision with the ion, the electron loses some energy. A steady-state condition is reached, when a finite value of *drift velocity* (V_d) is obtained. Thus, we can say that drift current is due to the influence of electric field. The drift velocity is in a direction opposite to that of the applied electric field.

Now, let us obtain an expression for the drift velocity in a metal.

We know that force experienced by the electron due to applied electric field.

$$F = q \cdot E$$

where,

E = Applied electric field

q = Charge of an electron

and acceleration,

$$a = \frac{\text{Force}}{\text{Mass}} = \frac{F}{m} = \frac{q.E}{m}$$

$$\therefore \quad \text{Drift velocity, } V_d = a\,.\,t = \frac{q.E}{m} \cdot t = \mu E$$

where μ is equal to $\frac{q.t}{m}$ and is known as mobility of an electron. The mobility is expressed in m^2/(volt-sec) and its value is different for different metals.

1.10.4. Electric Current in a Conductor (Metal)

Consider a piece of a metal or conductor in which electrons are uniformly distributed as shown in Fig. 1.20.

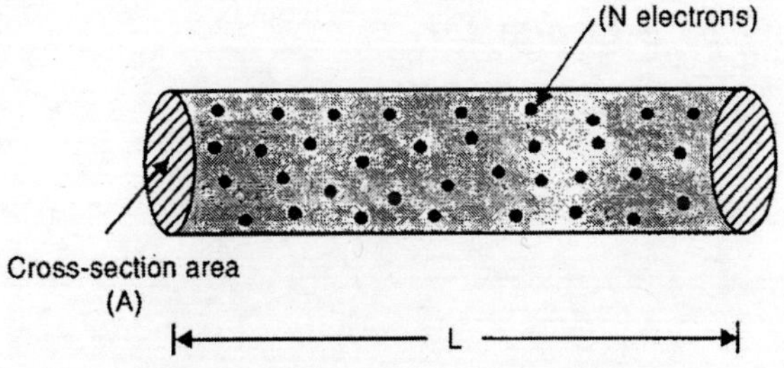

Fig. 1.20. *Metal or conductor.*

Let, N = No. of free electrons distributed in the conductor

L = Length of the conductor

A = Cross-sectional area of the conductor

We know that when an electric field is applied to the conductor, the electrons start moving in the semiconductor. Therefore, the average velocity of the electrons

$$v = \frac{L}{T}$$

where, T is the time taken by the electrons to travel through a distance.

The number of electrons passing through any area per second,

$$= \frac{N}{T}$$

∴ Total charge passing through any area per second (*i.e.*, current)

= charge × number of electrons per second passing through any area

$$= q \times \frac{N}{T}$$

$$= q \times \frac{N}{T} \cdot \frac{L}{L}$$

$$= \frac{q.N.v}{L}$$

Thus, the current in a conductor,

$$I = \frac{q.N.v}{L} \qquad \left\{ \because \text{ current density } \left(J = \frac{I}{A} \right) \right\}$$

$$J = \frac{I}{A} = \frac{q.N.v}{L.A}$$

$$J = q \cdot \left(\frac{N}{LA} \right) \cdot v \qquad \left\{ n = \frac{N}{LA} = \text{Concentration of electrons per unit volume} \right\}$$

$$\boxed{J = q \,.\, n \,.\, v}$$

1.10.5. Conductivity of a Conductor

From the relation of current density,

$$J = q \,.\, n \,.\, v$$

Also current density J (*i.e.*, electric current per unit area) is given by the relation.

$$J = \sigma \,.\, E$$

where, σ = Conductivity of a metal

E = Applied electric field

So, $$q \,.\, n \,.\, v = \sigma \,.\, E$$

$$q \cdot n \cdot (E \cdot \mu) = \sigma \cdot E$$

or $$q \cdot n \cdot \mu = \sigma$$

or $$\boxed{\sigma = q \cdot n \cdot \mu}$$

Also, $$\boxed{\sigma = \frac{1}{\rho} = q \cdot n \cdot \mu}$$

where, ρ = Resistivity of a metal

μ = Mobility of a metal.

NOTE : **Electrical Conduction.** In metal and semiconductor it is represented by relation :

(*a*) **In Metal** $\sigma = nq\mu$

where, n = no. of charge carriers

q = charge

μ = mobility

(*i*) at $T = 0$, $\sigma = \sigma_{max}$ (*i.e.*, maximum conductivity)

(*ii*) n is independent of temperature (for metal $T = 10^4$ K)

so, $$\mu \propto \frac{1}{T}$$

or $$\sigma \propto \frac{1}{T}$$

Thus in the case of metal, conductivity is inversely proportional to temperature.

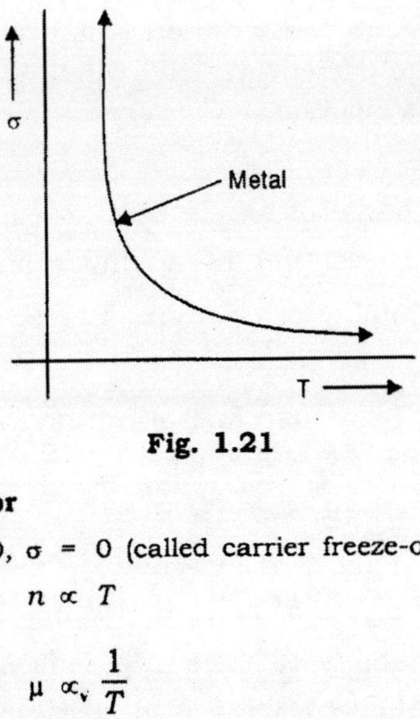

Fig. 1.21

(*b*) **In semiconductor**

(*i*) at $T = 0$, $\sigma = 0$ (called carrier freeze-out effect)

(*ii*) $$n \propto T$$

(*iii*) $$\mu \propto \frac{1}{T}$$

NOTE : In the case of semiconductor on increasing temperature there will be less change in mobility μ as compare to n.

so, $$\boxed{\sigma \propto T}$$

Thus in the case of semiconductor conductivity σ is directly proportional to the temperature.

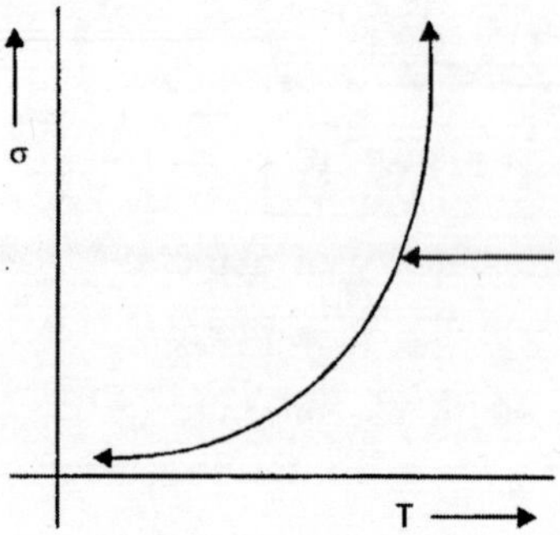

Fig. 1.22

1.10.6. Conduction in Intrinsic Semiconductors

We have already discussed that in an intrinsic semiconductor at room temperature (*i.e.*, at 300 K) the thermal energy is sufficient to produce a large number of electron hole pairs. When an electric field is applied, the current flows through the semiconductor. The flow of current, in a semiconductor, is due to the movement of free electrons in one direction and holes in the opposite direction.

As we know that electric current density in a metal (which is due to the movement of free electrons) is given by the relation,

$$J = q \cdot n \cdot \mu \cdot E$$

Using the above relation, we can write an equation for the current density in an intrinsic semiconductor. Thus, current density, which results due to the movement of electrons only is given by the relation,

$$J_n = q \cdot n \cdot \mu_n \cdot E$$

where,
q = Charge of an electron
n = No. of electron per unit volume
μ_n = Mobility of electrons in a semiconductor
E = Applied electric field.

A similar expression can be written for current density due to holes in an intrinsic semiconductor and is given by the relation,

$$J_p = q \cdot p \cdot \mu_p \cdot E$$

where,
q = Charge of a hole and is equivalent to that of an electron
p = Number of holes per unit volume
μ_p = Mobility of holes in a semiconductor
E = Applied electric field strength.

The total current density within an intrinsic semiconductor,

$$J = J_n + J_p$$
$$= q \cdot n \cdot \mu_n \cdot E + q \cdot p \cdot \mu_p \cdot E$$

$$= qE(n \cdot \mu_n + p \cdot \mu_p)$$
$$= \sigma \cdot E$$

where, $\sigma = q(n \cdot \mu_n + p \cdot \mu_p)$ called conductivity of a semiconductor.

We know that for an intrinsic semiconductor, the number of holes is equal to the number of electrons *i.e.*, $p = n = n_i$ where, n_i is called intrinsic carrier concentration. So, in the above equation substituting $n = p = n_i$, we get,

$$\boxed{J = q \cdot n_i (\mu_n + \mu_p) E}$$

and the conductivity of an intrinsic semiconductor,

$$\boxed{\sigma_i = q \cdot n_i (\mu_n + \mu_p)}$$

Thus, from above relation, it is clear that conductivity of an intrinsic semiconductor depends upon its intrinsic concentration (n_i) and the mobility of holes and electrons.

1.11. DIFFUSION

The term Diffusion may be defined as the transportation of charge carriers in a semiconductor due to result of non-uniform concentration gradient. Or in other words, we can say that in addition to a conduction current, the transport of charges in a semiconductor may be accounted by a mechanism called *diffusion,* not ordinarily encountered in metals. The essential features of diffusion are now discussed.

It is possible to have a non-uniform concentration of particles in a semiconductor. Fig. 1.23 shows a concentration $p(x)$ results in a diffusion current density for hole J_p, here the concentration p of holes varies with distance x in the semiconductor and there exists a concentration gradient, $\frac{dp}{dx}$, in the density of carriers.

Fig. 1.23 shows that the concentration gradient of $\left(\frac{dp}{dx}\right)$ of holes on one side of surface is larger than other side. The holes are in a random motion as a result of their thermal energy. This means that, holes will continue to move back and forth across this surface. We may then expect that, in a given time interval, more holes will cross

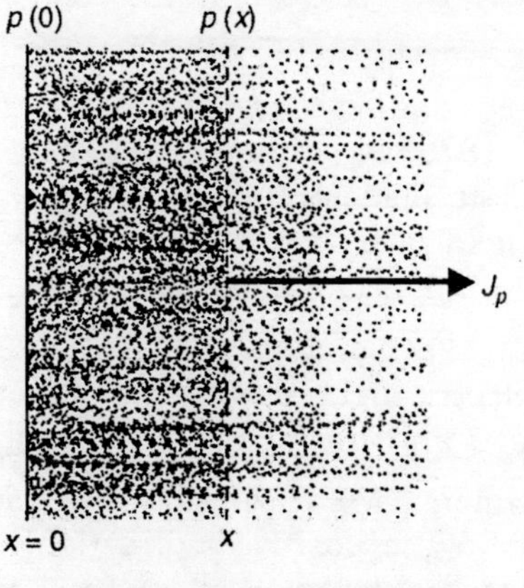

Fig. 1.23. *A non-uniform concentration p (x) results in a diffusion current J_p.*

the surface from the side of greater concentration than in the reverse direction. It should be noted that this net transport of charges is not the result of mutual repulsion among charges of like sign, but is simply the result of a statistical phenomenon. This diffusion is exactly analogous to that which occurs in a neutral gas if concentration gradient exists in the gaseous container. The diffusion hole-current density J_p is proportional to the concentration gradient and is given by

$$\boxed{J_p = -q\,D_p \frac{dp}{dx}}$$

where, D_p = diffusion constant for holes (square meter per second)

Since, p in the above figure decreases with increasing x, then $\frac{dp}{dx}$ is negative and the – ve sign is taken here. A similar equation exists for diffusion electron–current density here p is replaced by n and minus sign is replaced by plus sign. Thus diffusion electron-current density is given by the relation

$$\boxed{J_n = q \cdot D_n \frac{dn}{dx}}$$

1.11.1. Total Current

It is possible for both a potential gradient and concentration gradient to exist simultaneously within a semiconductor. In such a situation the total electron current is the sum of drift current (due to influence of applied electric field strength) and diffusion current, and given by the relation,

J_n = drift current density + diffusion current density

or

$$\boxed{J_n = q\,.\,n\,.\,\mu_n\,.\,E + q\,.\,D_n\,.\,\frac{dn}{dx}}$$

Similarly, the net hole current is

$$\boxed{J_p = q\,.\,p\,.\,\mu_p\,.\,E - q\,.\,D_p\,.\,\frac{dp}{dx}.}$$

1.12. HALL EFFECT

When a current carrying conductor is placed in a magnetic field (B) a transverse effect (discovered by Hall in 1879) is pointed. This effect is called the Hall effect. Hall found that : "When a magnetic field is applied at right angles to the direction of the electric current, an electric field is set up which is perpendicular to both the direction of electric current (I) and the applied magnetic field (B)."

In other words, "when any specimen carrying a current I is placed in the transverse magnetic field B, then an electric field E is induced in the specimen in the direction perpendicular to both B and I and the phenomenon known as Hall effect."

Fig. 1.24 shows a specimen bar carrying current I in the position-x direction. Let magnetic field B_z be applied in the positive-z direction. Then according to Hall effect, a force (due of electric field set up) gets exerted on the charge carriers (whether electrons or holes) in the negative-y direction. This current I may be due to holes moving in the positive-x direction or due to free electrons moving in the negative-x direction through the semiconductor specimen. Hence, irrespective of the nature of charge carriers, whether holes or electrons, these charge carriers get pressed towards face 1 of the specimen shown in Fig. 1.24.

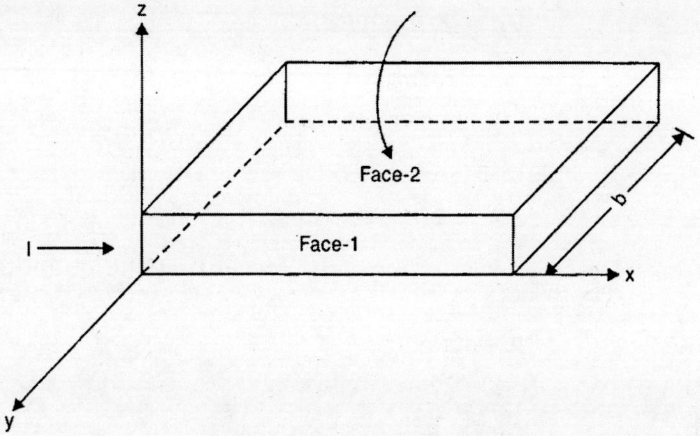

Fig. 1.24. *Current carrying semiconductor bar subject to transverse magnetic field.*

The current, in an n-type specimen is carried almost fully by electrons. These electrons, as a result of Hall effect, accumulate on face-1. As we know that atoms as whole electrically neutral therefore face-2 gets negatively charged. Consequently, a potential difference develops between face-1 and face-2 and is called the 'Hall Voltage' (V_H).

This Hall voltage for an n-type semiconductor is positive at face-2 and negative for an p-type semiconductor at face-1. These two results have been verified experimentally.

The polarity of Hall voltage enables us to determine experimentally whether the semiconductor specimen is of n-type or p-type.

1.12.1. Application of Hall Effect

The Hall effect may be used for :

(*i*) Determining whether a semiconductor is n-type or p-type.

(*ii*) Determining the carrier concentration.

(*iii*) Calculating the mobility, having measured the conductivity.

1.12.2. Experimental Determination of Mobility

The equilibrium condition of the specimen shown in Fig. 1.24 requires that the electrical field intensity due to Hall effect exerts a force on the carrier equal in magnitude to that exerted by the field.

$$q.E = qvB \qquad ...(1.13)$$

But

$$E = \frac{V_H}{d} \qquad ...(1.14)$$

Also current density J is given by

$$J = \frac{I}{b \times d} = \rho v \quad \text{...(1.15)}$$

By combining equations (1.13), (1.14) and (1.15), we get

$$V_H = E.d = vBd$$

or $$V_H = \frac{JBd}{\rho}$$

or $$V_H = \frac{BI}{\rho b} \quad \text{...(1.16)}$$

Thus the value of ρ can be determined by measuring the quantities V_H, B, I and b for a specimen.

The Hall co-efficient R_H is defined as

$$R_H = \frac{1}{\rho} \quad \text{...(1.17)}$$

Equation (1.16) yields

$$R_H = \frac{V_H . b}{BI} \quad \text{...(1.18)}$$

where,
I = Current carried by the specimen
J = Current density
B = Magnetic field
E = Electric field
V_H = Hall voltage
R_H = Hall co-efficient
σ = Conductivity
q = Magnitude of charge of an electron or hole
v = Drift velocity
ρ = Charge density
d = Distance between face-1 and face-2
b = Width of the specimen along the magnetic field.

The conductivity, in any extrinsic semiconductor, is due to primarily charges of only one sign and given as

$$\sigma = \rho\mu \quad \text{...(1.19)}$$

If both R_H and σ are measured, the mobility may be defined from the following relation,

$$\mu = \sigma R_H \quad \text{...(1.20)}$$

In the above discussion it is assumed that all charge carriers move with mean drift speed v. In fact, the charge carriers have a random motion. Taking this distribution into consideration, we find that equation (1.18) is valid provided that R_H is defined as $\frac{3\pi}{8\rho}$ instead of simply as $\frac{1}{\rho}$, so, equation (1.20) gets modified as below :

$$\mu = \left(\frac{8\sigma}{3\pi}\right) R_H \qquad ...(1.21)$$

1.13. INTRODUCTION OF JUNCTION DIODE

We have already discussed about the P-type and N-type semiconductors in this chapter. As a matter of fact, the P-type and N-type semiconductors taken separately, are of little use in actual practice. If we join a piece of P-type semiconductor to a piece of N-type semiconductor such that the crystal structure remains continuous at the boundary as shown in Fig. 1.25 (*a*), a P-N junction is formed.

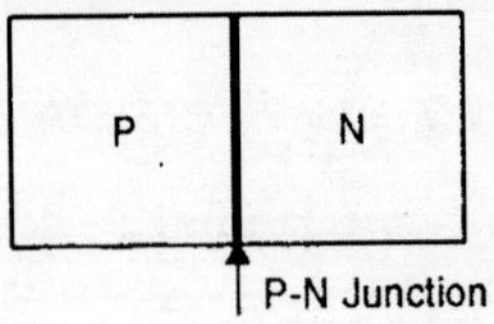

Fig. 1.25 (*a*). *P-N Junction.*

Such a P-N junction forms a very useful device and is called a semiconductor diode, PN junction diode or simply a crystal diode.

The symbol of P-N junction diode is shown in Fig. 1.25 (*b*).

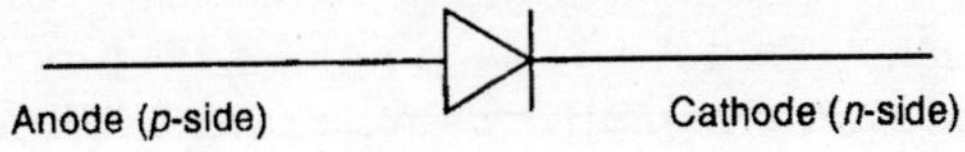

Fig. 1.25 (*b*). *Diode symbol.*

It will be interesting to know that a P-N junction cannot be formed by simply joining or welding the two pieces together, because it would produce a discontinuous crystal structure. Special fabrication techniques are used to prepare P-N junction.

1.14. P-N JUNCTION

Fig. 1.26 shows the general structure of P-N junction. Here in the given figure donor ion is represented by a plus sign because, after the impurity atom 'donates' an electron becomes a positive ion.

Whereas the acceptor ion is indicated by a minus sign because, after this atom 'accepts' an electron, it becomes a negative ion. Initially, there are nominally only P-type carriers to the left of the junction and only N-type carriers to the right. P-type posses holes as a majority carriers and electrons as a minority carriers whereas N-type posses electrons as a majority carriers and holes as a minority carriers.

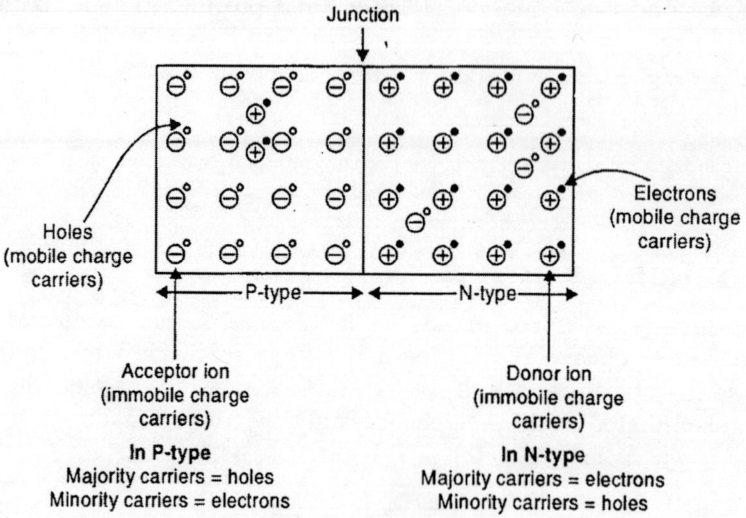

Fig. 1.26. *PN Junction with their charge carriers.*

Since the junction diode is a two terminal device, the application of voltage across its terminals leaves three possibilities :

1. No bias. (*i.e.*, $V_D = 0\ V$)
2. Forward bias. (*i.e.*, $V_D > 0\ V$)
3. Reverse bias. (*i.e.*, $V_D < 0\ V$)

1.14.1. No Bias Across the PN Junction (Formation of Depletion Layer in a PN Junction)

Fig. 1.27 shows a semiconductor consisting of a PN junction which has been just formed. It may be noted that this entire sample is a single crystal. Its left half is a P-type and the right half is N-type. Holes and electrons are mobile charge carriers but acceptor ions and Donor ions are immobile charge carriers. The sample as a whole is electrically neutral and so are the P-region and N-region considered separately. Therefore in the P-region, the charges of moving holes equal the total charges on its free electrons and immobile ions. Similarly, in the N-region, the negative charge of its majority carriers is compensated by the charge of its minority carriers and immobile ions.

It may be noted that no external voltage has been connected to the PN junction of Fig. 1.27. As soon as junction is formed, the conduction and valence bonds of P- and N-type material overlap. As a result of this the following processes are initiated :

(*a*) The holes from the P-region diffuses into the N-region. They then combine with the free electrons in the N-region.

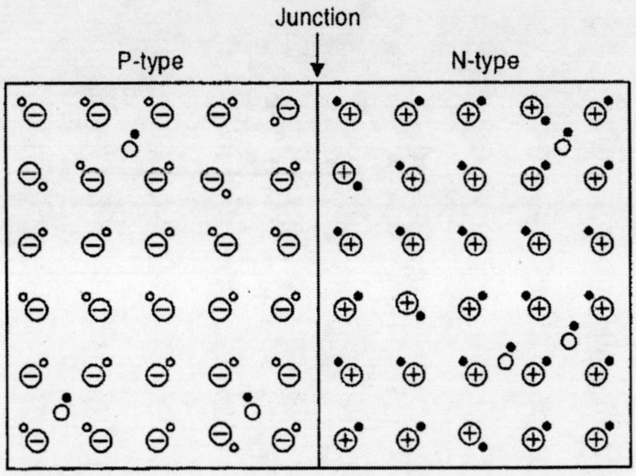

Fig. 1.27. *A P N-Junction when just formed.*

(*b*) The free electrons from the N-region diffuses into the P-region. These electrons combine with the holes.

The process (*a*) and (*b*) can be easily understood by Fig. 1.28 as shown below:

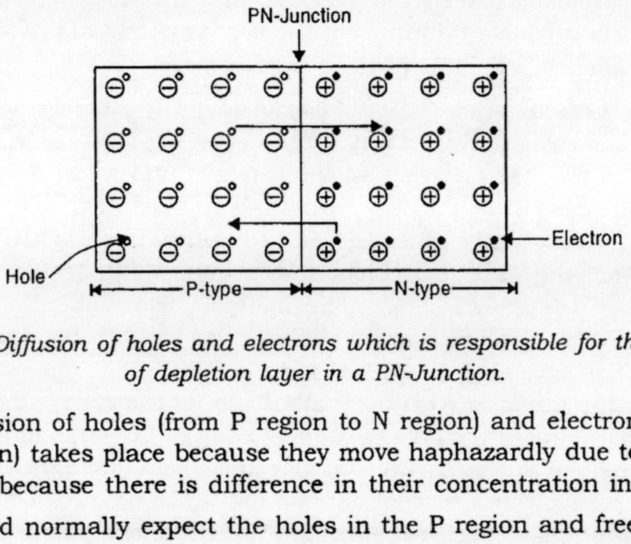

Fig. 1.28. *Diffusion of holes and electrons which is responsible for the formation of depletion layer in a PN-Junction.*

(*c*) The diffusion of holes (from P region to N region) and electrons (from N region to P region) takes place because they move haphazardly due to thermal energy, and also because there is difference in their concentration in the two regions.

(*d*) One would normally expect the holes in the P region and free electrons in the N region to flow towards each other and combine. But in actual practice this does not occur. The diffusion of holes and electrons across the junction occurs for a very short time. After a few recombination of holes and free electrons, in the vicinity of the junction, a restraining force is automatically set up. This force is produced due to *depletion region* or the *space-charge region* or the *transition region*. Which exists on either side of the junction. The thickness of this region is of the order of wavelength of visible light ($\cong$ 0. 5 μm). As a result of this, further diffusion of holes and free electrons from one region to the other is stopped by this *depletion region (layer)*. It may be noted that depletion region consists of only uncompensated acceptor ions and donor ions. Fig. 1.29 shows the formation of depletion layer.

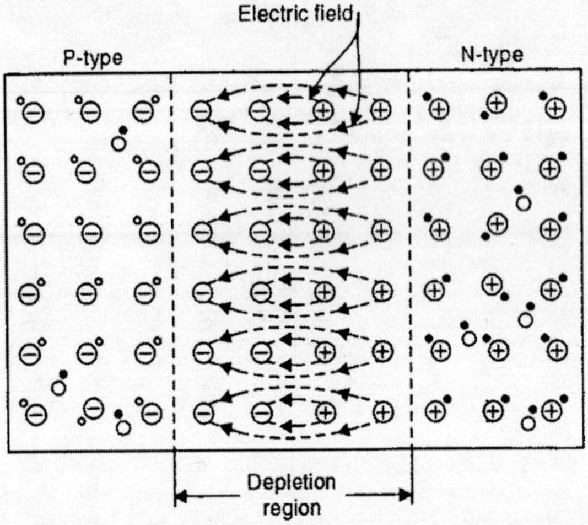

Fig. 1.29. *Space-charge region or depletion region is formed in the vicinity of the junction.*

The electric field between the acceptor and the donor ion is called a *barrier potential.* For a silicon PN junction, the barrier potential is about 0.7 V, whereas for a germanium PN junction it is approximately 0.3 V.

(*e*) The barrier discourages the diffusion of majority carriers across the junction. But here one important question arises—what happens to the minority carriers ? There are a few free electrons in the P region and a few holes in the N region. The barrier helps these minority carriers to drift across the junction. The minority carriers are constantly generated due to thermal energy. It will be interesting to know that the width of depletion layer depends upon the doping level of the impurity in N-type or P-type semiconductor. The higher the doping level, the thinner will be depletion layer and vice versa. This is due to the fact that potential barrier exerts a repelling force on the mobile charge carriers, trying to crossover the junction. This force stops the mobile charge carriers to crossover the junction, unless the energy is supplied from an external force. Does it mean there would be a current due to the movement of these minority carriers ? Certainly not. Electric current cannot flow since no circuit has been connected to the PN-junction. The drift of minority carriers across the junction is counter balanced by the diffusion of the same number of majority carriers across the junction.

Thus, we conclude that a barrier voltage is developed across the PN-junctions even if no external battery is connected.

NOTE : The barrier potential of a silicon diode decreases by 2 mV for each degree celsius rise

i.e., $\frac{dV}{dt} = -2 \text{ mV/°C} = \frac{\Delta V}{\Delta T}$.

1.14.2. PN-Junction with Forward Bias (*i.e.* $V_D > 0$ V)

If we connect a battery to the PN-junction diode such that the positive terminal of the battery is connected to the P-side and negative terminal to the N-side, as shown in Fig. 1.30 (*a*). In this situation the PN-junction is said to be forward-biased.

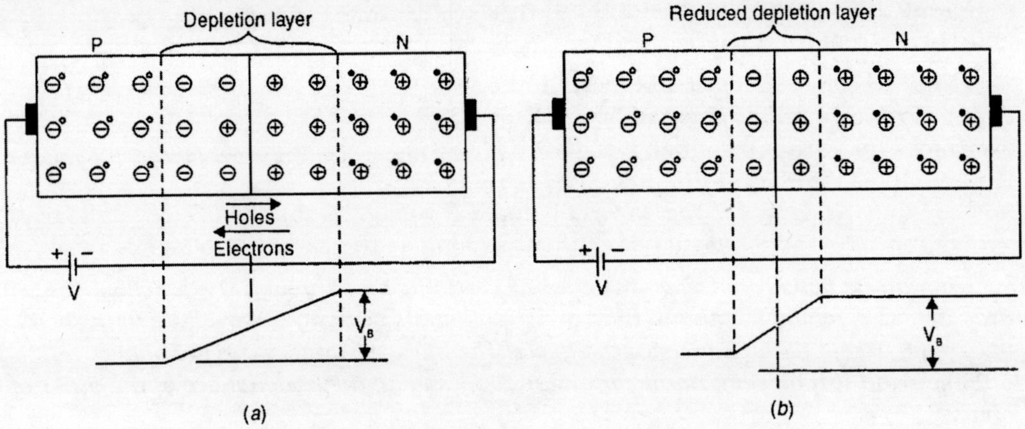

Fig. 1.30. *PN-Junction with forward bias.*

When the PN-junction is forward-biased, the holes are repelled from the positive terminal of the battery and are compelled to move towards the junction. Similarly, the electrons are repelled from the negative terminal of the battery and drift towards the junction. Because of their acquired energy (from the voltage source), some of the holes and electrons enter the depletion layer and recombine themselves. *This reduces the width as well as height of the potential barrier* (V_B) as shown in Fig. 1.30 (*b*). As a result of this, more majority carriers diffuse across the junction. Therefore it causes a large current to flow through the PN-junction.

It may be noted that for each recombination of free electron and hole, which occurs, an electron from the negative terminal of the voltage source enters the N-type region. Then it moves towards the junction. Similarly, in the P-type region near the positive terminal of the voltage source, an electron breaks a covalent bond in the crystal and enters the positive terminal of the voltage source. Thus for each electron, which break its bond, a hole is created. This hole drift towards the junction. The current through the external circuit, is due to the movements of electrons only. On the other hand, the current within the PN-junction is the sum of electron current and hole current.

Thus, the current in the external circuit continues to flow as long as the voltage source is present in the circuit. The current increases with increase in applied voltage and is of the order of several milliamperes. The maximum value of current depends upon the actual resistance, called *bulk resistance* of the semiconductor material. The bulk resistances are ohmic resistances of P-type and N-type semiconductor materials. Usually the values of bulk resistances are order of few ohms.

Effect of Barrier Potential on the Forward Biased PN-junction

The barrier potential of the depletion layer can be considered as a small battery, which opposes the external d.c. voltage as shown in Fig. 1.31. The resistance R_p and R_n represent the bulk resistances of the P-type and N-type semiconductors.

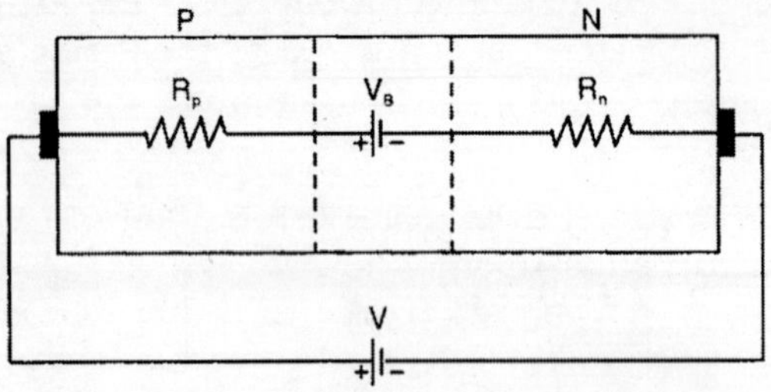

Fig. 1.31. *Forward biased PN-junction.*

The PN-junction does not permit the current to flow, until the external bias voltage overcomes the barrier potential, (*i.e.*, ($V \ngtr V_B$). For example, the *silicon* PN junction, does not conduct as long as the external applied voltage is below 0.7 V. Similarly, the germanium PN junction does not conduct as long as the external voltage is below 0.3 V.

It may be noted, that once the current starts flowing through the junction, the voltage drop across it remains equal to the barrier potential. It changes very little with the change in current, except for bulk resistance effects. The bulk resistances are the ohmic resistances of P-type and N-type semiconductor materials. Usually, their values are of the order of few ohms.

1.14.3. PN-Junction with Reverse Biased (*i.e.*, $V_D < 0$ V)

If we connect a voltage source to a PN-junction, such that positive terminal of the voltage source is connected to the N-region and negative to the P-region, then the PN-junction is said to be reversed biased. Fig. 1.32 shows a reverse biased PN-junction.

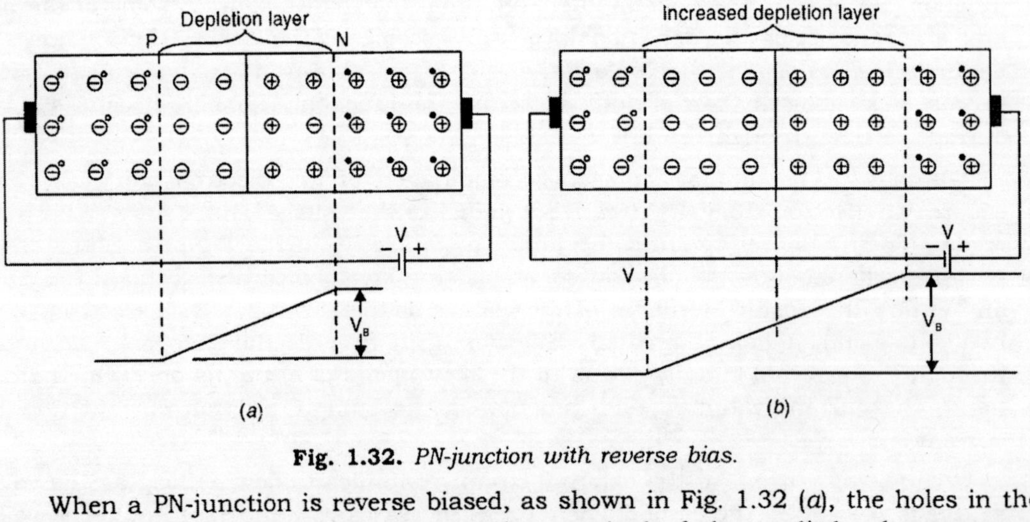

Fig. 1.32. *PN-junction with reverse bias.*

When a PN-junction is reverse biased, as shown in Fig. 1.32 (*a*), the holes in the P-region are attracted towards the negative terminal of the applied voltage source, whereas the electrons in the N-region are attracted to the positive terminal of the voltage source. Thus the majority carriers are drawn away from the junction. *This widens the depletion layer and increases the barrier potential* as shown in Fig. 1.32 (*b*).

The increased barrier potential makes it very difficult for the majority carriers to diffuse across the junction. Thus there is no current due to majority carriers in a reverse bias PN-junction. Or in other words we can say that the junction offers very high resistance under reverse biased condition.

However, the barrier potential helps the minority carriers in crossing the junction. As a matter of fact, as soon as a minority carrier is generated, it is swept (*i.e.*, drifted) across the junction because of the barrier potential. Hence a small amount of current (called reverse saturation current or leakage current or current due to minority charge carriers) does flow through the reverse biased PN-junction. The amount of this current depends upon the generation of minority carriers diffusing across the junction.

NOTE :

(*i*) It may be noted that generation of minority carriers depends upon the temperature and is independent of the applied reverse voltage. Therefore the current due to flow of minority carriers remains the same whether the applied voltage is increased or decreased. Because of this reason, the current is known as reverse saturation current.

(*ii*) The reverse saturation current is of the order of nanoamperes (*n*A) for silicon and microamperes (μA) for Germanium PN-junctions. The reverse saturation current for, Si devices is smaller than that of Ge because silicon has a large energy gap (≈ 1.1 eV) between the conduction and valence bands compare to the energy band gap of (≈ .7 eV). As a result of this, less minority carriers are thermally generated at any temperature in silicon than that of germanium. This is the reason why Si is mostly preferred over Ge.

(*iii*) **Temperature dependency of reverse saturation current.** It has been experimentally found that the value of reverse saturation current (I_0) is about double in magnitude for every 10°C rise in temperature. If $I_0 = I_0$, at $T = T_1$, then at temperature T, I_0 is given by

$$I_0\ (T) = I_0 \times 2^{\frac{(T - T_1)}{10}} \qquad \text{...(1.22)}$$

For example : if for silicon I_0 is 5 μA at 25°C, then it is approximately 10 *nA* at 35°C.

1.15. V-I CHARACTERISTICS OF A PN-JUNCTION DIODE

In order to know how a device responds when it is connected in an electrical circuit, *V-I* characteristics of PN-junction diode is analysed. It is the graph between the voltage applied across its terminals and the current that flows through it. The *V-I* characteristic of an ideal PN-junction diode is shown in Fig. 1.33. For a PN-junction, the current *I* is related to the voltage *V* by the equation.

$$I = I_0\left(e^{\frac{V}{\eta V_T}} - 1\right) \qquad \text{...(1.23)}$$

where I = Diode current (+ve value of I means current flows from p to n side)

I_0 = Reverse saturation current

V = Applied voltage $\begin{cases} \text{+ve for forward biased} \\ \text{−ve for reverse biased} \end{cases}$

$V_T \cong$ Volt equivalent by temperature

$$\cong \frac{T}{11{,}600} \text{ volt.}$$

$$\eta = \begin{cases} 1 \text{ for germanium, and} \\ 2 \text{ for silicon at rated current} \end{cases}$$

At room temperature (T = 300° K), V_T = 26 mV eqn. (1.23) represents the general form of V-I characteristics.

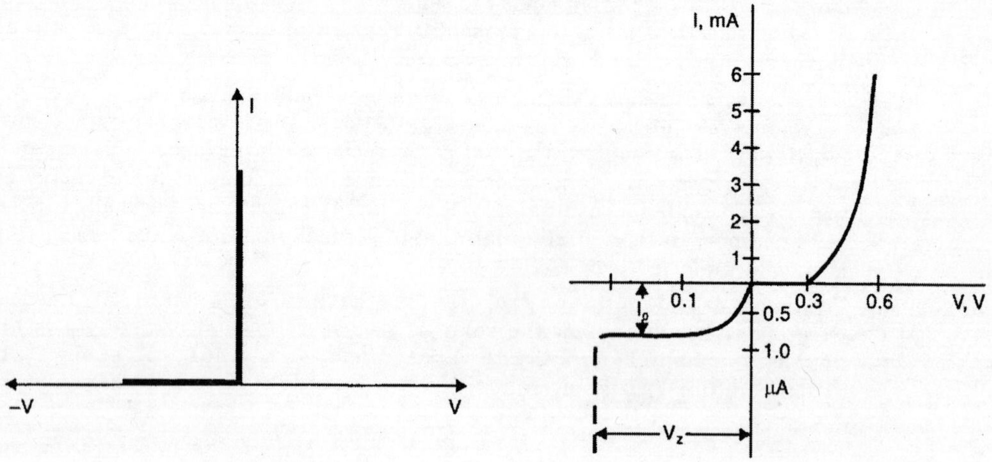

(*a*) The volt-ampere characteristic of an ideal PN diode.

(*b*) The volt-ampere characteristic for a practical germanium diode. The dashed portion indicates breakdown at V_Z.

Fig. 1.33.

When the voltage V is positive and several times V_T (*i.e.*, $V >> V_T$) eqn. (1.23) may be written as :

$$I = I_0 \left(e^{\frac{V}{\eta V_T}} \right) \quad \text{... (1.24)}$$

When the voltage V is negative (*i.e.*, diode is reverse biased) and $|V|$ is several times V_T ($|V| >> V_T$), then the eqn. (1.23) may be written as :

$$I \cong -I_0 \quad \text{... (1.25)}$$

The reverse current is therefore constant, independent of the applied reverse bias. Therefore, I_0 is called as reverse saturation current.

As we have already discussed that temperature is main determining factor of leakage current. On increasing the value of temperature the value of cut in voltage continuosly decreases, this can be better understood by Fig. 1.33 (*c*).

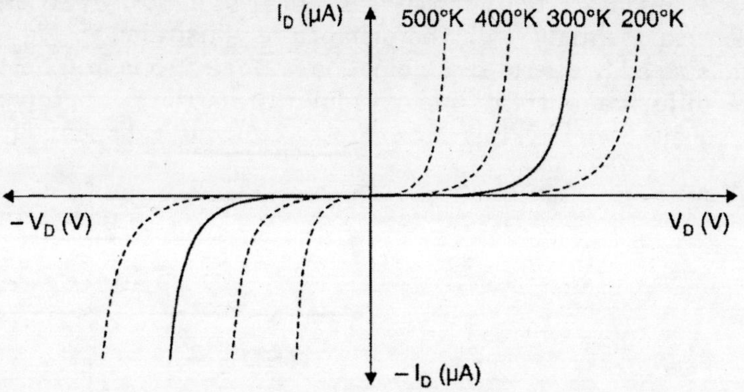

Fig. 1.33 (c). *Vartion in V-I characteristic of silicon diode due to temperature changes.*

Practical PN-junction Diode and Concept of Cutin Voltage (V_r)

From Fig. 1.33 it is clear that for small value of applied voltage V there will be a definite value of the diode current I. But in actual practice it is not possible, as there is some voltage (equal to cut in voltage V_r) is drop across the depletion layer of the junction, or in other words we can say that diode does not conduct well until the external voltage (*i.e.* V) overcome the barrier potential. Cut is voltage (V_r) also called knee voltage.

For a silicon diode it is approximately 0.7 V for Si diode and approximately 0.3 V for Ge diode as shown in Fig. 1.34 (*a*). The cut in voltage of Si is equal to 0.7 V means

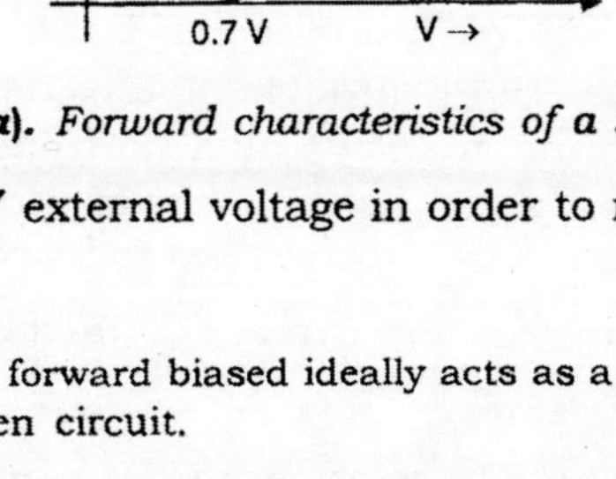

Fig. 1.34 (a). *Forward characteristics of a silicon diode.*

that we require minimum 0.7 external voltage in order to make the diode in conducting mode.

NOTE : PN-junction diode in forward biased ideally acts as a short circuit, whereas in reverse biased acts as a open circuit.

1.15.1. Current Composition in a P-N Junction Diode

Fig. 1.34 (*b*) shows a P-N junction diode with a forward bias with an external voltage V. Because of forward bias, there exists a potential gradient in P and N materials. Now, the holes from the P-region and electrons from the N-region are injected into the N-region and P-region respectively. The holes injected into the N-region and electrons injected into the *P*-region are the minority carriers. The minority carriers diffuse away from the junction exponentially with the distance as illustrated in Fig. 1.34 (c). Their concentration falls steadily due to recombination with electrons and holes respectively. Now, since the diffusion current due to minority carriers is proportional to the concentration gradient, therefore, this current must also vary exponentially with distance

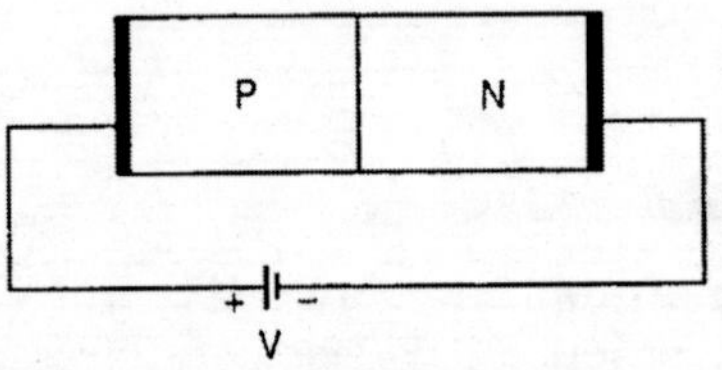

Fig. 1.34 (*b*)

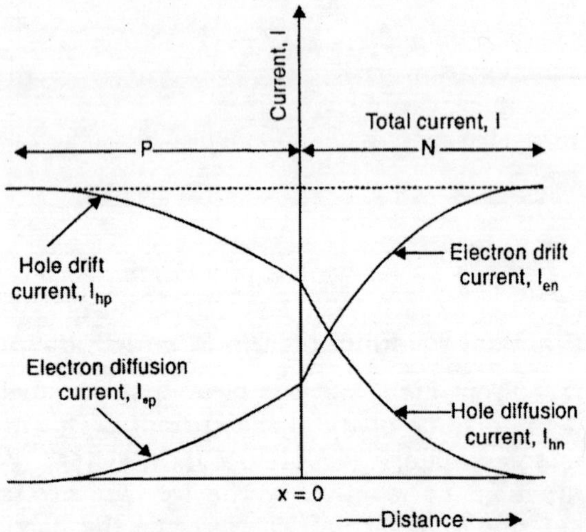

Fig. 1.34 (c). *Current components in a forward biased P-N junction diode*

There are two minority (*i.e.*, diffusion) current denoted as I_{ep} and I_{hm} as illustrated in Fig. 1.34 (*c*).

Let the symbols $I_{ep}(x)$ and $I_{hn}(x)$ represent the electron current in the *P*-region and the hole current in the *N*-region respectively.

Electrons crossing the junction at $x = 0$ from right to the left constitute a current in the same direction as the holes crossing the junction from left to right.

Hence, the total current *I* at junction (*i.e.*, x = 0) may be expressed as under :

$$I = I_{hn}(0) + I_{ep}(0)$$

Since the current is the same throughout a series circuit, therefore the total current *I* is independent of *x* and is shown as a horizontal line in Fig. 1.34 (*c*). Thus, in the *P*-region, there must be a second component of current I_{hp} which when combined with I_{ep} gives the total current *I*.

Therefore, the majority (*i.e.*, hole) current, I_{hp} in the *P*-region is expressed as

$$I_{hp}(x) = I - I_{ep}(x)$$

Similarly, the majority (*i.e.*, electron) current, I_{en} in the *N*-region is expressed as

$$I_{en}(x) = I - I_{hn}(x)$$

The currents I_{np} and I_{en} are plotted as a function of distance in Fig. 1.34 (*c*). Fig. 1.34 (*b*) is drawn for unsymmetrically doped diode, so that

$$I_{ep} \neq I_{hn}$$

Now, it may be noted that far away from the junction in *P*-region, the current is drift current I_{hp} of holes sustained by small electric field in the semi-conductor. As the holes approach the junction, some of them recombine with the electrons which have been injected into the *P*-region from the *N*-region. Hence, part of the current I_{hp} becomes a negative current just equal in magnitude of the diffusion current I_{ep}. The drift current I_{hp}, therefore, decreases towards the junction. The rate of decrease is such that the total current remains constant, independent of distance. The reduced hole current at the junction now enters the *N*-region and becomes hole diffusion current I_{hn}. A similar statement can be made with respect to electron drift current in the *N*-region I_{en}. Hence, in a forward-biased *P-N* diode, the current enters the *P*-region as a hole current and leaves the *N*-region as an electron current of the same magnitude.

From the above discussion, it may be concluded that the current in a *P-N* diode is bipolar in character (since it is made up of both positive and negative charge carriers) and the total current is constant throughout the device, but the proportion of current due to holes and that due to electrons varies with the distance as shown in Fig. 1.34 (*c*).

1.16. STATIC AND DYNAMIC RESISTANCE OF A PN-JUNCTION DIODE

One thing should be always kept in mind that no diode can act as an ideal diode. An actual diode does not behaves as a perfect conductor when forward-biased, and as a perfect insulator when reverse-biased. It does not offer zero resistance when forward-biased, also not infinite resistance when reverse biased.

Generally there are two types of resistance of a PN-junction diode is considered namely :

(*i*) Static resistance

(*ii*) Dynamic resistance.

(*i*) **Static Resistance.** Static resistance of a PN-junction diode is calculated when the diode is connected in a d.c. circuit. This resistance is also known as d.c. resistance or static resistance, or we can say it is simply the ratio of d.c. voltage across the diode to the d.c. current flowing through it. It is denoted by R_F.

(*ii*) **Dynamic Resistance.** Dynamic resistance of a PN-junction diode may be defined as the resistance offered by the diode to the a.c. signal is called dynamic or a.c. resistance. Dynamic resistance of a diode is equal to the slope of *V-I* characteristics $\left(\frac{dV}{dI} \text{ or } \frac{\Delta V}{\Delta I}\right)$ of the diode.

i.e.,

$$\boxed{r_f = \frac{\text{Change in voltage}}{\text{Resulting change in current}} = \frac{dV}{dI} = \frac{\Delta V}{\Delta I}}$$

Consider the eqn. (1.23)

$$I = I_0\left(e^{\frac{V}{\eta V_T}} - 1\right)$$

Differentiating eqn. (1.23) with respect to V we get dynamic conductance

$$g = \frac{dI}{dV} = \frac{I_0 e^{\frac{V}{\eta V_T}}}{\eta V_T}$$

or $$g = \frac{I + I_0}{\eta V_T} \qquad \text{(From equation (1.23))}$$

or $$r = \frac{\eta V_T}{I + I_0} \qquad \text{...(1.26)}$$

For a ϕ reverse biased junction (*i.e.*, $\frac{V}{\eta V_T}$ >> '). g is extremely small and r is very large. On the other hand, for a forward biased junction, $I > I_0$, so equation (1.26) becomes

$$\boxed{r \approx \frac{\eta V_T}{I}} \qquad \text{...(1.27)}$$

From above expression it is clear that dynamic resistance inversely proportional to the reverse saturation current.

1.17. DIODE CAPACITANCE

There are usually two types of diode capacitances namely :

(1) Depletion layer capacitance or Transition capacitance (C_T).

(2) Diffusion or Storage capacitance (C_D).

(1) Depletion Layer Capacitance or Transition Capacitance (C_T)

We have already discussed that when a PN-junction is formed, a layer of positive and negative ions called depletion layer is formed on either side of the junction. Since this capacitance is formed in the junction area, therfore, it is also called space charge capacitance. This capacitance is arises due to immobile charges at the junction varying with the applied voltage, that's why this is simply called junction capacitance. Fig. 1.35 shows the formation of depletion layer capacitance.

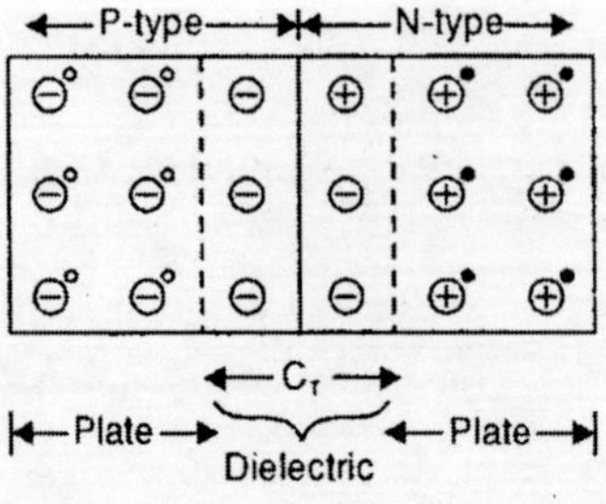

Fig. 1.35. *Depletion layer capacitance.*

The capacitance of a parallel plate capacitor is given by the equation $C = \frac{\epsilon_0 \cdot A}{d}$ where ϵ_0 is the permitivity of dielectric (insulator) between the plates of area A separated by a distance d. Since the depletion layer width (d) increases with the increase in reverse bias voltage, the resulting depletion layer capacitance will decrease with the increased reverse bias.

The depletion layer capacitance depends upon the nature of a PN-junction, semiconductor material and magnitude of the applied reverse voltage. It is given by the relation :

$$\boxed{C_T = \frac{C_0}{\left(1 + \frac{V_r}{V_T}\right)^{\eta}}} \qquad \text{... (1.28)}$$

where, C_0 = Capacitance at zero bias condition

V_r = Applied reverse voltage

V_T = Volt equivalent of temperature

$$\eta = \begin{cases} \frac{1}{2} & \text{for step or abrupt alloy junction} \\ \frac{1}{3} & \text{for linearly graded junction or diffused junction.} \end{cases}$$

It is evident from the above relation that the value of depletion layer capacitance (C_T) can be controlled by varying the applied reverse voltage. This property of variable capacitance, possessed by a reverse biased PN-junction is used in the construction of a device called *varicap* of *varacter*, which is chiefly used in FM circuits. Range of depletion layer capacitance is approximately in nF (nanofarad).

(2) Diffusion or Storage Capacitance (C_D)

The capacitance which exists in a forward-biased junction is called a diffusion or storage capacitance. It is different from the depletion layer capacitance, which exists in a reversed biased junction. The diffusion capacitance arises due to the arrangement of minority carrier density. The value of diffusion capacitance is much larger than the depletion layer capacitance (*i.e.*, $C_D >> C_T$). Its value for abrupt junction is given by the relation :

$$\boxed{C_D = \frac{\tau \cdot I}{\eta \cdot V_T}} \qquad \text{... (1.29)}$$

where τ = Mean life time of the carriers

I = Forward current

V_T = Volt equivalent of temperature

$$\eta = \text{Constant} = \begin{cases} 1 \text{ for Ge} \\ 2 \text{ for Si} \end{cases}$$

It is clear from eqn. (1.29) that, the diffusion capacitance is directly proportional to the forward current (I). Consider a forward-biased PN-junction carrying a current of

(I) amperes through it. Suppose the junction is suddenly reverse biased. It causes the forward current to reduce quickly to zero. But it leaves a number of majority charge carriers stored within the depletion region. This charge represents a stored charge in the reverse biased condition and must be removed out of the space charge region. The removal of stored charge takes a finite time. It represents an effect similar to the discharging of a capacitor. Thus the quantity of the stored charge represents the magnitude of diffusion capacitance. It may be noted that the effect of diffusion capacitance is negligible for a reverse-biased PN-junction.

1.18. DIFFERENT RATINGS OF DIODE

There are different types of rating in a diode given below :

(*i*) Maximum power rating

(*ii*) Peak inverse voltage rating (PIV)

(*iii*) Maximum forward current rating.

(*i*) **Maximum Power Rating.** Maximum power rating of a diode is defined as the maximum power that a *pn*-junction (or diode) can dissipate without damaging it is called its *maximum power rating.*

Usually, maximum power rating in specified by the manufacturer in its data sheet. The power dissipated at the junction is equal to the product of junction current and the voltage across the junction. If the power developed across the junction is more than the maximum power dissipated by it, the junction will be over-heated and may be destroyed.

(*ii*) **Peak Inverse Voltage Rating (PIV).** Peak inverse voltage rating of a diode may be defined as the maximum value of reverse voltage that a *p-n* junction (or diode) can withstand without damaging it is called the *peak inverse voltage rating* (*PIV*) of a diode.

This rating of a *p-n* junction or diode is also specified by the manufacturer in its data sheet. However, it may be noted that here if we increase the voltage across the junction at reverse bias condition beyond this specified value, the junction will be destroyed.

(*iii*) **Maximum Forward Current Rating.** The maximum forward current rating of a diode may be defined as the maximum value of forward current that a *pn*-junction or diode can carry without damaging itself called its *maximum forward current rating* of a diode.

? SELF-TEST (SEMICONDUCTOR MATERIALS)

1. Define Resistivity and Conductivity.
2. Explain hole Mechanism in brief.
3. Why energy bands are necessary ?
4. What is an insulator ? What is fermi level ? Why does a pure semiconductor behaves like an insulator at absolute zero ?
5. Differentiate between intrinsic and extrinsic semiconductors. Give examples.

6. Explain clearly the difference between acceptor and donor impurities and state what type of carriers are contributed to them.

7. (*i*) Differentiate between intrinsic and extrinsic semiconductors.

(*ii*) What is an insulator ?

(*iii*) What is a conductor ?

8. Define the terms conductivity, intrinsic concentration and energy gap of semi- conductors.

9. Differentiate semiconductors, conductors and insulators on the basis of band gap. What is the mechanism by which conduction takes place inside the semiconductor ?

10. What is meant by fermi level in semiconductors ? Where does the fermi level lie in an intrinsic semiconductor ? Prove that the fermi level in *n*-type material is much closer to the conduction band.

11. Which of the two semiconductor materials have Si or Ge has larger conductivity at room temperature and why ? What happens to the conductivity of the semiconductor with the rise in temperature ?

12. Sketch the energy-band picture for :

(*i*) an intrinsic semiconductor

(*ii*) an *n*-type semiconductor

(*iii*) an *p*-type semiconductor.

SOLVED EXAMPLES

Example 1. *A semiconductor wafer is 0.5 mm thick.*

(a) A potential of 100 mV is applied across the thickness, what is the electron drift velocity if their mobility is 0.2 m²/V-s ?

(b) How much time is required for an electron to move across the thickness.

Solution : (*a*) We know that,

$$\text{Mobility, } \mu = \frac{V_d}{E}$$

or

$$\text{Drift velocity, } V_d = \mu \times E$$

$$= \mu \times \frac{V}{d} \qquad \{\because V = E \times d\}$$

$$= 0.2 \times \frac{100 \times 10^{-3}}{.5 \times 10^{-3}} = 40 \text{ m/s.} \quad \textbf{Ans.}$$

(*b*) Time required,

$$T = \frac{d}{V_d} = \frac{.5 \times 10^{-3}}{40} = 12.5 \ \mu\text{S.} \quad \textbf{Ans.}$$

Example 2. *The mobilities of free electrons and holes in a pure germanium are 0.38 and 0.18 m²/V-s. Find the value of intrinsic conductivity. Assume* $n_i = 2.5 \times 10^{19}/m^3$ *at room temperature.*

Solution : Intrinsic conductivity for germanium

$$\sigma_i = q \cdot n_i (\mu_n + \mu_p)$$

or $$\sigma_i = 1.6 \times 10^{-19} \times 2.5 \times 10^{19} (0.38 + 0.18)\ (\Omega\text{-m})^{-1}$$

or $$\sigma_i = 2.24\ (\Omega\text{-m})^{-1}.$$ **Ans.**

Example 3. *Find the intrinsic carrier concentration of germanium, if its intrinsic resistivity at 300 K is 0.50 Ω-m. It is given that the electronic charge is* 1.6×10^{-19} *coulombs and that electron and hole mobilities at 300 K are 0.39 and 0.19 m²/V-s respectively.*

Solution : Given :

$$\rho = 0.50\ \Omega\text{-m}$$
$$q = 1.6 \times 10^{-19}\ \text{C}$$
$$\mu_n = 0.39\ \text{m}^2/\text{V-s}$$
$$\mu_p = 0.19\ \text{m}^2/\text{V-s}$$

Let n_i = Intrinsic carrier concentration of germanium.

We know that conductivity of a semiconductor,

$$\sigma = \frac{1}{\rho} = \frac{1}{.5} = 2\ (\Omega\text{-m})^{-1}$$

and the conductivity in intrinsic semiconductor (σ)

$$\sigma = q \cdot n_i (\mu_n + \mu_p)$$
$$2 = 1.6 \times 10^{-19} \cdot n_i (0.39 + 0.19)$$

or $$n_i = \frac{2}{1.6 \times 10^{-19} (0.58)}$$

$$= 2.155 \times 10^{19}/\text{m}^3.$$ **Ans.**

Example 4. *A sample of silicon is doped with phosphorous to a density of* $10^{21}/m^3$ *as well as with boron to a density of* $5 \times 10^{20}/m^3$*. What will be conductivity of the Si sample. The electron mobility in silicon is 0.18 m²/V-s.*

Solution : Given :

$$N_D = 10^{21}/\text{m}^3$$
$$N_A = 5 \times 10^{20}/\text{m}^3$$

Since, phosphorous atoms are donor and boron atoms are acceptor, therefore, net donor density.

$$N_D' = N_D - N_A = 10^{21} - 5 \times 10^{20} = 5 \times 10^{20}/\text{m}^3$$

∴ Number of free electrons, $n = N_D' = 5 \times 10^{20}/\text{m}^3$

We know that conductivity of silicon,

$$\sigma = q \cdot n \cdot \mu_n$$
$$= 1.6 \times 10^{-19} \times 5 \times 10^{20} \times 0.18$$
$$= 14.4\ (\Omega\text{-m})^{-1}.$$ **Ans.**

Example 5. *Find the conductivity of the copper if the mobility is given as 34.8 cm^2/V and density is given as 8.9 gm/cm^3. Assume one free electron per atom. Atomic weight of copper is 63.54.*

Solution : Concentration of electrons n can be given as

$$n = 6.023 \times 10^{23}(\text{atom/mole}) \times (1 \text{ mole}/63.54 \text{ gm}) \times 8.9 \text{ (gm/cm}^3)$$

or $$n = 8.44 \times 10^{22} \text{ electrons/cm}^3.$$

We know that

$$\sigma = nq\mu$$
$$= 8.44 \times 10^{22} \times 1.6 \times 10^{-19} \times 34.8$$
$$= 4.7 \times 10^5 \text{ mho/cm. } \textbf{Ans.}$$

Example 6. *Find the concentration of holes and electrons in a p-type germanium at 300 K, if the conductivity is 100 per ohm per cm [i.e., 100 $(\Omega\text{-m})^{-1}$]. Also find these values for N-type silicon, if the conductivity is 0.1 $(\Omega\text{-m})^{-1}$. Given for Germanium, n_i = 2.5 × $10^{13}/cm^3$; μ_n = 3800 cm^2/V-sec; μ_p = 1800 cm^2/V-sec and for silicon n_i = 1.5 × $10^{10}/cm^3$; μ_n = 1300 cm^2/V-sec and μ_p = 500 cm^2/V-sec.*

Solution : Given for Germanium,

$$\sigma = 100\ (\Omega\text{-m})^{-1}$$
$$n_i = 2.5 \times 10^{13}/\text{cm}^3$$
$$\mu_n = 3800 \text{ cm}^2/\text{V-sec}$$
$$\mu_p = 1800 \text{ cm}^2/\text{V-sec}$$

Let n = Concentration of electrons

p = Concentration of hole

Concentration in p-type Germanium :

We know that conductivity of a p-type Germanium (σ),

$$\sigma = q \cdot p \cdot \mu_p$$
$$100 = 1.6 \times 10^{-19} \times p \times 1800$$
$$p = \frac{100}{1.6 \times 10^{-19} \times 1800}$$

or $$p = 3.47 \times 10^{17}/\text{cm}^3. \ \textbf{Ans.}$$

Therefore, concentration of electrons,

$$n = \frac{n_i^2}{p} \qquad \{\because\ np = n_i^2 \text{ (mass-action law)}\}$$

$$n = \frac{(2.5 \times 10^{13})^2}{3.47 \times 10^{17}} = 1.8 \times 10^9/\text{cm}^3. \ \textbf{Ans.}$$

Concentration in n-type Silicon :

$$\sigma = q \cdot n \cdot \mu_n$$
$$0.1 = 1.6 \times 10^{-19} \times n \times 1300$$

$$n = \frac{0.1}{1.6 \times 10^{-19} \times 1300} = 4.8 \times 10^{14}/\text{cm}^3. \quad \textbf{Ans.}$$

and the concentration of holes,

$$p = \frac{n_i^2}{n} = \frac{(1.5 \times 10^{10})^2}{4.8 \times 10^{14}} = 4.7 \times 10^5/\text{cm}^3. \quad \textbf{Ans.}$$

Example 7. *In a germanium sample, a donor type impurity is added to the extent of 1 atom per 10^8 germanium atoms. Show that resistivity of the germanium sample drops to 3.7 ohm-cm. Given : μ_n = 3800 cm²/V-sec; μ_p = 1800 cm²/V-sec; $n_i = 2.5 \times 10^{13}/cm^3$; $N_{ge} = 4.41 \times 10^{22}/cm^3$ and $q = 1.602 \times 10^{-19}$ coulombs.*

Solution : Given : μ_n = 3800 cm²/V-sec; μ_p = 1800 cm²/V-sec; $n_i = 2.5 \times 10^{13}/\text{cm}^3$; $N_{ge} = 4.41 \times 10^{22}$ cm³ and $g = 1.602 \times 10^{-19}$ coulombs.

We know that the number of donor atoms,

$$N_D = \frac{N_{ge}}{10^8} = \frac{4.41 \times 10^{22}}{10^8} = 4.41 \times 10^{14}/\text{cm}^3$$

Since, the number of free electrons (n) is approximately equal to the donor atoms (N_D), therefore $n \approx 4.41 \times 10^{14}/\text{cm}^3$. We know that number of holes,

$$p = \frac{n_i^2}{N_D} = \frac{(2.5 \times 10^{13})^2}{4.41 \times 10^{14}} = 1.417 \times 10^{12}/\text{cm}^3$$

It may be noted that the number of electrons (4.41×10^{14}) is very large as compared to the number of holes (1.417×10^{12}) in a doped germanium. Therefore, conductivity of an N-type germanium,

$$\sigma = q \cdot n \cdot \mu_n = (1.602 \times 10^{-19}) \times (4.41 \times 10^{14}) \times 3800$$
$$= 0.268\ (\Omega\text{-cm})^{-1}$$

∴ Resistivity of a doped germanium,

$$\rho = \frac{1}{\sigma} = \frac{1}{0.268} = 3.73 \text{ ohm-cm.} \quad \textbf{Ans.}$$

Example 8. *Find the resistivity of intrinsic silicon. What will this change to when the silicon is doped with a pentavalent impurity to the extent of one impurity atom for each 50 million silicon atoms ? Following data may be used.*

Number of silicon atoms = $4.96 \times 10^{22}/cm^3$

Intrinsic carrier concentration = $1.52 \times 10^{10}/cm^3$

Electron charge = 1.6×10^{-19} coulomb

Electron mobility = 0.135 m²/Volt-sec

Hole mobility = 0.048 m²/Volt-sec.

Solution : Given : $N_{si} = 4.96 \times 10^{22}/\text{cm}^3$; $n_i = 1.52 \times 10^{10}/\text{cm}^2$; $q = 1.6 \times 10^{-19}$ C; μ_n = 0.135 m²/V-sec = 1350 cm²/V-sec and μ_p = 0.048 m²/V-sec = 480 cm²/V-sec.

Resistivity of intrinsic silicon :

We know that conductivity of an intrinsic silicon,

$$\sigma = q \cdot n_i (\mu_n + \mu_p) = (1.6 \times 10^{-19}) \times (1.52 \times 10^{10}) \times (1350 + 480)\ (\Omega\text{-cm})^{-1}$$

$$= 4.5 \times 10^{-6}\ (\Omega\text{-cm})^{-1}$$

$\therefore$ Resistivity of intrinsic silicon,

$$\rho = \frac{1}{\sigma} = \frac{1}{4.5 \times 10^{-6}} = 2.22 \times 10^{5}\ \Omega\text{-cm.} \quad \textbf{Ans.}$$

Resistivity of doped silicon :

We know that if the silicon is doped with a pentavalent impurity to the extent of one impurity atom for each 50 million (50×10^6) silicon atoms, then the number of donor atoms,

$$N_D = \frac{N_{si}}{50 \times 10^6} = \frac{4.96 \times 10^{22}}{50 \times 10^6} = 9.92 \times 10^{14}/\text{cm}^3$$

$\therefore$ Number of free electrons in a semiconductor,

$$n = N_D = 9.92 \times 10^{14}/\text{cm}^3$$

We also know that number of holes in a doped silicon,

$$p = \frac{n_i^2}{n} = \frac{(1.52 \times 10^{10})^2}{9.92 \times 10^{14}} = 2.33 \times 10^{5}/\text{cm}^3$$

Since, the number of electrons (equal to 9.92×10^{14}) in a doped silicon is much larger than the number of holes (equal to 2.33×10^5), therefore, conductivity of a doped silicon,

$$\sigma = q \cdot n \cdot \mu_n = (1.6 \times 10^{-19}) \times (9.92 \times 10^{14}) \times 1350 = 0.21\ (\Omega\text{-cm})^{-1}$$

$\therefore$ Resistivity of a doped silicon,

$$\rho = \frac{1}{\sigma} = \frac{1}{0.21} = 4.67\ \Omega\text{-cm.} \quad \textbf{Ans.}$$

Example 9. *In a n-type silicon, the donor concentration is 1 atom per 2×10^8 atoms. Assuming the effective mass of the electron is equal to true mass, find the value of the temperature at which the Fermi level coincides with the edge of the conduction band.*

Solution :

$$N_D = \frac{5 \times 10^{28}}{2 \times 10^8} = 2.5 \times 10^{20}\ \text{m}^{-3}$$

The Fermi level, $E_F = E_C$ if $N_C = N_D$

We know, $N_C = 4.82 \times 10^{21}\ \text{T}^{3/2}\ \text{m}^{-3}$

$$2.5 \times 10^{20} = 4.82 \times 10^{21} \times \text{T}^{3/2}$$

$$\text{T} = 0.139\ \text{K.} \quad \textbf{Ans.}$$

Example 10. *In a semiconductor at room temperature (300°K), the intrinsic carrier concentration and resistivity are $1.5 \times 10^{16}/\text{m}^3$ and $2 \times 10^3\ \Omega$-m respectively. It is converted* *to an extrinsic semiconductor with a doping concentration of $10^{20}/m^3$. For the extrinsic semiconductor, calculate the (a) minority carrier concentration (b) resistivity (c) shift in Fermi level due to doping and (d) minority carrier concentration when its temperature is increased to a value at which the intrinsic carrier concentration n_i doubles.*

Assume (1) The mobility of majority and minority carriers to be the same.

(2) kT = 26 meV at room temperature.

Solution :

(a) Minority carrier concentration $= \dfrac{n_i^2}{\text{Doping concentration}}$

$$= \frac{(1.5 \times 10^{16})^2}{10^{20}} = 2.25 \times 10^{12} \text{ atoms/m}^3$$

$$(\mu_n + \mu_n) = \frac{1}{2 \times 10^3 \times 1.6 \times 10^{-19} \times 1.5 \times 10^{16}}$$

$$2\mu_n = \frac{1}{4.8}$$

$$\mu_n' = \frac{1}{9.6} = 0.1042 \text{ m}^2/\text{volt-sec.} \quad \textbf{Ans.}$$

(b) Resistivity, $\sigma = qn(\mu_n)$

As doping concentration >> minority concentration, only doping is to be considered.

$$\therefore \quad \sigma = 1.6 \times 10^{-19} \times 10^{20} \times 0.1042$$

$$= 1.6672$$

Resistivity = 0.5998 Ω-m. **Ans.**

(c) Shift in fermi level due to doping is given by relation :

$$E_F - E_i = kT \ln \frac{n_0}{n_i}$$

$$= 0.026 \ln \frac{10^{20}}{10^{16} \times 1.5}$$

$$= 0.026 \ln 66.67 \times 10^2$$

$$= 0.229 \text{ eV}$$

E_F lies 0.229 eV above fermi level E_i. **Ans.**

(d) Minority carrier concentration $= \dfrac{(1.5 \times 2 \times 10^{16})^2}{10^{20}}$

$$= \frac{9 \times 10^{32}}{10^{20}} = 9 \times 10^{12} \text{ atoms/cm}^3. \quad \textbf{Ans.}$$

Example 11. *Determine the conductivity and resistivity of an intrinsic sample of Si at normal room temperature at 300 K.*

$$\mu_n = 1350 \ cm^2/V\text{-}sec$$
$$\mu_p = 480 \ cm^2/V\text{-}sec$$
$$n_i = 1.52 \times 10^{10}/cm^3 \text{ at } 300 \text{ K}$$
$$q = 1.6 \times 10^{-19} \text{ C.}$$

Solution : We know that conductivity of an intrinsic sample

$$\sigma = q \cdot n_i (\mu_p + \mu_n)$$

or
$$\sigma = 1.6 \times 10^{-19} \times 1.52 \times 10^{10} (1350 + 480)$$

or
$$\sigma = 4.451 \times 10^{-6} (\Omega\text{-cm})^{-1} \textbf{ Ans.}$$

and
$$\text{Resistivity} = \frac{1}{\text{Conductivity}}$$

$$\rho = \frac{1}{\sigma} = \frac{1}{4.451 \times 10^{-6}} = 2.25 \times 10^5 \text{ ohm-m.} \quad \textbf{Ans.}$$

Example 12. *Assuming standard values for silicon, find the resistivity if a donor type impurity is added to the extent of 1 in 10^8 atoms.*

Solution : Density of atoms in Si = 5.0×10^{22} atoms/cm^3

$$\therefore \quad N_D = \frac{5 \times 10^{22}}{10^8} = 5 \times 10^{14} \text{ (i.e., no. of donor atoms)}$$

By using law of mass-action,

$$np = n_i^2$$
$$N_D \cdot p = n_i^2$$

or
$$p = \frac{n_i^2}{N_D} = \frac{(1.45 \times 10^{10})^2}{5 \times 10^4} = 4.2 \times 10^5/\text{cm}^3$$

as $n >> p$, it is an n-type material with electron as majority carrier.

$$\therefore \quad \sigma = q \cdot \mu_n \cdot n_i$$
$$= 1.6 \times 10^{-19} \times 1300 \times 5 \times 10^{14}$$
$$= 0.104 \ (\Omega\text{-cm})^{-1}$$

or
$$\text{Resistivity} = \frac{1}{\sigma} = \frac{1}{0.104} = 9.62 \text{ ohm-cm.} \quad \textbf{Ans.}$$

Example 13. *A sample of Germanium is doped to the extent of 10^{14} donor atoms/cm^3 and 5×10^{13} acceptor atoms/cm^3. At 300 K, the resistivity of the intrinsic germanium is 60 Ω-cm. If the applied electric field is 2 V/cm. Find the total conduction current density.*

Assume, $\dfrac{\mu_p}{\mu_n} = \dfrac{1}{2}$ *and* $n_i = 2.5 \times 10^{13}/cm^3$ *at 300 K.*

Solution : Given :
$$N_D = 10^{14} \text{ atoms/cm}^3$$
$$N_A = 5 \times 10^{13} \text{ atoms/cm}^3$$

$$\rho = 60\ \Omega\text{-cm}$$
$$E = 2\ \text{V/cm}$$
$$J = ?$$

We know that,

$$\sigma = q \cdot n_i (\mu_n + \mu_p)$$

Since given, $$\frac{\mu_p}{\mu_n} = \frac{1}{2} \Rightarrow \mu_n = 2\mu_p$$

$$\frac{1}{\rho} = 1.6 \times 10^{-19} \times 2.5 \times 10^{13} (2\mu_p + \mu_p)$$

or $$\mu_p = \frac{1}{60 \times 1.6 \times 10^{-19} \times 2.5 \times 10^{13} \times 3}$$

$$\mu_p = 1.3889 \times 10^3 = 1389$$

$$\mu_n = 2\mu_p = 2 \times 1389 = 2778$$

We know, $$n_p = n_i^2 \qquad \text{...(1)}$$

also, the charge neutrality equation,

$$n + N_A = p + N_D \qquad \text{...(2)}$$

or $$p - n = N_A - N_D$$

or $$p - n = 5 \times 10^{13} - 10^{14}$$

or $$p - n = -5 \times 10^{13}$$

or $$n = p + 5 \times 10^{13} \qquad \text{...(3)}$$

Equation on putting in eqn. (1) we get,

$$(p + 5 \times 10^{13}) \cdot p = n_i^2 = (2.5 \times 10^{13})^2$$

$$p^2 + 5 \times 10^{13} p = 6.25 \times 10^{26}$$

$$p^2 + 5 \times 10^{13} p - 6.25 \times 10^{26} = 0$$

On solving the quadratic equation for p, we get,

$$p = 1.0356 \times 10^{13}$$

and $$n = 6.0356 \times 10^{13}$$

So, total conduction current density is given by the relation, $J = q(n\mu_n + p\mu_p) E$

$$J = 1.6 \times 10^{-19} (6.0356 \times 10^{13} \times 2778 + 1.0356 \times 10^{13} \times 1389) \times 2$$

$$= 3.2 \times 10^{-6} (18205.34)$$

$$= .0583\ \text{amp/m}^2.$$ **Ans.**

Example 14. *A Si sample is doped with 10^{17} arsenic atoms/cm^3*

(i) find the minority carrier concentration at room temperature.

(ii) find the location of fermi level with respect to the intrinsic fermi level.

Given that $n_i = 1.5 \times 10^{10}$ electrons/cm^3.

Solution : (*i*) Since from given data,

$$N_D >> n_i \qquad \{\because 10^{17} >> 1.5 \times 10^{10}\}$$

So, by charge neutrality equation :

$$n = N_D = 10^{17} \text{ electrons/cm}^3$$

$$p = \frac{n_i^2}{n} = \frac{(1.5 \times 10^{10})^2}{10^{17}} = 2.25 \times 10^3 \text{ holes/cm}^3. \quad \textbf{Ans.}$$

(*ii*) As we know that,

$$E_F - E_i = kT \ln \frac{N_D}{n_i}$$

$$= 0.026 \ln \frac{10^{17}}{1.5 \times 10^{10}}$$

$$= .407 \text{ eV.} \quad \textbf{Ans.}$$

The resulting energy band diagram is :

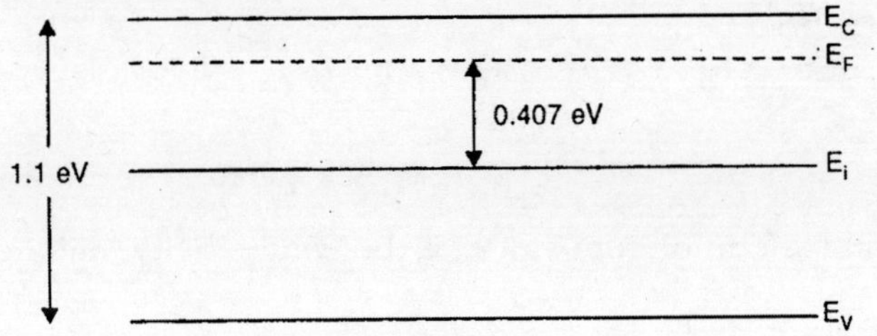

Fig. P (1.1).

Example 15. *In N-type Si bar is 2 cm long and has a cross-section of 2 mm × 2 mm. When a 1 V battery is connected across it, a current of 8 mA flows. Find*

(a) doping level

(b) drift velocity.

Solution : (*a*) As we know that,

$$R = \frac{\rho L}{A}; \quad R = \frac{V}{I} = \frac{1}{8 \times 10^{-3}} = 125 \ \Omega$$

$$R = \frac{1}{\sigma}\frac{L}{A}$$

$$\sigma = \frac{L}{RA} = \frac{2}{125 \times .2 \times .2} = .4 \text{ (ohm-cm)}^{-1}$$

As conductivity (σ) is much higher than the intrinsic conductivity (= 4.3×10^{-6}). This is heavily doped sample and so we can use approximate formula,

$$n = N_D$$

and $$\sigma = nq\mu_n \qquad \{\because \mu_n = 1300 \text{ cm}^2\text{/V-sec}\}$$

$$.4 = N_D \cdot 1.6 \times 10^{-19} \times 1300$$

or $$N_D = \frac{.4}{1300 \times 1.6 \times 10^{-19}}$$

or $$N_D = 192 \times 10^{13}/\text{cm}^3. \quad \textbf{Ans.}$$

(b) Drift velocity, $V_d = \mu . E$

or $$V_d = \mu_n . \left(\frac{V}{d}\right)$$

or $$V_d = 1300 \times \frac{1}{2}$$

or $$V_d = 650 \text{ cm/sec.} \quad \textbf{Ans.}$$

Example 16. *(a) Show that the conductivity of a semiconductor is minimum when it is lightly doped with p-type impurity such that*

$$p = n_i \left[\frac{\mu_n}{\mu_p}\right]^{1/2}$$

(b) Show that minimum conductivity is $2n_i[\mu_n \mu_p]^{1/2} q$.

Solution : (a) We know that,

$$\sigma = (n\mu_n + p\mu_p) q \qquad ...(1)$$

$$\sigma = \left(\frac{n_i^2}{p} \cdot \mu_n + q\,\mu_p\right) q$$

In order to find the minimum value of the conductivity, equate the derivative of σ to zero.

$$\frac{d\sigma}{dp} = \left(-\frac{n_i^2}{p^2} \cdot \mu_n + \mu_p\right) q = 0$$

We get, $$\mu_p = \frac{n_i^2}{p^2} \mu_n$$

or $$p = n_i \left[\frac{\mu_n}{\mu_p}\right]^{1/2} \quad \textbf{(Proved).}$$

(b) To calculate minimum value, put,

$$p = n_i \left[\frac{\mu_n}{\mu_p}\right]^{1/2} \text{ in eqn. (1)}$$

$$\sigma_{\text{min}} = \left[\frac{n_i^2 \mu_n}{n_i \left(\frac{\mu_n}{\mu_p}\right)^{1/2}} + n_i \left(\frac{\mu_n}{\mu_p}\right)^{1/2} \mu_p\right] q$$

$$= [n_i (\mu_n \mu_p]^{1/2} + n_i (\mu_n \mu_p)^{1/2}] q$$

$$= 2n_i (\mu_n \mu_p)^{1/2} . q. \quad \textbf{(Proved).}$$

Example 17. *Find the following for a copper wire whose diameter is 1.03 mm and the resistance is given as 6.51 Ω/1000 ft.*

(i) *Conductivity,*

(ii) *Mobility,*

(iii) *Drift velocity.*

The concentration of electrons is 8.4 × 10^{28} electrons/m^3 and the current flowing in the wire is 2 amp.

Solution : (*i*) 1000 ft = 1000 × 12 × 2.54 cm = 3.05 × 10^4 cm.

So, the resistivity ρ is given by

$$\rho = \frac{6.51\,\Omega}{3.05 \times 10^4 \text{ cm}} = 2.14 \times 10^{-4}\ \Omega/\text{cm}$$

As we know that the conductivity is the reciprocal of the resistivity.

$$\sigma = \frac{1}{\rho} = \frac{1}{2.14 \times 10^4} = 4.68 \times 10^3 \text{ mho/cm}$$

$$= 4.68 \times 10^5 \text{ mho/m.} \quad \textbf{Ans.}$$

(*ii*) From, $\sigma = nq\mu$

$$\text{Mobility, } \mu = \frac{\sigma}{nq} = \frac{4.68 \times 10^5}{8.4 \times 10^{28} \times 1.6 \times 10^{-19}}$$

or $$\mu = 3.48 \times 10^{-5} \text{ m}^2/\text{V-s}.$$

(*iii*) Drift velocity, $V = \dfrac{T}{nqA}$

But cross-section area, $A = \dfrac{\pi D^2}{4} = 8.33 \times 10^{-7} \text{ m}^2$.

or $$V = \frac{2}{8.4 \times 10^{28} \times 1.6 \times 10^{-19} \times 8.33 \times 10^{-7}} = 1.79 \times 10^{-4} \text{ m/s.} \quad \textbf{Ans.}$$

Example 18. *A piece of silicon at room temperature is doped with 2 × 10^{16}/cm^3 concentration of Boron atoms and 5 × 10^{16}/cm^3 of phosphorous atoms.*

(*i*) *Find the hole and electron concentration in this material.*

(*ii*) *Is the silicon p- or n-type?*

Solution : Acceptor atoms added is N_A and donor atoms added are N_D. So, $N_A = 2 \times 10^{16}/\text{cm}^3$ and $N_D = 5 \times 10^{15}/\text{cm}^3$.

Since the donor concentration is less than the acceptor concentration, the material is likely to be *p*-type. Hence the concentration of holes should be given as:

$$N_D - N_A = p_p = 2 \times 10^{16} - 5 \times 10^{15} = 15 \times 10^{15}/\text{cm}^3$$

From Mass action law, the concentration of the electrons can be determined.

$$n_p = n_i^2/p_p = \frac{(10^{12})^2}{15 \times 10^{15}} = \frac{1}{15} \times 10^9/\text{cm}^3.$$

Ans. (*i*) $n_p = \dfrac{1}{15} \times 10^9/\text{cm}^3$

$p_p = 15 \times 10^{15}$

(*ii*) Material is of *p*-type.

Example 19. *The resistivity of semiconductor material was known to be 0.005 Ω-m at room temperature. The flux density in the Hall model was 0.45 wb/m²*

Calculate the Hall angle for a Hall co-efficient of 3.55 × 10⁻⁴ m³/coulomb.

Solution : Given that,

$$\rho = 0.005\ \Omega\text{-m}$$
$$B_z = 0.48\ \text{wb/m}^2$$
$$R_H = 3.55 \times 10^{-4}\ \text{m}^3/\text{C}$$
$$\theta_H = ?$$

From Fig. P (1.2)

$$\text{Resistivity, } \rho = \frac{E_x}{J_x}$$

$$0.005 = \frac{E_x}{J_x}$$

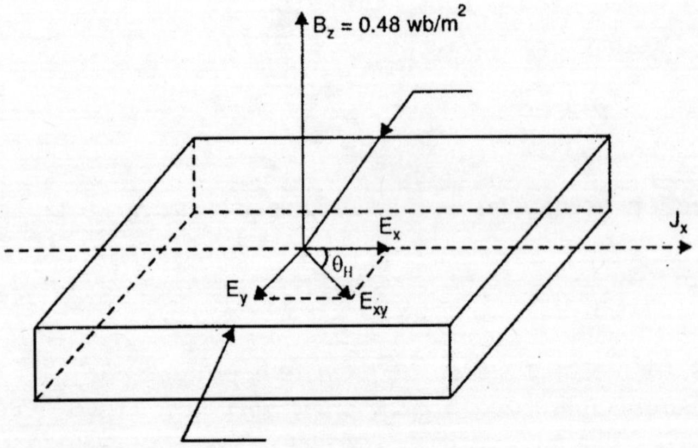

Fig. P (1.2)

or $$E_x = 0.005\ J_x \qquad ...(a)$$

and we know that,

$$R_H = \frac{E_y}{B_z J_x}$$

$$3.55 \times 10^{-4} = \frac{E_y}{0.48 \times J_x}$$

∴ $$E_y = 1.704 \times 10^{-4}\ J_x \qquad ...(b)$$

Now, $$\tan \theta_H = \frac{E_y}{E_x}$$

$$= \frac{1.704 \times 10^{-4} J_x}{5 \times 10^{-3} J_x}$$

Example 20. *Fig. P (1.3) shows a specimen of a silicon doped semiconductor having the Hall co-effieicnt of 3.55 × 10⁻⁴ m³/coulomb. Calculate the voltage between contacts when a current of 15 mA is flowing.*

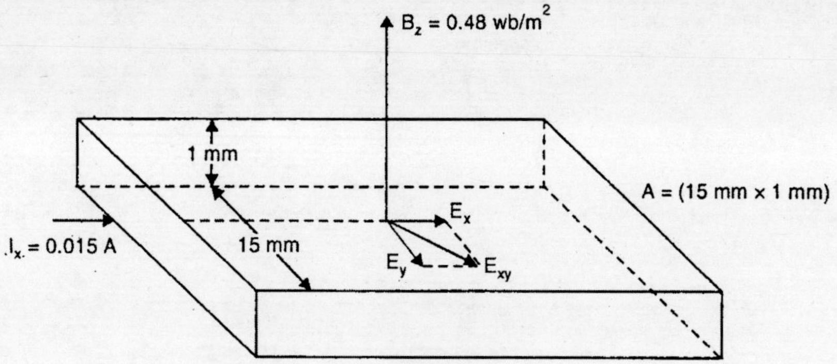

Fig. P (1.3)

Solution : Hall co-efficient, $R_H = 3.55 \times 10^{-4}$ m³/C

Current, $I_x = 15$ mA $= 0.015$ A

Area, $A = 15$ mm × 1 mm

$= 15 \times 10^{-6}$ m²

Flux density, $B_z = 0.48$ Wb/m²

Voltage between contacts = ?

$$\text{Now current density, } J_x = \frac{I_x}{A} = \frac{0.015}{15 \times 10^{-6}}$$

$= 1000$ A/m²

Hall co-effieicnt is given by the relation :

$$R_H = \frac{E_y}{B_z J_x}$$

$$3.55 \times 10^{-4} = \frac{E_y}{0.48 \times 1000}$$

$\therefore$ $E_y = 3.55 \times 10^{-4} \times 0.48 \times 1000 = 0.1704$ V/m

and *voltage between contacts*

$= 0.1704 \times (15 \times 10^{-3})$

$= 0.002556$ V. **Ans.**

Example 21. *A silicon crystal having a cross-sectional area of 0.001 cm² and a length of 20 μm is connected to its ends to a 20 V battery. At T = 300 K, we want a current of 100 mA in crystal. The concentration of donor atoms to be added is.*

Solution : Given, $A = 0.001$ cm²

$l = 20$ μm $= 20 \times 10^{-4}$ cm

$$V = 20 \text{ V}$$

$$I = 100 \text{ mA} = 100 \times 10^{-3} \text{ A}$$

From relation,

$$R = \frac{V}{I}$$

or $$R = \frac{20}{100 \times 10^{-3}} = 200\ \Omega$$

and we know that,

$$R = \frac{1}{\sigma} \cdot \frac{l}{A}$$

or $$\sigma = \frac{l}{RA} = \frac{20 \times 10^{-6}}{200 \times .001 \times 10^{-4}} = 0.01\ (\Omega\text{-cm})^{-1}$$

and $$\sigma = n.q.\mu_n$$

For Silicon, $\mu_n = 1350 \text{ cm}^2/\text{V-sec}$

Now, $$n = \frac{\sigma}{q.\mu_n} = \frac{0.01}{1.6 \times 10^{-19} \times 1350} = 4.6 \times 10^{13} \text{ cm}^{-3}$$

Ans.

Since, $n >> n_i$ (intrinsic concentration)

It means, $n = N_D$.

Example 22. *A thin film resistor is to be made from a GaAs film doped n-type. The resistor is to have a value of 2 kΩ. The resistor length is to be 200 μm and area is to be 10^{-6} cm². The doping efficiency is known to be 90%. The mobility of electrons is 8000 cm²/V-s. The doping needed is*

Solution : Given data,

$$R = 2 \text{ k}\Omega = 2000\ \Omega$$

$$L = 200\ \mu\text{m} = 200 \times 10^{-4} \text{ cm}$$

$$A = 10^{-6} \text{ cm}^2$$

$$n = 0.9 \text{ N}_D$$

Now from relation,

$$R = \frac{\rho l}{A}$$

or $$R = \frac{1}{nq\,\mu_n} \times \frac{l}{A}$$

or $$\rho = \frac{L}{Rq\,\mu_n A}$$

or $$qN_d = \frac{20 \times 10^{-4}}{2000 \times 1.6 \times 10^{-19} \times 8000 \times 10^{-6}}$$

or $$N_d = 8.7 \times 10^{15} \text{ cm}^{-3}$$ **Ans.**

Example 23. *A silicon semiconductor sample at T = 300 K is doped with phosphorus atoms at a concentrations of 10^{15} cm^{-3}. The position of the Fermi level with respect to the intrinsic Fermi level is.*

Solution : We know that,

$$E_F - E_{Fi} = KT\ ln\left|\frac{N_d}{n_i}\right|$$

$$= 26 \times 10^{-3}\ ln\left|\frac{10^{15}}{1.5 \times 10^{10}}\right|.eV$$

$$= 0.287 \text{ eV} \quad \textbf{Ans.}$$

Example 24. *A silicon sample contains acceptor atoms at a concentration of $N_a = 5 \times 10^{15}$ cm^{-3}. Donor atoms are added forming and n-type compensated semiconductor such that the Fermi level is 0.215 eV below the conduction band edge. Calculate the concentration of donors atoms added.*

Solution : We know that,

$$n = N_d - N_a = N_c\, e^{-\frac{(E_C - E_F)}{KT}}$$

For Si, $N_c = 2.8 \times 10^{19}$ cm^{-3}

Given, $N_a = 5 \times 10^{15}$ cm^{-3}

Now, $$N_d = N_a + N_c\, e^{-\frac{(E_C - E_F)}{KT}}$$

or $$N_d = 5 \times 10^{15} + 2.8 \times 10^{19}\, e^{-\left(\frac{0.215 \text{ eV}}{26 \times 10^{-3} \text{ eV}}\right)}$$

or $$N_d = 1.19 \times 10^{16} \text{ cm}^{-3}$$

[∴ given that Fermi level is 0.215 eV below the conduction band *i.e.,* $E_C - E_F = 0.215$ eV]

Ans.

Example 25. *A silicon sample doped n-type at 10^{18} cm^{-3} have a resistance of 10 Ω. The sample has an area of 10^{-6} cm^2 and a length of 10 μm. The doping efficiency of the sample is (μ_a = 800 cm^2/V–s).*

Solution : Given data,

$N_d = 10^{18}$

$R = 10\ \Omega$

$A = 10^{-6}$ cm^2

$L = 10$ mm $= 10 \times 10^{-4}$ cm, $\mu_n = 800$ cm^2/V–s

$$\% \text{ doping efficiency} = \frac{n}{N_d} \times 100 = ?$$

Now, $$n = \frac{L}{q.\mu_n.A.R}$$

or $$n = \frac{10^{-4}}{1.6 \times 10^{-19} \times 800 \times 10^{-6} \times 10}$$

$$= 7.81 \times 10^{17} \text{ cm}^{-3}$$

$$\% \text{ doping efficiency} = \frac{7.81 \times 10^{17}}{1 \times 10^{18}} \times 100$$

$$= 78.1\%.$$ **Ans.**

? SELF-TEST (DIODES)

1. What do you mean by *p-n* junction ?

2. Explain the phenomenon of the formation of depletion layer in the *p-n* junction.

3. What will be the effect on the depletion layer of a *p-n* junction diode under different biased condition ?

4. Differentiate between transition capacitance and diffusion capacitance of a *p-n* junction diode.

5. What is an ideal diode ? How can it be represented as a switch ? Draw the equivalent circuit and its characteristics.

6. Define diffusion current in a semiconductor.

7. (*i*) Define cut-in voltage of a *p-n* junction diode.

(*ii*) Draw the reverse bias characteristics of a *p-n* junction.

8. Explain why the energy levels of an atom become energy bands in a solid.

9. Define a hole in a semiconductor. Explain why a semiconductor acts as an insulator at 0°K and why its conductivity increases with increasing temperature.

10. Draw the energy band diagram for a *p-n* junction under open-circuit condition. Clearly indicate various energy levels in *p*-region, space charge region and *n*-region. How will it be modified if *p-n* junction is forward biased ?

11. For a semiconductor diode, define static and dynamic resistance.

12. Derive an expression for the dynamic resistance of a diode in forward bias condition and hence interrupt graphically its variation with (*i*) current and (*ii*) temperature.

13. Explain the construction and *V-I* characteristics of a *p-n* junction diode.

14. Is a hole a fundamental particles in an atom ? Between electron and hole which one has a greater mobility ? Give a reason for the difference.

SOLVED EXAMPLES

Example 26. *The current flowing in a certain PN-junction diode at room temperature is 2 × 10^{-7} A, when the large reverse voltage is applied. Calculate the current flowing, when 0.1 V forward bias is applied at room temperature.*

Solution : Given $I_0 = 2 \times 10^{-7}$ A

$V = 0.1$ volt

We know that the current flowing through the diode under forward bias,

$$I = I_0 (e^{40V} - 1)$$
$$= 2 \times 10^{-7} (e^{40 \times 0.1} - 1)$$
$$= 1.16\ \mu A. \quad \textbf{Ans.}$$

Example 27. *The current voltage characteristic of a PN-junction diode is given by the relation :*

$$I = I_0 (e^{qV/\eta . k . T} - 1)$$

The diode current is 0.5 mA at V = 340 mV and 15 mA at V = 440 mV. Determine the value of η. Assume kT/q = 25 mV.

Solution : Given : $I_1 = 0.5$ mA $= 0.5 \times 10^{-3}$ A; $V_1 = 340$ mV; $I_2 = 15$ mA $= 15 \times 10^{-3}$ A; $V_2 = 440$ mV and $k . T/q = 25$ mV.

We know that the diode current (I_1)

$$0.5 \times 10^{-3} = I_0 (e^{qV_1/\eta . k . T} - 1) = I_0 . e^{qV_1/\eta . k . T}$$
$$= I_0 . e^{340/25\eta}$$
$$= I_0 . e^{13.6/\eta} \quad ...(1)$$

Similarly, the diode current (I_2)

$$15 \times 10^{-3} = I_0 . e^{440/25\eta} = I_0 . e^{17.6/\eta} \quad ...(2)$$

Dividing equation (2) by (1),

$$30 = e^{(17.6 - 13.6)/\eta} = e^{4/\eta}$$

Taking natural logarithms on both sides,

$$\log_e 30 = \log_e(e^{4/\eta}) = \frac{4}{\eta} . \log_e e = \frac{4}{\eta} \quad (\because \log_e e = 1)$$

$$3.4 = 4/\eta \text{ or } \eta = 4/3.4 = 1.18 \quad \textbf{Ans.}$$

Example 28. *An ideal germanium diode at a temperature of 125°C has a reverse saturation current of 30 μA. At a temperature of 125°C, find the dynamic resistance for a 0.2 V bias*

(a) in the forward direction

(b) in the reverse direction.

Solution : (*a*) We know that the diode current is given by the relation

$$I = I_0\left(e^{\frac{V}{\eta V_T}} - 1\right)$$

$$\frac{dI}{dV} = \frac{I_0}{\eta V_T} \cdot e^{\frac{V}{\eta V_T}}$$

$$\frac{dI}{dV} = \frac{I_0}{V_T} e^{\frac{V}{V_T}} \quad (\eta = 1 \text{ for germanium})$$

We know that dynamic resistance,

$$r_f = \frac{dV}{dI}$$

$$= \frac{V_T}{I_0\, e^{\frac{V}{V_T}}}$$

$$= \frac{T}{11600} \times \frac{1}{30 \times 10^{-6}} \times \frac{1}{e^{\frac{0.2}{V_T}}} \qquad \left(\because V_T = \frac{T}{11600}\right)$$

$$= \frac{398}{11600} \cdot \frac{1}{30 \times 10^{-6}} \cdot \frac{1}{e^{\frac{0.2}{398} \times 11600}} \qquad (T = 273 + 125 = 398°K)$$

$$= 3.36\ \Omega. \quad \textbf{Ans.}$$

(*b*) Dynamic resistance in reverse direction

$$r_f = \frac{398}{11600} \times \frac{1}{30 \times 10^{-6}} \times \frac{1}{e^{-\frac{0.2}{398} \times 11600}}$$

or $$r_f = .389\ M\Omega. \quad \textbf{Ans.}$$

Example 29. *(a) Find the voltage at which the reverse current in a Ge diode at room temperature will reach 90% of its saturation value.*

(b) Find the ratio of diode current with a forward bias of 0.05 V to the current with a reverse bias of 0.05 V.

(c) The reverse saturation current of a Ge diode is 10 μA. Find current under forward biases of 0.1 V, 0.2 V, 0.3 V.

Solution : (*a*) From diode equation given by,

$$I = I_0\,(e^{V/\eta V_T} - 1)$$

Saturation value of $I = -I_0$

$\therefore$ Putting $I = -0.9\, I_0$ in above equation we get

$$-0.9\, I_0 = I_0\,(e^{V/\eta V_T} - 1)$$

$$\therefore \quad e^{V/\eta V_T} - 1 = -0.9$$

$$\therefore \quad \frac{V}{\eta V_T} = \ln 0.1 = -2.3$$

$$\therefore \quad V = \eta V_T \times (-2.3) = -1 \times 0.026 \times 2.3$$

$$= -59.8\ \text{mV}. \quad \textbf{Ans.}$$

(*b*) $$\frac{I_f}{I_r} = \frac{I_0\,(e^{0.05/0.026} - 1)}{I_0\,(e^{-0.05/0.026} - 1)} = \frac{5.84\, I_0}{-0.854\, I_0} = -6.84 \quad \textbf{Ans.}$$

(*c*) $$I_1 = 10\,(e^{0.1/0.026} - 1) = 10 \times 45.8 = 458\ \mu A$$

$$I_2 = 10\,(e^{0.2/0.026} - 1) = 10 \times 21900$$

or $I_2 = 21900\ \mu A = 21.9\ mA$

$I_3 = 10\ (e^{0.3/0.026} - 1)$

$= 10 \times 1.02 \times 10^5\ \mu A = 1.02$ Amps. **Ans.**

Example 30. *A Si diode operates at a fixed forward bias of 0.4 V. Calculate the factor by which the current will get multiplied when its temperature is raised from 25°C to 150°C.*

Solution : In a practical diode, I_0 doubles for every 10° rise in temperature.

$$\therefore \quad I_0\ (150) = I_0\ (25) \times 2^{(150-25)/10}$$

$$= I_0\ (25)\ 2^{12.5} = 5800\ I_0(25)$$

Thus rev. sat. current is multiplied 5800 times

$$V_T\ (150) = kT = 8.62 \times 10^{-5} \times (273 + 150) = 0.0364\ V$$

Using diode equation

$$\therefore \quad \frac{I(150)}{I}(25) = \frac{I_0\ (150)(e^{V/\eta V_T} - 1)}{I_0\ (25)(e^{V/\eta V_T} - 1)}$$

$$= 5800 \times \frac{e^{0.4/(2 \times 0.0364)}}{e^{0.4/(2 \times 0.026)}}$$

$$= \frac{5800 \times 241}{2191} = 638.$$ **Ans.**

Example 31. *In the circuit shown below, the diode is used has a threshold voltage of 0.5 V and a forward resistance of 1 Ω. Find the operating point of the diode.*

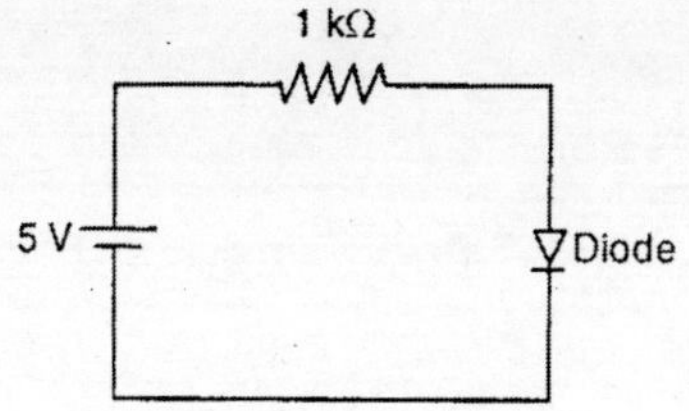

Fig. P (1.4).

Solution : The redrawn circuit is shown in Fig. P (1.5) with diode having 1 Ω resistance.

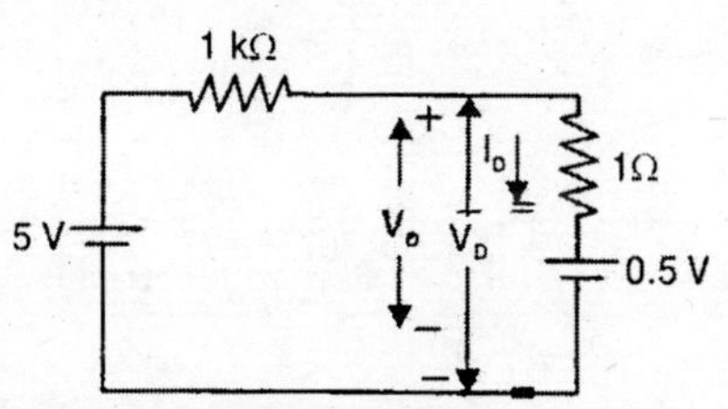

Fig. P (1.5).

In order to calculate the value of operating point we have to calculate the value of I_D and V_D.

Applying KVL in the following circuit

$$5 - I_D \times 1000 - I_D . 1 - 0.5 = 0$$

$$I_D = \frac{4.5}{1001}$$

or $$I_D = 4.49 \text{ mA. } \textbf{Ans.}$$

and $$V_D = I_D \times 1\ \Omega + 0.5$$

or $$V_D = 4.49 \times 10^{-3} \times 1 + 0.5$$

or $$V_D = .5045 \text{ V. } \textbf{Ans.}$$

Example 32. *(a) Two ideal and identical (ideality factor η = 1) junction diodes are connected in series as shown in figure. Show that exp.* $\left(\frac{eV_1}{kT}\right)$ *+ exp.* $\left(\frac{-eV_2}{kT}\right)$ *= 2 where* V_1 *and* V_2 *are the voltage drops across the diodes.*

(b) Assuming that the reverse biased diode is saturated at I_0. *Calculate the voltage drop across the forward biased diode. (Assume* $\frac{kT}{q}$ *= 26 meV).*

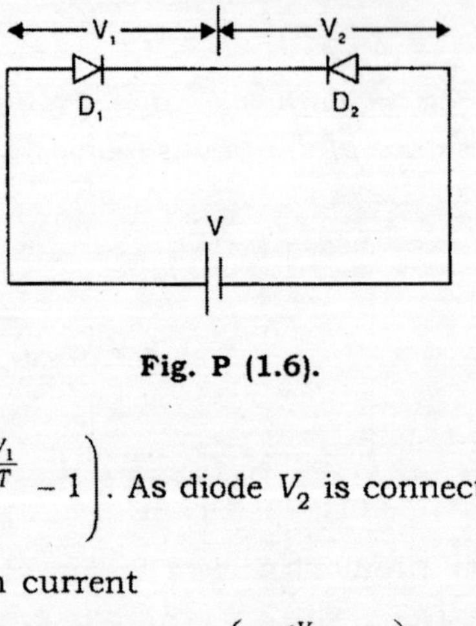

Fig. P (1.6).

Solution : (a) $I = I_0\left(e^{\frac{eV_1}{kT}} - 1\right)$. As diode V_2 is connected in the reverse-bias mode $I = I_0$, the reverse saturation current

$$I = I_0\left(e^{\frac{eV_1}{kT}} - 1\right) \text{ for } D_1 \text{ and } I = -I_0\left(e^{\frac{-eV_2}{kT}} - 1\right) \text{ for } D_2.$$

$$I_0\left(e^{\frac{eV_1}{kT}} - 1\right) = -I_0\left(e^{\frac{-eV_2}{kT}} - 1\right)$$

$$e^{\frac{eV_1}{kT}} + e^{\frac{-eV_2}{kT}} = 2$$

or $$\exp\left(\frac{eV_1}{kT}\right) + \exp\left(\frac{-eV_2}{kT}\right) = 2. \textbf{ Ans.}$$

(*b*)
$$e^{\frac{eV_1}{kT}} = 2$$
$$kT = 26 \text{ meV}$$
$$e^{eV_1/26e \times 10^{-1}} = 2$$
$$\frac{V_1}{26 \times 10^{-3}} = \ln 2 = 0.693; \; V_1 = 26 \times 10^{-3} \times 0.693$$
$$= 18 \text{ mV} = 0.018 \text{ V.} \quad \textbf{Ans.}$$

Example 33. *Find the space charge capacitance of germanium diode whose area is 1 mm² and whose space charge thickness is 2 × 10⁻⁴ cm. Permitivity for germanium is*

$$\epsilon_{Ge} = \frac{16}{36\pi \times 10^{11}} \; F/cm.$$

Solution : As we know that $C_T = \frac{\epsilon_0 A}{d} = \frac{\epsilon_{Ge} A}{d}$

therefore, $$C_T = \frac{\left(\frac{16}{36\pi \times 10^{11}}\right) \times 10^{-2}}{(2 \times 10^{-4})} = 70.73 \text{ pF} \;\textbf{Ans.}$$

Example 34. *Find C_T, if conductivity of p-material is 0.286 $(\Omega \text{ cm})^{-1}$, barrier potential is 0.35 V and cross-section area is circular with diameter of 0.12 cm. The applied reverse voltage is 5 V.*

Solution : Given, $$A = \frac{\pi D^2}{4} = \frac{\pi \times (0.102)^2}{4} \text{ cm}^2 = 8.17 \times 10^{-3} \text{ cm}^2$$

$$\sigma_p = q\mu_p N_A$$

or $$N_A = \frac{\sigma_p}{q\mu_p} = \frac{0.286}{1.6 \times 10^{-19} \times 500} = 3.575 \times 10^{15} \text{ atom/cm}^3$$

Therefore, the transition capacitance, C_T is

$$C_T = 2.92 \times 10^{-4} \left(\frac{N_A}{V_b}\right)^{1/2} \times A \text{ pF/cm}^2$$

$$= 2.92 \times 10^{-4} \left(\frac{3.575 \times 10^{15}}{5 + 0.35}\right)^{1/2} \times 8.17 \times 10^{-3}$$

$$= 61.25 \text{ pF/cm}^2 \;\textbf{Ans.}$$

Example 35. *The transition capacitance C_T is 15 pF for a reverse bias of 8V. What will be the value of C_T for a reverse bias of 12V.*

Solution : Transition Capacitance is inversely proportional to square root of junction potential

$$C_T \propto \sqrt{\frac{1}{V_b}}$$ where, V_b = Junction potential or barrier potential

Therefore,

$$\frac{C_{T_1}}{C_{T_2}} = \left(\frac{V_{b_2}}{V_{b_1}}\right)^{1/2}$$

$$C_{T_2} = C_{T_1} \cdot \left(\frac{V_{b_1}}{V_{b_2}}\right)^{1/2}$$

$$= 15\left(\frac{8}{12}\right)^{1/2} = 12.28 \text{ pF } \textbf{Ans.}$$

Example 36. *At what voltage, the reverse current flowing through a diode (Ge) will reach 90% of its saturation value at room temperature (300°K).*

Solution : $V_T = 0.026$ V at 300°K

Using diode equation, we get

$$T = I_0\left(e^{\frac{V}{\eta V_T}} - 1\right)$$

or $$I = I_0\left(e^{\frac{V}{\eta(0.026)}} - 1\right)$$

For germanium, $\eta = 1$

So, $I = 90\%$ of $I_0 = -0.9\ I_0$

$$-0.9\ I_0 = I_0\left(e^{\frac{V}{(0.026)}} - 1\right)$$

or $$-0.9 = e^{\frac{V}{(0.026)}} - 1$$

or $$e^{\frac{V}{0.026}} = 1 - 0.9 = 0.1$$

Taking log both side

or $$\log_e e^{\frac{V}{0.026}} = \log_e .1$$

or $$\frac{V}{0.026} = (-2.3)$$

or $$V = -59.8 \text{ mV.}$$

NOTE : $I = -0.9\ I_0$, because direction of diode current is positive while flowing from p to n-side but reverse saturation current flows from n-side to p-side due to flow of minority charge carriers.

Example 37. *Determine I_D for each circuit given in Fig. P (1.7)*

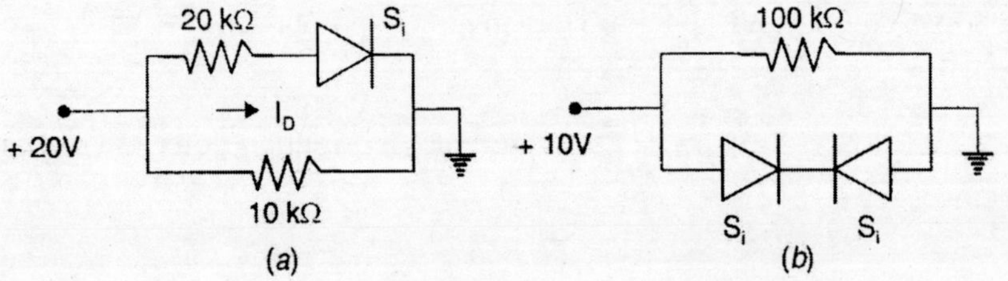

Fig. P (1.7).

Solution : (*a*) Given $V_{in} = 20$ V

V_γ = Cut in voltage = 0.7 V for Si

Apply KVL,

$$20 - I_D \, 20 - 0.7 = 0$$

$$I_D = \frac{20 - 0.7 \text{ V}}{20 \text{ k}\Omega} = 0.965 \text{ mA } \textbf{Ans.}$$

(*b*) Given $V_{in} = 10$ V

$V_\gamma = 0.7$ V **Ans.**

NOTE : Here, the arrangement of diode is back to back. hence, one will be reverse biased and other will be forward biased. But due to reverse biased diode, no current will flow through the branch containing diodes. The total voltage will be impressed on resistance of 100 kΩ, since it is parallel to the branch containing diodes. Therefore,

$$I_D = \frac{10 \text{ V}}{100 \text{ k}\Omega} = 0.1 \text{ mA } \textbf{Ans.}$$

Example 38. *Calculate V_0 and I_D for the circuits given in Fig. P (1.8).*

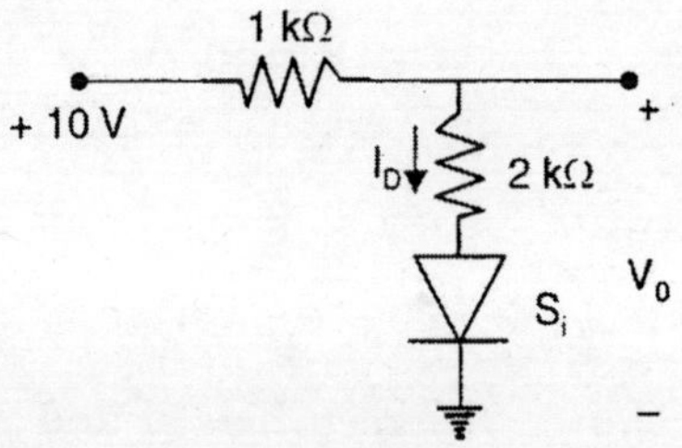

Fig. P (1.8).

Solution : Apply KVL in the given loop, we get

$$10 - (1 \text{ k}\Omega)\, I_D - (2 \text{ k}\Omega)\, I_D - 0.7 = 0$$

$$I_D\,(1 + 2) \text{ k}\Omega = 10 - 0.7$$

or $$I_D = \frac{9.3 \text{ V}}{3 \text{ k}\Omega}$$

or $$I_D = 3.1 \text{ mA } \textbf{Ans.}$$

and, $$V_0 = I_D\,(2 \text{ k}\Omega) + 0.7$$

$$= 3.1 \times 10^{-3} \times 2 \times 10^3 + 0.7$$

$$= 6.2 + 0.7$$

$$= 6.9 \text{ V } \textbf{Ans.}$$

Example 39. *For each of the ckt as shown in Fig. P (1.9), determine the state of the diode—ON or OFF and calculate the current in each branch. Assume ideal diodes.*

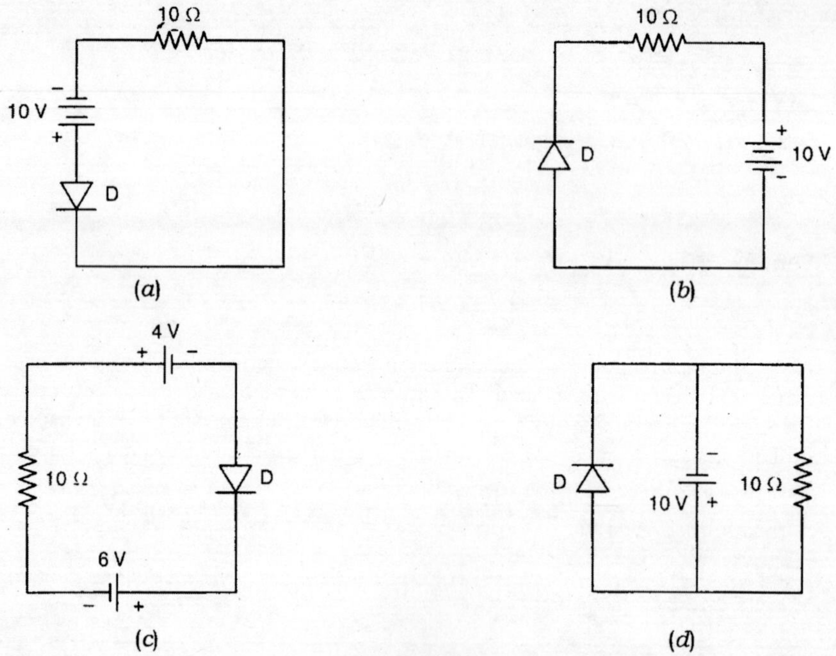

Fig. P (1.9).

Solution :

(*a*) The diode D is in forward biased so if will be in ON-State and current through resistance will be $I = \frac{V}{R} = \frac{10}{10} = 1$ amp.

(*b*) The diode in ckt-D will be reversed biased and since diode is ideal so reverse bias resistance will be of. So the current through 10 resistance will be 0.

D-OFF.

(*c*) Here 4 V and. 6 Volt batteries are connected in series with reverse polarity in total voltage across diode will become : 2 V; so diode will be OFF and current through this will be zero.

(*d*) The diode will be ON and current through this $= \frac{10}{10} = 1\text{A}$ (*i.e.*, $\frac{V}{R}$).

Example 40. *In the ckt of Fig. P (1.10), diode has a threshold voltage of 0.5 V and* $R_F = 1\ k\Omega$. *Find the operating point of the diode.*

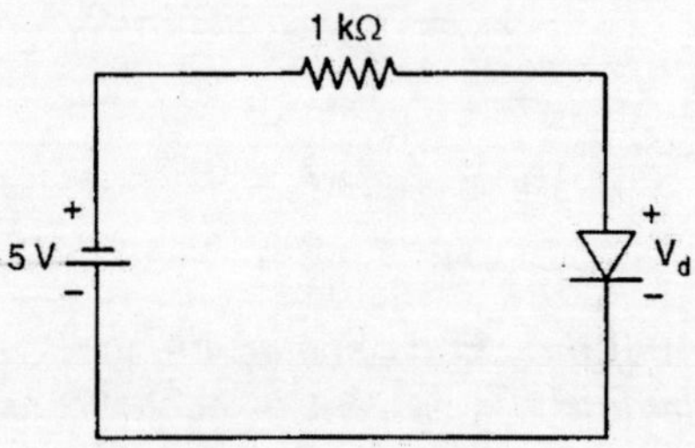

Fig. P (1.10).

Solution : Apply KVL for loop we get,

$$5 - V_d - R_i I_i = 0 \quad ...(1)$$

When $\qquad I_i = 0$

$\therefore \qquad V_d = 5 \text{ V}$

When $\qquad V_d = 0$

$$I_i = \frac{5}{R_F} = \frac{5}{1 \text{ k}\Omega}$$

or $\qquad I_i = 5 \text{ mA}$

Now from equation (1)

$$5 - V_d - R_i I_i = 0$$

$$I_i = \frac{5 - 0.5}{R_F}$$

$$= \frac{4.5}{1 \text{ k}\Omega} = 4.5 \text{ mA}$$

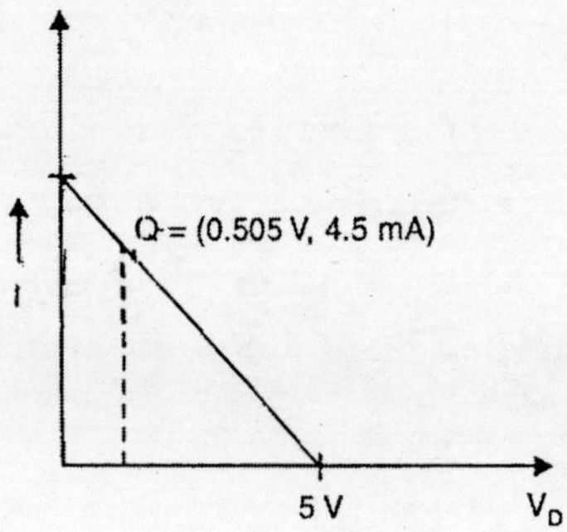

Fig. P (1.11).

The load line is shown in Fig. P (1.11)

The operating point, Q = (0.505 V, 4.5 mA)

Example 41. *For the circuit shown in Fig. P (1.12), the input* V_i *has positive and negative swings. Find the output voltage,* V_0.

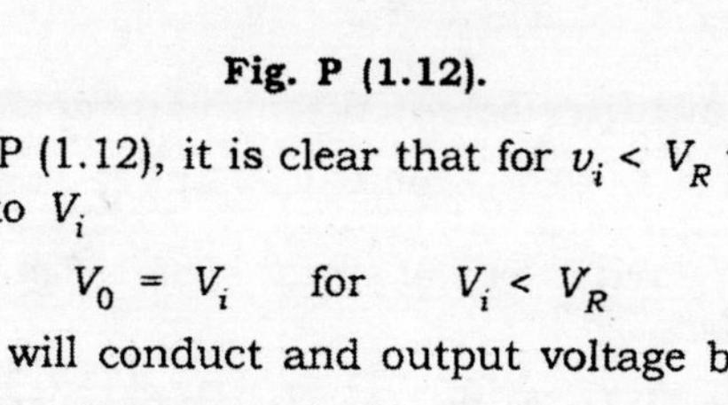

Fig. P (1.12).

Solution : From Fig. P (1.12), it is clear that for $v_i < V_R$ the diode will not conduct so output voltage equals to V_i

i.e., $\qquad V_0 = V_i \quad \text{for} \quad V_i < V_R$

and for $V_i > V_R$ the diode will conduct and output voltage becomes equal to V_R.

i.e., $\qquad V_0 = V_R \quad \text{for} \quad V_i > V_R$ (neglecting the voltage drop across R and cut in voltage of diode)

Example 42. *For the circuit shown in Fig. P (1.13). The output voltage V_0.*

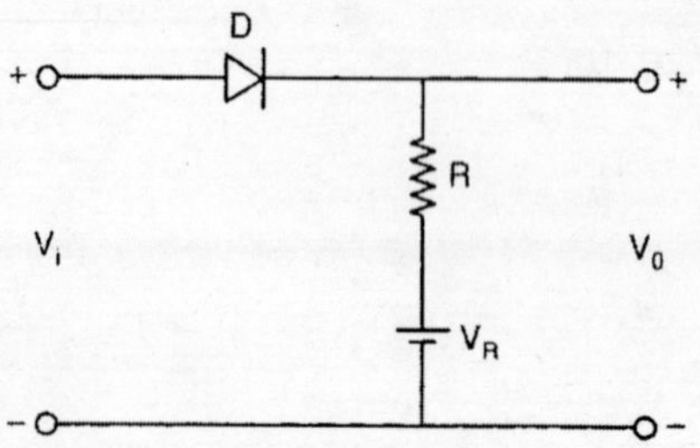

Fig. P (1.13).

Solution : Try yourself.

Example 43. *Thevoltages at V_1 and V_2 of the arrangement shown in the Fig. P (1.14) will be respectively. (Assume cut-in voltage of diode is .6 V).*

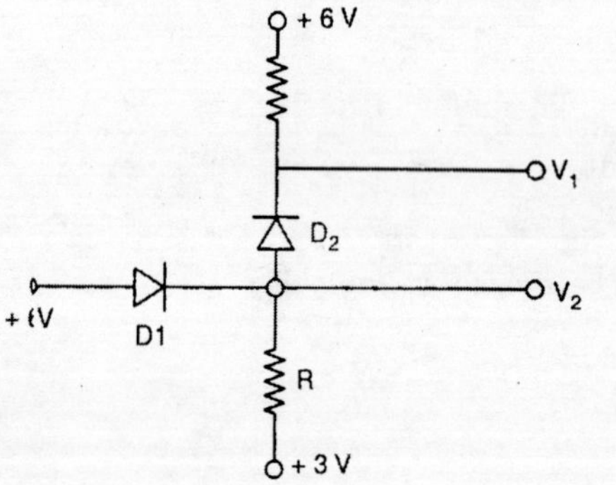

Fig. P (1.14).

Solution : From given Fig P (1.14), for the diode D_2 cathode is at higher potential than anode. Hence diode D_2 vill not conduct, *i.e.*, no current in upper RΩ resistance, so $V_1 = 6$ V.

But anode of D_1 is at higier potential than cathode so the diode D_1 will conduct.

$V_2 = 6 - V_r = 6 - 0.6 = 5.4$ V (where V_r = cut in voltage of diode)

Example 44. *Find I_1 and I_2 for the circuit shown Fig. P (1.15) terms of V_S assume $V_T = 0.7$ volt for all diode.*

Solution : From Fig. P (1.15) we see that here current I_1 and I_2 depends upon voltage V_x.

$$I_1 = \frac{5 - V_x}{2\ \text{k}\Omega} \qquad \text{If } 0 < V_S < 0.7 \text{ V}$$

Also,
$$V_x = V_S + 0.7$$

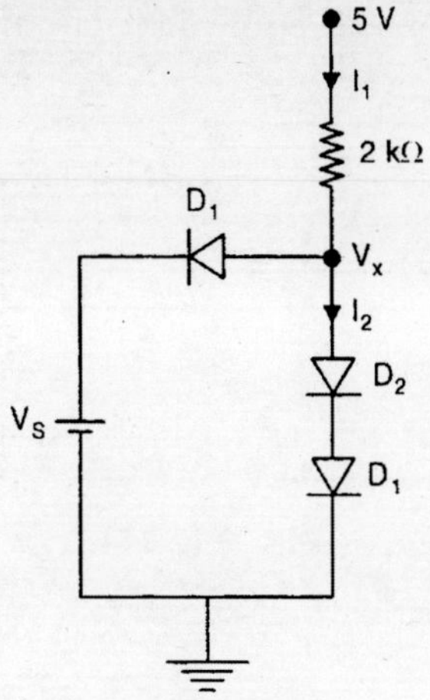

Fig. P (1.15).

If $V_S > 0.7$, the diode D_1 will be in reverse biased, so, in this situation, $I_1 = I_2 = \frac{5-1.4}{2\ k\Omega} = 1.8$ mA. **Ans.**

Where, 1.4 V is the voltage drop across diode D_2 and D_3.

Example 45. *Consider the circuit shown in Fig. P (1.16). If the diode used here has the V-I characteristic as in Fig. P (1.17). Find the output waveform* V_O.

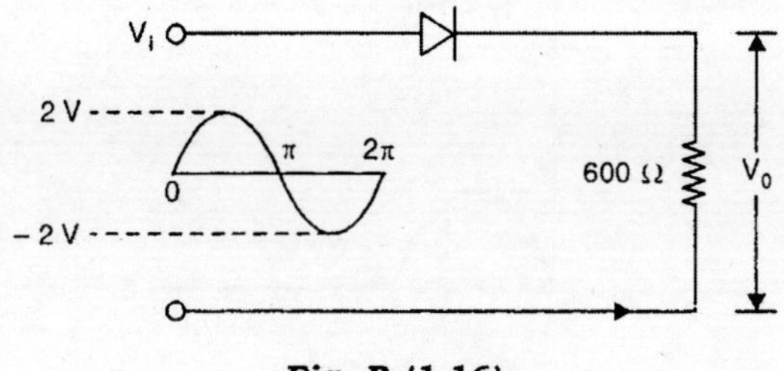

Fig. P (1.16)

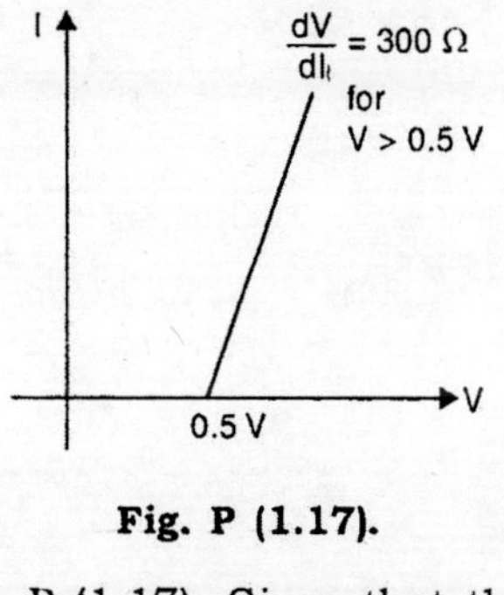

Fig. P (1.17).

Solution : From given Fig. P (1.17). Given that the resistance of diode, *i.e.*, $R_f = \frac{dV}{dI} = 300\ \Omega$ and the diode will conduct for input voltage greater than 0.5 V.

So, for the given information the equivalent circuit of Fig. P (1.16) can be redrawn as shown in Fig. P (1.18).

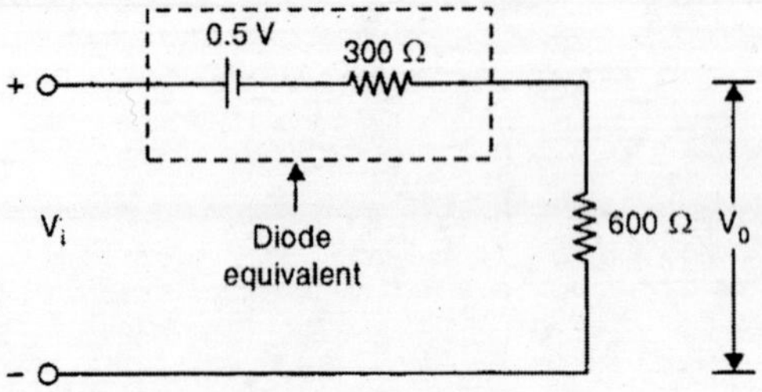

Fig. P (1.18).

From Fig. P (1.17), for positive half-cycle

So, $$V_0 = (V_i - 0.5)\left(\frac{600}{600+300}\right)$$

or $$V_0 = 1.5 \times \frac{600}{900}$$

or $$V_0 = 1.0 \text{ V}$$

However, for negative half cycle of the input voltage V_i, the diode is in the non-conduction state and we get zero output.

Finally the output wave form of Fig. P (1.16) is shown in Fig. P (1.19)

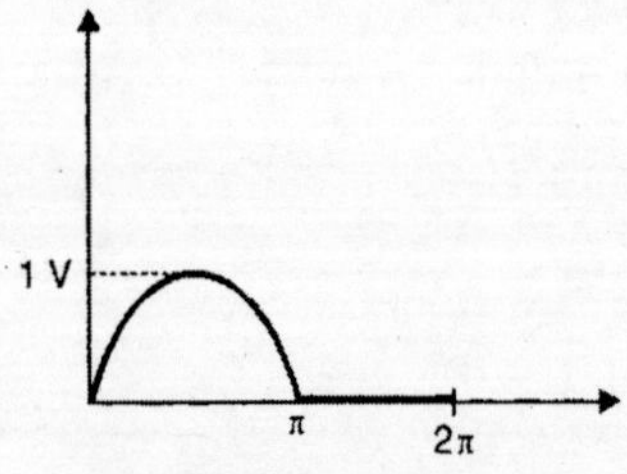

Fig. P (1.19).

□□□

2

DIODE APPLICATIONS

2.1. INTRODUCTION

Diode plays a very important role in various application. The most widely use of diode is in the field of rectification, (rectification is the process of a.c. to d.c. conversion) without diode rectification is not possible. The a.c. may be either in voltage or current form. a.c. power is easily produced in bulk form through different methods, but generally in many power control circuits and in other industrial applications d.c. power is very much required. Hence a.c. power necessarily has to be converted into d.c. power by means of electronic rectifier which is simple cheaper and highly efficient compared to rotatory converters or motor generators, and this process is achieved by the use of diode. The output obtained by the diode rectification is not pure d.c. there is some ripples in the output wave (current or voltage). So in order to get pure d.c. output filter circuit becomes necessary requirement. There are different filter circuits are used in order to filter the a.c. components.

A rectifier is a diode circuit which facilitates the a.c. current to flow only in one direction. A single diode-rectifier circuit cuts off the negative portion of the a.c. current and allows only one positive portion of the signal. This operation can be termed as half wave rectification (HWR). Similarly, for both parts of the cycle, we require a two diode rectifier circuit. This operation can be termed as full wave rectification.

The main application area of diodes can be listed as below :

* rectifiers
* clippers or limiters
* clampers
* voltage multipliers
* logic gates etc.

2.2. RECTIFIERS

If we look around, we find that most of electronic devices, such as TV, radio transistor, stereo, VCD/DVD players and computers etc., require the d.c. supply to work properly. But we have the a.c. mains supply from the power grid at a higher voltage level *i.e.*, 230 V, 50 Hz (Single phase) and 440 V, 50 Hz (Three phase) for domestic purpose.

Now, we have two options to supply the d.c. voltage to the electronic equipments, either we use batteries or some systems, which can convert the a.c. mains supply into d.c. supply. The system for this purpose, used in power supply section of electronic equipments is known as **Rectifier,** and the process of a.c. to d.c. conversion is known as **rectification.**

There are three types of rectifier circuit configurations :

* Half wave rectifier.
* Full wave rectifier with center tapped transformer.
* Full wave bridge rectifier.

These are discussed in detail below.

2.3. HALF-WAVE RECTIFIER (HWR)

The basic circuit diagram of half-wave rectifier with a pure resistive load, is shown in Fig. 2.1.

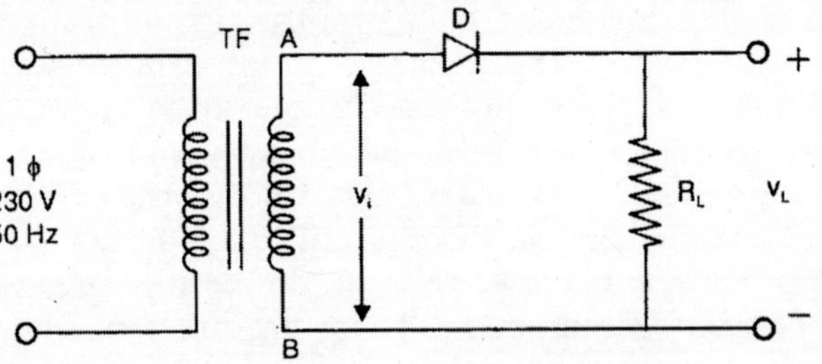

Fig. 2.1. *Half wave rectifier circuit.*

In the circuit, TF is the step down transformer. The resistance of the diode is assumed to be R_f in forward conduction or ON state and the d.c. resistance of secondary winding is R_2.

2.3.1. Operation of the Circuit

This circuit utilizes the unidirectional conduction property of the *p-n* junction, *i.e.*, the *p-n* junction allows to flow the current through it only when it is forward biased and restricts when it is reverse biased. In a.c. mains supply voltage, there are two half cycles, one is positive and second is negative. The operation of the circuit, can be well understood from the Fig. 2.2, and the waveforms are shown in Fig. 2.3.

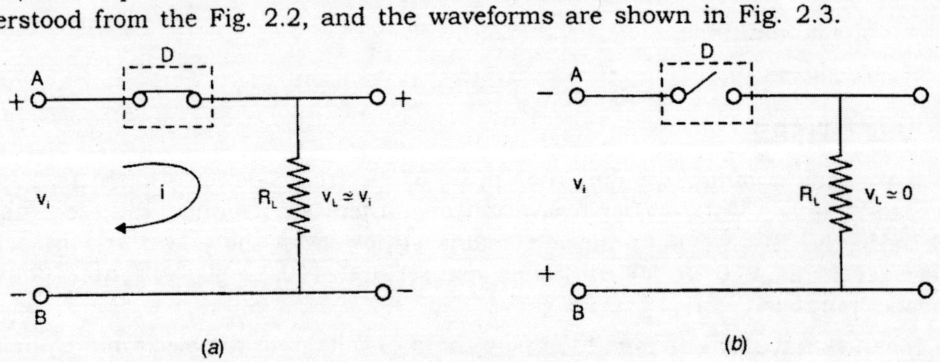

Fig. 2.2.

During the positive half cycle, $(0 \leq \omega t \leq \pi)$ when terminal A of secondary is at higher potential than that of terminal B, diode D will get forward biased and hence possesses a low resistance path (often with a very low resistance R_f) to the current. This will make the load voltage positive and almost equal to the instantaneous input (supply) voltage.

[Fig. 2.2 (*a*)] Here, the effect of cut-in voltage of diode, *i.e.*, V_D, is ignored assuming that $V_D << V_m$, (V_m is maximum value of input voltage), but when input is a small signal it can't be ignored, the effect of V_D is clear in examples.

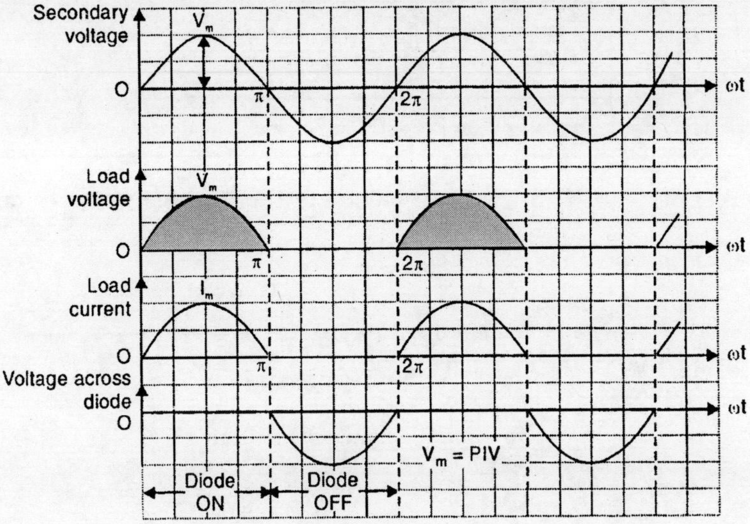

Fig. 2.3. *Waveforms of the half wave rectifier.*

During the negative half cycle, ($\pi \le \omega t \le 2\pi$), potential at B, becomes higher than that at A and diode becomes reverse biased and hence acts as an open switch. Here it is to be noted that a minute current is still flowing through the diode, which is known as **reverse saturation current** but ignored here for the simplification. The load voltage here is almost zero as shown in Fig. 2.3.

It is clear from above discussion that diode D conducts only for one half cycle or in other words we get only one half cycle of the input across the load. The output voltage obtained from this circuit is not of constant magnitude but pulsating in nature. Its magnitude is continuously changing with time but only in one direction (either positive or negative).

2.3.2. Analysis of Half Wave Rectifier Parameters

As, we assume here a sinusoidal input, hence we can write,

$$v_i = V_m \sin \omega t \qquad \text{... (2.1)}$$

If we neglect the drop across the diode D, we have output voltage (*i.e.*, load voltage) from Fig. 2.3,

$$v_L = v_i = V_m \sin \omega t \quad ; 0 \le \omega t \le \pi$$
$$= 0 \quad ; \pi \le \omega t \le 2\pi \qquad \text{... (2.2)}$$

Similarly,

$$i = I_m \sin \omega t \quad ; 0 \le \omega t \le \pi$$
$$= 0 \quad ; \pi \le \omega t \le 2\pi \qquad \text{... (2.3)}$$

where

$$I_m = \frac{V_m}{(R_2 + R_f + R_L)} \qquad \text{... (2.4)}$$

(*i*) Average (d.c.) Load Current

Mathematically, average or mean value of any function $f(x)$ can be calculated between the limits l and u as ;

$$F_{mean} = \frac{1}{u-l}\int_{l}^{u} f(x)\,dx \quad \text{... (2.5)}$$

where, l = lower limit
u = upper limit

Therefore from eqn. (2.5), we can calculate the average value of current i for one cycle as ;

$$I_{av} \text{ or } I_{dc} = \frac{1}{2\pi}\int_{0}^{2\pi} i\,d(\omega t)$$

Substituting the value of i here

$$I_{dc} = \frac{1}{2\pi}\left[\int_{0}^{\pi} I_m \sin(\omega t)\,d(\omega t) + 0\right]$$

$$= \frac{1}{2\pi} I_m \int_{0}^{\pi} \sin(\omega t)\,d(\omega t)$$

$$= \frac{I_m}{2\pi}[-\cos(\omega t)]_0^{\pi}$$

or
$$I_{dc} = \frac{I_m}{\pi} \quad \text{... (2.6)}$$

or
$$I_{dc} = \frac{V_m}{\pi(R_f + R_2 + R_L)}$$

where, R_f = Forward diode resistance
R_2 = Transformer secondary (d.c.) resistance
R_L = Load resistance.

(*ii*) RMS (a.c.) Load Current

Mathematically, **R**oot **M**ean **S**quare (RMS) value of a function $f(x)$ is defined as

$$F_{rms} = \sqrt{\frac{1}{u-l}\int_{l}^{u} [f(x)]^2\,dx} \quad \text{... (2.7)}$$

Using the relation of eqn. (2.7) we can calculate the RMS value of the current i, *i.e.*,

$$I_{rms} = \left[\frac{1}{2\pi}\int_{0}^{2\pi} i^2\,d(\omega t)\right]^{1/2} = \left[\frac{1}{2\pi}\int_{0}^{\pi} I_m{}^2 \sin^2(\omega t)\,d(\omega t)\right]^{1/2}$$

$$= \left[\frac{I_m^2}{2\pi} \int_0^{\pi} \frac{[1 - \cos(2\,\omega t)]}{2}\, d(\omega t) \right]^{1/2}$$

$$= \left[\frac{I_m^2}{2\pi \times 2} \left| \omega t - \frac{\sin(2\omega t)}{2} \right|_0^{\pi} \right]^{1/2}$$

or $$I_{rms} = \left[\frac{I_m^2}{4\pi} \times \pi \right]^{1/2}$$

or $$\boxed{I_{rms} = \frac{I_m}{2}} \quad \text{... (2.8)}$$

or $$I_{rms} = \frac{V_m}{2\,(R_f + R_2 + R_L)}$$

(*iii*) Form Factor

Form factor for any periodic wave is defined as the ratio of RMS and average value, *i.e.*,

$$\text{F.F.} = \frac{\text{RMS value}}{\text{DC value}} \quad \text{... (2.9)}$$

Therefore, **form factor** for the half wave rectifier is :

$$\text{F.F.} = \frac{(I_{m/2})}{(I_{m/\pi})} = \frac{\pi}{2}$$

or $$\boxed{\text{F.F.} = 1.57}$$

(*iv*) Average (d.c.) Load Voltage

We can determine the d.c. load voltage using the Ohm's law, *i.e.*,

$$V_{L_{dc}} = I_{dc} \times R_L \quad \text{... (2.10)}$$

$$= \frac{V_m}{\pi\,(R_f + R_2 + R_L)} \times R_L$$

If $R_L >>> (R_f + R_2)$, the general case, we get

$$\boxed{V_{L_{dc}} \simeq \frac{V_m}{\pi}} \quad \text{... (2.11)}$$

(*v*) RMS Load Voltage

Similarly, the RMS value of load voltage can be find as,

$$V_{L_{rms}} = I_{rms} \times R_L$$

$$= \frac{V_m}{2\,(R_f + R_2 + R_L)} \times R_L$$

Making the assumption, $R_L >>> (R_f + R_2)$, we get

$$\therefore \quad \boxed{V_{L_{rms}} \simeq \frac{V_m}{2}} \qquad \text{... (2.12)}$$

(*vi*) Ripple Factor

The rectifier output consists of a.c. as well as d.c. The ripple factor measures the percentage of a.c. component in the rectified output. The ideal value of ripple factor should be zero, *i.e.*, output should be pure d.c. Ripple factor is defined as :

$$\text{Ripple factor,} \quad \boxed{\gamma = \frac{\text{RMS value of the a.c. component of output}}{\text{d.c. or Average value of output}}}$$

As the instantaneous value of a.c. fluctuations is measured with respect to the d.c. level (*i.e.*, the difference of instantaneous total value and the d.c. value). Thus, the instantaneous a.c. value is given as :

$$i_{ac} = i - I_{dc} \qquad \text{... (2.13)}$$

Therefore, the RMS value of a.c. components is given as :

$$I_{ac_{rms}} = \left[\frac{1}{2\pi}\int_0^{2\pi}(i - I_{dc})^2\, d(\omega t)\right]^{1/2}$$

$$= \left[\frac{1}{2\pi}\int_0^{2\pi}(i^2 + I_{dc}^2 - 2\, i\, I_{dc})\, d(\omega t)\right]^{1/2}$$

$$= \left[\frac{1}{2\pi}\int_0^{2\pi} i^2\, d(\omega t) + \frac{1}{2\pi}\int_0^{2\pi} I_{dc}^2\, d(\omega t) - 2\, I_{dc}\int_0^{2\pi} i\, d(\omega t)\right]^{\frac{1}{2}}$$

$$= \left[I_{rms}^2 + I_{dc}^2 - 2I_{dc}^2\right]^{\frac{1}{2}}$$

$$I_{ac_{rms}} = \left[I_{rms}^2 - I_{dc}^2\right]^{\frac{1}{2}} \qquad \text{... (2.14)}$$

Therefore, **ripple factor,**

$$\gamma = \frac{I_{ac_{rms}}}{I_{dc}} = \frac{[I_{rms}^2 - I_{dc}^2]^{\frac{1}{2}}}{I_{dc}}$$

$$= \left[\left(\frac{I_{rms}}{I_{dc}}\right)^2 - 1\right]^{\frac{1}{2}}$$

$$\boxed{\gamma = [(\text{F.F.})^2 - 1]^{\frac{1}{2}}} \qquad \text{... (2.15)}$$

For half wave rectifier, ripple factor can be calculated by substituting the value of form factor into eqn. (2.15), therefore,

$$\gamma = [(1.57)^2 - 1]^{\frac{1}{2}}$$

or

$$\boxed{\gamma = 1.21}$$

(*vii*) Voltage Regulation

Under d.c. conditions, the rectifier circuit can be modelled as shown in Fig. 2.4, where R_f is the forward resistance of diode, R_2 is the d.c. resistance of transformer's secondary winding and R_L is the load resistance.

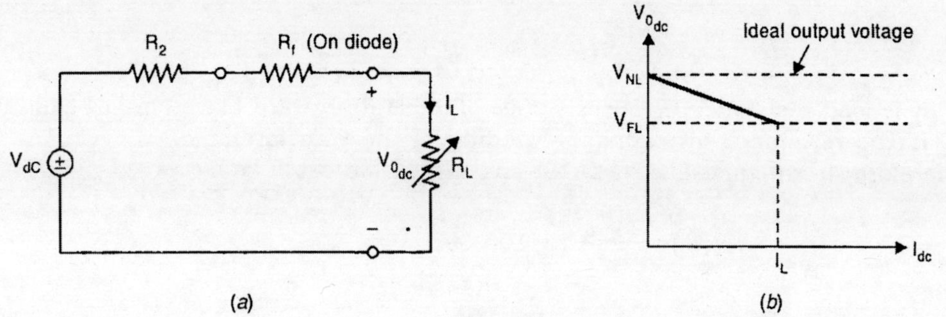

Fig. 2.4.

The voltage at output terminals can be calculated by voltage dividing rule as :

$$V_{0_{dc}} = \frac{V_{dc}}{(R_2 + R_f + R_L)} \times R_L = I_{dc} \times R_L$$

The value of load current depends on the value of the load resistance, *i.e.*, lower the value of R_L and higher will be the current (I_{dc}). Since, the secondary voltage V_{dc} is constant, the output voltage $V_{0_{dc}}$ will decrease with decreasing the value of R_L. This variation in output voltage from no-load to full load is measured by voltage regulation, *i.e.*,

$$\% \text{ V.R.} = \frac{V_{NL} - V_{FL}}{V_{FL}} \times 100 \qquad \text{... (2.16)}$$

where, V_{NL} = Voltage at no load

V_{FL} = Voltage at full load

The **voltage regulation** is also termed as **load regulation.** The ideal value of load regulation should be zero.

(*viii*) Rectification or Power Conversion Efficiency

The efficiency of a rectifier can be defined as the ratio of output d.c. power available at the load to the input a.c. power from the mains, and can be represented mathematically as

$$\text{Rectification efficiency} = \frac{\text{Output d.c. power available at the load}}{\text{Input a.c. power from the mains}}$$

i.e.,
$$\boxed{\eta\ (\%) = \frac{P_{o_{D.C.}}}{P_{i_{A.C.}}} \times 100.} \qquad \text{... (2.17)}$$

Now,

The d.c. power available at the load

$$P_{o_{d.c.}} = I_{dc}^2 R_L$$
$$= \frac{I_m^2}{\pi^2} R_L \qquad \text{... (2.18)}$$

And the input a.c. power from the mains,

$$P_{i_{a.c.}} = I_{rms}^2 (R_L + R_f + R_2)$$
$$= \frac{I_m^2}{4} (R_L + R_f + R_2) \qquad \text{... (2.19)}$$

where, R_L is load resistance, R_f is resistance of diode in forward biased or ON condition and R_2 is the resistance of secondary winding of the transformer.

Therefore, from eqns. (2.17), (2.18) and (2.19), we get

$$\eta(\%) = \frac{P_{o_{D.C.}}}{P_{i_{A.C.}}} \times 100\ \%$$
$$= \frac{(I_{m/\pi})^2 R_L}{(I_{m/2})^2 (R_L + R_f + R_2)} \times 100\ \%$$
$$= \frac{4 R_L}{\pi^2 (R_L + R_f + R_2)} \times 100\ \%$$
$$= \frac{(0.4053) \times 100}{1 + \left(\dfrac{R_f + R_2}{R_L}\right)} \%$$
$$= \frac{40.53}{1 + \dfrac{R_f + R_2}{R_L}} \%$$

As per assumption, if $R_L >> (R_f + R_2)$, we have

$$\boxed{\eta = 40.53\%} \qquad \text{... (2.20)}$$

It is clear from eqn. (2.20) that maximum achievable **rectification** and **conversion efficiency** with the half wave rectifier circuit is only approximately 41%. In practical case, where $(R_f + R_2)$ is comparable to the load resistance R_L, the efficiency is very less than this maximum value.

(*ix*) Transformer Utilization Factor (TUF)

Transformer Utilization Factor (TUF) can be defined as the ratio of output d.c. power available at the load to the power rating (Volt Ampere) of the transformer used in the rectifier circuit. Mathematically,

$$\text{TUF} = \frac{\text{Output d.c. power available at the load}}{\text{Power rating (VA) of the transformer}}$$

$$\boxed{\text{TUF} = \frac{P_{0_{d.c.}}}{\text{VA rating of transformer}}} \qquad \text{... (2.21)}$$

The a.c. power rating or VA rating of transformer can be calculated with the RMS voltage developed across the winding and RMS current flowing through the winding, *i.e.*,

$$\text{VA rating of transformer} = V_{rms} \times I_{rms}$$

As the voltage developed across the secondary winding is full sinusoidal cycle but the current flows only for the half cycle (for which diode conducts), so

$$V_{rms} = \frac{V_m}{\sqrt{2}}$$

and

$$= I_{rms} = \frac{I_m}{2}$$

Therefore,

$$\text{VA rating of transformer} = \frac{V_m}{\sqrt{2}} \times \frac{I_m}{2} = \frac{V_m I_m}{2\sqrt{2}}$$

but

$$I_m = \frac{V_m}{(R_L + R_f + R_2)}$$

Therefore,

$$\text{VA rating of transformer} = \frac{I_m^2 (R_L + R_f + R_2)}{2\sqrt{2}} \qquad \text{... (2.22)}$$

Hence, using eqns. (2.18) and (2.22) we can calculate the TUF as :

$$\text{TUF} = \frac{I_m^2 R_L}{\pi^2} \times \frac{2\sqrt{2}}{I_m^2 (R_L + R_f + R_2)}$$

$$= \frac{2\sqrt{2}}{\pi^2} \times \frac{1}{1 + \left(\dfrac{R_f + R_2}{R_L}\right)}$$

If $(R_f + R_2) << R_L$, we get

$$\text{TUF} = \frac{2\sqrt{2}}{\pi^2} = 0.287$$

or

$$\boxed{\text{TUF} = 28.7\%} \qquad \text{... (2.23)}$$

Ideal value of TUF should be 100% and practically it should be as high as possible.

(*x*) Peak Inverse Voltage (PIV)

This is the maximum voltage, which appears across the diode during the non-conducting or OFF state.

The PIV for the diode used in the rectifier circuit can be calculated using the circuit of Fig. 2.5. As shown in Fig. 2.5, PIV appears across the diode when the negative

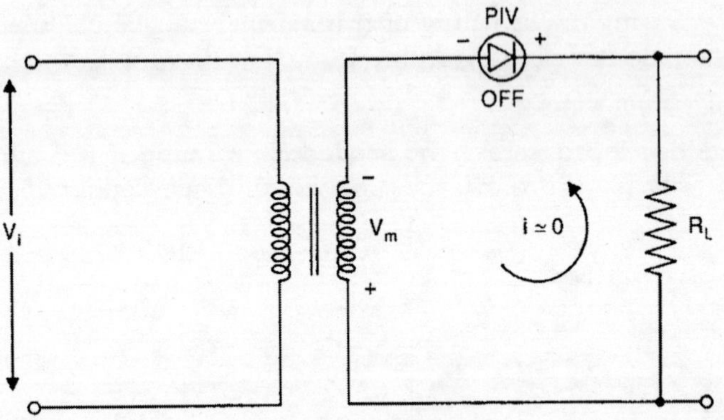

Fig. 2.5. *Calculation of PIV.*

peak occurs at the secondary of transformer. At this instant the current in the circuit is zero (only a leakage current flows), hence there is no voltage developed across the load resistance. Now applying KVL in this closed loop, we get

$$V_m - PIV = 0$$

or

$$\boxed{PIV = V_m} \quad \text{... (2.24)}$$

PIV is useful in selecting the diode for the rectifier. The diode must be selected for any rectifier circuit such that the PIV does not exceed the diode's maximum reverse voltage withstand capability, otherwise the diode will be damaged due to break down.

(*xi*) Ripple Frequency

The frequency of the pulsating load voltage is known as **Ripple Frequency.** It is equal to the frequency of supply voltage is Half Wave Rectifier, *i.e.*,

$$\boxed{f_r = f_s}$$

Disadvantages of Half Wave Rectifier

(*a*) Low efficiency, only 40%.

(*b*) High ripple factor, 121%.

(*c*) Low Transformer Utilization Factor (TUF), only 28%.

(*d*) Low d.c. output voltage and current.

(*e*) Possibility of transformer core saturation due to unidirectional current flow.

2.4. FULL WAVE RECTIFIER (FWR)

Unlike the HWR, in the full wave rectifier, we get the output across the load for the full cycle of the input signal. According to the circuit configuration the full wave rectifiers (FWRs) are classified into two :

* Full wave rectifier with center-tapped transformer
* Full wave bridge rectifier.

2.5. FWR WITH CENTER TAPPED TRANSFORMER

The basic circuit diagram for the FWR with center tapped transformer is shown in Fig. 2.6. In this circuit, two diodes D_1 and D_2 are used as the switching elements, one diode acts for one half cycle and second for a othelr half cycle. The transformer used is a center tapped transformer.

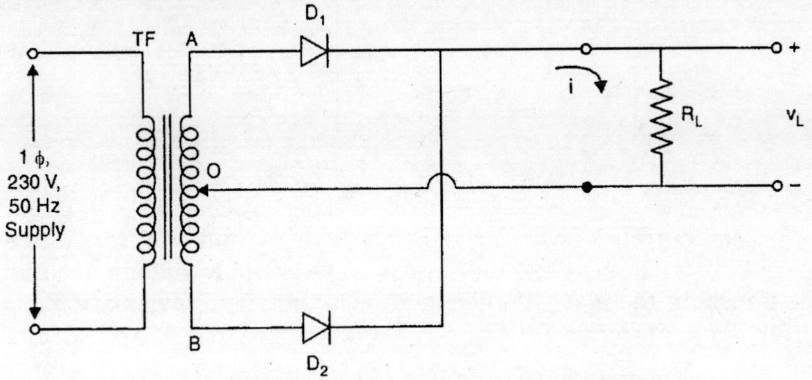

Fig. 2.6. *FWR with center tapped transformer.*

2.5.1. Operation of the Circuit

In the positive half cycle of the input a.c. supply, the polarities of secondary voltages are as shown in Fig. 2.7 (*a*), *i.e.*, terminal A of the winding is at a higher potential than the center terminal O and the terminal B is more negative, with respect to terminal O.

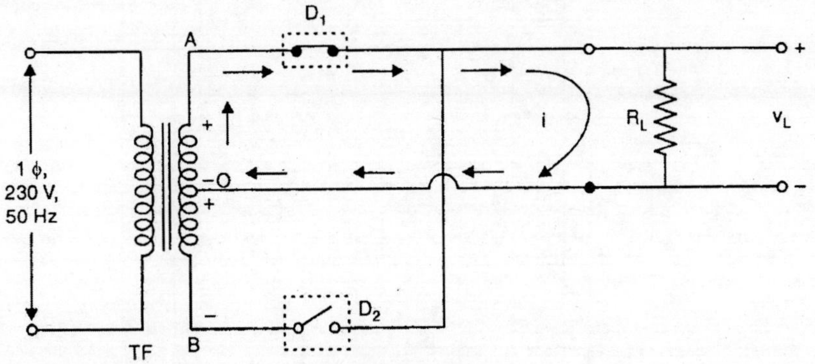

Fig. 2.7(*a*).

This makes the forward biasing for the diode D_1 and reverse biasing for the diode D_2. Therefore in this positive half cycle diode D_1 conducts and D_2 remains off. Thus load current flows through the diode D_1 and load R_L producing a load voltage (*i.e.*, output voltage) across R_L. The path of current is shown in figure.

In the negative half cycle ($\pi \leq \omega t \leq 2\pi$), terminal B of transformer secondary winding becomes positive to the center terminal O and A becomes negative with respect to the terminal. Therefore, diode D_2 conducts for this duration due to forward bias and D_1 is OFF [Fig. 2.7 (*b*)].

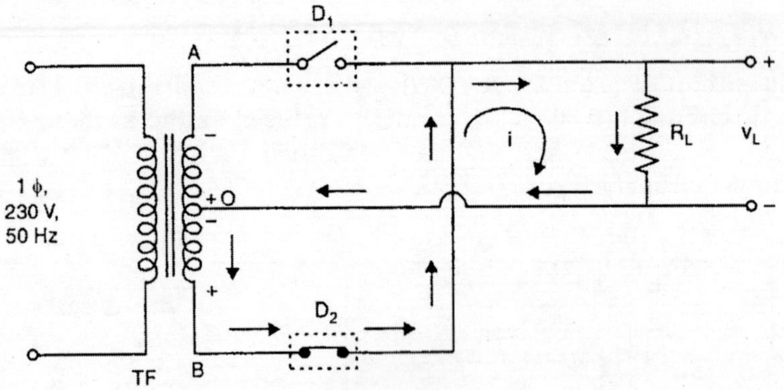

Fig. 2.7(*b*).

Now, the load currents flows through the diode D_2 and load resistance R_L. It is to be noted here that for both of the half cycles of a.c. supply voltage, load current flows in the same direction, therefore producing an equidirectional pulsating d.c. voltage. The waveforms for diode currents, load current and load voltage is shown in Fig. 2.8.

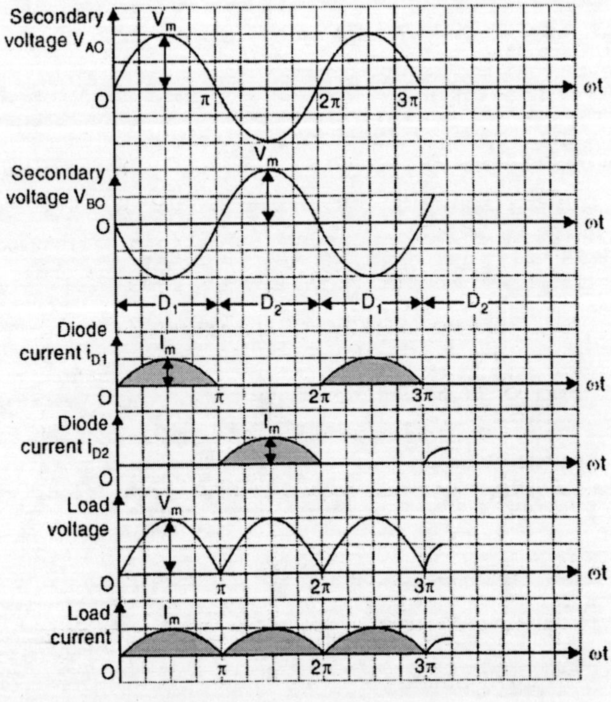

Fig. 2.8. *Voltage and current waveforms for a FWR.*

2.5.2. Analysis of Centre Tapped FWR Parameters

Similar to the previous section,

$$V_i = V_m \sin \omega t$$

Now, if we neglect the drop of a diode in forward bias we have load voltage,

$$V_L = \begin{cases} V_m \sin \omega t & ; \text{for } 0 \le \omega t \le \pi \\ -V_m \sin \omega t & ; \text{for } \pi \le \omega t \le 2\pi \end{cases} \quad \text{... (2.25)}$$

Similarly load current

$$i = \begin{cases} I_m \sin \omega t & ; \text{for } 0 \le \omega t \le \pi \\ -I_m \sin \omega t & ; \text{for } \pi \le \omega t \le 2\pi \end{cases} \quad \text{... (2.26)}$$

where,

$$\boxed{I_m = \frac{V_m}{(R_L + R_f + R_2)}} \quad \text{... (2.27)}$$

where R_f is the forward resistance of diode and R_2 is the d.c. resistance of secondary winding.

(*i*) Average (d.c.) Load Current

Putting the value of current i into eqn. (2.5) from eqn. (2.26), we have,

$$I_{av} \text{ or } I_{dc} = \frac{1}{2\pi}\left[\int_0^{2\pi} i \, d(\omega t)\right]$$

$$= \frac{1}{2\pi}\left[\int_0^{\pi} I_m \sin(\omega t)\, d(\omega t) + \int_{\pi}^{2\pi} (-I_m \sin \omega t)\, d(\omega t)\right]$$

$$= \frac{I_m}{2\pi}\left[(-\cos \omega t)\Big|_0^{\pi} + (\cos \omega t)\Big|_{\pi}^{2\pi}\right]$$

$$= \frac{I_m}{2\pi}[-\cos \pi + \cos 0 + \cos 2\pi - \cos \pi]$$

or

$$\boxed{I_{dc} = \frac{2I_m}{\pi}} \quad \text{... (2.28)}$$

or

$$I_{dc} = \frac{2}{\pi}\,\frac{V_m}{(R_L + R_f + R_2)}$$

(*ii*) RMS (a.c.) Load Current

Substituting the value of current i from eqn. (2.26) into eqn. (2.7), we get

$$I_{rms} = \left[\frac{1}{2\pi}\left\{\int_0^{\pi} I_m^2 \sin^2 \omega t \, d(\omega t) + \int_{\pi}^{2\pi} I_m^2 \sin^2 \omega t \, d(\omega t)\right\}\right]^{\frac{1}{2}}$$

$$= \left(\frac{I_m}{2\pi}\right)^{\frac{1}{2}} \left[\int_0^{\pi}\left(\frac{1-\cos 2\omega t}{2}\right) d(\omega t) + \int_{\pi}^{2\pi}\left(\frac{1-\cos 2\omega t}{2}\right) d(\omega t)\right]^{\frac{1}{2}}$$

After solving we get

$$\boxed{I_{rms} = \frac{I_m}{\sqrt{2}}} \quad \ldots (2.29)$$

or

$$I_{rms} = \frac{1}{\sqrt{2}} \frac{V_m}{(R_L + R_f + R_2)}$$

(*iii*) Form Factor (F.F.)

As we have from the definition of **Form Factor** that

$$\text{F.F.} = \frac{\text{RMS value}}{\text{Average value}}$$

We have for full wave rectifier,

$$\text{F.F.} = \frac{(I_m/\sqrt{2})}{(2\, I_m/\pi)} = \frac{\pi}{2\sqrt{2}}$$

or

$$\boxed{\text{F.F.} = 1.11}$$

(*iv*) Average (d.c.) Load Voltage

We have from Ohm's Law, d.c. load voltage,

$$V_{L_{dc}} = I_{dc} \times R_L = \frac{2\, I_m}{\pi} \times R_L$$

or

$$V_{L_{dc}} = \frac{2}{\pi} \frac{V_m}{(R_L + R_f + R_2)} \times R_L$$

$$= \frac{2}{\pi} \frac{V_m}{1 + \left(\frac{R_f + R_2}{R_L}\right)}$$

If $(R_f + R_2) << R_L$, then,

$$\boxed{V_{L_{dc}} \simeq \frac{2\, V_m}{\pi}} \quad \ldots (2.30)$$

(*v*) RMS Load Voltage

The RMS value of load voltage is given by

$$V_{L_{rms}} = I_{rms} + R_L$$

Substituting the value of I_{rms}, we get,

$$V_{L_{rms}} = \frac{I_m}{\sqrt{2}} \times R_L$$

or

$$V_{L_{rms}} = \frac{1}{\sqrt{2}} \frac{V_m}{(R_L + R_f + R_2)} \times R_L$$

$$= \frac{1}{\sqrt{2}} \frac{V_m}{1 + \left(\dfrac{R_f + R_2}{R_L}\right)}$$

Assuming $R_f + R_2 << R_L$, then

$$\boxed{V_{L_{rms}} \simeq \frac{V_m}{\sqrt{2}}} \quad \text{... (2.31)}$$

(*vi*) Ripple Factor (γ)

From eqn. (2.15), we have,

$$\gamma = [(FF)^2 - 1]^{1/2}$$

Substituting the value of form-factor for the full wave rectifier, we get

$$\gamma = [(1.11)^2 - 1]^{1/2}$$

$$= 0.48$$

or

$$\boxed{\gamma = 48\%}$$

Therefore, the quality of the d.c. voltage obtained from full-wave rectifier is better than that of half wave rectifier, because the ripples present in the rectified output of full wave rectifiers are just 48% which are much less than the 121% of half wave rectifier (HWR).

(*vii*) Rectification Efficiency

Rectification efficiency is defined as,

$$\% \, \eta = \frac{P_{O_{DC}}}{P_{i_{AC}}} \times 100$$

Where

$$P_{O_{DC}} = I_{dc}^2 \times R_L$$

$$= \frac{4 \, I_m^2}{\pi^2} \times R_L \quad \text{... (2.32)}$$

and

$$P_{i_{AC}} = I_{rms}^2 \, (R_L + R_f + R_2)$$

$$= \frac{I_m^2}{2} \, (R_L + R_f + R_2) \quad \text{... (2.33)}$$

Therefore, from eqns. (2.32) and (2.33), we have,

$$\% \eta = \frac{P_{O_{DC}}}{P_{i_{AC}}} \times 100$$

or $$= \frac{(4I_m^2/\pi^2)R_L}{(I_m^2/2)(R_L + R_f + R_2)} \times 100$$

or $$= \frac{4I_m^2}{\pi^2} \times \frac{2}{I_m^2} \times \frac{R_L}{R_L + R_f + R_2} \times 100$$

or $$= \frac{8}{\pi^2} \times \frac{1}{1 + \left(\frac{R_f + R_2}{R_L}\right)} \times 100$$

$$\% \eta = 81.06 \times \frac{1}{1 + \left(\frac{R_f + R_2}{R_L}\right)} \% \qquad \text{... (2.34)}$$

If we assume $(R_f + R_2) << R_L$ then eqn. (2.34) reduces to

$$\boxed{\eta \simeq 81.06\%}$$

Thus the maximum theoretical rectification efficiency, which can be achieved with a full wave rectifier is 81.06%. It is much larger than that of half wave rectifier circuit, *i.e.*, 40.53%.

(*viii*) Transformer Utilization Factor (TUF)

The method to calculate the TUF for a full wave rectifier (with center tapped transformer) is slightly different to that for half wave rectifier. For this circuit the TUF for primary and secondary windings is calculated separately, and the average of these two is the TUF for the circuit. This is because the half of secondary supplies the power to the load only for the half cycle duration, while the currents flows in the primary winding for the full cycle of input voltage. Now,

$$\text{Secondary TUF} = \frac{P_{O_{DC}}}{\text{VA rating of transformer}}$$

$$= \frac{I_{DC} \times V_{L_{DC}}}{I_{rms} \times V_{i_{rms}}} = \frac{(2I_m/\pi) \times (2V_m/\pi)}{(I_m/\sqrt{2})(V_m/\sqrt{2})}$$

$$= \frac{8}{\pi^2}$$

$$\boxed{\text{Secondary TUF} = 81.1\%} \qquad \text{... (2.35)}$$

and TUF for primary winding, *i.e.*,

$$\text{Primary TUF} = 2 \times \text{TUF or HWR}$$

$$= 2 \times 0.287$$

$$\boxed{\text{Primary TUF} = 0.574} \qquad \ldots (2.36)$$

Thus, the TUF for a full wave rectifier with center tapped transformer is given as :

$$\text{TUF} = \frac{\text{Primary TUF} + \text{Secondary TUF}}{2}$$

$$= \frac{0.811 + 0.574}{2}$$

$$\boxed{\text{TUF} = 69.25\%} \qquad \ldots (2.37)$$

(*ix*) Peak Inverse Voltage (PIV)

PIV for the center tapped transformer, full wave rectifier circuit can be calculated using the circuit of Fig. 2.9, where diode D_2 is reverse biased at the time of peak of the input voltage, and D_1 conducts, therefore, the maximum voltage V_m also appears across the load resistance R_L.

Now applying the KVL in loop (2), we have

$$V_m + V_m - PIV = 0$$

or

$$\boxed{PIV = 2V_m} \qquad \ldots (2.38)$$

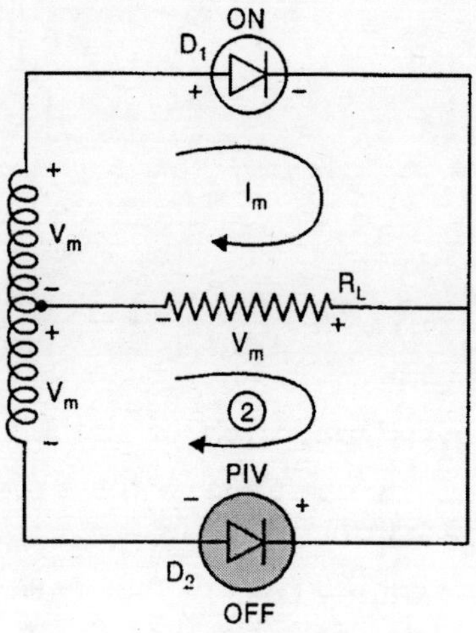

Fig. 2.9. *Calculation of PIV.*

(*x*) Ripple Frequency

The ripple frequency in a full-wave rectifier is twice that of the supply voltage, *i.e.*,

$$\boxed{f_r = 2f_s} \qquad \ldots (2.39)$$

Advantages of FWR with Center Tapped Transformer

(*a*) Low ripples as compared to HWR.

(*b*) Better rectification efficiency.

(*c*) Better TUF.

(*d*) Higher values of d.c. load current and voltage.

(*e*) No possibility of transformer core saturation.

Disadvantages of FWR with Center Tapped Transformer

(*a*) The output maximum voltage is half that of maximum induced voltage at secondary winding.

(*b*) PIV is twice that of HWR, therefore diodes are required of higher rating.

(*c*) Increased cost due center tapped transformer.

2.6. FULL WAVE BRIDGE RECTIFIER

The main disadvantages of the center-tapped transformer full wave rectifier were the high PIV and essence of the **center tapped transformer.** These can be overcome simply by using a **bridge of diodes** as shown in Fig. 2.10. Figure shows the basic circuit of the bridge rectifier. TF is a simple **step down** transformer, if needed, the four diodes D_1, D_2, D_3 and D_4 are connected in the form of a bridge, two diode, either D_1 and D_2 or D_3 and D_4 conduct at a time, this gives the full wave rectification.

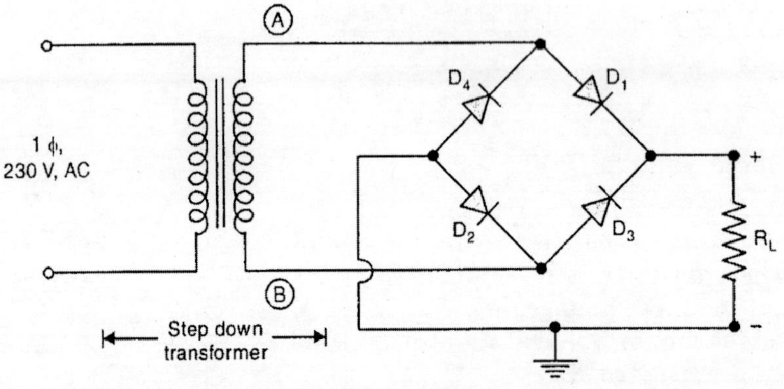

Fig. 2.10. *A bridge rectifier circuit.*

2.6.1. Operation of the Circuit

The operation of the circuit can be understood with the help of Fig. 2.11(*a*) and Fig. 2.11(*b*).

During positive half cycle (*i.e.*, $0 \le \omega t \le \pi$) of the supply voltage, the terminal A is at higher potential than terminal B, thus providing the forward biasing to diodes D_1 and D_2, whereas D_3 and D_4 are reverse biased Fig. 2.11(*a*). Here the load current and load voltage, both are positive.

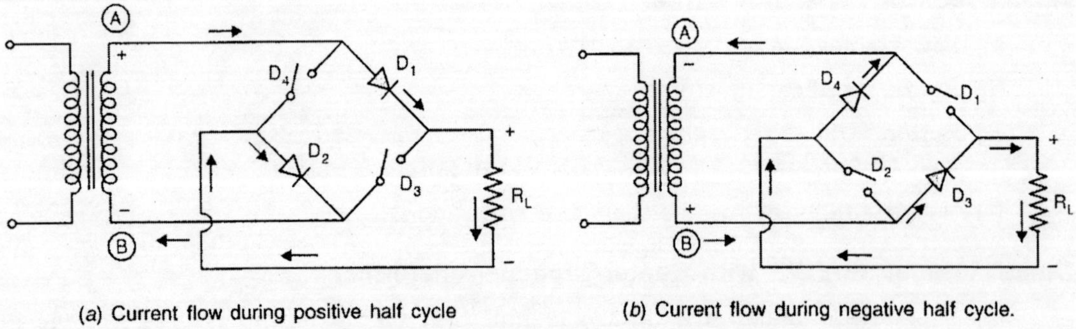

Fig. 2.11.

During negative half cycle (*i.e.*, $\pi \le \omega t \le 2\pi$), of the supply voltage, terminal B becomes more positive than terminal A, therefore, it forward biases the diode D_3 and D_4 and D_1 and D_2 are OFF during this period Fig. 2.11(*b*). The waveforms are shown in Fig. 2.12. Therefore, the direction of the current and hence load voltage is same during both half cycles of the supply voltage.

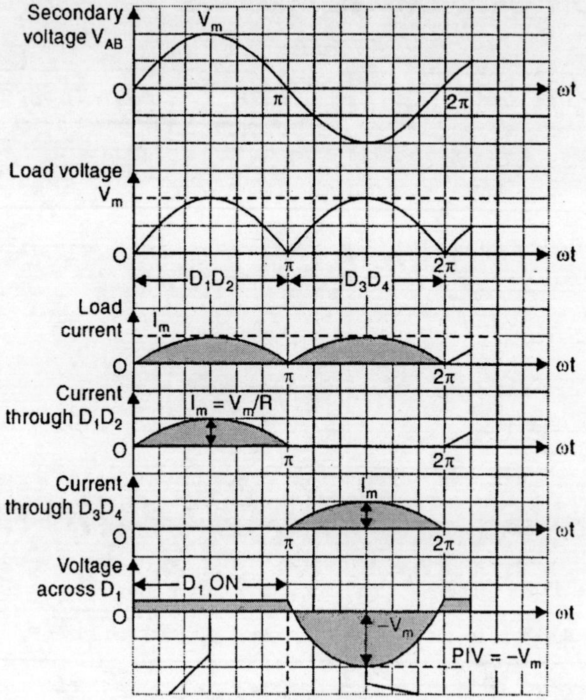

Fig. 2.12.

2.6.2. Analysis

The analysis for performance parameters of the Bridge rectifier is same as for the full wave with center tapped transformer rectifier circuit, but the diode forward resistance is now twice, because two diodes conduct in series, thus contributing total ON resistance of $2R_f$. The *PIV* of this circuit is found equal to the maximum value (peak value) of the supply voltage, *i.e.*, V_m.

The TUF should be calculated for this secondary winding only, which is equal to the calculated in previous section for full wave rectifier, eqn. (2.35), *i.e.*,

$$\boxed{\text{TUF} = 81.1\%}$$

Advantages

(*a*) Transformer is not compulsory.

(*b*) Circuit is suitable for high voltage applications because PIV is less, *i.e.*, equal to V_m.

(*c*) Core saturation does not take place.

Disadvantage

The only disadvantage of this circuit is the need of four diodes, but it is not major one compared to the advantages of the circuit.

2.7. COMPARISON OF RECTIFIER CIRCUITS

Table 2.1

S.No.	Parameters	HWR	Center Tapped FWR	Bridge FWR
1.	d.c. Load Current ($I_{L_{DC}}$)	$\frac{I_m}{\pi}$	$\frac{2I_m}{\pi}$	$\frac{2I_m}{\pi}$
2.	No load d.c. Load Voltage ($V_{L_{DC}}$)	$\frac{V_m}{\pi}$	$\frac{2V_m}{\pi}$	$\frac{2V_m}{\pi}$
3.	RMS Load Current (I_{rms})	$\frac{I_m}{2}$	$\frac{I_m}{\sqrt{2}}$	$\frac{I_m}{\sqrt{2}}$
4.	RMS Load Voltage ($V_{L_{RMS}}$)	$\frac{V_m}{2}$	$\frac{V_m}{\sqrt{2}}$	$\frac{V_m}{\sqrt{2}}$
5.	d.c. Load Power ($P_{O_{DC}}$)	$\left(\frac{I_m}{\pi}\right)^2 R_L$	$\left(\frac{2I_m}{\pi}\right)^2 R_L$	$\left(\frac{2I_m}{\pi}\right)^2 R_L$
6.	Maximum Efficiency (η%)	40%	81.2%	81.2%
7.	Ripple Factor (γ)	121%	48%	48%
8.	Ripple Frequency	fs	$2fs$	$2fs$
9.	TUF (Turn utilization factor)	28.7%	69.24%	81.2%
10.	Number of Diodes	1	2	4
11.	Center tap transformer	Not required	Required	Not required
12.	Transformer core saturation	Possible	Not possible	Not possible
13.	PIV (Peak inverse voltage)	V_m	$2V_m$	V_m

2.8. FILTERS

Need of Filters

Upto now we have discussed about the rectifier circuit. As we have seen that the output of any rectifier circuit is not purely d.c. The output is pulsating in nature because of *ripples* present with the d.c. output. The ripples are unwanted a.c. signals which if not removed before applying to the load produces distortions in the form of *hum*, so in order to get a smooth d.c. output for perfectly operating to any electronic

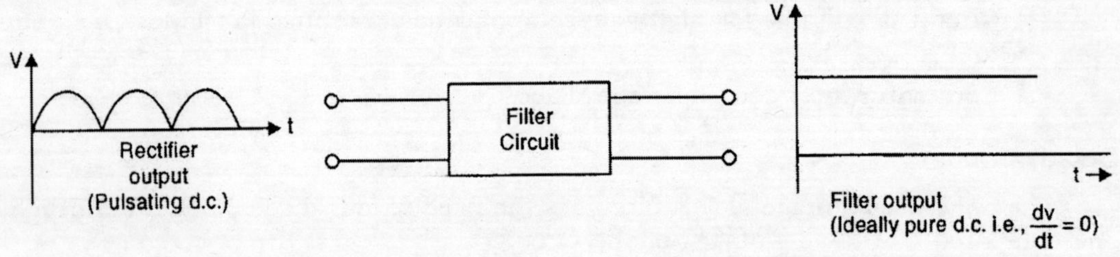

Fig. 2.13. *Block diagram representation of filter.*

circuit we require an extra circuit between rectifier output and load. *Thus we can say that filters are the devices which convert pulsating d.c. output of a rectifier to steady d.c. level.*

Before discussing the different types of filters it is very important to know about the main component used in the filter circuits *i.e.*, inductor and capacitor.

Usually all filter circuits employ inductors or capacitors or both as basic frequency selective components. Depending on the arrangement of these component with respect to load, filters have been classified in following categories :

(*i*) Series inductor filter

(*ii*) Shunt capacitor filter

(*iii*) Choke input or *L*-section filter

(*iv*) Capacitor input or π-section filter

The block diagram showing the function of filter is shown in Fig. 2.13.

Why do we connect an inductor in series and a capacitor in shunt in a filter circuit ?

Inductor. The term inductance may be defined as the opposition offered by the magnetic field of an inductor (an inductor is just a choke *i.e.*, a coil having large number of turns with small resistance wound). The resistance offered is called inductive reactance or inductance. Its value can be calculated by using the relation :

$$\boxed{X_L = 2\pi fL}$$

where L = Inductance of choke in Henry

f = Operating frequency in Hertz

From the above relation (*i.e.*, $X_L = 2\pi fL$) it is clear that, for d.c., the value of inductive reactance is zero, because for d.c., $f = 0$ (*i.e.*, $X_L = 2\pi.0.L = 0$) and hence it provides a least resistance path to d.c. component. However, for a.c. the value of inductive reactance is high. It means it provides high opposition to the a.c. components and blocks it.

Thus, an inductor is connected in series with the circuit so that it blocks the a.c. component and provides an easy path to the d.c. component to reach the destination (i.e., load).

Capacitor. The term capacitance may be defined as the opposition offered by a capacitor due to electric field. (Generally, an electrolytic capacitor at large capacitance is used in the filter circuits). The resistance offered is called capacitive reactance or capacitance and its value is determined by the relation

$$\boxed{X_c = \frac{1}{2\pi f\, C}}$$

Therefore the value of X_c, for d.c. (*i.e.*, $X_c = \dfrac{1}{2\pi.0.C.} = \infty$) and hence it provides very high impedance (*i.e.*, as open circuit) for d.c. However, it provides an easy path (low impedance) for a.c.

Thus a *capacitor is connected in parallel to the circuit so that it bypasses the a.c. component and blocks the d.c. component to reach at the load.*

2.8.1. Shunt Capacitor Filter

This is the most simple form among all the filter circuits and in this arrangement a high value capacitor C is placed directly across the load terminals as illustrated in Fig. 2.14. During the conduction period capacitor gets charged and stores up energy in the electrostatic field and discharges through the load resistance R_L during the non-conduction period. In this process, the time duration during which current flows through the load resistance gets prolonged (tedius) and a.c. component or ripples get considerably reduced. Here it is very interesting to note that the capacitor C get charged to the peak value of input voltage quickly because charging time constant and is almost zero. It is so because there is no resistance (except the negligible forward resistance of diode) in the charging path.

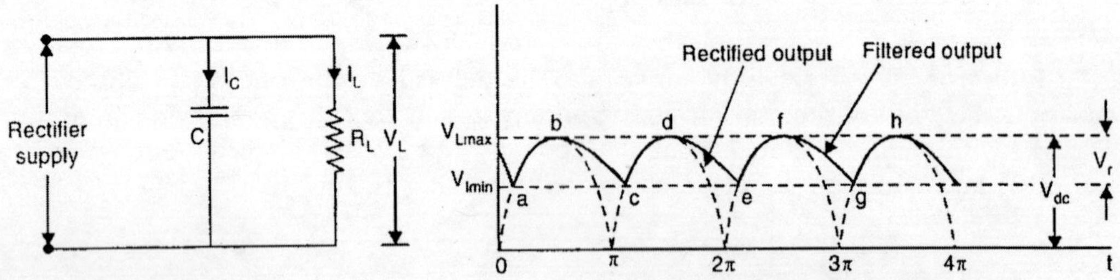

(*a*) Circuit diagram

(*b*) Rectified and filtered output voltage waveform for full-wave rectifier with shunt capacitor filter.

Fig. 2.14. *Shunt capacitor filter.*

2.8.2. Series Inductor Filter

Circuit diagram of series inductor filter is shown in Fig. 2.15(*a*). In this arrangement (filter) a high value inductor (choke) L is connected in series with the load resistance and rectifier supply (output). Due to the unique property of inductor, its filtering action depends upon its property of opposing any change in the current flowing through it when the output current of the rectifier increases above a certain value, energy is stored in it in the form of magnetic field and energy is given up when the output current falls below the average value. So, placing a inductor in series with the rectifier output and load, any sudden change in current that might have occurred in the circuit without an inductor is smoothed out by the presence of the inductor L.

Again, the inductor uses the property that it opposes any change of current flowing through it. Inductor stores energy when the input signal is positive and delivers the energy when the input signal is negative. So a lower magnitude of voltage is obtained at the output. As the input decreases from zero, the inductor again opposes the decrease in current and supply energy to the load. Thus, the current at the load remains continuous with reduced pulsating nature.

We know that the choke offers a high impedance to the a.c. component, but offers almost zero resistance to the derived d.c. components. Thus ripples are removed upto a large extent. Nature of the output voltage without filter and with choke filter are shown in Fig. 2.15(*b*).

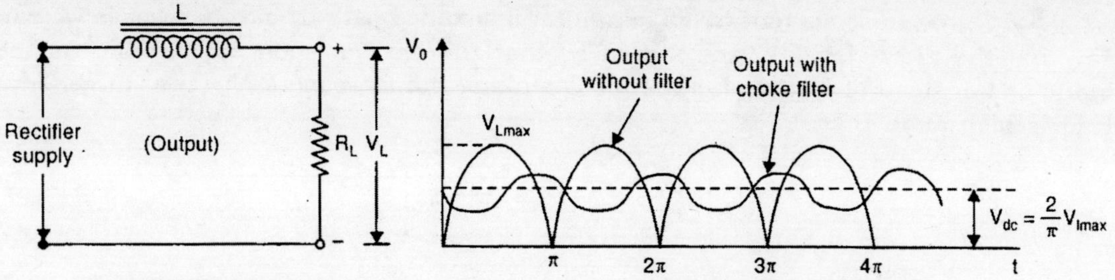

(*a*) Circuit diagram of series inductor filter

(*b*) The output voltage waveforms

Fig. 2.15. *Full wave rectifier with series inductor filter.*

2.8.3. Choke Input or *L*-section Filter

Upto now we have studied about simple series inductor filter and simple shunt capacitor filter. The former reduces both the peak and effective values of the output current and output voltage while the later reduces the ripple voltage but at the same time increases the diode current which may cause the diode to be damaged. So, individual application of these filters in the power supply system adds certain disadvantages.

In order to avoid the drawbacks of these two filter circuits and to add up their advantages, another type of filter is introduced which is referred to as the choke input or *L*-section filter. Fig. 2.16 illustrates the arrangement of combination of inductor and capacitor in the fiiter circuit.

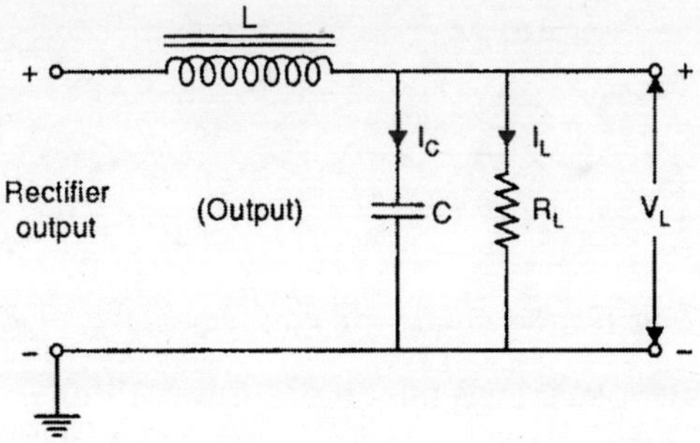

Fig. 2.16. *Circuit diagram of L-section filter.*

The circuit diagram in Fig. 2.15 consists of a inductor in series and capacitor in shunt with a load R_L. The name *L*-section of this filter circuit has been derived from the basic '*L*' shaped structure of the circuit.

The input is fed through the inductor so it is also known as the choke—input filter. Here, the inductor plays its role as a current smoothing element and capacitor as the voltage stabilizing element.

Here, it is notable that several L-section in cascade can be connected in order to get more smooth filter output.

The choke L on the input side of the circuit easily allows d.c. to pass ($2\pi f L = 0$ for d.c.) but restricts the a.c. components to flow. Now, if any component of a.c. *i.e.*, pulsating d.c. is still present then it is effectively passed by the capacitor in the shunt ($\therefore X_C << R_L$) because current takes minimum resistance path or in other words we can say that the filter elements are chosen so that the a.c. variation appears across the inductor L instead of load and capacitor C making the filter more effective. In general, load is high such that.

$$\boxed{R_L >> 2\omega L >> \frac{1}{2\,\omega C}}$$

where, R_L = Load Resistance

ω = Frequency

C = Capacitor

L = Inductor

In actual practice we can't get pure d.c. across the load because some ripples have not yet been filtered, however, in most of the applications it is practically avoidable. The rectified and filtered output for a *L*-section filter is given in Fig. 2.17.

From Fig. 2.17, we can see that the filtered output starts traversing from zero instant and readily goes towards negative side i.e., below the constant d.c. level output. The reason behind this phenomenon in that the shunt capacitor when charges, takes the voltage from the d.c. level and the filtered output is diminished. When capacitor discharges during its operation, it adds up the current to the load R_L and voltage level gets slightly increased.

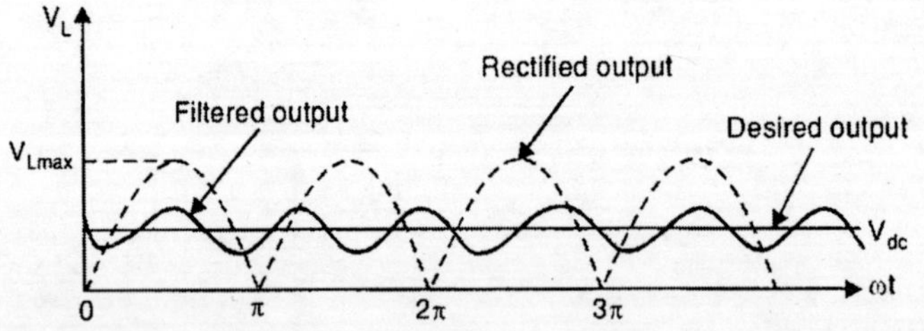

Fig. 2.17. *Rectified and filtered output voltage waveform for FWR for L-section filter.*

The expression for ripple factor for choke input filter can be expressed as :

$$\text{Ripple factor, } \gamma = \frac{V_{ac\,rms}}{V_{dc}} = \frac{\sqrt{2}}{3}\frac{X_C}{X_L}$$

or

$$\gamma = \frac{\sqrt{2}}{3}\frac{1}{2\omega C}\cdot\frac{1}{2\omega L} = \frac{1}{6\sqrt{2}\omega^2 LC}$$

$$\boxed{\gamma = \frac{1}{6\sqrt{2}\omega^2 LC}}$$

2.8.4. Capacitor Input or π-Filter

In this class of configuration the output from a rectifier is first fed across the capacitor and named as capacitor input filter. The circuit diagram for capacitor input filter is given in Fig. 2.18.

Basically a π-filter is a combination of two capacitors and one inductor. It consist of two stages:

(*i*) Capacitor shunt filter,

(*ii*) Choke input filter.

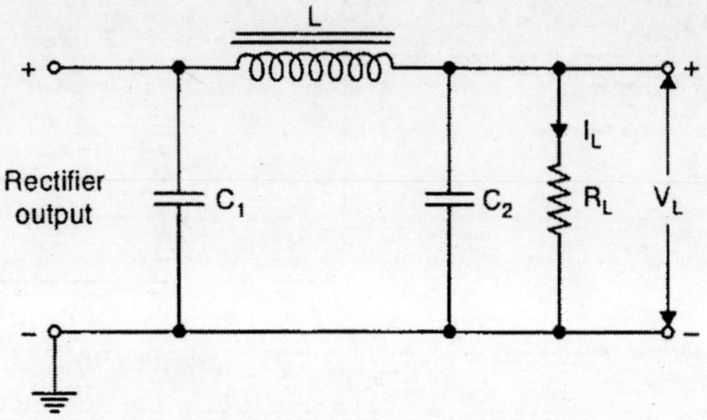

Fig. 2.18. *Circuit diagram of a π-Filter.*

Fig. 2.18 consists of a shunt capacitor C_1 at the input terminal. An induction L in them connected to C_1 in series. This inductor L is again shunted by another capacitor C_2 and finally load is connected across the C_2. This type of arrangement of filter circuit is often referred to as π-filter just because of its shape which resembles to the symbol π (*Pie*).

π-filter is result of need of higher output voltages at low magnitude loads. To form of π-filter, a input capacitor is shunted to a choke input filter.

The most important feature of this filter is that it can be used for a half wave rectifier circuit because rectifier output is directly connected to the capacitor.

In general, both the capacitors are confined in a single metal container and the metal acts as a common ground for both the capacitors.

It is notable here that in filter circuits we employ electrolytic capacitors.

The main filtering action is performed by the input capacitor C_1 in this case. The filtering action of capacitor used not to be explained again because it has been given in detail in the beginning of this article. The ripple factor of π-filter is product of the ripple factor of capacitor shunt filter stage and ripple factor of choke input filter stage. If compared with *L*-section filter, π-filter have a higher output voltage but the voltage regulation is poor. If any ripples are still present in the output C_1, series inductor L and shunt capacitor C_2 again smoothen the output and a desirable destabilized d.c. output gets associated with the load. The main disadvantage of π-filter is that the output voltage of π-filter falls off rapidly with the increase in load.

Comparison between *L*-section and π-Filter

1. In a π-filter ripples are less in comparison to shunt capacitor or *L*-section filter.
2. A π-filter requires an inductor of relatively low magnitude than that used in *L*-section filter.
3. Voltage regulation in case of π-filter is poor, so π filters are generally employed with fixed loads, while *L*-section filters suit better for varying loads.
4. PIV is larger in case of a π-filter than in case of *L*-section filter.
5. Current and therefore voltage-regulation for this filter circuit is usually poor. The rectified output and filtered-output in case of capacitor input filter is as given under here :

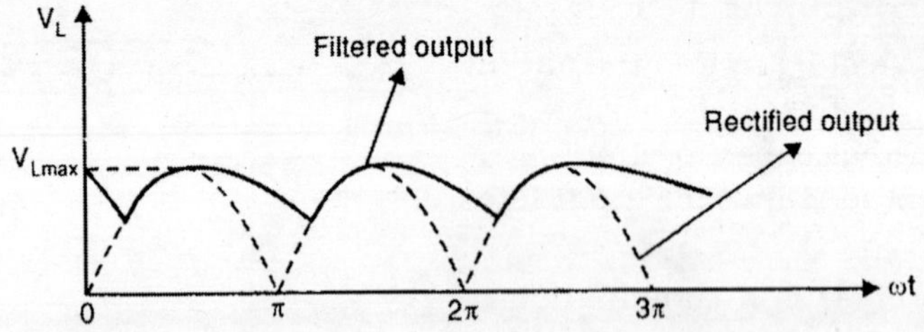

Fig. 2.19. *Rectified and filtered output voltage waveforms for a full wave rectifier with capacitor input-filter.*

6. π-filter provides better d.c. output voltage than that obtained in the case of *L*-section filter under similar input conditions.

Expression for ripple factor is given under here :

Ripple factors, $$\gamma = \frac{V'_{ac\ rms}}{V_{dc}} = \frac{\sqrt{2}\, I_{dc} X_{C_1} X_{C_2}}{V_{dc}\ X_L} \qquad (\because V_{dc} = I_{dc} \cdot R_L)$$

$$= \frac{\sqrt{2}\, I_{dc}\, X_{C_1}\, X_{C_2}}{I_{dc} R_L\, X_L} = \frac{\sqrt{2}\, X_{C_1}\, X_{C_2}}{R_L\, X_L}$$

or $$\boxed{\gamma = \frac{\sqrt{2}}{R_L} \cdot \frac{1}{\omega C_1}\, \frac{1}{\omega C_2} \cdot \frac{1}{\omega L} = \frac{\sqrt{2}}{\omega^3 C_1 C_2 L R_L}}$$

Advantages of π-filter

(*i*) Reduction in the ripples.
(*ii*) Increase in the average load voltage.
(*iii*) Simple circuitary.

Disadvantages of π-filter

(*i*) Ripple factor is dependent on the load.
(*ii*) Regulation is relatively poor.
(*iii*) Diodes-handle large peak currents.

2.9. CLIPPING CIRCUITS

The circuit, in which the waveform is shaped by removing (or clipping) a certain portion of the input voltage wave above or below a certain specific level, is called **clipping** circuit, **amplitude limiter,** or **amplitude slicer** etc.

The clipper circuits are classified as :

(*i*) positive clipper
(*ii*) negative clipper
(*iii*) biased clipper
(*iv*) combination clipper.

2.9.1. Positive Clipper

A clipper circuit, which clips off the positive half cycles of the input signals, is called the **positive clipper.** The clipping circuits use the switching property of the diode, similar to the half wave rectifier.

The positive clippers are again classified into two :

(*a*) Positive series clipper and

(*b*) Positive shunt clipper,

according to the output taken in the circuit.

(*a*) **Positive Series Clippers.** The circuit configuration for the positive series clipper is shown in Fig. 2.20. For the positive half cycle of the input signal, the diode is reverse biased and as such no current flows through the circuit. Thus the output voltage

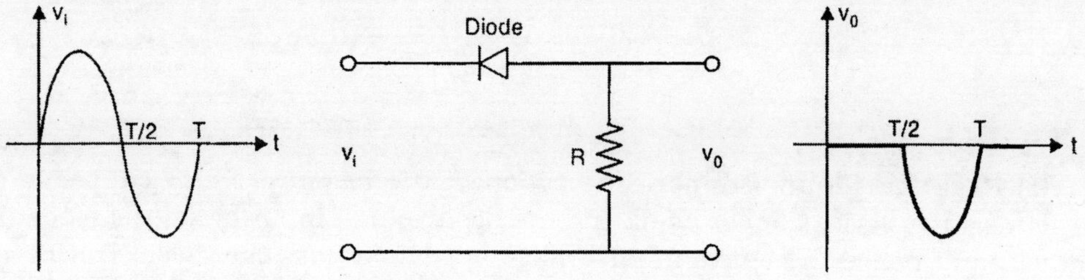

Fig. 2.20. *Positive series clipper.*

is zero. During the negative half cycle of the input signal, the diode becomes forward biased and current flows through the circuit, thereby the output across the resistor *R* is half sinusoidal waveform. In this discussion, the diode is assumed to be ideal, but for practical region, **reverse saturation current** I_s flows through the circuit, hence, there also a minute voltage appears across the resistance *R*.

(*b*) **Positive Shunt Clipper.** The circuit for the positive shunt clipper is shown in Fig. 2.21. Here, the output is taken across the diode. During positive half input cycle the output is nearly zero, since, the forward biased diode resistance is very small. And during negative half cycle of the input the output is a negative half sinusoidal wave, as shown in Fig. 2.21.

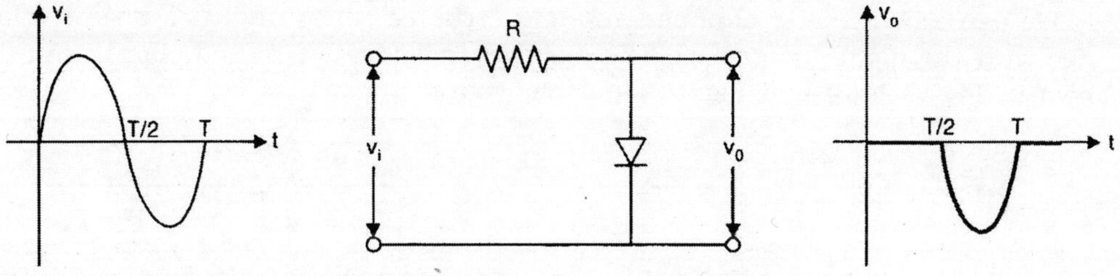

Fig. 2.21. *Positive shunt clipper.*

2.9.2. Negative Clipper

This is used to clip the negative half of the input signal. Like the positive clipper, these are also classified as series and shunt clipper circuits.

(*a*) **Negative Series Clipper.** The circuit shown in Fig. 2.22 acts as negative clipper. The input signal applied, appears across the output only for the positive half cycle, because during the negative half cycle of input the diode is reverse biased, and no current flows, while for positive half cycle of input, the diode is forward biased and applies very low resistance to the input. In actual practice, some saturated current I_s does flows through the diode when diode is reverse biased.

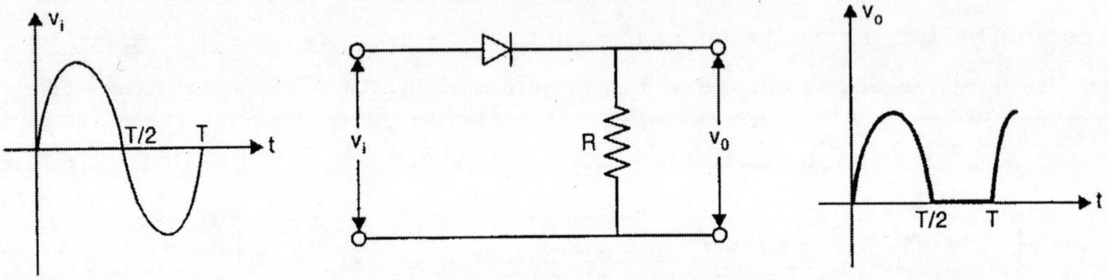

Fig. 2.22. *Negative series clipper.*

(*b*) **Negative Shunt Clipper.** The action of the negative shunt clipper is quite to that of negative series clipper as during the positive half input cycle, the diode does not conduct, and hence the entire input voltage appears across the diode. During negative half cycle of the input, diode is forward biased, and current flows through it, but the voltage drop is nearly zero, hence the output voltage is zero, for ideal case.

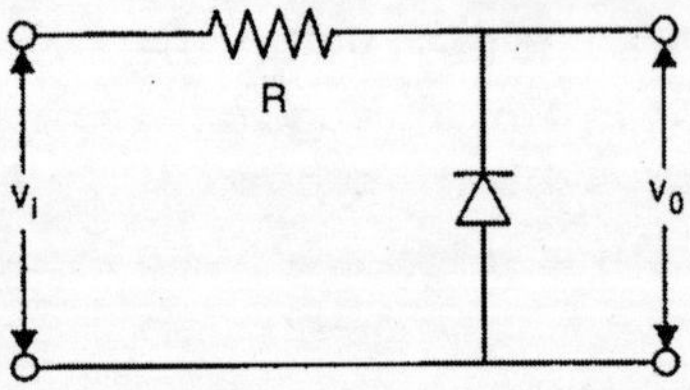

Fig. 2.23. *Negative shunt clipper.*

2.9.3. Biased Clipper

We can adjust the level of clipping to a certain level by adding a biased voltage in series with the diode or resistor. The positive and negative biased clipper circuits are shown in Fig. 2.24(*a*) and Fig. 2.24(*b*), respectively.

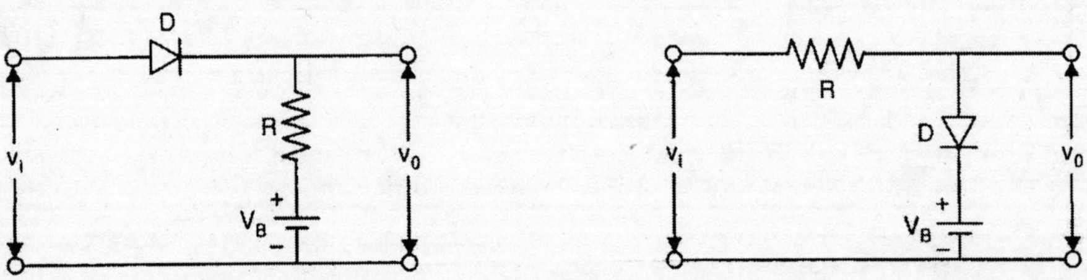

(*a*) Biased series positive clipper (*b*) Biased shunt positive clipper

Fig. 2.24. *Biased clippers.*

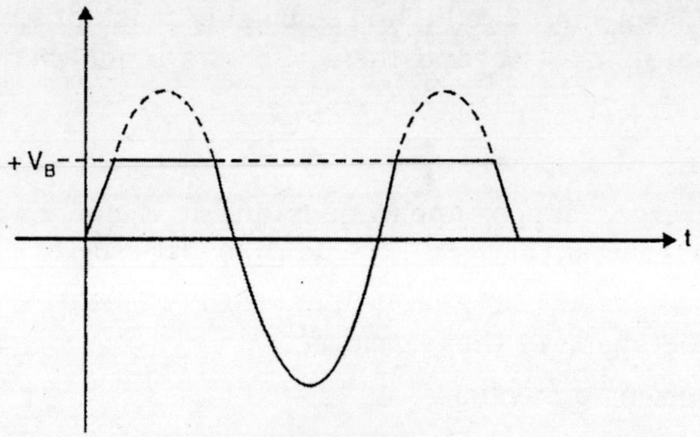

Fig. 2.25.

In both the circuits, clipping takes place during the positive half cycle, when the input voltage is more than bias voltage V_B, as shown in Fig. 2.25.

During the half cycle when the input voltage, is more than the bias voltage V_B, the diodes conduct and the output voltage is at a constant level in **shunt biased clipper.**

During the negative half cycle the diode does not conduct in series clipper and thus the all the input appears across output. If the battery connection in the above circuit are reversed then the circuit is known as biased negative clipper. In this situation clipping takes place during the negative half cycle when input voltage $V_i \leq (-)\ V_B$. This level can be changed by adjusting the bias voltage V_B.

2.9.4. Combination Clipper

It is a combination of the *biased positive clipper* and *biased negative clipper* when connected together as shown in Fig. 2.26 is called a combination clipper.

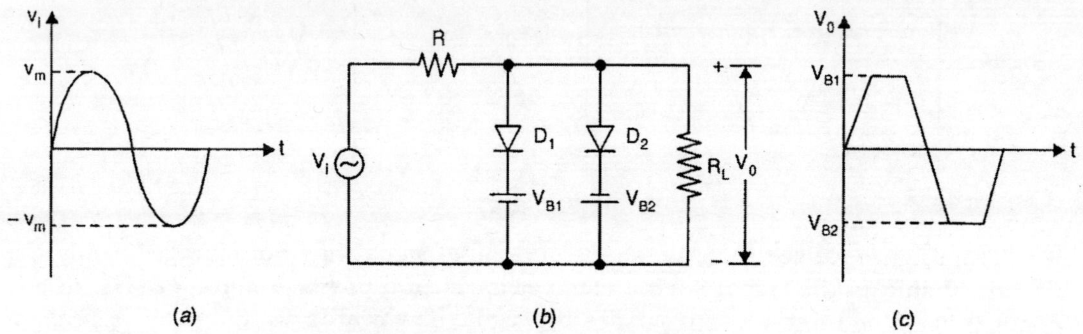

Fig. 2.26. *Combination clipper.*

Here the biasing voltage V_{B_1} and V_{B_2} are unequal in magnitude. When the input voltage signal goes positive the diode D_1 acts as short circuit and D_2 is reversed biased; so D_1 act as short circuit and D_2 as open circuit. The output voltage waveform follows the input voltage wave till it attains a value $(+)V_B$, after which it remains constant since $v_i > V_{B_1}$ till the input voltage is below V_{B_1}. Now it follows the input waveform. During negative half cycle of the input voltage, the output voltage waveform follows.

The input voltage waveform till it attains a value $-V_{B_1}$ since diode D_1 act as open circuit and diode D_2 as a short circuit, and it remains constant for interval during which input voltage is less than $-V_{B_2}$ and then it follows input then again the process is repeated.

2.9.5. Applications of Clippers

There are numerous clipper applications and it is not possible to discuss all of them. However, in general, clippers are used to perform one of the following two functions.

(*i*) Changing the shape of the waveform.

(*ii*) Circuit transient protection.

2.10. CLAMPING CIRCUITS

A circuit that places either the positive or negative peak of signal at a desired d.c. level is known as a *clamping circuit.* On the other hand we can say that a clamping circuit (or a clamper) essentially adds a d.c. component to the signal.

Clamping circuits broadally divided into two groups namely

(*i*) Positive clamper.

(*ii*) Negative clamper.

Fig. 2.27 shows the key idea behind clamping. The input signal is a sine wave having a peak-to-peak value of 20 V. The clamper adds the d.c. component and pushes the signal upwards so that the negative peaks fall on the zero level.

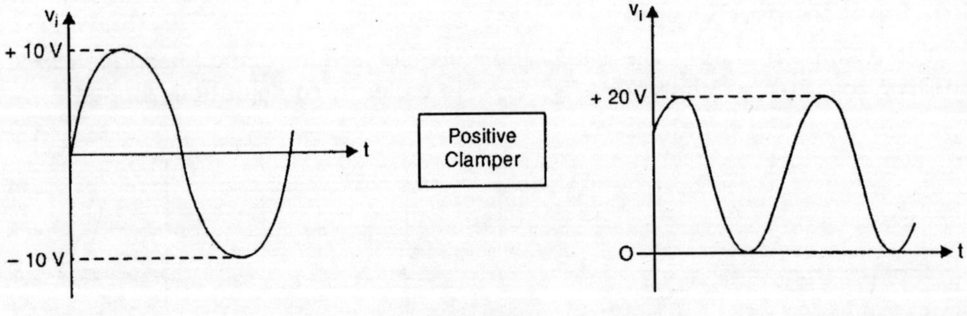

Fig. 2.27.

From figure we see that the shape of original signal has not changed : only there is vertical shift in the signal such a clamper is called a **positive clamper.** The negative clamper does the reverse *i.e.*, it pushes the signal downwards so that positive peaks fall on the zero level.

Things to be remember about the clamper circuit.

(*i*) *The clamping circuit does not change the peak-to-peak or rms value of the waveform.* From Fig. 2.27, the input waveform and clamped output has the same peak-to-peak value *i.e.*, 20V in this case.

(*ii*) *A clamping circuit changes the peak and average value of a waveform.* From the above circuit it is clear that input waveform has a peak value of 10V and average value over a cycle is zero. The clamped output voltage varies between 20V and 0V. Therefore the peak value of clamped output is 20V and average value (or d.c. value $= \frac{20 + 0}{2} = 10V$) is 10V. Hence we conclude that a clamper changes the peak value as well as the average value of a waveform.

2.10.1. Positive Clamper

Fig. 2.28 shows a circuit of a positive clamper.

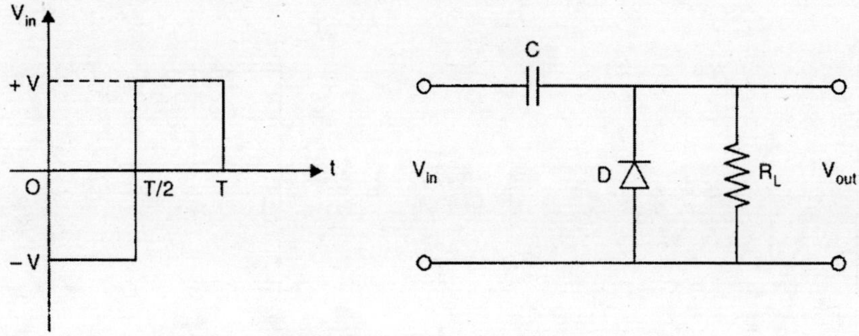

Fig. 2.28. *Positive clamper.*

Here the input signal is assumed to be a square wave with time period *T*. The clamped output is obtained across R_L. One things should be remember that for the proper operation of the circuit the charging time ($\tau_c = R_f \times C$) where R_f is the forward resistance of the diode and *C* is the capacitor connected. Between the diode and input supply is small as compared to the discharging time ($\tau_d = R_L \times C$). Where R_L is the value of the load resistance, this condition is based on the fact that voltage across the capacitor will not discharge significantly during the interval of diode in non-conducting. Therefore the discharging time is deliberately made much greater than the charging time.

Operation

(*i*) During the negative half cycle of the input signal, the diode is forward biased. Therefore the diode behaves as a short circuit (S.C.) as shown in Fig. 2.29. Under this situation the capacitor will charge to *V* volts very quickly. From figure it is clear that during this interval, the output voltage is directly across the short circuit. Therefore V_{out} = 0V.

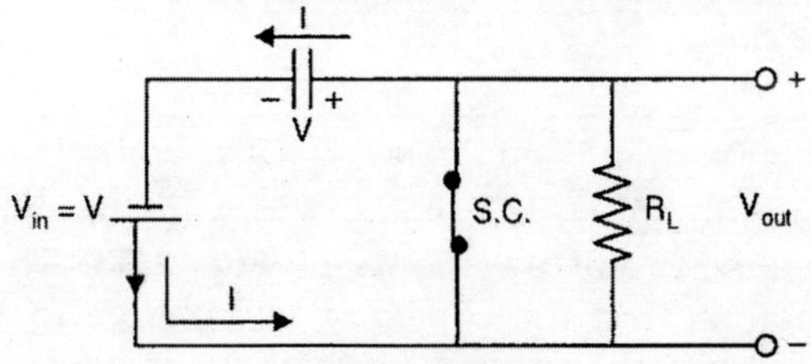

Fig. 2.29.

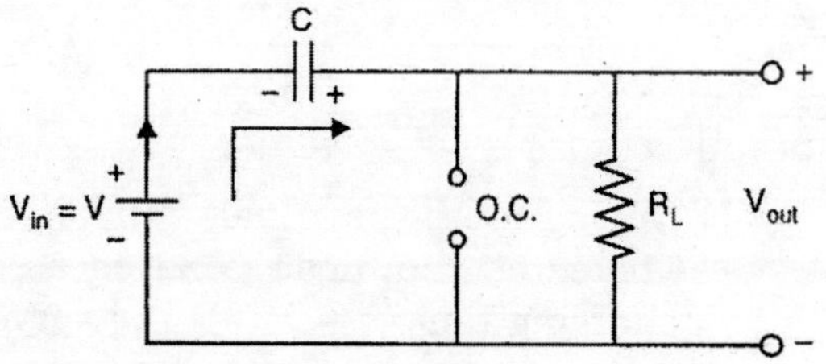

Fig. 2.30.

(*ii*) During the positive half-cycle, the diode is reverse biased and behaves as an open circuit (O.C.). Since the discharging time ($\tau = CR_L$) is much greater than the time period of the input signal, the capacitor remains almost fully charged to V volts during the off time of diode. Now, applying KVL to the input loop, we have

$$V + V - V_{out} = 0$$

or

$$V_{out} = 2V.$$

The resulting waveform is shown in Fig. 2.31. It is clear that it is a positively clamped output. Or in other words we can say that input signal has been pushed upward by V volts, so that negative peaks fall on the zero level.

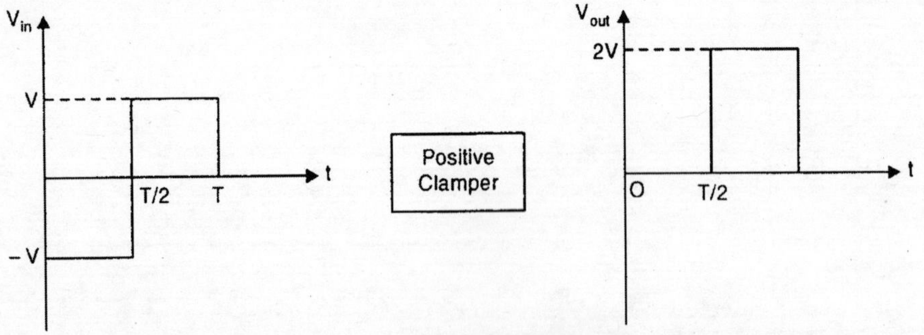

Fig. 2.31. *Input and output waveform of a positive clamper.*

2.10.2. Negative Clamper

Fig. 2.32 shows the circuit of a negative clamper. The main difference between the positive clamper and negative clamper is that in the case of negative clamper the terminal (*p* and *n*) are reversed as compare to the connection of diode terminal in positive clamper.

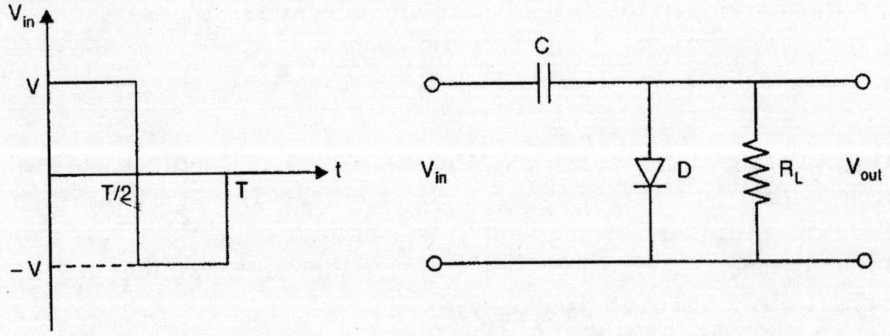

Fig. 2.32.

Operation

(*i*) During the positive half-cycle of the input signal, the diode is forward biased. Therefore, the diode behaves as a short circuit as shown in Fig. 2.33. The charging time ($\tau = R_f \times C$) is very small so that the capacitor will charge to V volts very quickly. From the figure, it is clear that, the output voltage is directly across the short circuit. Therefore, $V_{out} = 0$ volt.

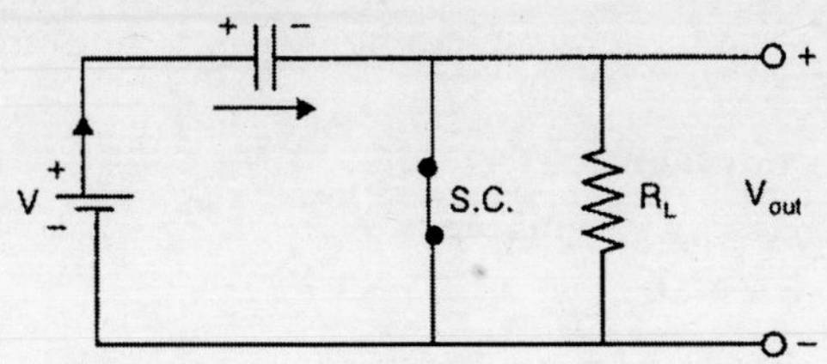

Fig. 2.33.

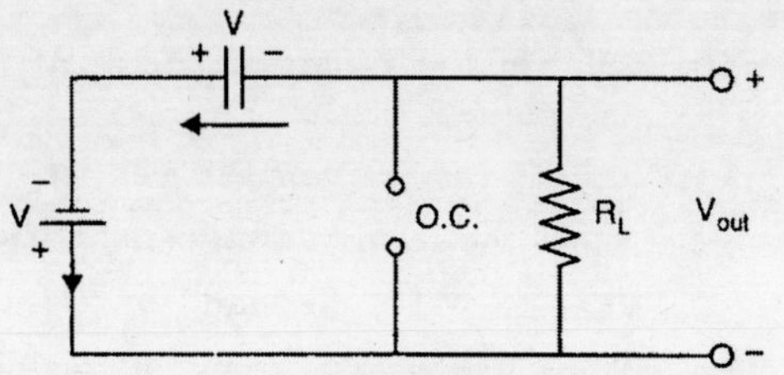

Fig. 2.34.

(*ii*) During the negative half cycle the diode is reverse biased and behaves as an open circuit as shown in Fig. 2.34. Since the discharging time ($\tau = R_L \times C$) is much greater than the time period of input signal, the capacitor almost remains fully charged to V volts during the off time of the diode. Applying KVL in Fig. 2.34, we have

$$-V - V - V_{out} = 0$$

$$V_{out} = -2\text{ V}$$

The resulting waveform is shown in Fig. 2.35.

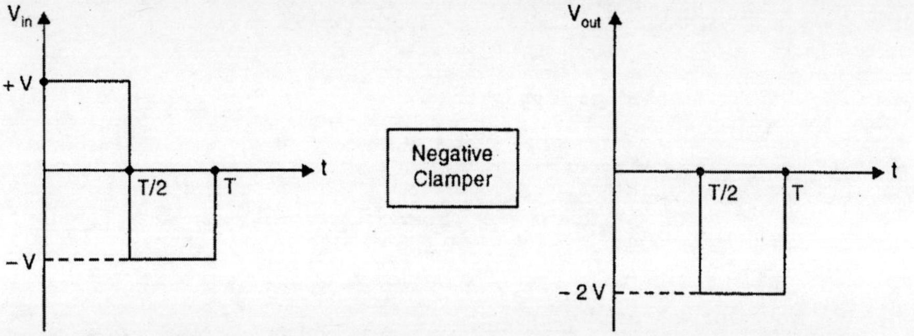

Fig. 2.35. *Input and output waveforms of a negative clamper.*

? SELF-TEST [HALF WAVE (RECTIFIER)]

1. What are the main application areas of semiconductor diodes ?

2. What is the need of rectifiers ?

3. Define the term **"rectification"** ?

4. Why the diodes are used in rectifier circuits ?

5. What is the need of transformer in rectifiers ?

6. Draw the circuit and output waveform for the half wave rectifier.

7. Is the output voltage is **exactly zero** for the negative half cycle in the half wave rectifier circuit ?

8. If an a.c. voltmeter is connected in the output of the circuit across the R_L, which value will it read ?

9. What is ripple factor ? What is its significance in rectifier circuits ?

10. What is the difference between instantaneous value of rectifier output and total instantaneous value ?

11. What is the relation between ripple factor and form factor ?

12. Define the term **"voltage regulation"**. What is its significance ?

13. What do you understand by **"rectification efficiency"** ?

14. Define TUF and also give its significance.

15. Define PIV.

SOLVED EXAMPLES

Example 1. *For a half wave rectifier circuit, the transformer used has a turn ratio* $N_2 : N_1 = 4 : 1$ *and load resistance* $R_L = 1$ kΩ. *Calculate the average current and load voltage, RMS current and voltage, efficiency of rectification, TUF and ripple factor, if the input to primary is,* $V_i = 10 \sin \omega t$, *assume ideal diode.*

Solution : Given for an HWR :

$$N_2 : N_1 = 4 : 1$$

$$R_L = 1 \text{ k}\Omega$$

$$V_i = 10 \sin \omega t$$

Ideal diode,

$$I_{av} \text{ and } V_{av} = ? \; I_{rms} \text{ and } V_{rms} = ? \; \eta_l = ?, \text{ TUF} = ?$$

$$\gamma = ?$$

(*i*) As we have for a transformer,

$$\frac{N_2}{N_1} = \frac{V_2}{V_1}$$

Therefore induced voltage at the secondary winding when v_1 is applied,

$$v_2 = \frac{N_2}{N_1} v_1 = 4 \times 10 \sin \omega t$$

$$v_2 = 40 \sin \omega t$$

$$\Rightarrow \quad V_m = 40 \text{ volts.}$$

(*ii*) Average load voltage,

$$V_{L_{av}} = \frac{V_m}{\pi} = V_{L_{DC}} = \frac{40}{\pi}$$

or $$V_{L_{DC}} = 12.73 \text{ V.} \qquad \textbf{Ans.}$$

and, average load current,

$$I_{DC} = \frac{I_m}{\pi} = \frac{V_m}{R_L \pi} = \frac{12.73}{1 \text{ K}}$$

$$\Rightarrow \quad I_{DC} = 12.73 \text{ mA} \qquad \textbf{Ans.}$$

(*iii*) RMS load voltage,

$$V_{L_{rms}} = \frac{V_m}{2} = \frac{40}{2}$$

$$\Rightarrow \qquad V_{L_{rms}} = 20 \text{ V.} \qquad \textbf{Ans.}$$

and,

$$I_{rms} = \frac{V_{L_{rms}}}{R_L}$$

or

$$I_{rms} = 20 \text{ mA} \qquad \textbf{Ans.}$$

(*iv*) **Rectification efficiency (η)**

$$\eta = \frac{P_{O_{DC}}}{P_{i_{AC}}} \times 100\%$$

$$= \frac{I_{L_{dc}}^2 \; R_L}{I_{rms}^2 \; R_L} \times 100 = \frac{(12.73)^2 \times 1}{(20)^2 \times 1} \times 100$$

$$\eta = 40.51\% \qquad \textbf{Ans.}$$

(*v*) **Transformer utilization factor (TUF) :** As defined,

$$\text{TUF} = \frac{P_{O_{DC}}}{\text{VA rating of transformer}}$$

$$= \frac{P_{O_{DC}}}{V_{2_{rms}} \times I_{2_{rms}}} = \frac{(12.73)^2 \times 1}{\left(\frac{40}{\sqrt{2}}\right) \times 20}$$

or

$$\text{TUF} = 28.65\% \qquad \textbf{Ans.}$$

(*vi*) **Ripple factor (γ)**

$$\therefore \qquad \gamma = \frac{\left[V_{L_{rms}}^2 - V_{L_{DC}}^2\right]^{\frac{1}{2}}}{V_{L_{DC}}}$$

Substituting the values we have :

$$\gamma = \frac{\left[(20)^2 - (12.73)^2\right]^{\frac{1}{2}}}{12.73}$$

$$\gamma = 121\% \qquad \textbf{Ans.}$$

Example 2. *A diode whose internal resistance is 10 Ω, is to supply power to a 1 kΩ load from a 230 V (rms) source of supply. Calculate input a.c. power, output d.c. power, efficiency and percentage regulation from no load to full load.*

Solution : Given,

$$R_f = 10\ \Omega$$
$$R_L = 1\ \text{k}\Omega$$
$$V_i = 230\ \text{V}\ (rms)$$

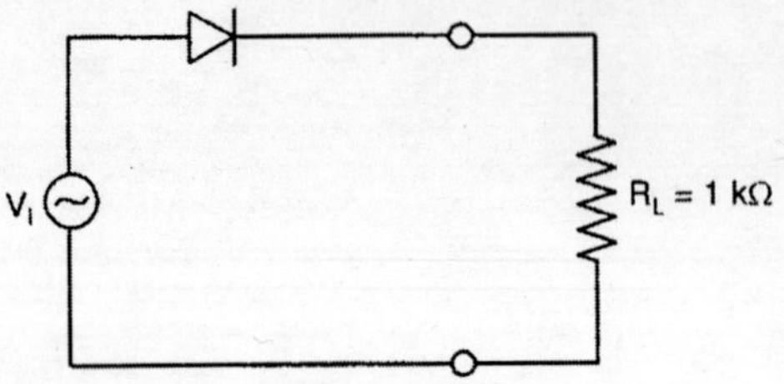

Fig. P (2.1).

As we have,

$$I_{DC} = \frac{I_m}{\pi}$$

$$= \frac{V_m}{(R_L + R_f)\,\pi} = \frac{230 \times \sqrt{2}}{(10 + 1000)\,\pi}$$

$$= 102.5\ \text{mA}$$

or $$I_{DC} = 102.5\ \text{mA}$$

and also $$I_{rms} = \frac{V_m}{(R_L + R_f) \times 2} \qquad \left[\because I_{rms} = \frac{I_m}{2}\right]$$

$$I_{rms} = 161.02\ \text{mA}$$

(*i*) Input a.c. power,

$$P_{iAC} = I_{rms}^2\ (R_s + R_f) = (0.161)^2 \times 1010$$

$$P_{iAC} = 26.9\ \text{watt} \qquad \textbf{Ans.}$$

(*ii*) Output d.c. power,

$$P_{O_{DC}} = I_{DC}^2\ R_L$$

$$= (0.103)^2 \times 1000$$

$$P_{O_{DC}} = 10.5\ \text{watt} \qquad \textbf{Ans.}$$

(*iii*) **Efficiency of rectification** (η)

As we have,

$$\eta = \frac{P_{O_{DC}}}{P_{i_{AC}}} \times 100\%$$

Substituting the values, we get

$$\eta = \frac{10.5}{26.9} \times 100\%$$

$$\eta = 39.03\% \qquad \textbf{Ans.}$$

(*iv*) **Load regulation**

We have, $$R = \frac{V_{NL} - V_{FL}}{V_{FL}} \times 100\%$$

here, $$V_{NL} = \frac{V_m}{\pi}$$

$$V_{FL} = I_{DC}\,R_L = \frac{V_m}{\pi\,(R_f + R_L)} \times R_L$$

$$\Rightarrow \qquad R = \frac{\dfrac{V_m}{\pi} - \dfrac{V_m\,R_L}{\pi(R_f + R_L)}}{\dfrac{V_m}{\pi}\dfrac{R_L}{(R_L + R_f)}} \times 100\%$$

$$= \frac{(R_f + R_L) - R_L}{R_L}\,\%$$

$$= \frac{R_f}{R_L} \times 100\%$$

$$= \frac{10}{1000} \times 100\%$$

or $$R = 1\% \qquad \textbf{Ans.}$$

Example 3. *An HWR circuit is connected to a 230 V mains supply by a transformer of 1 : 4 turn ratio, calculate the output d.c. voltage and current for a load resistance of 1 kΩ and assuming the Si diode.*

Solution : Given for an HWR circuit

$$v = 230 \text{ V } (rms)$$

$$N_2 : N_1 = 1 : 4$$

$$R_L = 1 \text{ k}\Omega$$

Drop across the diode = 0.7 V.

Therefore average load voltage :

$$V_{L_{DC}} = \frac{1}{2\pi}\int_0^{\pi} (V_m - 0.7)\,\sin \omega t\, d\omega t$$

$$= \left(\frac{V_m - 0.7}{2\pi}\right)(-\cos \omega t)_0^{\pi}$$

$$V_{L_{DC}} = \frac{V_m - 0.7}{\pi} \text{ volts.}$$

Now, the voltage induced at the secondary winding is,

$$V_2 = V_1\frac{N_2}{N_1} = 230 \times \frac{1}{4}$$

$$= 57.5 \text{ volt } (rms)$$

$$V_m = \sqrt{2}\,V_{rms} = \sqrt{2}\,V_2$$

or $$= \sqrt{2} \times 57.5$$

or $$V_m = 81.32 \text{ volts}$$

Substituting the value of V_m, we get

$$V_{L_{DC}} = \frac{81.32 - 0.7}{\pi}$$

$$V_{L_{DC}} = 25.66 \text{ V} \qquad \textbf{Ans.}$$

and the load current,

$$I_{L_{DC}} = \frac{V_{L_{DC}}}{R_L} = \frac{25.66 \text{ V}}{1 \text{ K}}$$

$$I_{L_{DC}} = 25.66 \text{ mA} \qquad \textbf{Ans.}$$

Example 4. *Find the voltage drop across 2 kΩ resistance in the following circuit (Assuming diode to be ideal).*

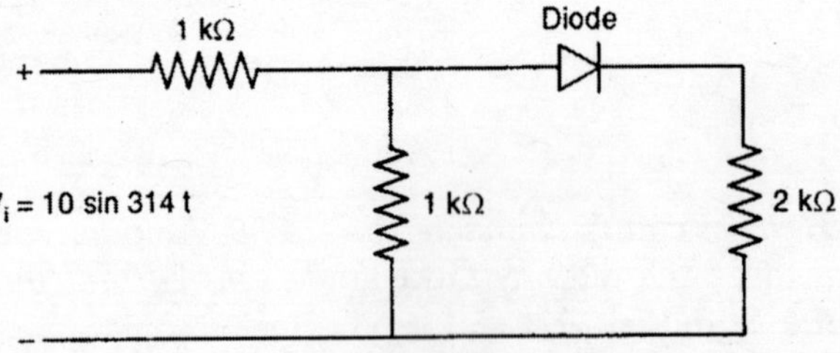

Fig. P (2.2).

Solution : During positive half cycle, diode D is forward biased and it behaves as a short circuit (Assuming diode to be ideal) in this condition, above circuit can be redrawn as :

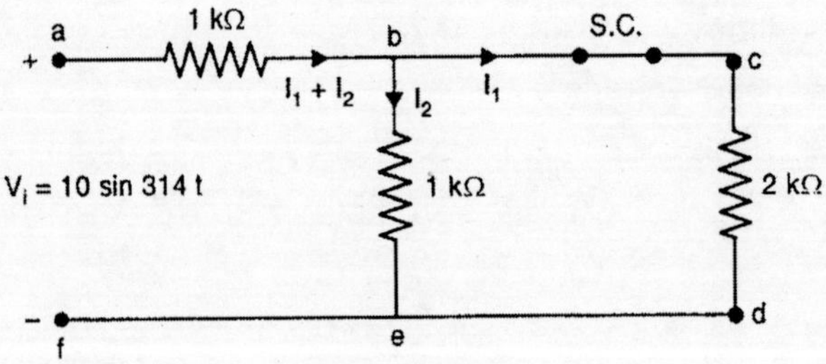

Fig. P (2.3).

By KVL along *abcdefa*

$$v_i = 1 \times 10^3 (I_1 + I_2) + 2 \times 10^3 I_1$$

$$10 \sin 314\, t = 3 \times 10^3 I_1 + 1 \times 10^3 I_2 \qquad ...(1)$$

By KVL along path *bcdeb*

$$2 \times 10^3 I_1 = 1 \times 10^3 I_2$$

or

$$2I_1 = I_2 \qquad ...(2)$$

From eqn. (1) and eqn. (2) we have

$$10 \sin 314\, t = 3 \times 10^3 \times I_1 + 1 \times 10^3 \times 2\, I_1$$

$$10 \sin 314\, t = 5 \times 10^3 I_1$$

or
$$I_1 = \frac{10 \sin 314\, t}{5 \times 10^3}$$
$$I_1 = 2 \times 10^{-3} \sin 314\, t \text{ A}$$
$$I_1 = 2 \sin 314\, t \text{ mA}$$

During negative half cycle, diode will not conduct. It behaves as an open circuit.

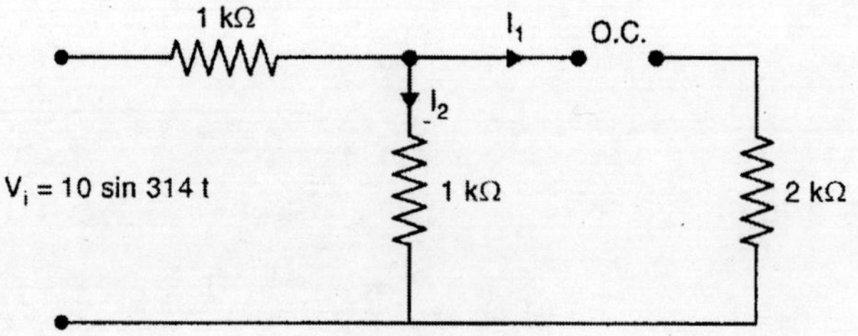

Fig. P (2.4).

From Fig. P (2.4), we have $I_1 = 0$ and voltage drop across 2 kΩ resistance = 0.

Thus, we have
$$I_1 = \begin{cases} 2 \sin 314\, t & 0 < \omega t < \pi \\ 0 & \frac{\pi}{\omega} < \omega t < 2\pi \end{cases}$$

R.M.S. value of current for a half wave rectifier is $\frac{I_m}{2}$
$$\therefore \qquad I_1,\ rms = \frac{2}{2} = 1$$

Voltage drop across 2 kΩ = $1 \times 10^{-3} \times 2 \times 10^3 = 2$ V. **Ans.**

Example 5. *Sketch the output voltage waveform for the following circuit. Assume the diode is ideal.*

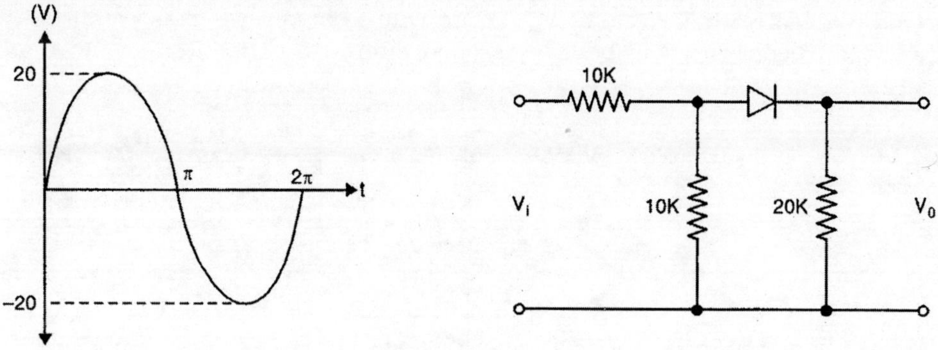

Fig. P (2.5).

Solution : During the positive half cycle the ideal diode will work as closed switch. Hence from 0 → π circuit will be as below.

This circuit can be reduced to the following circuit

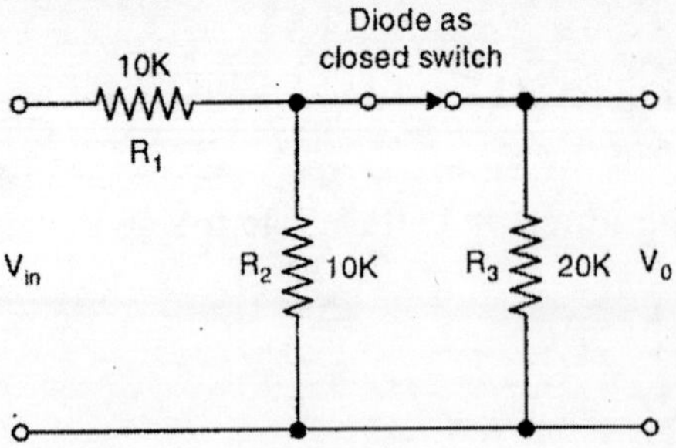

Fig. P (2.6).

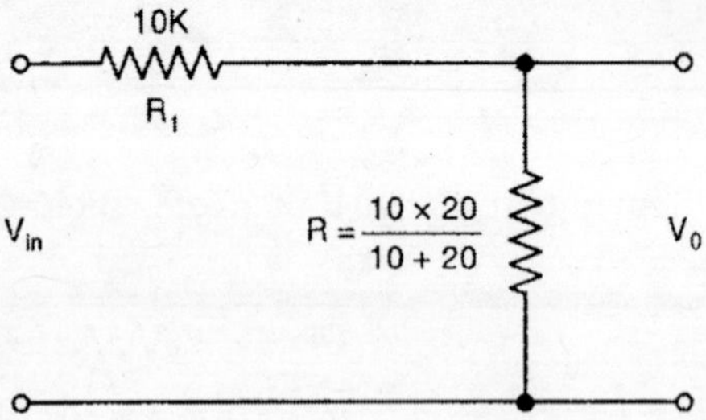

Fig. P (2.7).

Here,
$$R = \frac{R_2 R_3}{R_2 + R_3} = \frac{10 \times 20}{10 + 20} \text{K} = \frac{20}{3} \text{K}$$

This circuit is simply a voltage divider circuit.

So,
$$V_0 = \frac{R}{R + R_1} V_{in} = \frac{\frac{20}{3}}{\frac{20}{3} + 10} V_{in}$$

$$= \frac{20}{50} V_{in} = \frac{2}{5} V_{in}$$

where,
$$V_{in} = 20 \sin \omega t$$

So,
$$V_0 = \frac{2}{5} \times 20 \sin \omega t$$

$$= 8 \sin \omega t$$

Now from 0 → 2π

This is negative half cycle of V_{in}. Diode will be in reverse bias condition. Because it is an ideal diode so it will behave as open circuit. So the circuit will be as follow :

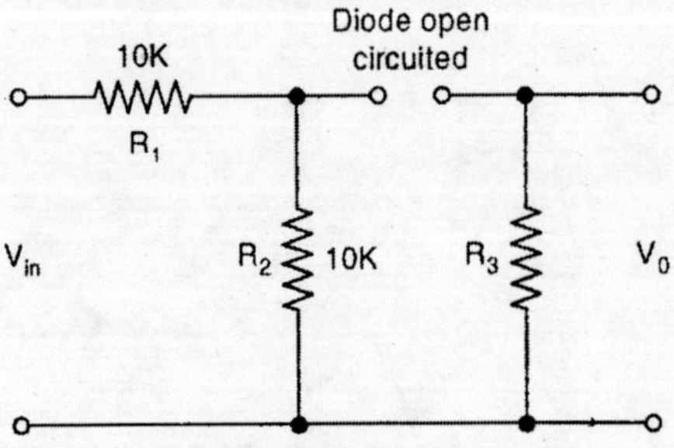

Fig. P (2.8).

It is clear from the circuit that from $\pi \rightarrow 2\pi$

$$V_0 = 0$$

So the output waveform will be as in Fig. P (2.9)

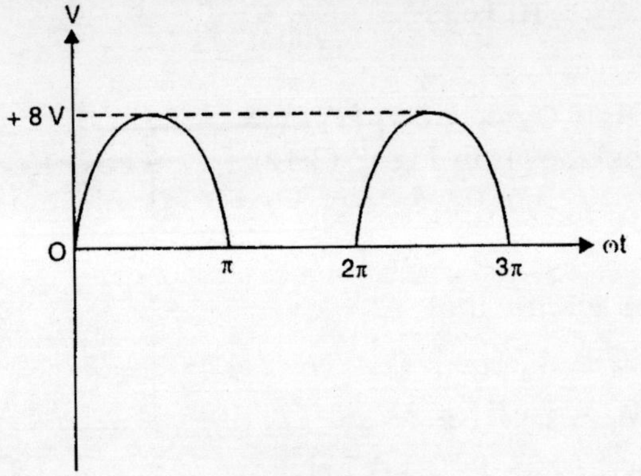

Fig. P (2.9).

Example 6. *Determine the output waveform for the network of Fig. P (2.10) and calculate the output d.c. level and the required PIV of each diode.*

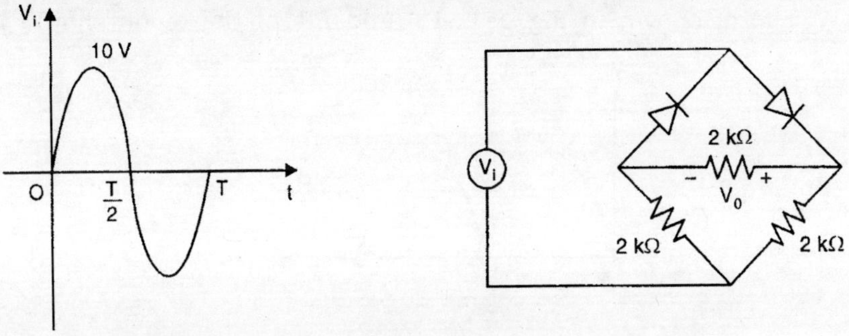

Fig. P (2.10).

Solution : During Positive Half Cycle of V_i (0 to $T/2$). The circuit will be as shown in Fig. P (2.11) as D_2 is forward biased and D_1 is reverse biased.

It is clear from the circuit that

$$V_0 = V_{AB}$$
$$= 2k \times I.$$

If we calculate the value of I then

$$I = \frac{V_i}{4k}.$$

So $$V_0 = 2k \times \frac{V_i}{4k}$$

$$= \frac{V_i}{2}.$$

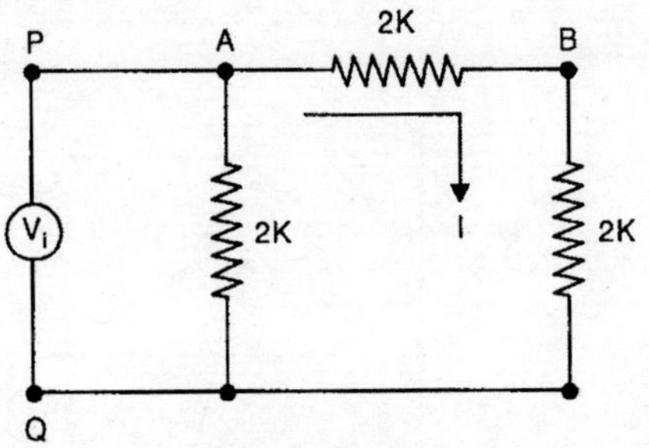

Fig. P (2.11).

The maximum value of V_0 during $0 \rightarrow T/2$ will be at $T/4$

$$V_{0\,max} = \frac{10 \text{ volts}}{2} = 5 \text{ volts.}$$

During Negative Half Cycle of V_i ($T/2$ to T). The circuit will be as shown in Fig. P (2.12).

The diode D_1 will be forward biased and D_2 will be reverse biased.

It is clear from the circuit that

$$V_0 = V_{AB}$$

or $$V_0 = (-I') \times 2k$$

as the circuit is similar to Fig. P (2.11) here

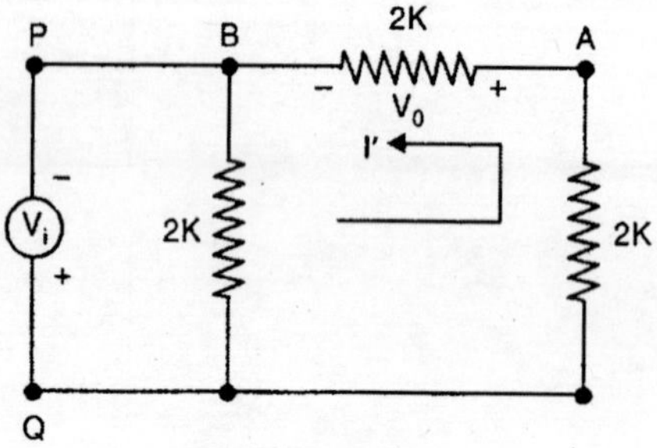

Fig. P (2.12).

or $$V_0 = \frac{-V_i}{2}$$

as the V_i has become negative so we get V_0 again positive and same as in the positive half cycle. Waveform shown in Fig. P (2.13) and *PIV* of the diode should be 5 volts.

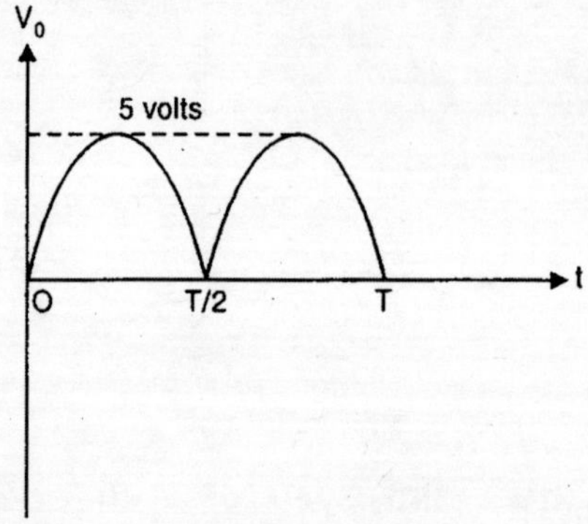

Fig. P (2.13).

Example 7. *For a circuit shown in Fig. P (2.14), sketch V_0 and determine PIV rating of each diode.*

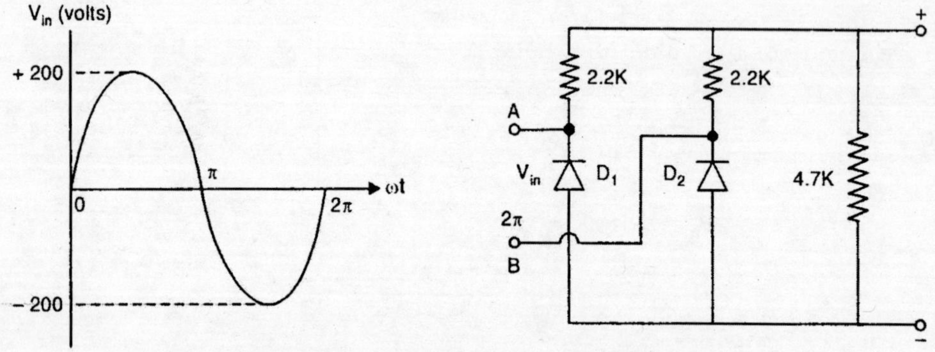

Fig. P (2.14). *Given circuit and waveform.*

Solution : During positive half cycle of input *i.e.*, $\omega t = 0 \rightarrow \pi$ the circuit will be as shown in Fig. P (2.15). Here point A is positive with respect to point B. Hence diode D_1 is reverse biased (open circuited) and diode D_2 is forward biased (short circuited). D_1 is shunted by $V_{AB} \simeq V_{in}$ so max voltage of V_{in} appears across diode D_1 which is in reverse bias state. So *PIV* for D_1 is + 200 V. Fig. P (2.16) is clearly observed as potential divider network as in Fig. P (2.16).

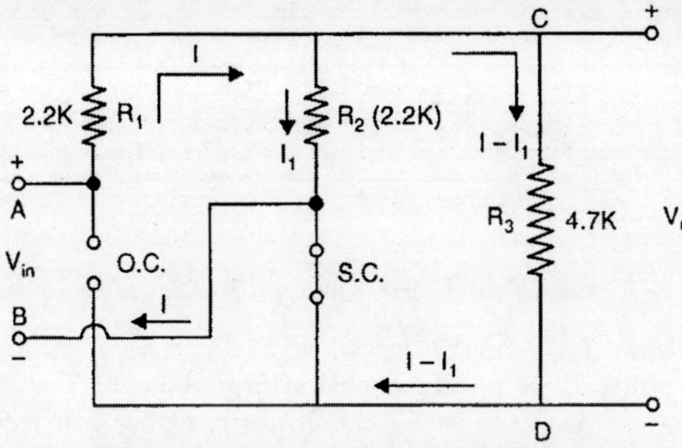

Fig. P (2.15). *Circuit during $0 \rightarrow \pi$.*

$$V_0 = \frac{R}{R_1 + R} V_{in}$$

where $$R = R_2 \parallel R_3$$

$$= \frac{R_2 R_3}{R_2 + R_3} = \frac{2.2 \times 4.7}{2.2 + 4.7} = 1.5 \text{ K.}$$

So $$V_0 = \frac{1.5}{2.2 + 1.5} \times V_{in}$$

$$V_0 = 0.405\ V_{in}$$

$$V_{0,\ max} = 0.405 \times 200$$

$$V_{0,\ max} = 81 \text{ volts.}$$

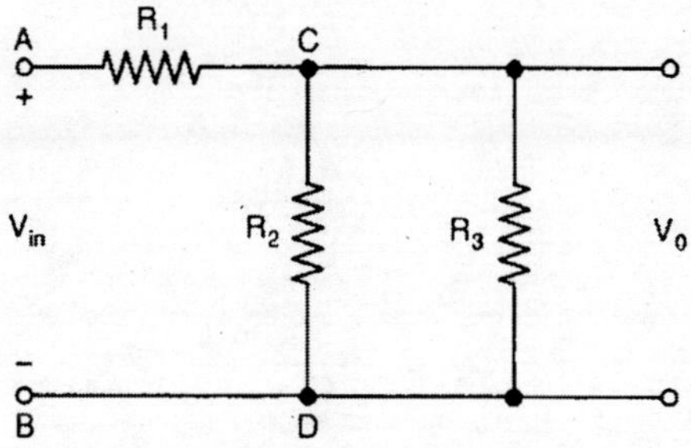

Fig. P (2.16). *Equivalent for Fig. P (2.15).*

The current through R_3 is from C to D so positive voltage appears across R_3.

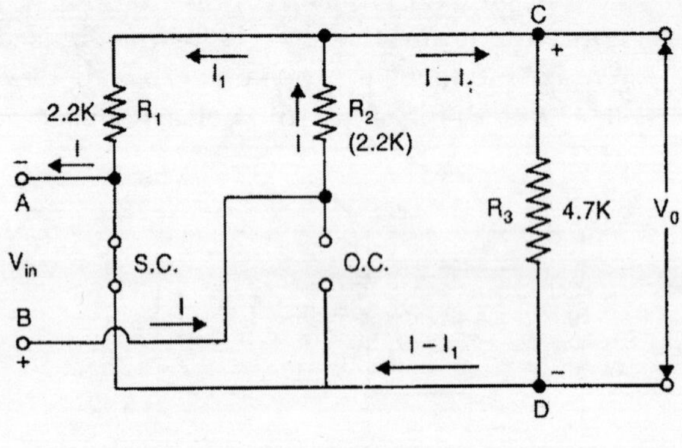

Fig. P (2.17). *Circuit during π – 2π.*

During negative half cycle of input *i.e.*, $\omega t = \pi \rightarrow 2\pi$ the circuit will be as shown in Fig. P (2.17). Here point B is positive with respect to A. Hence diode D_1 is forward biased (short) and diode D_2 is reverse biased (open). D_2 is shunted by $V_{BA} \simeq V_{in}$ so max voltage of V_{in} appears across diode D_2 which is in reverse bias state. So PIV for D_2 is + 200 V. Fig. P (2.17) is clearly observed as potential divider network as in Fig. P (2.18).

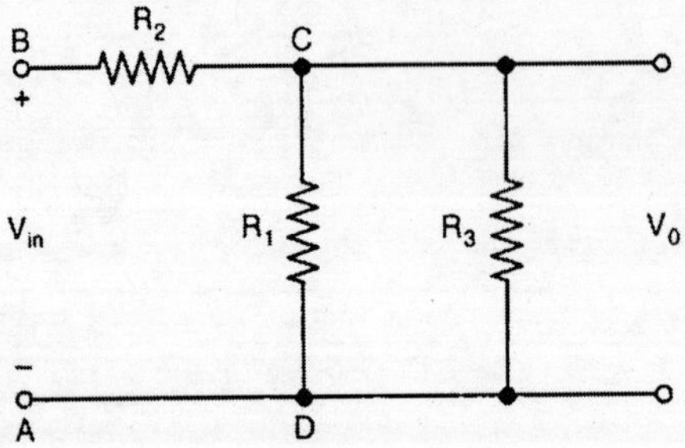

Fig. P (2.18). *Equivalent for Fig. P (2.17).*

$$V_0 = \frac{R'}{R_2 + R'}(-V_{in})$$

where $$R' = R_1 \parallel R_3$$

$$= \frac{R_1 R_3}{R_1 + R_3} = \frac{2.2 \times 4.7}{2.2 + 4.7} = 1.5 \text{ K.}$$

So $$V_0 = \frac{1.5}{2.2 + 1.5}(-V_{in})$$

$$= 0.405\ (-V_{in})$$

$$V_{0,\,max} = 0.405 \times 200$$

$$= 81 \text{ volts.}$$

The current through R_3 is from C to D so positive voltage appears across R_3.

Output waveform for V_0 is shown in Fig. P (2.19).

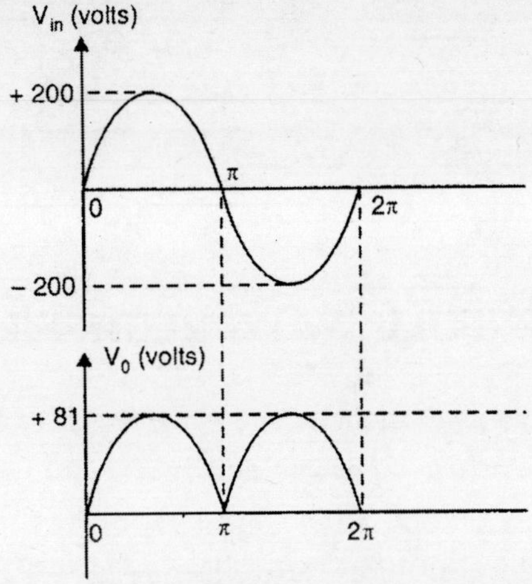

Fig. P (2.19).

Example 8. *Draw the output waveform for the following circuits.*

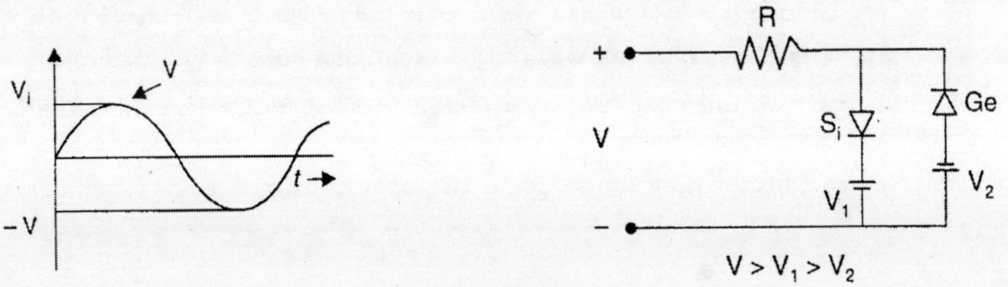

Fig. P (2.20).

Solution : Output waveform of the circuit is given below in Fig. P (2.21).

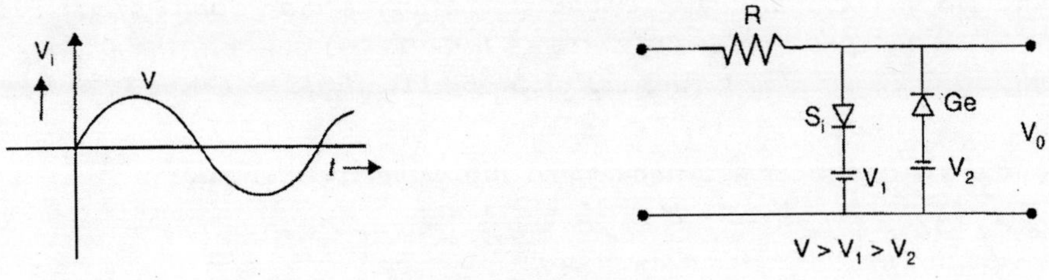

Fig. P (2.21).

As we know that Si diode has threshold voltage 0.7 and Ge have 0.3 we get the output waveform when +ve half cycle is input Ge diode is cut-off and Si diode is also cut-off till the $V_1 = V_i$ when $V_i > (V_1 + .7)$ it get on and current flows through it similarly for negative half cycle of Ge.

Fig. P (2.22).

? SELF-TEST (FULL WAVE RECTIFIER)

1. What are the two configurations of full wave rectifiers ?
2. What is the role of center tapped transformer in full wave rectifier ?
3. Why we say the d.c. output, although we get alternating half sinusoidal pulses ?
4. Why do ripples reduce in full wave rectifier ?
5. How the full wave rectifier is superior to half wave rectifier ?
6. What are the main disadvantages of full wave rectifiers ?
7. How the TUF of center-tapped FWR differs from HWR and bridge rectifiers ?
8. What are the main advantages of bridge rectifier over center-tapped full wave rectifier ?
9. Explain the working of a full wave-bridge rectifier. Find the values of PIV of each diode.
10. Explain the working of a full wave rectifier using centre tapped transformer. What is the PIV of this rectifier ?
11. With a neat circuit diagram, explain the operation of a bridge rectifier.

SOLVED EXAMPLES

Example 9. *For a full wave rectifier circuit (with center tapped transformer), find the average, RMS and peak values of currents through the diode, if the voltage across half of secondary is 15 sin 314 t. Also calculate the PIV of diode. Given R_L = 1 kΩ and ideal diodes.*

Solution : Given for a outer-tapped full wave rectifier circuit :

$$v_i = 15 \sin 314\, t$$

$$\Rightarrow \therefore \quad V_m = 15 \text{ V}$$

(*i*) **Average current through the diodes :**

$$I_{D_{av}} = I_{dc} = \frac{I_m}{\pi}$$

but $$I_m = \frac{V_m}{R_L} \quad \text{(as } R_f \text{ and } R_2 \text{ are assumed zero)}$$

$$\Rightarrow \qquad I_{D_{av}} = \frac{15}{\pi \times 1} = 4.77 \text{ mA}$$

$$I_{D_{av}} = 4.77 \text{ mA.} \qquad \textbf{Ans.}$$

(*ii*) **RMS current :**

$$I_{D_{rms}} = \frac{I_m}{2}$$

as one diode conducts only for one half cycle.

$$\Rightarrow \qquad I_{D_{rms}} = \frac{15}{2 \times 1} = 7.5 \text{ mA}$$

$$\therefore \qquad I_{D_{rms}} = 7.5 \text{ mA.} \qquad \textbf{Ans.}$$

(*iii*) **Peak diode current :**

$$I_{D_{peak}} = I_m = \frac{V_m}{R_L}$$

or

$$I_{D_{peak}} = \frac{15}{1}$$

or

$$I_{D_{peak}} = 15 \text{ mA.} \quad \textbf{Ans.}$$

(*iv*) **Peak inverse voltage :**

As we know the *PIV*, appears across the diode in center-tapped full wave rectifier is given by :

$$PIV = 2V_m = 2 \times 15$$

or

$$PIV = 30 \text{ V.} \quad \textbf{Ans.}$$

Example 10. *For a center-tapped transformer full wave rectifier, the turn ratio of the transformer between primary and half the secondary is 4 : 1. The resistance of each half of the secondary is 2Ω and the forward resistance of each diode is 1Ω. Calculate,*

(a) Average load current (b) Average load voltage at no load (c) Average load voltage at full load (d) Percentage load regulation (e) Rectification efficiency.

Given load resistance R_L = 1 kΩ ; and mains supply at primary is 240 V, 50 Hz.

Solution : Given, $R_f = 1\ \Omega$, $R_2 = 2\ \Omega$, $R_L = 1\ \text{k}\Omega$

$$N_1 : N_2 = 4 : 1$$

(*a*)

$$I_{L_{DC}} = \frac{2V_m}{\pi(R_2 + R_f + R_L)}$$

Where

$$V_m = \sqrt{2} \times V_{2_{rms}}$$

$$V_{2_{rms}} = \frac{N_2}{N_1} \times V_{1_{rms}} = \frac{1}{4} \times 240$$

or

$$V_{2_{rms}} = 60 \text{ V}$$

$$\therefore \quad V_m = \sqrt{2} \times V_{2_{rms}}$$

$$= \sqrt{2} \times 60$$

$$V_m = 84.85 \text{ volt}$$

Therefore
$$I_{L_{DC}} = \frac{2 \times 84.85}{\pi (2 + 1 + 1000)}$$

$$I_{L_{DC}} = 53.86 \text{ mA.} \quad \textbf{Ans.}$$

(*b*) **Average load voltage at no load :**

$$V_{NL} = \frac{2V_m}{\pi} = \frac{2 \times 84.85}{\pi}$$

$$V_{NL} = 54.02 \text{ V.} \quad \textbf{Ans.}$$

(*c*) **Average load voltage at full load :**

$$V_{L_{DC}} = V_{L_{DC}} \times R_L = 53.86 \times 1$$

$$V_{L_{DC}} = 53.86 \text{ volt.} \quad \textbf{Ans.}$$

(*d*) **Load regulation :**

$$\% \text{ load regulation} = \frac{V_{NL} - V_{FL}}{V_{FL}} \times 100\%$$

$$= \frac{54.02 - 53.86}{53.86} \times 100$$

$\therefore$ % load regulation

$$= 0.297\% \quad \textbf{Ans.}$$

(*e*) **Rectification Efficiency** (η)

Input a.c. power
$$P_{i_{AC}} = I_{rms}^2 (R_2 + R_f + R_l)$$

$$= \frac{I_m^2}{2} (R_2 + R_f + R_l)$$

but
$$I_m = \frac{V_m}{(R_L + R_2 + R_f)}$$

$$\Rightarrow \quad P_{i_{AC}} = \frac{V_m^2}{2(R_L + R_2 + R_f)} = \frac{(84.85)^2}{2\,(1000 + 2 + 1)}$$

$$P_{i_{AC}} = 3.6 \text{ watts}$$

and output d.c. power

$$P_{O_{DC}} = I_{L_{DC}}^2 R_L$$

$$= (53.86 \times 10^{-3})^2 \times 1000$$

$$= 2.9 \text{ watts}$$

As we know, rectification efficiency :

$$\eta = \frac{P_{O_{DC}}}{P_{i_{AC}}} = \frac{2.9}{3.6}$$

$$\eta = 80.56\%$$ **Ans.**

Example 11. *Determine the output waveform for the network of Fig. P (2.23) and calculate the output d.c. level and the required PIV of each diode.*

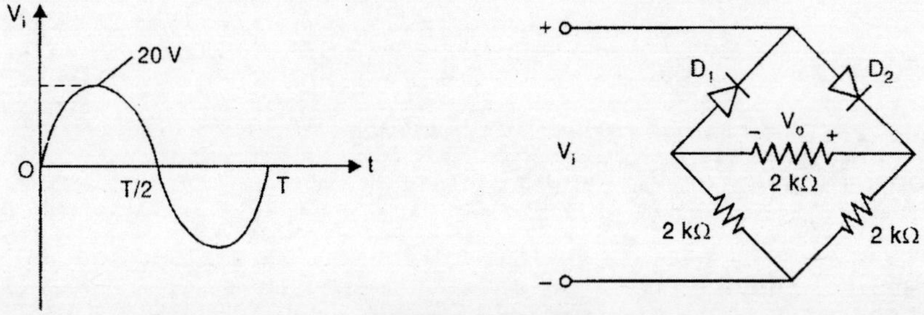

Fig. P (2.23).

Solution : The Network shown in Fig. P (2.23). Since all the resistances of the same value here, therefore the output will be same for both positive and negative cycle. For positive half cycle diode D_2 conduct while D_1 does not. Now for easy to understand we will draw its equivalent circuit as shown in Fig. P (2.24).

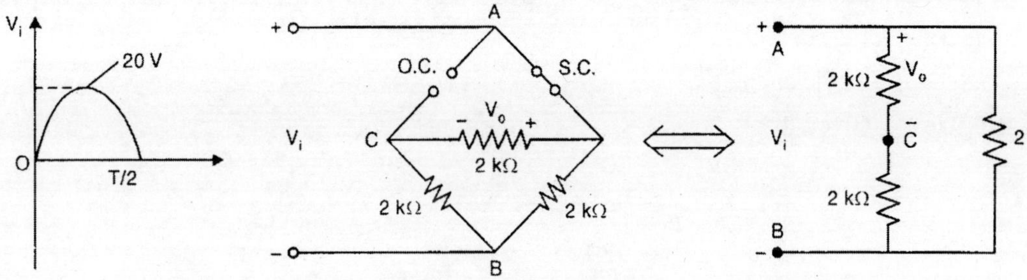

Fig. P (2.24). *Network.*

Applying potential divider method between terminal A and B.

$$V_{AC} = v_0 = V_{AB} \times \frac{2}{2+2} = 20 \times \frac{2}{4} = 10 \text{ V}$$

PIV for each diode = V_m, *i.e.*, is equal to the maximum voltage across R_L (*i.e.*, 2 kΩ), which is 10 V. **Ans.**

Example 12. *A half wave rectifier uses a diode with an equivalent forward resistance of 0.3 kΩ. If the input a.c. voltage is 10 V (rms) and the load is a resistance of 2.0 Ω, calculate I_{dc} and I_{rms} in the load.*

Solution : Given that $V_{rms} = 10$ V

$r_f = 0.3$ Ω

$R_L = 2.0$ Ω

So, $V_m = \sqrt{2}\, V_{rms} = 10\sqrt{2}$ V

Peak value of current in load, I_{max} or I_m

$$= \frac{V_m}{R_L + r_f} = \frac{10\sqrt{2}}{2 + 0.3} = 6.15 \text{ A}$$

d.c. output current, $I_{dc} = \frac{I_m}{\pi} = \frac{6.15}{\pi} = 1.958$ A. **Ans.**

RMS value of output current,

$$I_{rms} = \frac{I_m}{2} = \frac{6.15}{2} = 3.075 \text{ A.} \quad \textbf{Ans.}$$

Example 13. *What is the ripple 5 V on average of 50 V ?*

Solution : Given V_{av} or $V_{dc} = 50$ V

$V_{rms} = 5$ V

We know that ripple factor,

$$\gamma = \frac{V_{rms}}{V_{dc}} = \frac{5}{50} = 0.10 \quad \textbf{Ans.}$$

Example 14. *A single phase full-wave rectifier uses two diodes, the internal resistance of each being 20 Ω. The transformer rms secondary voltage from center tap to each end of secondary is 50 V and Load Resistance is 980 Ω. Find :*

(i) The mean load current.

(ii) rms load current.

(iii) Output efficiency.

Solution : Given that

$V_{rms} = 50$ V

$r_f = 20\ \Omega$

$R_L = 980\ \Omega$

So, $V_m = \sqrt{2}\, V_{rms} = 50\sqrt{2}$

(*i*) Mean load current

$$I_{dc} = \frac{2I_m}{\pi} = 2\frac{2.V_m}{\pi\,(R_L + r_f)} = \frac{2}{\pi} \cdot \frac{50\sqrt{2}}{(980 + 20)}$$

$$= 45 \text{ mA.} \quad \textbf{Ans.}$$

(*ii*) rms load current

$$I_{rms} = \frac{I_m}{\sqrt{2}} = \frac{V_m}{\sqrt{2}(R_L + r_f)} = \frac{50\sqrt{2}}{\sqrt{2}\,(980 + 20)} = 50 \text{ mA.}$$

Ans.

(*iii*) Output efficiency, $\eta = \frac{.812}{1 + \frac{r_f}{R_L}} = \frac{.812}{1 + \frac{20}{980}} = 79.58\%$. **Ans.**

Example 15. *For the circuit shown in Fig. P (2.25) a symmetrical 5 kHz square wave whose output varies between + 10 V to – 10 V is applied. Find out the output for a complete cycle.*

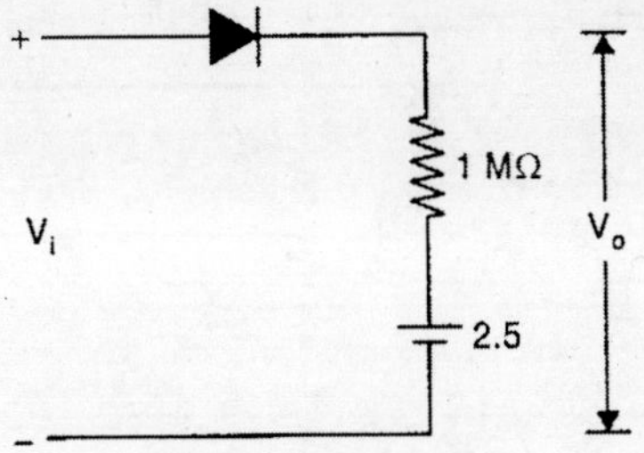

Fig. P (2.25).

Solution : In the given problem, input is look like as shown in Fig. P (2.26).

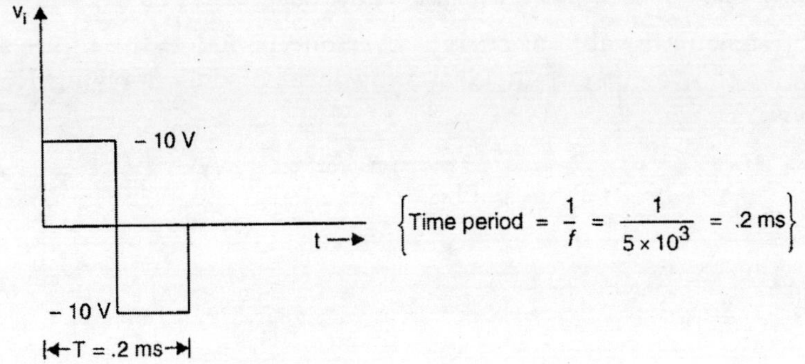

Fig. P (2.26).

From the input waveform it is clear that during positive half cycle diode will be in conducting stage *i.e.*, in forward biased. To calculate the output for positive half cycle we will draw its equivalent diagram in Fig. P (2.27).

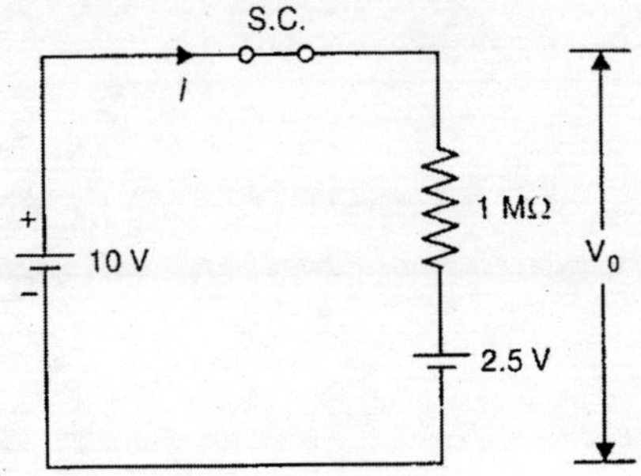

Fig. P (2.27).

$$i = \frac{10 - 2.5 \text{ V}}{1 \text{ M}\Omega} = 7.5 \ \mu\text{A}$$

$$V_{o_1} = iR + 2.5 = 7.5 \times 10^{-6} \times 1 \times 10^6 + 2.5 = 10 \text{ V}. \quad \textbf{Ans.}$$

Now, for negative half cycle, diode will be in reverse biased, therefore, there is no current flowing through the circuit due to open circuit. So in this case

$$V_{o_2} = 2.5 \text{ V}. \quad \textbf{Ans.}$$

Example 16. *In the circuit shown in Fig. P (2.28) calculate current over one period of the input voltage. Assuming that the diodes to be ideal.*

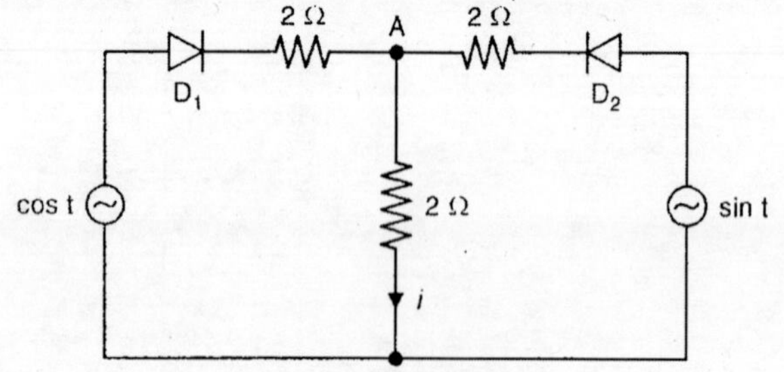

Fig. P (2.28).

Solution : From the given Fig. P (2.28), it is clear that both the diodes are in forward biased because both sin t and cos t are positive.

To calculate value of current over one period of the input signal first of all we will draw its equivalent circuit (assuming ideal diode). Applying KCL at node A,

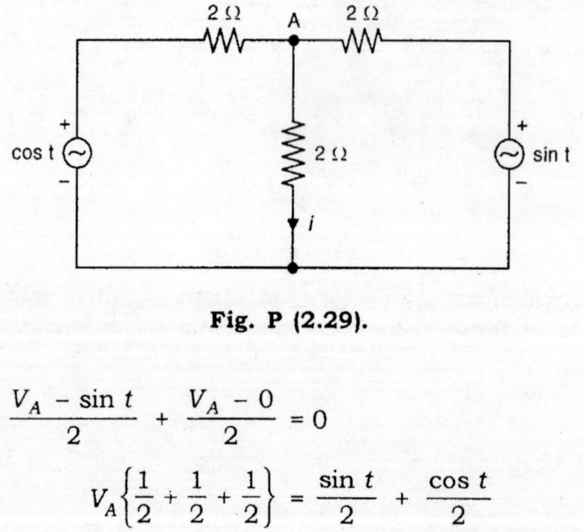

Fig. P (2.29).

KCL at node A

$$\frac{V_A - \cos t}{2} + \frac{V_A - \sin t}{2} + \frac{V_A - 0}{2} = 0$$

$$V_A\left\{\frac{1}{2} + \frac{1}{2} + \frac{1}{2}\right\} = \frac{\sin t}{2} + \frac{\cos t}{2}$$

$$V_A = \left(\frac{\sin t + \cos t}{3}\right)$$

$$i = \frac{V_A}{2} = \frac{(\sin t + \cos t)}{3.2} = \frac{\sin t + \cos t}{6}\text{A.}$$

Ans.

However, during the negative half cycle both the diodes in the reverse biased. Therefore current i in this case will be zero. **Ans.**

Example 17. *For the given centre-tapped full wave rectifier. Calculate :*

(a) d.c. output voltage

(b) PIV

(c) Rectification efficiency.

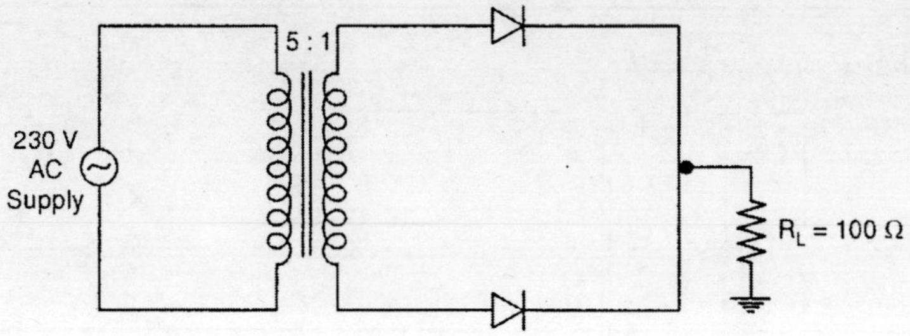

Fig. P (2.30).

Solution : Given that $V_1 = 230$ V

$$\frac{N_2}{N_1} = \frac{1}{5} = 0.2$$

$$R_L = 100\ \Omega$$

(*a*) **d.c. output voltage**

We know that the rms value of secondary voltage,

$$V_{rms} = V_2 = V_1 \times \frac{N_2}{N_1} = 230 \times 0.2 = 46 \text{ V}$$

and half of the secondary voltage

$$V_s = \frac{V_2}{2} = \frac{46}{2} = 23 \text{ V}$$

∴ Maximum value of the half of the secondary voltage,

$$V_m = \sqrt{2} \times V_s = \sqrt{2} \times 23 = 32.5 \text{ V}$$

and d.c. output voltage,

$$V_{dc} = \frac{2V_m}{\pi} = \frac{2 \times 32.5}{\pi} = 20.7 \text{ V.} \quad \textbf{Ans.}$$

(*b*) **Peak inverse voltage of a diode**

For centre tapped FWR, we know that peak inverse voltage of a diode,

$$PIV = 2\ V_m$$
$$= 2 \times 32.5$$
$$= 65 \text{ V.} \quad \textbf{Ans.}$$

(*c*) **Rectification Efficiency**

We also know that rectification efficiency of a full-wave rectifier,

$$\eta = 0.812 = 81.2\% \quad \textbf{Ans.}$$

Example 18. *For the circuit shown below, determine*

(a) d.c. output voltage

(b) Rectification efficiency

(c) Peak inverse voltage (PIV)

(d) Output frequency.

Assume the diodes is ideal.

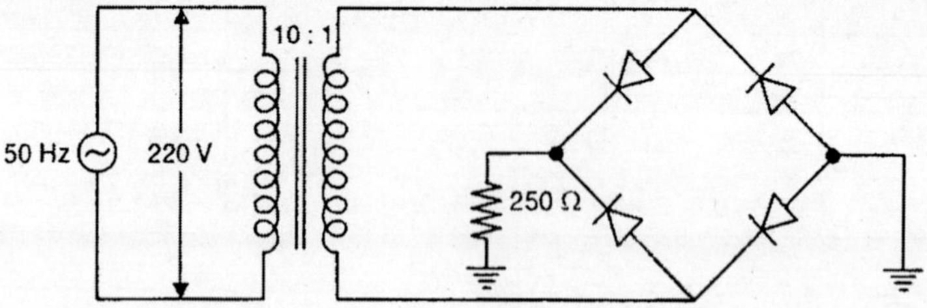

Fig. P (2.31).

Solution : Given that,

$$V_1 = 220 \text{ V}$$

$$\frac{N_2}{N_1} = 1/10 = 0.1$$

$$R_L = 250 \ \Omega$$

(*a*) **d.c. output voltage**

rms value of secondary voltage,

$$V_2 = V_1 \times \frac{N_2}{N_1} = 220 \times 0.1 = 22 \text{ V}$$

∴ Maximum value of secondary voltage to be rectified

$$V_m = \sqrt{2}\, V_2$$

$$= \sqrt{2} \times 22 \text{ V}$$

$$= 31.11 \text{ V}$$

Now,

$$I_m = \frac{V_m}{R_L} \quad (\because \text{ diode is assumed to be ideal } i.e., \ r_f = 0)$$

$$= \frac{31.11}{250} = 124.44 \text{ mA}$$

$$I_{dc} = I_{av} = \frac{2I_m}{\pi} \qquad \text{(For FWR)}$$

$$= \frac{2 \times 124.44}{\pi} = 79.22 \text{ mA}$$

∴ d.c. output voltage, $V_{dc} = I_{dc} \times R_L$

$$= 79.22 \times 250 \times 10^{-3}$$

$$= 19.80 \text{ V.}$$ **Ans.**

(*b*) **Rectification efficiency**

d.c. output power, $P_{dc} = I_{dc}^2 \, R_L$

$$= (79.22 \times 10^{-3})^2 \times 250$$

$$= 1.56 \text{ W}$$

a.c. input power, $P_{ac} = I_{rms}^2 \times R_L$

$$= \left(\frac{I_m}{\sqrt{2}}\right)^2 R_L$$

$$= \left(\frac{124.44 \times 10^{-3}}{\sqrt{2}}\right)^2 \times 250$$

$$= 1.935 \text{ W}$$

Now, rectification efficiency $\eta = \dfrac{P_{dc}}{P_{ac}}$

$$= \frac{1.56}{1.935} = 0.805$$

$\therefore$ $\% \eta = 0.805 \times 100$

$= 80.5\%$ **Ans.**

(*c*) **Peak inverse voltage**

We know that *PIV* for a bridge Network full wave rectifier is

$$PIV = V_m = 31.11 \text{ V} \quad \textbf{Ans.}$$

(*d*) **Output frequency**

A full wave rectifier doubles the input frequency

$$f_{out} = 2 \times f_{in}$$

$$= 2 \times 50$$

$$= 100 \text{ Hz} \quad \textbf{Ans.}$$

Example 19. *Calculate the output d.c. level and the required PIV for each diode in the circuit shown :*

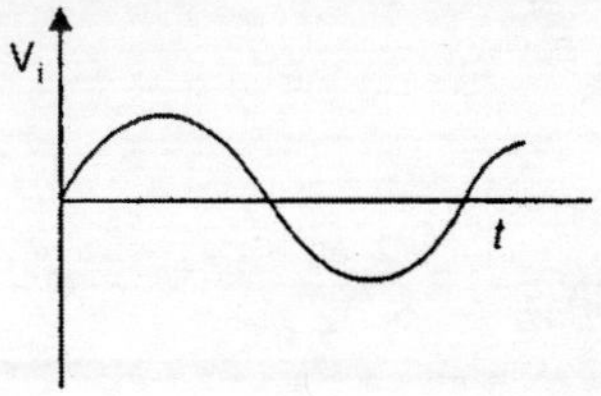

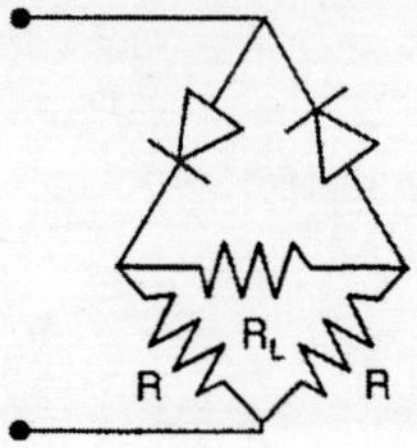

Fig. P (2.32).

Solution : Given circuit diagram is as follows :

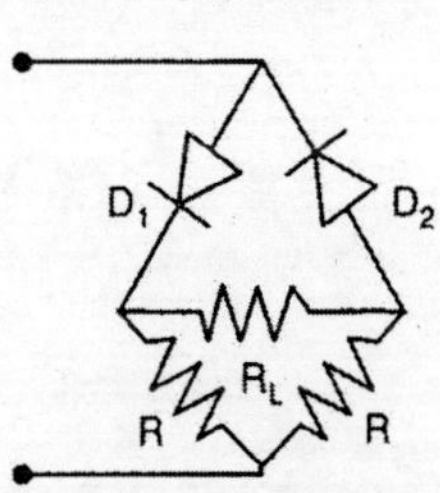

Fig. P (2.33).

It may be shown that output when diode D_1 is conducting as follows :

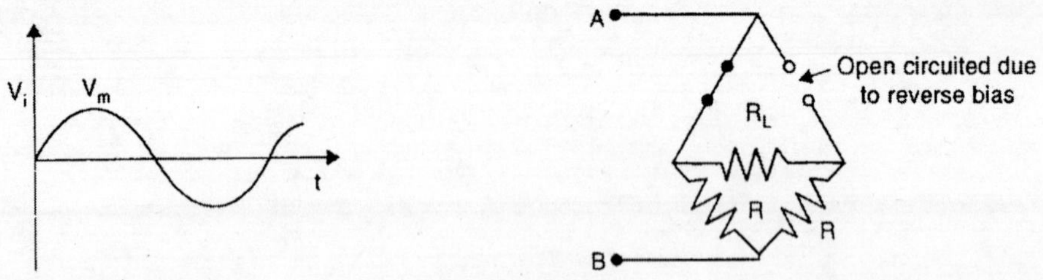

Fig. P (2.34).

and we get the above diagram as follow :

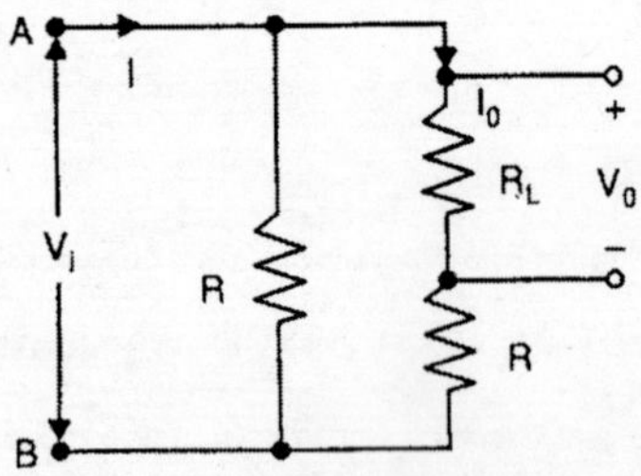

Fig. P (2.35).

So output voltage may be calculated as follow :

$$I = \frac{V_i}{\text{Req}} = \frac{V_i}{(R_L + R) \parallel R}$$

and $$I_0 = \frac{IR}{R_L + R + R} = \frac{V_i}{R_L + R}$$

and we have output voltage

$$V_0 = I_0 \cdot R_L = \frac{V_i}{R_L + R} \times R_L = \frac{V_i R_L}{R_L + R}$$

and PIV of diodes may be calculated as follow :

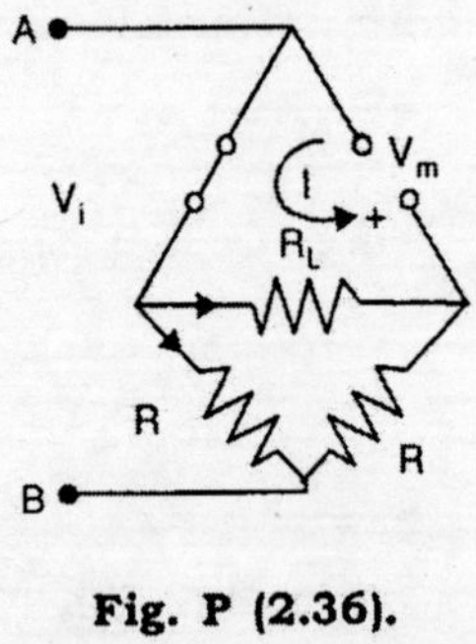

Fig. P (2.36).

Now considering the loop we get the V_D as follows :

$$V_m - I_0 R_L = 0$$

or $$V_m = I_0 R_L$$

or $$V_m = \frac{V_i R_L}{R_L + R}.$$ **Ans.**

This is the required PIV of each diode.

Example 20. *A full wave rectifier circuit is fed from a transformer whose secondary is centre-tapped with a line-to-line voltage of 60 V (rms). Assuming the total resistance of the diode and half of the transformer secondary winding to be 10 Ω and load resistance to be 1000 Ω, determine :*

(a) The no-load Dc voltage,

(b) The dc current through load,

(c) The dc load voltage,

(d) The percentage regulaltion from no load to full load,

(e) The ac ripple voltage across the load and the ripple frequency,

(f) The PIV across each diode,

(g) Power-conversion efficiency, and

(h) Transformer-Secondary ruting.

Solution :

(a) Line-to-line secondary voltage $= \frac{2V_m}{V^2} = 60$ V

Hence, $V_m = \frac{60\sqrt{2}}{2} = 42.4$ V

No-load DC voltage $= \frac{2V_m}{\pi} = 0.636\ V_m = 29.96$ V

(b) DC load current $= \frac{0.636\ V_m}{R_f + R_S + R_L} = \frac{26.96}{1010}$

$= 0.027$ A

$= 27$ mA

(c) DC load voltage, $Vdc = \frac{2V_m}{\pi} - I_{dc}(R_f + R_S)$

$= 26.96 - 0.027 \times 10$

$= 26.7$ V

(d) Regulation from no load to full load

$$= \frac{R_f + R_S}{R_L} \times 100\%$$

$$= \frac{10}{1000} \times 100\% = 1\%$$

(*e*) AC ripple voltage across load

$= 0.482 \times V_{dc}$

$= 0.482 \times 26.7 = 12.86$ V

Ripple frequency = 2 × mains frequency = 100 Hz

(*f*) PIV across each diode $= 2V_m = 2 \times 42.4 = 84.8$ V

(*g*) Power-conversion efficiency = 81.2%

(*h*) Transformer secondary rating

$$= \frac{P_{dc}}{TUF}$$

$$= \frac{0.713}{0.693} = 1.03 \text{ VA}$$

Example 21. *A full wave rectifier using a capacitor fitter has to supply 30 V dc to a 1 kΩ load. Assuming the diode and transformer winding resistance to be negligible, estimate the input voltage required and the value of the filter capacitor for a ripple factor of 0.01.*

Solution : Given

$$R_F = 0.01$$

$$V_{dc} = 30 \text{ V}$$

$$R_L = 1 \text{ k}\Omega$$

Hence,
$$I_{dc} = \frac{30 \text{ V}}{1 \text{ k}\Omega} = 30 \text{ mA}$$

Also,
$$V_{dc} = V_m - \frac{5000\, I_{dc}}{C}$$

In order to determine V_m, we must estimate C first

$$\text{Ripple factor} = \frac{2900}{C.R_L} = 0.01$$

Hence,
$$\therefore \quad C = \frac{2900}{0.01 \times 1000} = 290 \text{ F}$$

$$V_m = V_{dc} + \frac{5000\, I_{dc}}{C} = 30 + \frac{5000 \times 0.03}{290}$$

$$= 30 + 0.862$$

$$\therefore \quad V_m = 30.862 \text{ V}$$

Line-to-line secondary voltage

$$= \frac{2\,V_m}{\sqrt{2}} = \frac{2 \times 30.862}{\sqrt{2}}$$

$$= 43.65 \text{ V}$$

? SELF-TEST (FILTER)

1. What is the need of filters ?

2. List different types of filter used.

3. What is basic function of inductor and capacitor in the filter ?

4. What are the advantages of π-filter over choke input filter ?

5. Compare different types of filter under different load condition.

SOLVED EXAMPLES

Example 22. *Design a filter for full wave circuit with LC filter to provide an output voltage of 2.5 V with a load current of 100 mA and its ripple is limited to 3%.*

Solution : Design of filter means to calculate the value of L and C used in LC filter.

Given that, $V_L = 25$ V and $I_L = 200$ mA

So load resistance $$R_L = \frac{V_L}{I_L} = \frac{25\text{ V}}{200\text{ mA}} = 125\ \Omega$$

We know that ripple factor for a LC filter is given by

$$\gamma = \frac{1}{6\sqrt{2}\omega^2 LC}$$

Given that $\gamma = 3\% = .03$

$$.03 = \frac{1}{6\sqrt{2} \times (2\pi f)^2 LC}$$

$$.03 = \frac{1}{6\sqrt{2} \times 4\pi^2 \times 50^2\ LC} \qquad (\because f = 50\text{ Hz})$$

$$LC = \frac{1}{.03 \times 836614.8} = \frac{1}{25098.4} \qquad \ldots(1)$$

We know that

$$L = \frac{R_L}{3\omega} = \frac{R_L}{3.(2\pi f)}$$

or $$L = \frac{125}{3 \times 2 \times 3.14 \times 50}$$

or $L = .1326$ H **Ans.**

$$C = \frac{1}{.1326 \times 25098.4} = \frac{1}{3330.46} = 300.25\ \mu\text{F}. \quad \textbf{Ans.}$$

Example 23. *(a) Find the output voltage, current and ripple for the circuit shown in Fig. P (2.37).*

(b) What is the maximum value of R_L ?

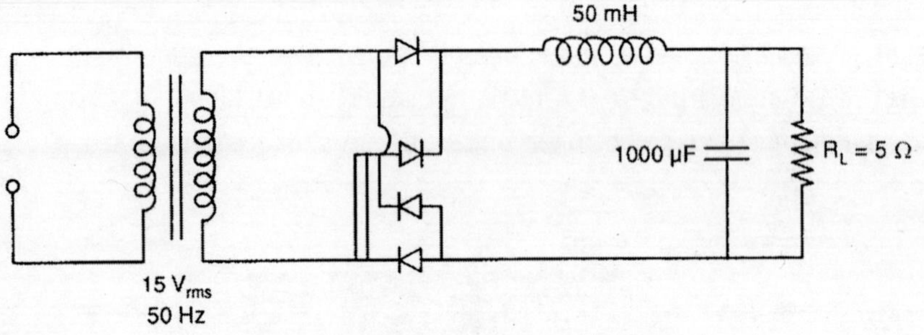

Fig. P (2.37).

Solution : The circuit of *L-C* filter is shown in Fig. P (2.37).

(*a*)

$$V_{dc} = \frac{2V_m}{\pi} = \frac{2\sqrt{2}V_m}{\pi} = \frac{2\sqrt{2} \times 15}{\pi} = 13.5 \text{ V}$$

$$I_{dc} = \frac{V_{dc}}{R_L} = \frac{13.5}{5} = 2.7 \text{ A}$$

$$\gamma = \frac{1}{6\sqrt{2}\omega^2 LC} = \frac{1}{6\sqrt{2}(2\pi f)^2 \, LC}$$

$$= \frac{1}{6\sqrt{2}\,(314)^2 \,.\, 50 \times 10^{-3} \times 1000 \times 10^{-6}}$$

$$\gamma = \frac{1}{41.83} = .024 \quad \textbf{Ans.}$$

(*b*) Peak a.c. current through inductor

$$I_m \cong \frac{V_m}{X_L}$$

$$\cong \frac{15\sqrt{2}}{2\pi f L} \cong \frac{15\sqrt{2}}{2 \times 3.14 \times 50 \times 50 \times 10^{-3}}$$

$$\cong 1.35 \text{ A.}$$

∴ Minimum d.c. current to avoid spiking (I_{Lmin}) = 1.35 A

$$R_{Lmax} = \frac{V_{dc}}{I_{L_{min}}} = \frac{13.5}{1.35} = 10 \ \Omega \quad \textbf{Ans.}$$

Example 24. *A single phase full-wave rectifier makes use of π-section filter with two 10 μF capacitors and a choke of 104. The secondary voltage is 280 V_{rms} with respect to centre tap output. If the load current is 100 mA, determine the percentage ripple. Given that V_{dc} = 346 V.*

Assume supply frequency of 50 Hz.

Solution : Given $V_{rms} = 280 \text{ V}$

$$V_{L\max} = V_m = \sqrt{2}\,V_{rms} = \sqrt{2} \times 280 = 396 \text{ V}$$

Load current, $I_{dc} = 100 \text{ mA}$

$$C_1 = C_2 = 10 \text{ μF}$$

$$f = 50 \text{ Hz}$$

Ripple factor in case of π-section filter is given by the relation

$$\gamma = \frac{\sqrt{2}}{8\omega^3 C_1 C_2 L R_L} \qquad \left\{\therefore R_L = \frac{V_{dc}}{I_{dc}} = \frac{346}{100 \times 10^{-3}} = 3460\right\}$$

or
$$\gamma = \frac{\sqrt{2}}{8(2\pi \times 50)^3 \times 10 \times 10^{-6} \times 10 \times 10^{-6} \times 104 \times 3460}$$

or
$$\gamma = .00165$$

Percentage ripple = .165% **Ans.**

? SELF-TEST (CLIPPER CLAMPER)

1. What do you mean by clipper ?
2. List different types of clipper.
3. List the application of the clipper circuit.
4. What do you mean by clamper ?
5. List different types of clamper.
6. What is basic difference between the clipper and clamper circuit ?
7. What are the main features of clamper circuit, with reference to the time constant ?
8. What do you mean by biased clipper ?
9. Compare biased clipper with the positive and negative clipper.
10. Does a clamper circuit changes the peak-to-peak (*i.e.*, *rms* value) ? Justify.

SOLVED EXAMPLES

Example 25. *The negative series clipper shown in Fig. P (2.38). What is the peak output voltage from the circuit ? Assume diode is ideal.*

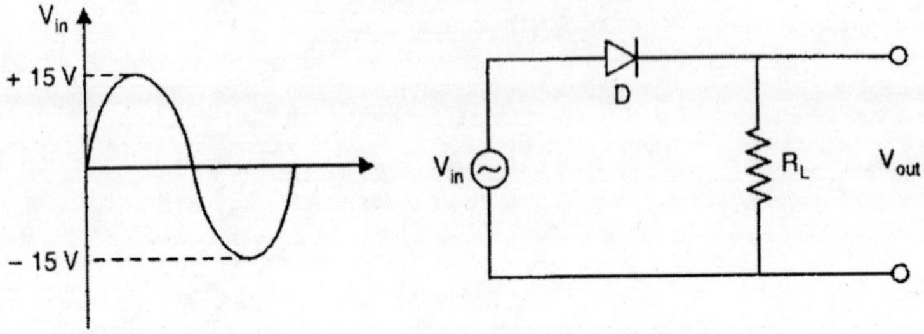

Fig. P (2.38).

Solution : When the diode is connected in series with the load, it is called a series clipper. Since it is a negative clipper, it will remove negative portion of the input a.c. signal. During the positive half cycle of the input signal diode is forward biased. As a result, the diode will conduct the output voltage. Fig. P (2.39) shows the output waveform.

$$V_{out} = V_{in} = 15 \text{ V}$$

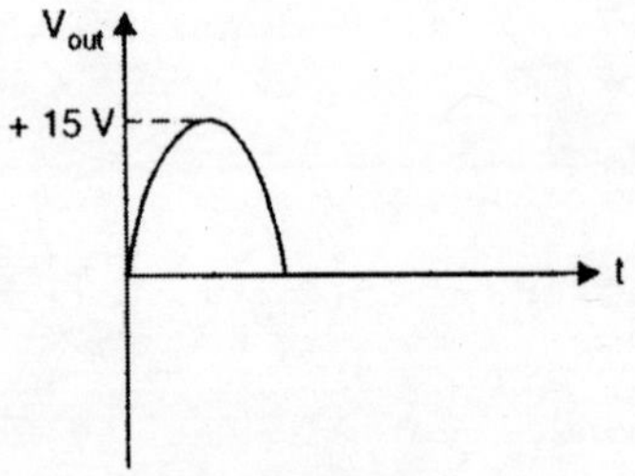

Fig. P (2.39).

During the negative half cycle of the input signal, the diode is reverse biased and consequently it will not conduct. Therefore V_{out} = 0 volt.

Example 26. *The negative shunt clipper shown in Fig. P (2.40) has a peak input voltage of + 5V. What is the peak output voltage from this circuit.*

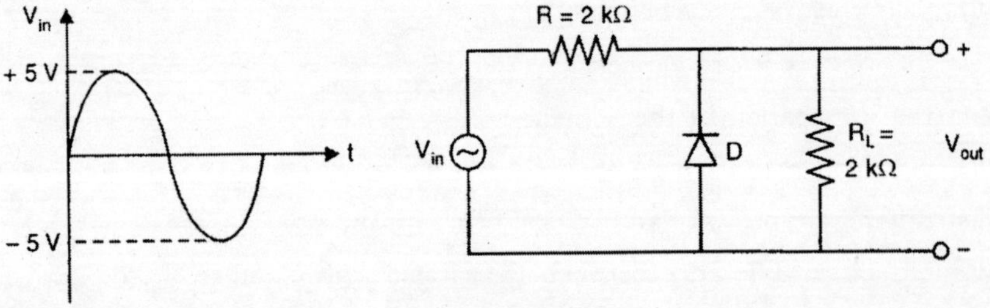

Fig. P (2.40).

Solution : During the positive half-cycle of input a.c. signal, the diode is reverse biased and it will behave as an open circuit. Output voltage in this case can be calculated by applying potential divider method.

$$V_{out} = \frac{R_L}{R + R_L} \times V_{in}$$

$$= \frac{2}{2 + 2} \times 5 = 2.5 \text{ volt.} \quad \textbf{Ans.}$$

The output waveform is shown in Fig. P (2.41).

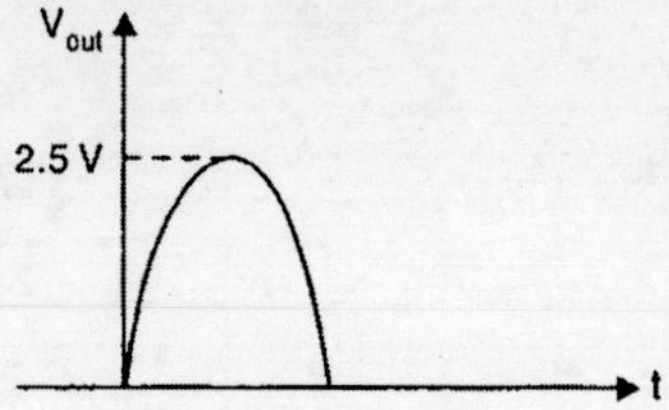

Fig. P (2.41).

However during the negative cycle the diode will conduct and output equal to zero volt.

Example 27. *Sketch L_R and V_o for the following circuit.*

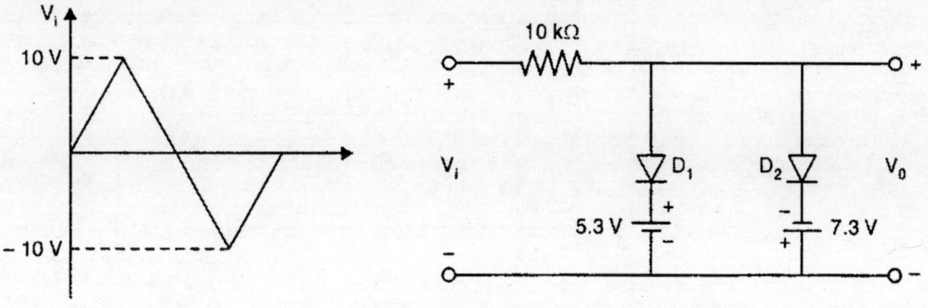

Fig. P (2.42).

Solution : We know that when a diode is forward biased then a voltage V_f = 0.7 V drops across the diode (D_1 and D_2). When a diode in reverse biased it behaves like an open circuit and all the voltage drops across it.

To make diode D_1 forward biased V_i should be positive and 0.7 V more than 5.3 V.

i.e., 6V (5.3 + 0.7). Once D_1 is forward biased the current i_R flows through 10K resistor and diode D_1 and 5.3 V battery. At this time V_0 becomes constant at 6 V and extra voltage of V_i drops across the resistor. Maximum positive current i_R will flow when V_i = + 10 V

$$i_{R_{max}} = \frac{10\text{ V} - 6\text{ V}}{10\text{ K}} = 0.4\text{ mA.}$$

Similarly when v_1 is negative, diode D_2 will be forward biased when input voltage V_i is 0.7 V more negative than (– 7.3 V) *i.e.*, (– 7.3 – 0.7 = – 8 V). Once V_i becomes more negative than – 8 V, a negative current i_R flows in resistance R and V_0 becomes constant at (– 8 V) and extra voltage drops across R.

Maximum negative current will flow when V_i = – 10V

$$i_R = \frac{-10\text{ V} + 8\text{ V}}{10\text{ K}} = -0.2\text{ mA.}$$

Required waveforms are shown in Fig. P (2.43)

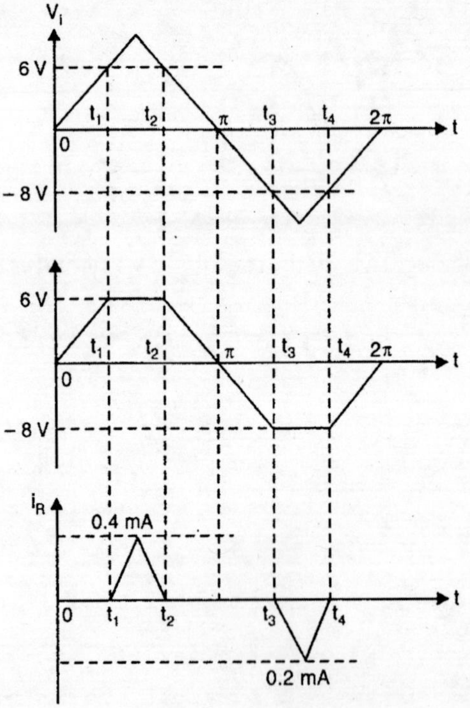

Fig. P (2.43).

Example 28. *For the given circuit, what will be the output voltage and voltage across R when the input voltage is – 12 V ?*

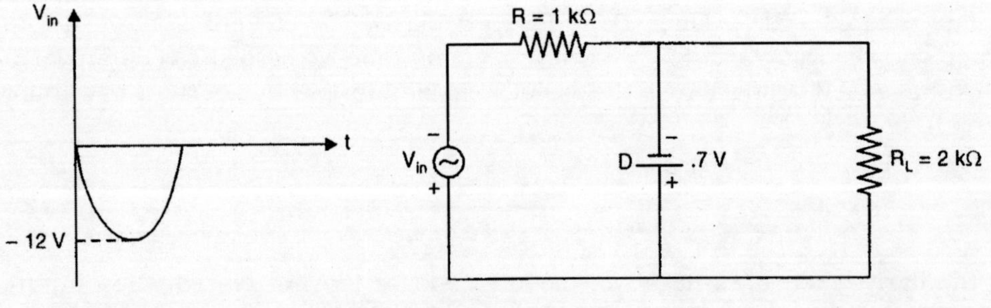

Fig. P (2.44).

Solution : During the negative half-cycle of input signal, the diode is forward biased

$$V_{out} = -\ 0.7\ \text{V}$$

Voltage across R, $V_R = -\ 12 - (-\ .7)$

or $V_R = -\ 11.3\ \text{V}$

The waveform for the output voltage is shown in Fig. P (2.45).

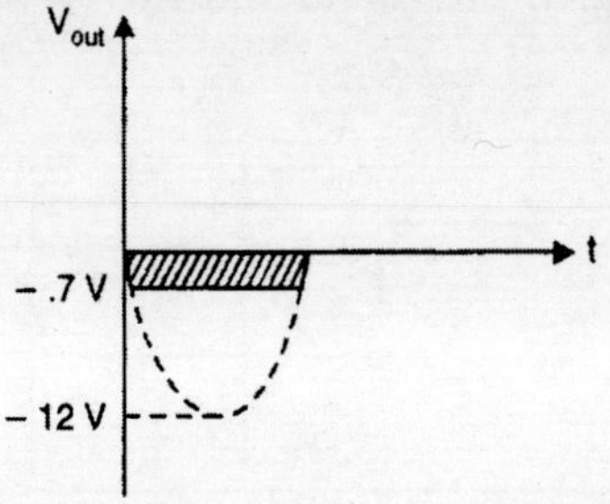

Fig. P (2.45). *Output waveform for the above circuit.*

Example 29. *Sketch the output waveform for the circuit shown in Fig. P (2.46). It is given that discharging time is much greater than the time period of charging time.*

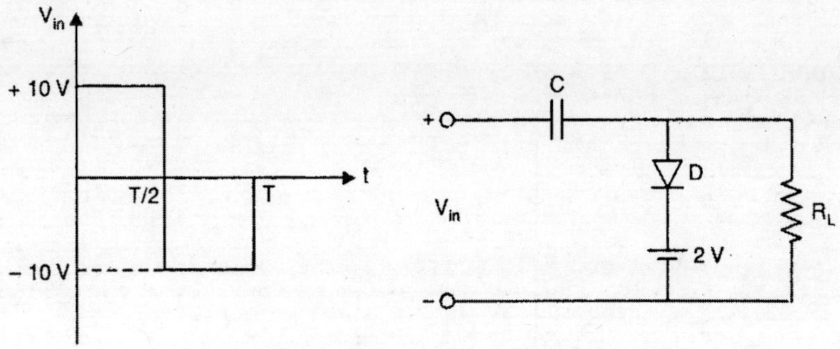

Fig. P (2.46).

Solution : During positive half-cycle of the input signal, the diode is forward biased. The network under this situation is shown in Fig. P (2.47).

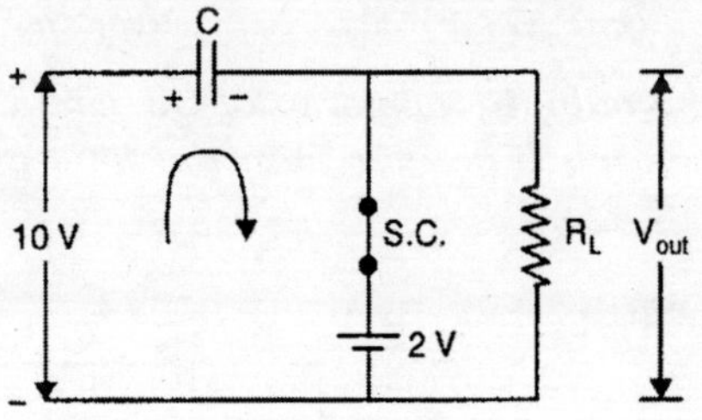

Fig. P (2.47).

Applying KVL, we have

$$10 - V_C - 2 = 0$$

$$V_C = 8 \text{ V (Therefore, the capacitor will charge upto 8 V)}$$

and $$V_{out} = V_{in} - V_C$$

or $$V_{out} = 10 - 8 = 2 \text{ V}.$$

During the negative half-cycle of the input signal, the diode is reverse biased and the diode will behave as an open circuit as shown in Fig. P (2.48).

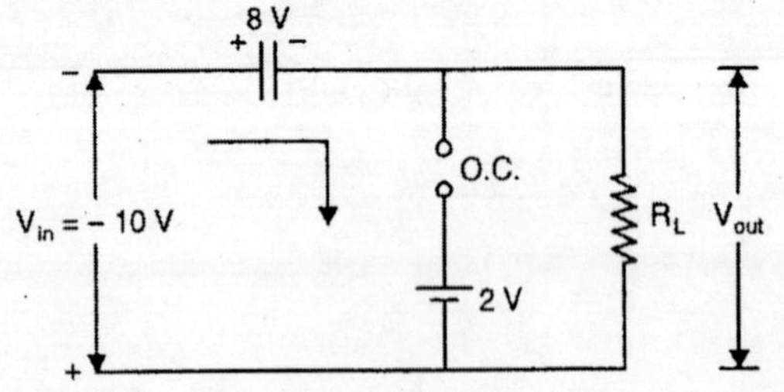

Fig. P (2.48).

Applying KVL to the loop, we have

$$-10 - 8 - V_{out} = 0$$

$$V_{out} = -18 \text{ V}$$

The complete output waveform is shown in Fig. P (2.49).

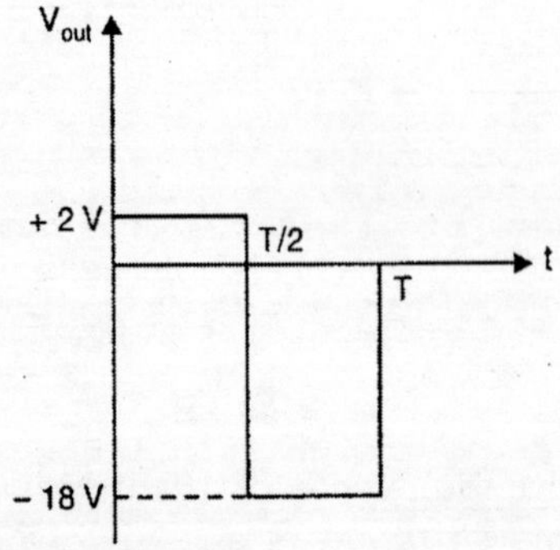

Fig. P (2.49). *Outwave waveform.*

Example 30. *Sketch the output waveform for the circuit shown in Fig. P (2.50). It is given that discharging time is much greater than the time period of input wave.*

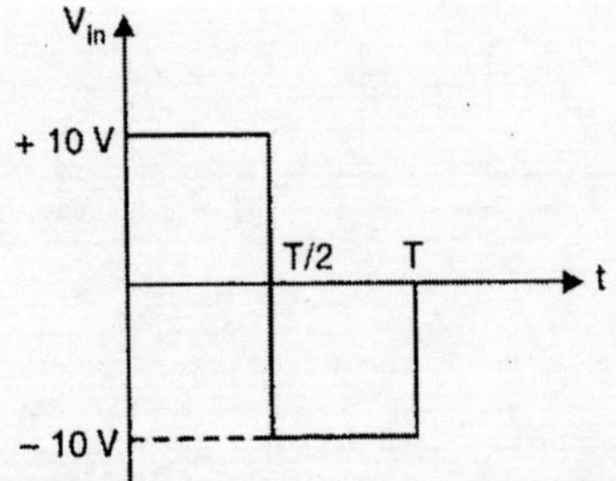

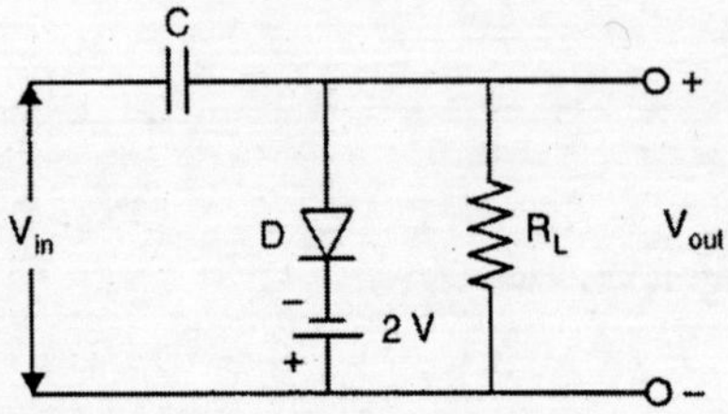

Fig. P (2.50).

Solution : During the positive half cycle, the diode is forward biased. The circuit becomes like Fig. P (2.51).

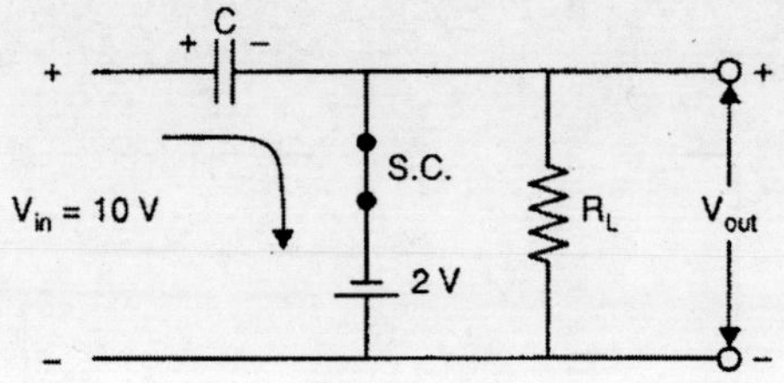

Fig. P (2.51).

Applying KVL to the loop, we have

$$10 - V_C + 2 = 0$$

$$V_C = 12 \text{ V}$$

and $$V_{out} = V_{in} - V_C$$

or $$V_{out} = 10 - 12$$

or $$V_{out} = -2 \text{ V} \qquad \textbf{Ans.}$$

During the negative half cycle the above circuit looks like as shown in Fig. P (2.52).

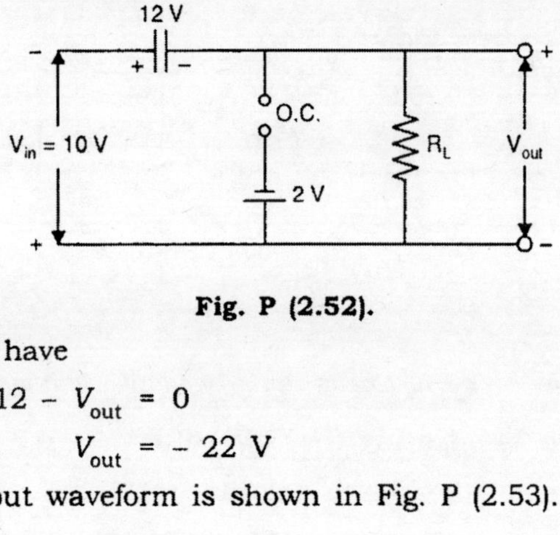

Fig. P (2.52).

Applying KVL, we have

$$-10 - 12 - V_{out} = 0$$

or $$V_{out} = -22 \text{ V}$$

The complete output waveform is shown in Fig. P (2.53).

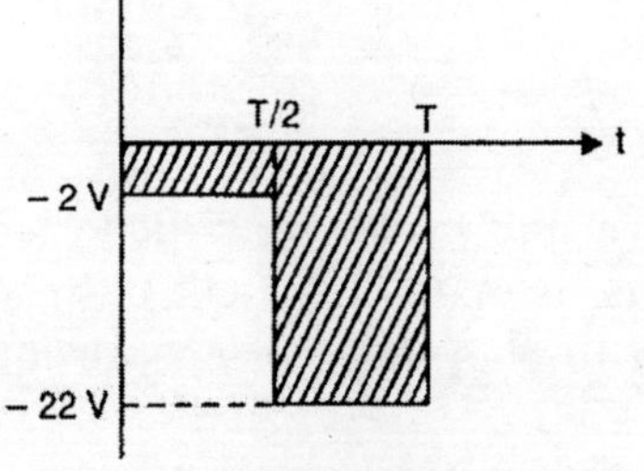

Fig. P (2.53).

Example 31. *Sketch V_0 for the following circuit in Fig. P (2.54).*

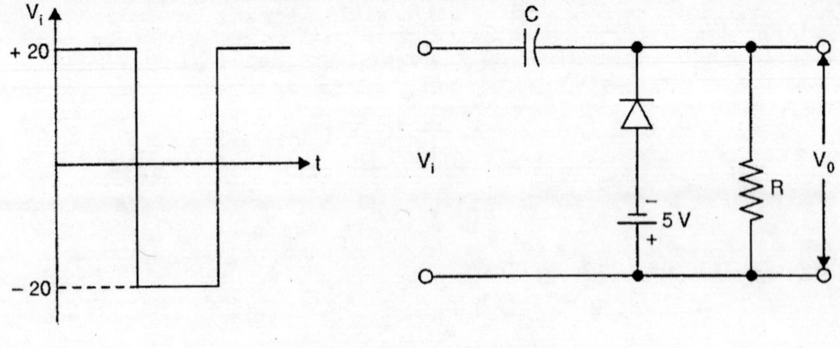

Fig. P (2.54). *Given circuit.*

Solution : First of all that condition is considered when diode will be forward biased and will behave as almost short circuit. This will happen when negative half cycle occurs and circuit behaves like Fig. P (2.55).

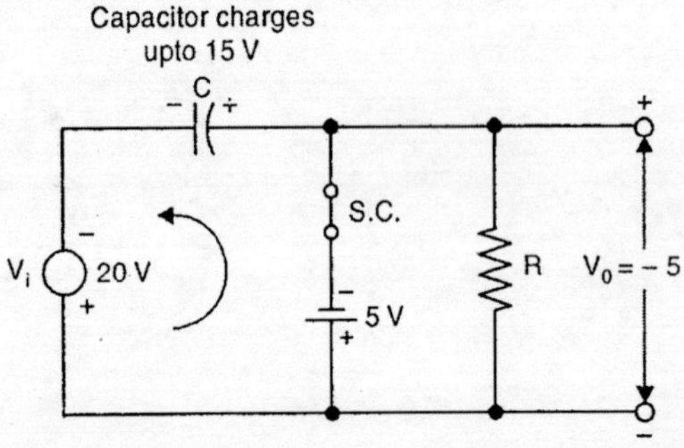

Fig. P (2.55). *Circuit during negative half cycle.*

As the input voltage is opposing the 5 V battery.

So capacitor charges in the shown polarity upto

$$V_c = V_i - 5 \text{ V}$$
$$= 20 - 5 = 15 \text{ volts.}$$

During this period diode is almost short circuited hence battery voltage (−5 V) appears across the output.

So $$V_0 = -5 \text{ V.}$$

Now consider the positive half cycle. Now diode is reverse biased and the circuit is shown in Fig. P (2.56). Diode is open circuited, battery is disconnected. Input voltage V_i and capacitor voltage (due to charging during previous negative half cycle) come in series and output voltage.

$$V_0 = V_i + V_c$$
$$= 20 + 15 = 35 \text{ volts.}$$

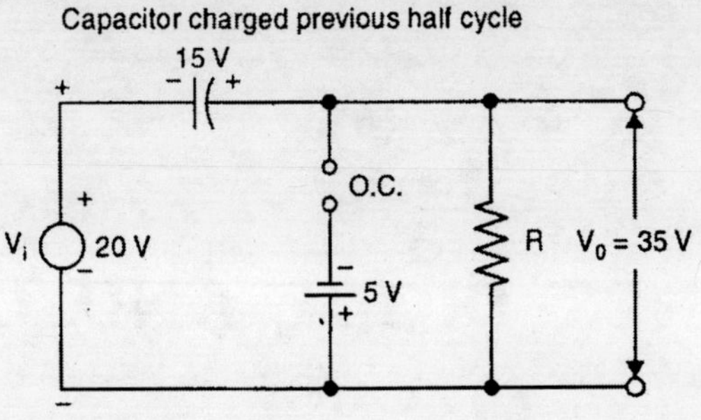

Fig. P (2.56). *Circuit during positive half cycle.*

Waveform for input and V_0 is shown in Fig. P (2.57).

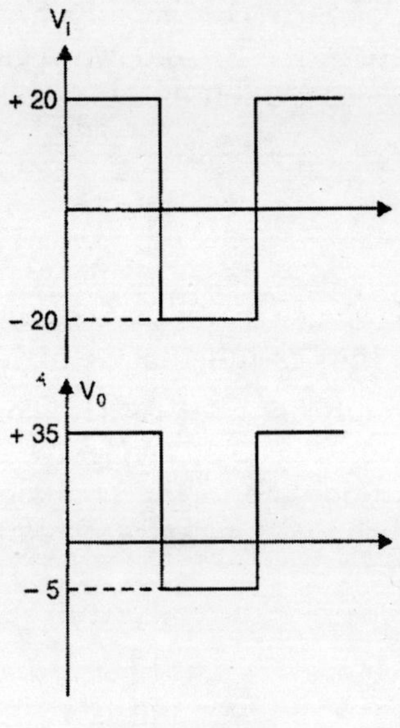

Fig. P (2.57).

□□□

3

ZENER DIODE AND LED

3.1. INTRODUCTION

We have already discussed about rectifiers and PN-junction diode. Now, we shall discuss the Breakdown diodes, which are used for specific purpose like voltage regulation, short circuit protection etc. Diodes which are designed with adequate power-dissipation capabilities to operate in the Breakdown region may be employed as voltage-reference or constant-voltage devices. Such diodes are known as avalanche, breakdown or zener diodes.

3.2. BREAKDOWN MECHANISM

The Breakdown mechanism of a zener diode made up of a silicon PN-junction devices, differs from a rectifier diode, in the sense, that it is operated in the reverse break down region (*i.e.,* when the PN-junction is highly reverse biased).

The reverse-voltage characteristic of a semiconductor diode, including the Breakdown region is shown as in Fig. 3.1. Diodes which are designed with adequate power-dissipation capabilities to operate in the breakdown region may be employed as voltage-reference or constant-voltage devices.

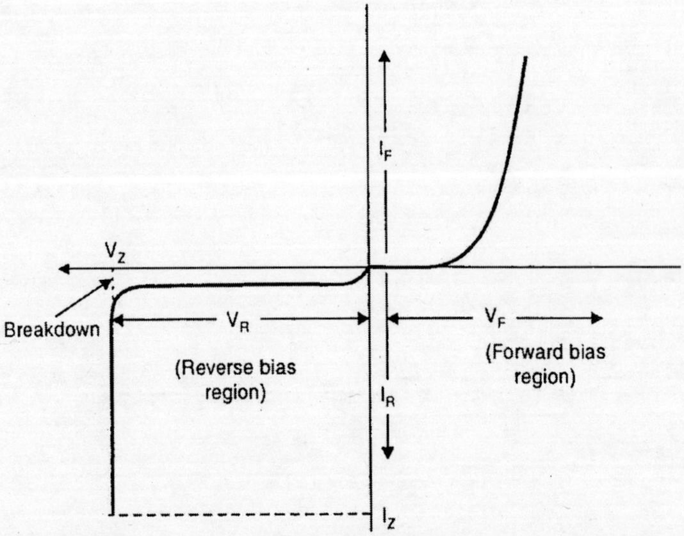

Fig. 3.1. *The V-I characteristic of a zener diode.*

Fig. 3.2 shows the symbol of zener diode which is quite different from the simple diode.

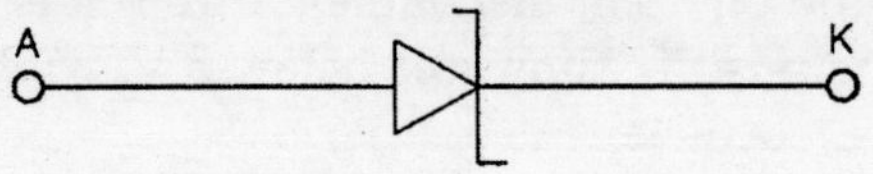

Fig. 3.2. *Symbol of zener diode.*

Two mechanism of diode Breakdown for increasing reverse voltage are namely :

(*i*) Zener Breakdown, and

(*ii*) Avalanche Breakdown.

One thing should be always kept in mind that the Zener Breakdown and avalanche Breakdown are usually differentiated on the basis of doping concentration. Zener Breakdown occur when the PN-junction is highly doped while avalanche Breakdown occurs only when the PN-junction is very lightly doped.

The Zener Breakdown occurs when the electric field across the junction, produced due to reverse voltage is sufficiently high. This electric field exerts a force on the electrons in the outer most shell. This force is so high that the electrons are pulled away from their parent nuclei and become free carriers. This ionization, which occurs due to the electrostatic force of attraction is known as *zener effect.* Also known as *high-field emission.* One thing should be always kept in mind that zener effect will occur only when a diode is heavily doped. Because when the diode is heavily doped depletion layer becomes very narrow. Due to this the electric field across the depletion layer is very intense, when the field strength reaches approximately 300,000 V/cm. It causes an increase in the number of free carriers and hence an increase in the reverse current.

The zener diodes, with Breakdown voltages of less than 6 V, operate predominantly in Zener Breakdown. However, the Breakdown voltages greater than 6 V, operates predominantly in Avalanche Breakdown.

3.2.1. Avalanche Multiplication

A thermally generated carrier falls down the junction barrier and acquires energy from the applied potential. This carriers collides with a crystal ion and imparts sufficient energy to disrupt a covalent bond. In addition to the original carrier, a new electron-hole pair has now been generated. These carriers may also pick up sufficient energy from the applied field, collide with another crystal ion and create still another electron-hole pair. This cumulative process is referred to as avalanche multiplication. It results in large reverse saturation currents and the diode is said to be in the region of *avalanche breakdown.* However, Zener Breakdown does not involve collisions of carriers with the crystal ions as does avalanche multiplication.

3.2.2. Temperature Characteristics

The temperature coefficient of any device may be defined as the percentage in reference voltage per centrigate degree change in diode temperature. This coefficient may be either positive or negative and will normally be lie in the range ± 0.1 percent /°C.

Here with reference to the Breakdown diode, if the reference voltage (*i.e.*, applied reverse voltage) is above 6 V, where the physical mechanism involved is avalanche multiplication, the temperature coefficient is positive, however below 6 V, where true Zener Breakdown is involved, the temperature coefficient is negative.

Explanation of why Zener Breakdown have negative temperature coefficient while Avalanche Breakdown have positive temperature coefficient?

A junction having a narrow depletion-layer width and hence high field intensity, will breakdown by the zener mechanism. An increase in temperature increases the energies of the valence electrons, and hence makes it easier for these electrons to escape from the covalent bonds. Less applied voltage is therefore required to pull these electrons from their positions in the crystal lattice and convert them into conduction electrons. Thus, the Zener Breakdown voltage decreases with temperature *i.e.*, negative temperature coefficient.

However, if a diode having very highly doping which gives a broad depletion layer, and therefore low field intensity, will break down by avalanche mechanism. In this case, we rely on intrinsic carriers to collide with valence electrons and create avalanche multiplication. As the temperature increases, the vibrational displacement of atoms in the crystal of grows. This vibration increases the probability of collisions with the lattice atoms of the intrinsic particles as they cross the depletion width. The intrinsic holes and electrons thus have less of an opportunity to gain sufficient energy between collisions to start the avalanche process. Therefore, the value of the avalanche voltage must increases with increased temperature.

3.3. BREAKDOWN CHARACTERISTIC OF A ZENER DIODE

We have already discussed that a zener diode is operated in the reverse-bias region. That is why we shall examine its characteristic in this region only. Fig. 3.3 shows the reverse characteristic of a zener diode.

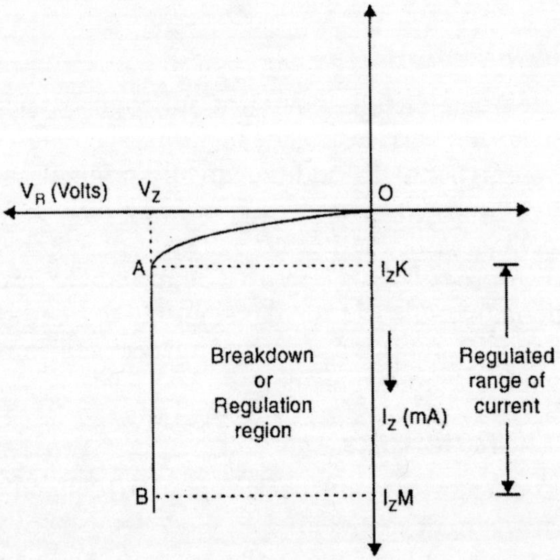

Fig. 3.3. *Reverse characteristic of a zener diode.*

It may be noted from Fig. 3.3 that as the reverse voltage (V_R) Starts to increase the reverse current (usually called zener current, I_Z) remains negligibly small up to the 'Knee' of the curve (Point A in Fig. 3.3). At this point, the effect of Breakdown process begins, at this point, the voltage is called zener Breakdown voltage or simply zener voltage V_Z, remains essentially constant. *This ability of diode is called regulating ability* and is important feature of zener diode. It maintains, an essentially a constant voltage across its terminals over a specified range of zener current values.

Two important point about the characteristics of zener diode :

1. There is a maximum value of zener current designated as I_{ZM} or $I_{Z(max)}$ above which diode may be damaged. The value of this current is given by the maximum power dissipation of zener diode. As long as the maximum power dissipation is not exceeded the diode will not be damaged.
2. There is a minimum value of zener current called breakover current designated as I_{ZK} or $I_{Z(min)}$ which must be maintained in order to keep the diode in breakdown (or regulation). When the current is reduced below the knee of the curve, the voltage changes drastically and the regulation is lost.

3.4. ZENER RESISTANCE

Fig. 3.4 (*a*) shows a ideal approximation of a zener diode in reverse breakdown. It shows that a zener diode is simply equivalent to a battery having a voltage equal to the zener voltage (V_Z).

Fig. 3.4 (*b*) shows a practical equivalent circuit of zener diode. This circuit shows that a zener diode is equivalent to a battery with voltage (V_Z), in series with a resistance (R_Z) called *zener resistance,* also called *dynamic resistance* or *a.c. resistance.*

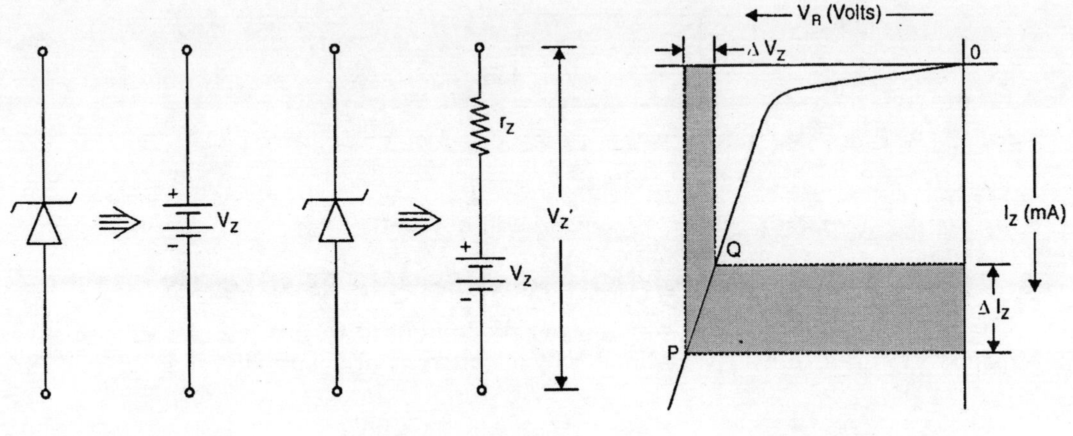

(*a*) Ideal zener equivalent diode circuit

(*b*) Practical zener diode equivalent circuit

(*c*) Reverse V.I. characteristic curve illustrating zener resistance

Fig. 3.4.

In order to calculate the value of zener resistance, consider any two points *P* and *Q* on the reverse *V-I* curve as shown in Fig. 3.4 (c).

Let, ΔV_Z = Change in the values of zener voltage between the points P and Q

ΔI_Z = Change in the values of zener current at the corresponding points

Now, the zener resistance is given by the relation :

$$r_Z = \frac{\Delta V_Z}{\Delta I_Z} \quad \text{... (3.1)}$$

3.5. ZENER DIODE SPECIFICATION

Any zener diodes are generally specified in terms of four factors namely :

(*i*) Zener voltage (V_Z)

(*ii*) Maximum power dissipation (P_{ZM} or $P_{D(max)}$)

(*iii*) Breakover current (I_{ZK}), and

(*iv*) Zener resistance (r_Z).

Zener diodes are available in market with Breakdown voltages ranging from 1.8 V to 2000 V. The power dissipation of a zener diode is the product of breakdown voltage (V_Z) and reverse current (I_Z). Mathematically, the power dissipation,

$$P_Z = V_Z \times I_Z \quad \text{... (3.2)}$$

The maximum value of power dissipation, which a zener can dissipate, without failure is called power rating and is designated by P_{ZM}.

The zener diodes are available with power rating from 150 mW to 50 W. The value of maximum zener current is given by the relation,

$$I_{ZM} = \frac{P_{ZM}}{V_Z} \quad \text{... (3.3)}$$

where, P_{ZM} = Maximum power rating of zener diode, and

V_Z = Breakdown voltage.

I_{ZM} or $I_{Z(max)}$ = Max. value of zener diode current.

3.6. ZENER DIODE APPLICATION AS A SHUNT REGULATOR

Fig. 3.5 shows a circuit in which zener diode is used to regulate the voltage across R_L against changes due to variations in load current and supply voltage.

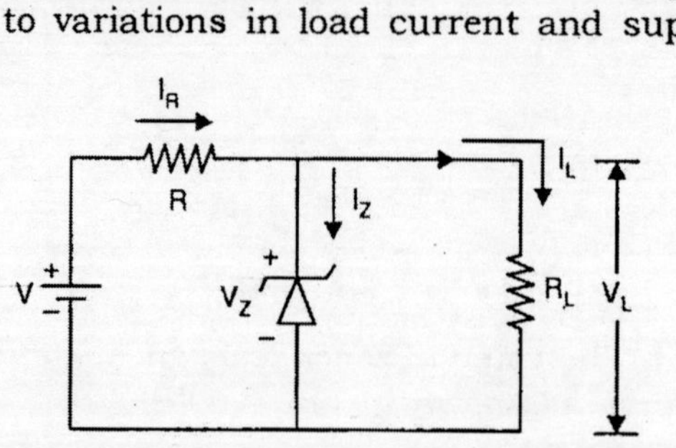

Fig. 3.5. *Application of zener diode as a shunt regulator.*

Working : The source voltage V and resistor R are selected, so that, initially, the diode is operating in the breakdown region. Here, the diode will now regulate the load voltage against variations in load current and again variations in supply voltage changes in diode voltage.

In order to better understanding. Consider :

Case 1 : When the input voltage increases. Since, the zener diode is equivalent to a battery, V_Z as shown in Fig. 3.3. It is clear that output voltage remain constant at V_Z (V_L). The excess voltage is dropped across the series resistance R. This will cause an increase in the value of total current I. The zener will conduct the increase of current in I while the load current remains constant. Hence, output voltage V_L remains constant irrespective of the changes in the voltage V.

Case 2 : When the input voltage (V) is constant but the Load Resistance R_L decreases. This will cause an increase in load current. The extra current cannot come from the source because drop in R will not change as the zener is within its regulating range. The additional load current will come from a decrease in zener current I_Z. Consequently, the output voltage stays at constant value.

Voltage drop across, $R = V - V_L$

Current through R, $I = I_Z + I_L$

$$R = \frac{V - V_L}{I_Z + I_L}$$

where, V = Input voltage

Moreover, as load current or supply voltage changes, the diode current will accommodate itself to these changes to maintain a nearly constant load voltage. The diode will continue to regulate until the circuit operation requires the diode current to fall to I_{ZK}, in the neighbourhood of the knee of the diode volt-ampere curve. The upper limit on diode current is determined by the power-dissipation rating of the diode.

A zener diode is chosen whose breakdown voltage V_Z is the same as output voltage V_L. Regulation takes place only for $V_i > V_Z$.

$$I_{Z(max)} = \frac{V_{i(max)} - V_Z}{R} - I_{L(min)}$$

$$I_{Z(min)} = \frac{V_{i(min)} - V_Z}{R} - I_{L(max)}$$

$$R_{min} = \frac{V_{i(min)} - V_Z}{I_{Z(max)} + I_{L(min)}}$$

$$R_{max} = \frac{V_{i(min)} - V_Z}{I_{Z(min)} + I_{L(max)}}$$

Other Applications of Zener Diodes

(*a*) As a fixed reference voltage in transistor biasing circuit.

(*b*) As peak clippers or limiters in waveshaping circuits.

(*c*) For meter protection against damage from accidental applications.

Advantages of Zener Diode Regulator

It is simple circuit, light weight, more reliable and provides regulation over a wide range of current.

Disadvantages of Zener Diode

(*a*) As there is power dissipation in series resistor and the diodes, it results in poor efficiency.

(*b*) The stablised output is determined by the zener breakdown voltage and cannot be varied.

3.7. LIGHT-EMITTING DIODE : (CONSTRUCTION AND WORKING PRINCIPLE)

The light emitting diode (LED) converts current to light. When a forward bias voltage is applied across a pn junction, electrons and holes flow across the space-charge region and become excess minority carriers. These excess minority carriers diffuse into the neutral semiconductors regions, where they recombine with majority carriers. If the semiconductor is a **direct bandgap material,** such as GaAs. The electron and hole can recombine with no change in momentum, and a photon or light wave can be emitted. Conversely, in an **indirect bandgap material,** such as silicon, when an electron and hole recombine, both energy and momentum must be conserved. Therefore, LEDs are fabricated from GaAs or other compound semiconductor material. In an LED, the diode current is directly proportional to the recombination rate, which means that the output light intensity is also propositional to the diode current *i.e.,* forward current.

Thus, higher the diode current (forward current), higher the light output. The schematic symbol and basic structure of a light emitting diode (LED) is shown in Fig. 3.6 (*a*). The arrows, pointing away from the diode symbol represent the light, which is being transmitted away from the junction.

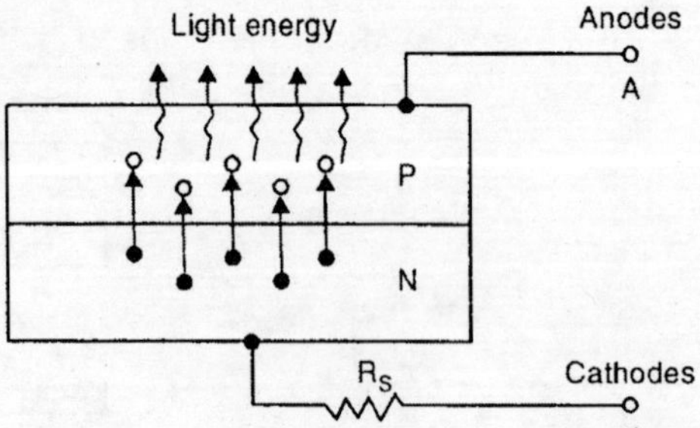

(*a*) Basic structure

Fig. 3.6. (*a*) *Light emitting diode.*

Here, an N-type layer is grown on a P-type substrate by a diffusion process. The metal connections to both the layers make anode and cathode terminals as shown. The light energy is released at the junction, when the recombination of electrons with holes take place.

It will be interesting to know that when the LED is forward biased, the electrons and holes move towards the junction and the recombination takes place. After recombination, the electrons, lying in the conduction bands of N-region, fall into the holes lying in the valence band of a P-region. The difference of energy between the conduction band and valence band is radiated in the form of light energy. In ordinary diodes, this energy is radiated in the form of heat.

The Semiconducting materials used for manufacturing light emitting diodes are gallium arsenide, gallium arsenide phosphide. The Si and Ge is not used for manufacturing LEDs because these are heat producing materials. Moreover, these materials are very poor in emitting light radiations.

The LEDs radiate light in different colours such as Red, Green, Blue, Orange etc. Some of the LEDs emit infrared (*i.e.*, invisible) light also. The colours of the emitted light, depends upon the type of the semiconductor used. It will be interesting to know that LEDs are not able to withstand reverse bias of even a very small voltages. For this reason, it is necessary to assure that reverse bias is never applied to an LED.

3.7.1. LED Voltage Drop and Current

The LEDs have also a certain forward voltage drop, when they are used in the circuit like an ordinary diode (which has a voltage drop equals to its Cut-in voltage). This voltage drop depends upon the LED current, colour of the emitted light etc.

Usually, the voltage drop has a typical value that varies from 1.5 to 2.5 V for current 10 to 50 mA.

Consider a LED circuit as shown in Fig. 3.6 (*b*). In this circuit, the LED is connected to the supply voltage through a current limiting resistor (R_S). The current through the LED is given by the relation.

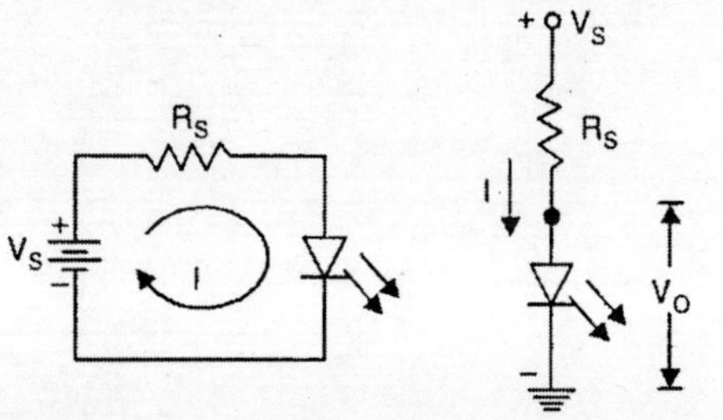

Fig. 3.6. (*b*) *LED circuit.*

$$I = \frac{V_S - V_D}{R_S}$$

where, V_S = Supply Voltage

V_D = Voltage drop across LED

R_S = Current limiting resistor.

The value of current limiting resistor is given by the relation,

$$R_S = \frac{V_{out(Peak)} - V_F}{I_F}$$

where, V_F = Peak output voltage

I_F = LED Current drop.

3.7.2. Radiative Recombination

The radiative recombination in semiconductors occurs predominantly via three different processes , namely,

- Interband transition
- Recombination via impurity centres, and
- Excitation recombination

The radiative recombination/transition takes place between injected electrons into the N-region and holes into the P-region as shown in Fig. 3.6 (*c*). In such transition the electrons move from conduction band to valence band and the holes move from valence band to conduction band. Fig. 3.6 (*d*) is a general energy level diagram which is used to describe the possible radiative and non-radiative electron transition involved in LEDs.

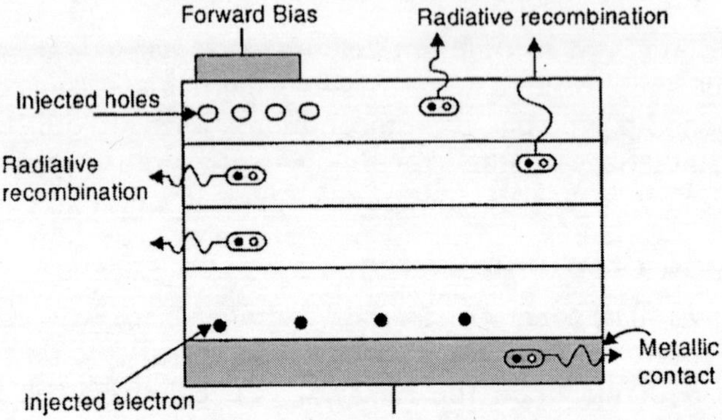

Fig. 3.6. (c) *Cross section of LED.*

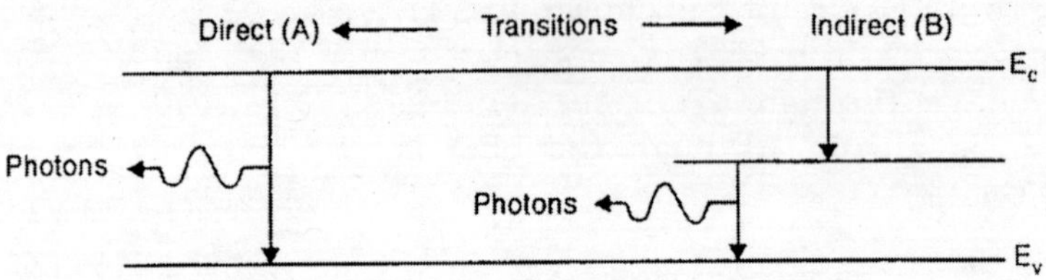

Fig. 3.6. (*d*) *Illustration of Direct & Indirect Transitions.*

In Si and Ge semiconductors the greater percentage of this energy is given in the form of heat and emitted light energy is so insignificant that it does not seem to emit light at all.

Two types of radiative recombination are commonly encountered in LEDs. They are :

- Direct Recombination, and
- Indirect Recombination.

These two types of recombination is dependent on the bandgap of the semiconductor material.

In either case of recombination, the wave length (λ) of radiation emitted is related with the bandgap energy (E_g) of semiconductor by

$$E_g = h\nu = \frac{hc}{\lambda}$$

or

$$\lambda = \frac{hc}{E_g} = \frac{12423}{E_g} \text{ Å} \cong \frac{1240}{E_g} \text{ nm} = \frac{1240}{h\nu\,(\text{eV})} \text{ nm}$$

where h = Plank's constant = 4.141×10^{-15} eV/sec

c = Velocity of light of EM radiation

= 3×10^8 m/sec

E_g = bandgap in eV

1 Å = 10^{-8} cm = 10^{-10} M

It is well known that the light given off is classified into two ways;

- Wavelength, and
- Intensity

The colour of the LED depends on the energy involved. Most of the LEDs are made of Gallium and Indium components. The LEDs are low power device, typically of the order of milliwatts. This suggests that LEDs are useful as an indicator but not especially good for illumination.

3.7.3. Light Source Material

The optical bandgaps of several materials are presented in table. Many group III-IV semiconductor possess E_G in the above range which makes them useful for LEDs. In addition, some group II-VI semiconductors also appear to be attractive for LEDs, but their technology being difficult, they are not exploited for this purpose.

The requirement of a good LED material are :

- The ability to fabricated a P-N junction to achieve good injection and illumination efficiencies at the same time.
- Extract a reasonable fraction of generated light.

Table 3.1. Optical Energy Bandgaps

Name	*Chemical name*	*Optical energy bandgap (eV)*
Gallium Arsenide	GaAs	1.4
Indium Arsenide	InAs	0.43
Gallium Phosphide	GaP	2.2

Thus, the main requirements for suitability of the LED materials are :

- It must have energy gap in the range of visible range *i.e.,* 1.75 eV to 3.15 eV. Generally, for visible radiation gap should be more than 2.0 eV.
- Both *n*-and *p*-type should have low resistivity. Unfortunately materials with large bandgap have high resistivity even after doping.
- Efficiency radiative path must be provided.

3.7.4. Direct Bandgap LEDs

In the direct bandgap semiconductors (GaAs, GaSb), the probability of recombination is high, because electrons and holes have the same value of momentum. Hence, in this case, the internal conversion efficiency *i.e.,* efficiency of converting electrical input to Electromagnetic (EM) radiation is high. Most of the electrons energy is given up in the form of radiant energy approximately equal to GaAs bandgap of 1.4 eV *i.e.*, the transition takes place all the way from conduction band to valence band.

Gallium Arsenide (GaAs) is one of the first and the best materials used for infrared LEDs (IREDs). Other popular direct bandgap semiconductors for LEDs are GaSb, InAs, InP and GaSb. But all these semiconductors have E_g below 1.75 eV. Therefore, the resulting emission of *p-n* junction of these materials is outside the visible spectrum *i.e.*, in IR region.

3.7.5. Indirect Bandgap LEDs

Popular indirect bandgap semiconductors such as GaP, GaN, and SiC are used to produce visible light LEDs. Indirect bandgap semiconductors have very poor efficiency. The indirect transition occurs, when the electron and hole momentum (Mass* velocity) differ. In such case, an intermediate transition must be made, and most of the energy is given off in the form of heat. The probability of recombination of electrons and holes in *p-n* junctions of these materials being low, impurity centers in controlled quantities are usually introduced during device fabrication to improve internal conversion efficiency of the LEDs. But, only limited number of impurities are suitable for this purpose. For example, the selective introduction of *N* in GaP *p-n* junction diode facilitates green emission with the internal efficiency (1%). Also, the use of Zn and O as impurities in GaP *p-n* junctions makes it possible to realize red emission with internal absorption of GaP devices (and in general of LEDs in the indirect bandgap material) is low, it leads to high external efficiency for the LEDs. Typically, the external efficiency of re-emitting GaP LEDs is 3% to 7%, whereas that of green-emitting GaP LEDs is about 0.6%.

3.7.6. Typical LEDs

It is of interest to note that red, green and yellow emissions are directly realizable, whereas blue emission is possible only through up conversion of IR emission. The importance of GaP LEDs in display technology is also clear from Fig. 3.7. For a junction area of 0.1 mm the values on the *X*-axis give the LED drive current in mA and the

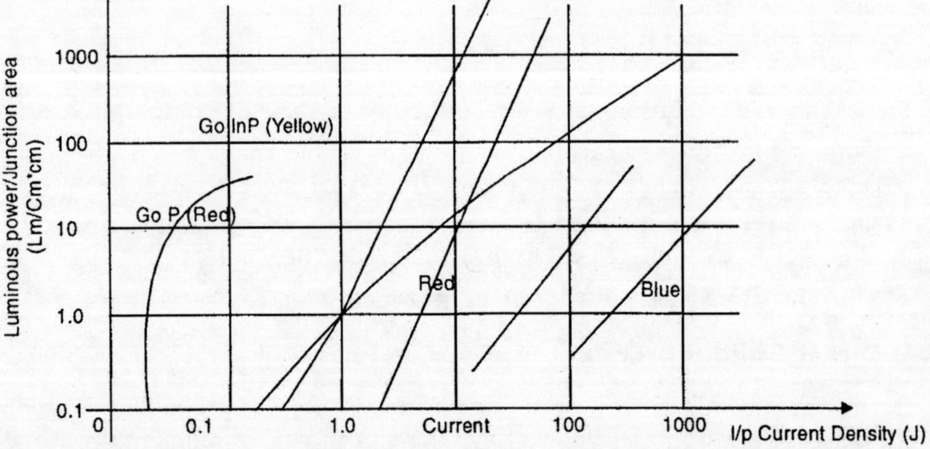

Fig. 3.7. *LED Performance in Different Materials.*

values on the Y-axis give the luminous output power of the LED in millilumens. A comparison of LEDs (Bergh and Dean loc. cit., Nuesses 1972) from the point of view of several importants.

While these LEDs are suitable for instrument displays, opto-electronics etc., their radiance (*i.e.*, optical power radiated into a solid-angle, per unit area of emitting surface) is low, generally below 0.25 $W/Sr/cm^2$.

The luminous efficiency of low-power visible LEDs varies from 50 to 1000 lumen/ watt (Lm/W). As the energy gap of the LED is increased (emission wavelength decreased), the probability of direct radiative transition decreased because the crystal flows provide intermediate energy levels through which two step radiationless transition can be made.

3.7.7. Quantum Efficiency and LED Power

The internal quantum efficiency in the active region is the fraction of the electron-hole pairs that recombine radiatively. If the radiative recombination rate is R_r and the non-radiative recombination rate is R_{nr}, then the internal quantum efficiency (η_{int}) is the ratio of the radiative recombination rate of the total recombination rate :

$$\eta_{int} = \frac{R_r}{R_r + R_{nr}} \qquad ...(3.4)$$

for exponential decay of excess carriers, the radiative recombination lifetime is $\tau_r = \frac{n}{R_r}$ and the non-radiative recombination lifetime is $\tau_{nr} = \frac{n}{R_{nr}}$.

Thus, the internal quantum efficiency can be expressed as

$$\eta_{int} = \frac{\eta/\tau_r}{\eta/\tau_r + n/\tau_{nr}}$$

$$\eta_{int} = \frac{1/\tau_r}{1/\tau_r + 1/\tau_{nr}}$$

$$\eta_{int} = \frac{\tau}{\tau_r} \qquad ...(3.5)$$

where τ is the bulk recombination lifetime, given by the expression :

$$\frac{1}{\tau} = \frac{1}{\tau_r} + \frac{1}{\tau_{nr}}$$

The power of radiated light as a function of forward current can be plotted in the form of a graph known as input-output characteristics of an LED, as shown in Fig. 3.8 (*a*). From Fig. 3.8 (*a*), it is clear that the greater the forward current, the greater the number of electrons that will be excited at conduction band and the greater the number of photons (light) that will be emitted. Hence, radiated light power of an LED is directly proportional to the forward current one expects the saturation effect as shown by dotted line in Fig. 3.8 (*a*). This takes at the point where all the available mobile electrons will be involved in radiation and further increasing the current value, will not produce additional Photons hence the emitted light, becomes constant.

Light Power, P (energy per second) is the number of photon (n_p) times the energy of an individual Photon, E_p.

$$P = \frac{n_p E_p}{t} \qquad \text{...(3.6)}$$

where N is the number of excited (injected) electrons and

$$n_p = N\eta_{int}$$

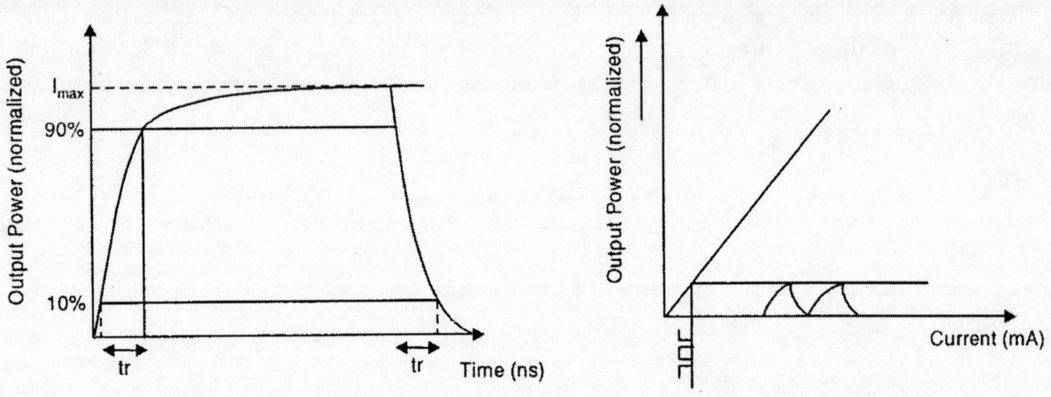

(*a*) Rise time and fall time (*b*) Modulation of pulse

Fig. 3.8. *Characteristics of an LED.*

Hence,
$$P = \frac{N\,\eta_{int} E_p}{t} \qquad \text{...(3.7)}$$

also, the current
$$I = \frac{Ne}{t} \qquad \text{...(3.8)}$$

Putting the value of N from Eq. (3.8) in Eq. (3.7).

Hence, the radiated light power is :

$$P = \left(\frac{It}{e}\right)\left(\frac{\eta_{int} E_p}{t}\right)$$

$$P = \left(\frac{\eta_{int} E_p}{e}\right) I \qquad \text{...(3.9)}$$

If E_p is measured in electron volts (eV) and I in milliamperes (mA), then

$$P(mw) = [\eta_{int}\ E_p\ (eV)]\ I\ (mA) \qquad \text{...(3.10)}$$

where, Light Power, P in milliwatts (mW)

Energy of individual photon in electron volt (eV)

Radiated current, I in milliampere (mA)

To find the emitted power, one needs to consider the external quantum efficiency (η_{ext}). This is defined as the ratio of the Photons emitted from the LED to the number of internally generated Photons. This can be expressed as

$$\eta_{ext} = \frac{1}{4\pi} \int_0^{\varphi_c} T(\varphi)(2\pi\pi \sin)\, d\varphi \quad ...(3.11)$$

where φ_c is the critical angle, φ is incidence angle, $T(\varphi)$ is the freshel transmission coefficient or resnel transmissivity.

The critical angle is given as

$$\phi_c = \sin^{-1}\left(\frac{n_2}{n_1}\right) \quad ...(3.12)$$

where n_1 is the refractive index of the semiconductor material and n_2 is the refractive index of the outside material.

For normal incidence, the freshel transmissivity is given as

$$T(0) = \frac{4\, n_1 n_2}{(n_1 + n_2)^2} \quad ...(3.13)$$

Assuming that the outside medium is air and *i.e.*, $n_2 = 1$ let $n_1 = n$

$$T(0) = \frac{4n}{(n+1)^2}$$

The external quantum efficiency is then approximately given by

$$\eta_{ext} = \frac{1}{n\,(n+1)^2} \quad ...(3.14)$$

The optical power emitted from the LED is given by

$$\boxed{P = \eta_{ext}\, P_{int}} \quad ...(3.15)$$

$$\boxed{P = \frac{P_{int}}{n\,(n+1)^2}} \quad ...(3.16)$$

3.7.8. Characteristics of LEDs

1. Lifetime, Rise and Fall Time

Lifetime τ of the charge carriers is the time between the moment they are excited (injected into a depletion region) and the moment they are recombined. This is also known as recombination lifetime and it ranges from nanoseconds to milliseconds. There are two recombination lifetimes : one is radiative (τ_r) and other is non-radiative (τ_{nr}). So, total carrier lifetime, τ is given as

$$\frac{1}{\tau} = \frac{1}{\tau_r} + \frac{1}{\tau_{nr}} \quad ...(3.17)$$

Rise/fall time, τ_r is defined as 10 to 90% of the maximum value of the pulse, as shown in Fig. 3.8 (*a*). For an LED, this characteristics shows how an output light pulse follows the electrical-modulating input pulse, which is illustrated in Fig. 3.8 (*b*).

Rise/fall time is determined by an LED's capacitance (C), input step current with amplitude (I_p) and the total recombination lifetime (τ).

$$\tau_r = 2.2\left[\tau + \frac{\left(1.7 \times 10^{-4} \times T^0 K \times C\right)}{I_p}\right] \quad ...(3.18)$$

where T^0K is absolute temperature.

3.7.9. Modulation Bandwidth

Bandwidth (BW), of an LED can be defined in either electrical or in optical terms. Normally, electrical terms are used since the bandwidth is actually determined via. the associated electrical circuitry. Thus, the optical bandwidth is defined as the point where the electrical signal power has dropped to half its constant value resulting from the modulated portion of the optical signal. This is the electrical –3dB point; that is frequency at which the output electrical power is reduced by –3dB with respect to the input electrical power as is illustrated in Fig. 3.9.

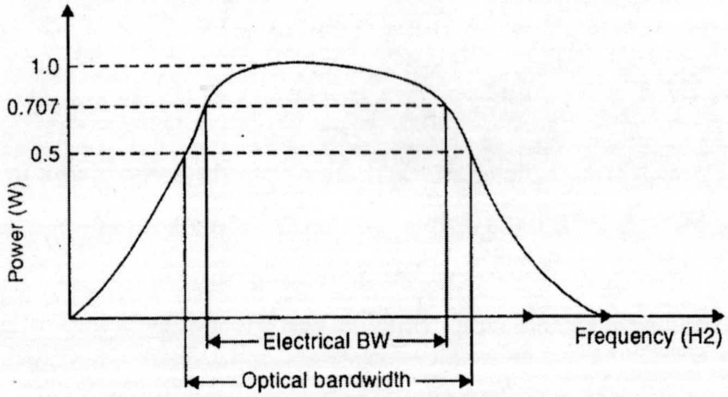

Fig. 3.9. *Frequency response of the optical source showing the electrical and optical bandwidth.*

In electronics, the general relationship between bandwidth and rise time is given as :

$$B\omega = \frac{0.35}{\tau_r} \quad ...(3.19)$$

But practical value of $B\omega$, does not match with the theoretical value, calculated from Eq. (3.19).

This discrepancy occurs because if the forward current is modulated at angular frequency, ω, an LED's output light intensity, $I(\omega)$, will vary as follows :

$$I(\omega) = I_0\,[1 + (\omega\tau)^2]^{-1/2} \quad ...(3.20)$$

where I_o is the LED's light intensity at constant current.

Detected electric power is proportional to I^2.

Taking $$\frac{I^2(\omega)}{I^2(0)} = \frac{1}{2}$$

Which is a –3dB decline. Hence, bandwidth becomes :

$$BW = \Delta\omega = \frac{1}{\tau} \quad ...(3.21)$$

From Eq. (3.21) an important conclusion can be made, that an LED's modulation bandwidth is limited by the recombination lifetime of the charge carriers.

Power-bandwidth product is another important characteristics of an LED. The product of LED's output power and its modulation bandwidth is constant.

$$\boxed{P \times BW = \text{Constant}} \quad ...(3.22)$$

This means that if we increase bandwidth, the output power will decrease and vice-versa.

3.7.10. Radiated Wavelength and Spectral Width

Radiated wavelength, referred as a peak position λ_p is determined by an energy gap, Eg, and spectral width, $\Delta\lambda$, is measured as fall width at half maximum (FWHM).

Both the Peak position and bandwidth are affected by the temperature. The peak moves to longer wavelength and the peak Broadens with rise in temperature. One can attribute the longer wavelength to a smaller energy gap which, in turn, can be attributed to less splitting between the bonding and antibonding orbitals as the atoms moves further apart.

The thermal peak broadening is due to a broadened thermal distribution of the electrons in the conduction band or the holes in the valence or both. The peak broadening is illustrated in Fig. 3.10.

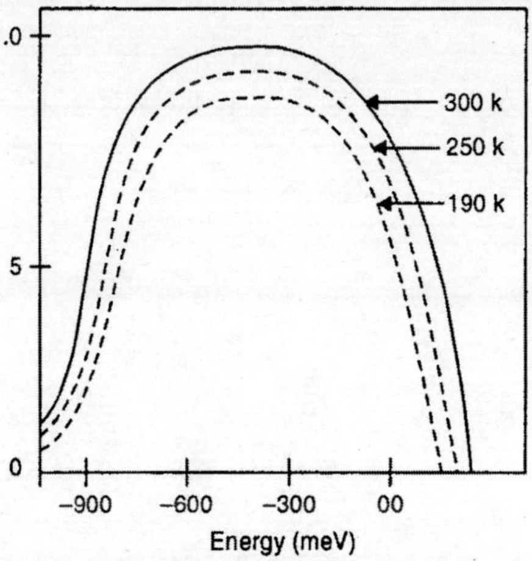

Fig. 3.10. *Thermal broadening (Relative intensity at various emission peak energy minus E_g at a given temperature).*

SELF-TEST (ZENER DIODE AND LED)

1. What do you mean by breakdown diode ?
2. List the different types of breakdown diode.
3. Describe the physical mechanism for zener breakdown diode.
4. What are the various breakdown mechanism in junction diode ? Give examples of at least one type, for each of these mechanism and elaborate their applications.
5. Qualitatively explain how mobility affects conductivity.
6. Distinguish between zener breakdown and avalanche breakdown.
7. What is the main application area of a zener diode ?
8. Breakdown diode with a breakdown voltage greater than 5 V generally have a positive temperature coefficient but those with lower breakdown voltage have a negative temperature coefficient. Do you agree with the statement ? Justify.

SOLVED EXAMPLES

Example 1. *When the reverse current in a particular zener diode increases from 20 mA to 30 mA, the zener voltage changes from 5.6 V to 5.75 V. What is the resistance of device ?*

Solution : Given :

$$\Delta I_Z = (30 - 20) \text{ mA} = 10 \text{ mA} = 10 \times 10^{-3} \text{ A}$$

and

$$\Delta V_Z = 5.75 - 5.6 = .15 \text{ V}$$

We know that the resistance of the zener diode,

$$r_Z = \frac{\Delta V_Z}{\Delta I_Z}$$

or

$$r_Z = \frac{0.15}{10 \times 10^{-3}}$$

or

$$r_Z = 15\ \Omega. \quad \textbf{Ans.}$$

Example 2. *A 4.7 V zener has a resistance of 15 Ω, what is the terminal voltage, when the current is 20 mA ?*

Solution : Given :

$$V_Z = 4.7 \text{ V}$$

$$r_Z = 15\ \Omega$$

$$I_Z = 20 \text{ mA}$$

$$V_Z' = ?$$

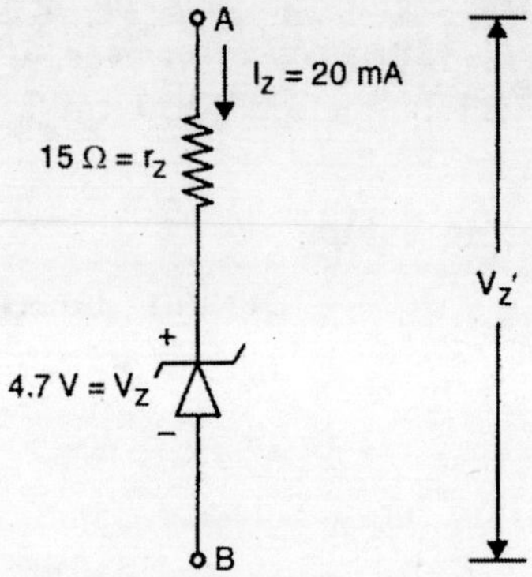

Fig. P (3.1).

The terminal voltage V_Z' is given by relation :

$$V_Z' = V_Z + I_Z r_Z$$

or $$V_Z' = 4.7 + 20 \times 10^{-3} \times 15$$

or $$V_Z' = 4.7 + 300 \times 10^{-3}$$

or $$V_Z' = 4.7 + .3$$

or $$V_Z' = 5 \text{ V.} \quad \textbf{Ans.}$$

Example 3. *A zener diode has a dc power dissipation rating of 500 mW and a zener voltage rating of 6.8 V. What is the value of I_{ZM} for the device ?*

Solution : Given : $P_{ZM} = 500 \text{ mW} = 500 \times 10^{-3}$ W

$V_Z = 6.8$ V

We know that the maximum current of zener diode is given by relation,

$$I_{ZM} = \frac{P_{ZM}}{V_Z}$$

$$= \frac{500 \times 10^{-3}}{6.8}$$

$$= 73.5 \times 10^{-3} \text{ A} = 73.5 \text{ mA.} \quad \textbf{Ans.}$$

NOTE : One thing always remember that while designing circuits using zener diodes, we must sure that the current through the zener diode does not exceed the I_{ZM} value.

Example 4. *A motorola's IN754 zener diode is being operated at 85°C. What is the maximum power dissipation for the device ? Assume that the dc power dissipation rating of the device is 1500 mW and the derating factor is 3.33 mW/°C above 60°C.*

Solution : Given : P_{ZM} = 1500 mW, and

Derating factor = 3.33 mW

We know that so long the zener diode is being operated at or below 60°C, its maximum power dissipation is 1500 mW. However, if it is so operated at 85°C, its maximum power dissipated will be below 1500 mW, because of the excessive heat generated inside.

We know that total derating value

$$= \text{Derating factor } (85 - 60)$$

$$= 3.33 \times (85 - 60)$$

$$= 83.25 \text{ mW}$$

Maximum power dissipation for the device,

$$P_{ZM} = 1500 - 83.25$$

$$= 1416.75 \text{ mW.} \quad \textbf{Ans.}$$

Example 5. *For the zener diode network of Fig. P (3.1), determine V_L, V_R, I_Z and P_Z.*

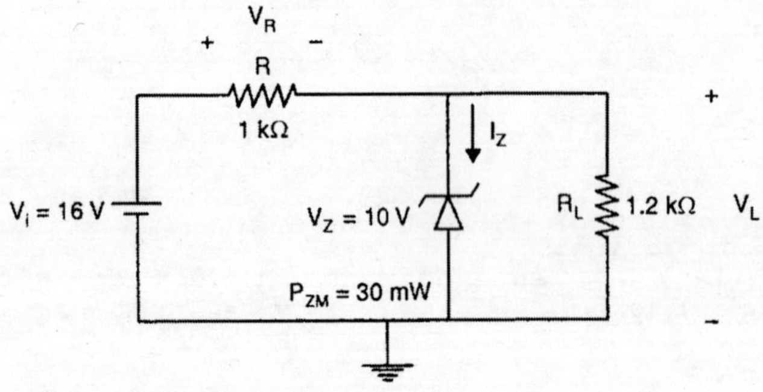

Fig. P (3.2).

Solution :

NOTE : Before solving this type of problem, consider two situation :

Situation 1 : Applying kVL to the given Fig. P (3.3).

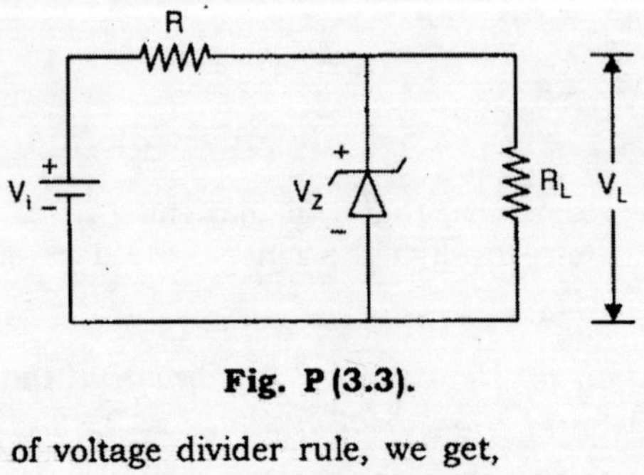

Fig. P (3.3).

Where the application of voltage divider rule, we get,

$$V_L = \frac{R_L . V_i}{(R_L + R)}$$

If $V_L > V_Z$, the zener diode is "on" and the equivalent model of Fig. P (3.2) can be substituted.

i.e., $$V_L = V_Z$$

Situation 2 : If $V_L \le V_Z$, the diode is "off" and the open circuit equivalent as shown in Fig. P (3.4) is substituted.

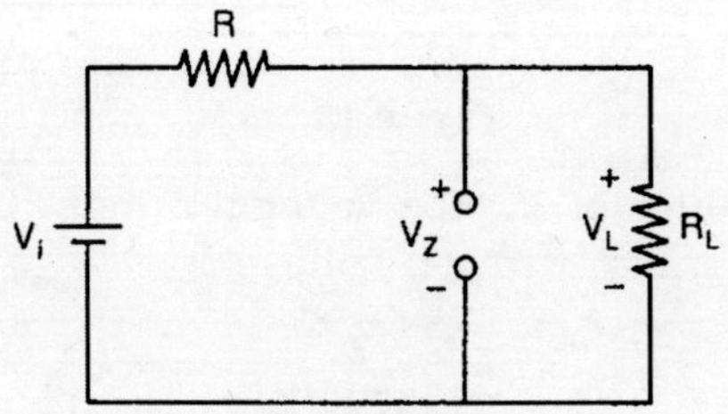

Fig. P (3.4).

In short, we can quickly memorize as shown below :

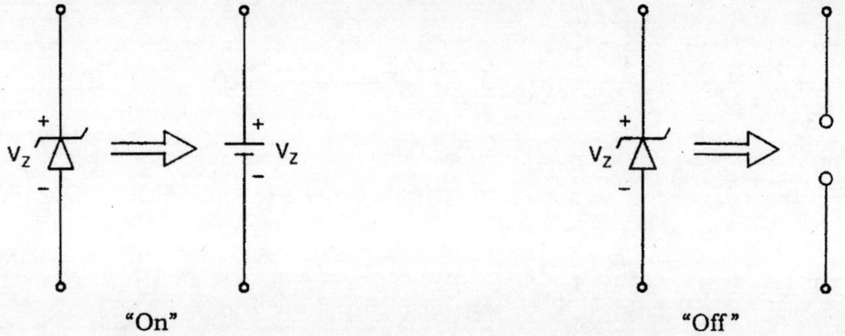

Now, from Fig. P (3.1), applying potential divider we get,

$$V_L = \frac{R_L . V_i}{R + R_L} = \frac{1.2\ k\Omega\ (16\ V)}{(1\ k\Omega + 1.2\ k\Omega)} = 8.73\ V$$

Since, $V = 8.73$ V is less than $V_Z = 10$ V, the diode is in "off" state. So, in order to calculate the parameter, substitute the open circuit equivalent.

$$V_L = 8.73\ V$$

$$V_R = V_i - V_L = 16 - 8.73 = 7.27\ V$$

$$I_Z = 0\ A$$

$$P_Z = V_Z I_Z = V_Z . 0 = 0\ W.$$ **Ans.**

Example 6. *Determine the range of value of V_i that will maintain the zener diode in the 'on' state as shown in Fig. P(3.5).*

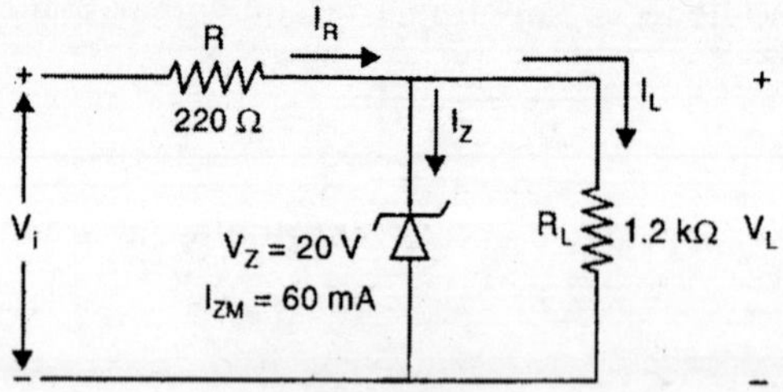

Fig. P (3.5).

Solution : For fixed values of R_L, the voltage V_i must be sufficiently large to turn the zener diode on. The minimum turn-on voltage $V_i = V_{i(min)}$ is determined by relation,

$$V_L = V_Z = \frac{R_L}{R_L + R} \cdot V_{i\,(min)}$$

or $$V_{i\,(min)} = \frac{(R_L + R)}{R_L} \cdot V_Z$$

or $$V_{i\,(min)} = \frac{(1.2 + .220)}{1.2} \cdot 20$$

or $$V_{i\,(min)} = 23.67 \text{ V}$$

However, the maximum value of V_i is limited by the maximum zener current I_{ZM}.

Since, $$I_{ZM} = I_R - I_L$$

$$I_{R\,(max)} = I_{ZM} + I_L$$

So, $$V_{i\,(max)} = V_{R\,(max)} + V_Z$$

or $$V_{i\,(max)} = I_{R\,(max)} \cdot R + V_Z$$

or $$V_{i\,(max)} = (I_{ZM} + I_L) \cdot R + V_Z$$

$$= \left(60 \text{ mA} + \frac{20 \text{ V}}{1.2 \text{ k}\Omega}\right) \cdot (0.22 \text{ k}\Omega) + 20$$

$$= 16.87 + 20 = 36.87 \text{ V. } \textbf{Ans.}$$

Example 7. *Check whether the zener diode of Fig. P (3.6) operating in the Breakdown region.*

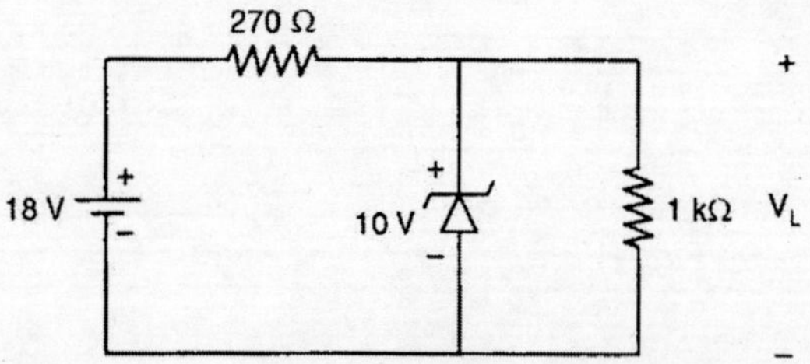

Fig. P (3.6).

Solution : The zener diode will be operated in the Breakdown region if $V_L > V_Z$. Applying potential divider method, we get,

$$V_L = \frac{1\ \text{k}\Omega}{(270\ \Omega + 1\ \text{k}\Omega)} \cdot 18\ \text{V}$$

or $$V_L = 14.2\ \text{V}$$

Since, V_Z (14.2 V) > V_Z (10 V)

Hence, the zener diode is operating in the Breakdown region. **Ans.**

Example 8. *Calculate the value of zener current as shown in Fig. P(3.7).*

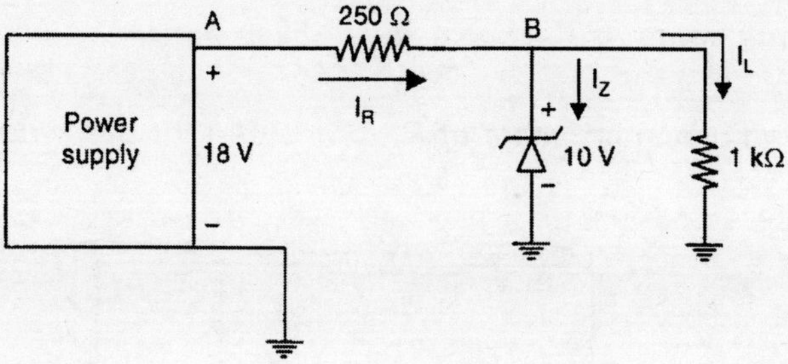

Fig. P (3.7).

Solution : From figure,

$$I_R = \frac{18-10}{250} = \frac{8}{250} = 32\ \text{mA}$$

$$I_L = \frac{10}{1\ \text{k}\Omega} = 10\ \text{mA}$$

Apply KCL at node *B*. We get,

$$I_R = I_L + I_Z$$

$$32 = 10 + I_Z$$

or $$I_Z = 32 - 10 = 22\ \text{mA.}\ \textbf{Ans.}$$

Example 9. *Find the current flowing through the resistance R as shown in Fig. P (3.8).*

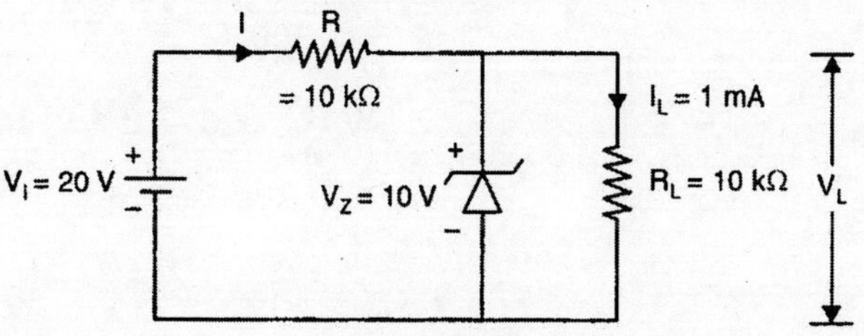

Fig. P (3.8).

Solution : First of all we will calculate V_L

$$V_L = \left(\frac{10}{10+10}\right) \cdot V_i = \frac{10}{20} \cdot 20 = 10 \text{ V}$$

Since, $V_L \geq V_Z = 10$ V, therefore, the zener diode is on. So,

$$I = \frac{V_I - V_Z}{R} = \frac{(20-10) \text{ V}}{10 \text{ k}\Omega} = 1 \text{ mA.} \quad \textbf{Ans.}$$

Example 10. *(a) Determine V_L, I_L, I_Z and I_R for the network shown in Fig. P(3.9), if $R_L = 300\ \Omega$.*

(b) Determine the value of R_L that will establish maximum power conditions for the zener diode.

(c) Determine the minimum value of R_L to ensure that the zener diode is in the "on" state.

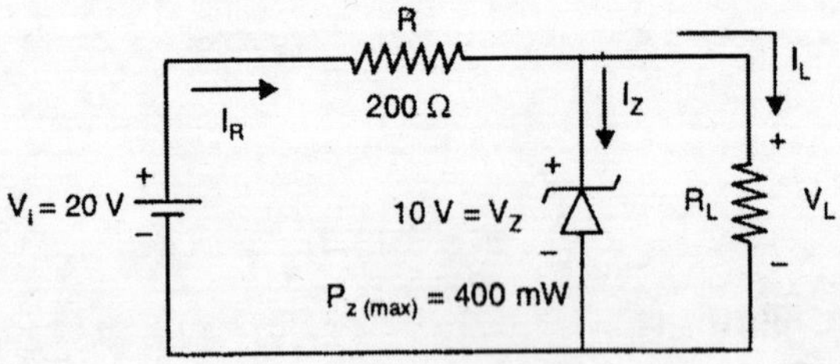

Fig. P (3.9).

Solution : Given : $R_L = 300\ \Omega$

(a) $$V_L = \frac{300}{300+200} \cdot 20$$

or $$V_L = \frac{3}{5} \cdot 20 = 12 \text{ V}$$

Since $V_L > V_Z$, it means zener diode is "on".

So, $$V_L = V_Z = 10 \text{ V}$$

$$I_L = \frac{V_L}{R_L} = \frac{10}{300} = 33.333 \text{ mA}$$

$$I_R = \frac{V_i - V_Z}{R} = \frac{20-10}{200} = \frac{10}{200} = 50 \text{ mA}$$

$$I_Z = I_R - I_L$$

or $$I_Z = 50 - 33.33$$

$$I_Z = 16.67 \text{ mA.} \quad \textbf{Ans.}$$

(*b*) In order to ensure that the zener diode is in the "on" state.

$$V_L \geq V_Z$$

$$\frac{R_L}{R_L + R} \cdot V_I \geq V_Z$$

$$\frac{R_L}{R_L + 200} \cdot 20 \geq 10$$

$$2R_L \geq R_L + 200$$

$$R_L \geq 200$$

$$(R_L)_{min} = 200\ \Omega. \quad \textbf{Ans.}$$

NOTE : From above result we conclude that the minimum value of R_L to ensure that the zener diode is in the "on" state is equal to the value of R.

Example 11. *The circuit diagram of a regulated power supply is shown in Fig. P (3.10). The zener diode characteristic is given by $V_Z = 8 + RI_Z$, where V_Z and I_Z are the diode voltage and current respectively and $R_L = 10$ ohms. If the maximum rated current of zener diode is 0.2 A. Calculate the minimum value of R_i.*

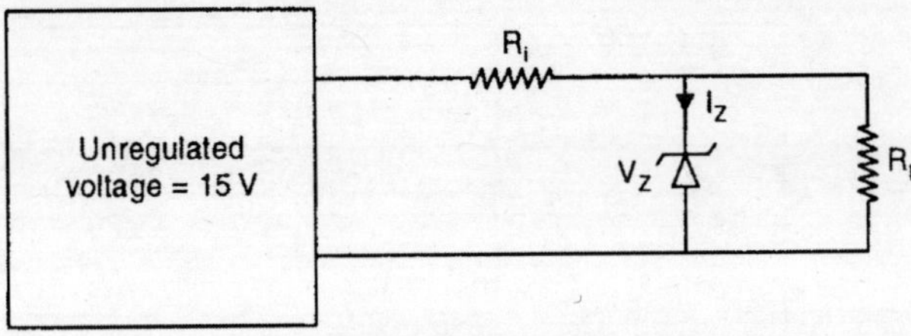

Fig. P (3.10).

Solution : For maximum current of 0.2 A, the voltage of zener diode is,

$$V_Z = 8 + RI_Z$$

$$= 8 + 10 \times 0.2$$

$$= 10 \text{ V}$$

This maximum current is drawn where the regulated voltage of 15 V is regulated to 10 V. So, the load current is

$$I_{L(min)} = \frac{10}{10} = 1 \text{ A.}$$

Given : $$I_{Z(max)} = 0.2 \text{ A}$$

so, $$R_{i(min)} = \frac{V_{i(max)} - V_Z}{I_{Z(max)} + I_{L(min)}}$$

$$= \frac{15 - 10}{0.2 + 1.0} = \frac{5}{1.2} = 4.7\ \Omega. \quad \textbf{Ans.}$$

Example 12. *For the circuit of the following figure, find (a) the output voltage, (b) voltage drop across RS and (c) the current through the zener diode.*

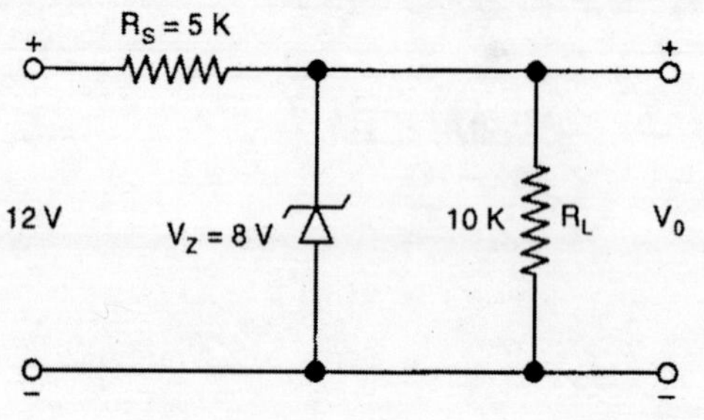

Fig. P (3.11).

Solution : While trouble shooting Zener regulator circuits, the first thing to check for its breakdown operation of the zener diode. Because of the load resistor, the Thevenin voltage driving the zener diode is less than the source voltage. See a zener regulator act as below :

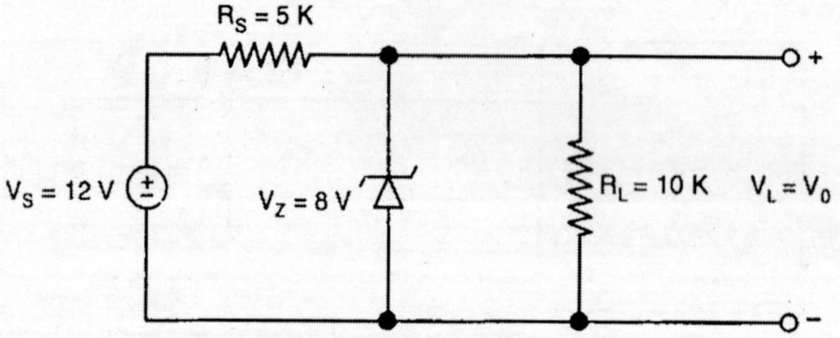

Fig. P (3.12).

Here, output voltage,

$$V_L = \frac{R_L}{R_s + R_L} \times V_S$$

or $$V_L = \frac{10}{5+10} \cdot 12$$

or $$V_L = 8 \text{ V}$$

For breakdown operation the zener diode must be greater than V_Z.

But, here, $$V_L = V_Z = 8 \text{ V}$$

So, zener diode will not be in the breakdown condition. Hence, no current will flow through it and it will behave as open circuited as shown in following circuit.

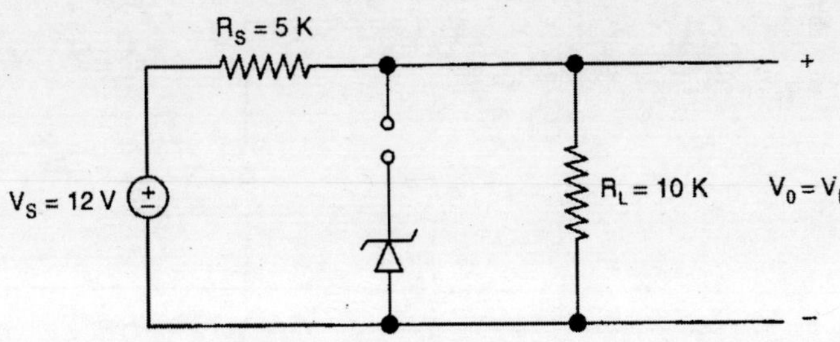

Fig. P (3.13).

So, (*a*) Output voltage $V_0 = V_L = 8$ V

(*b*) Voltage drop across $R_S = V_S - V_0 = 12 - 8 = 4$ V.

(*c*) Current through zener diode $I_Z = 0$. **Ans.**

Example 13. *Determine V_Z, I_Z, P_Z for the following circuit :*

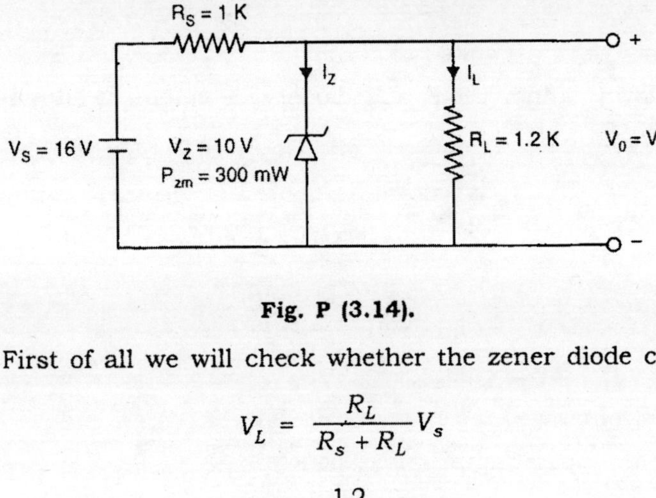

Fig. P (3.14).

Solution : First of all we will check whether the zener diode conduct or not.

$$V_L = \frac{R_L}{R_s + R_L} V_s$$

or $$V_L = \frac{1.2}{1+1.2} \times 16$$

or $$V_L = 8.72 \text{ V}$$

as $V_L < V_Z$, so diode will not conduct.

No, current will flow through diode.

i.e., $$I_Z = 0$$

$$P_Z = 0$$

$$V_0 = V_L = 8.72 \text{ V. } \textbf{Ans.}$$

Example 14. *For the circuit shown in Fig. P (3.15), determine V_L, I_L, I_Z and I_R with R_L = 200 ohm and R_L = 200 ohm. Comment on the operation of the circuit. V_{in} = 20 V, V_z = 10 V, $P_{Z, max}$ = 400 mW and R_S = 220 ohm.*

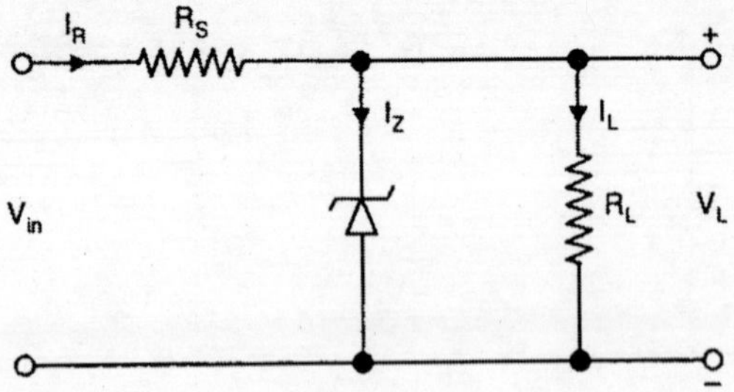

Fig. P (3.15).

Solution : When $R_L = 200$.

Checking for the operation of zener diode.

The diode is open circuited and V_L is calculated

$$V_L = \frac{R_L}{R_S + R_l} V_{in} = \frac{200}{220 + 200} \times 20 = 9.52 \text{ volts.}$$

as this value $V_L < V_z$ *i.e.*, 10 volts.

Zener breakdown cannot occur and diode will remain in off state so

$$I_Z = 0$$

and $$I_R = I_L = \frac{V_{in}}{R_S + R_L} = \frac{20 \text{ V}}{(220 + 200) \text{ ohms}} = \frac{20}{420} \text{ Amp.}$$

so, $$I_R = I_L = 47.62 \text{ m amp.}$$

$$V_L = 9.52 \text{ volts.}$$

$$I_Z = 0.$$

When $R_L = 50$ ohm.

Checking for the operating of zener diode

$$V_L = \frac{R_L}{R_S + R_L} V_{in} = \frac{50}{220 + 50} \times 20 = 3.7 \text{ volts}$$

Which is less than $V_Z = 10$ volts.

So, zener diode remains off. Hence, $I_Z = 0$

and $$I_R = I_L = \frac{V_{in}}{R_S + R_L} = \frac{20}{220 + 50} = 74.1 \text{ m amp.} \quad \textbf{Ans.}$$

$$I_Z = 0. \quad \textbf{Ans.}$$

$$V_L = 3.7 \text{ volts.} \quad \textbf{Ans.}$$

Example 15. *The circuit in Fig. P (3.16) shows a zener-regulated d.c. power supply. The zener diode voltage is 15V. The minimum value of R_L down to which the output voltage remains constant is*

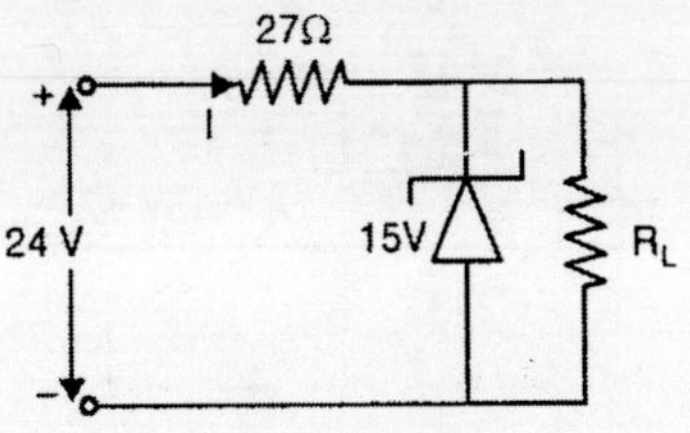

Fig. P (3.16).

Solution : From given Fig. P (3.16)

$$V_Z = 15 \text{ V}$$

$$I = \frac{24 - V_Z}{27} = \frac{24 - 15}{27} = \frac{1}{3} \text{A}$$

$$R_{L(min)} = \frac{V_Z}{I_{L(max)}} = \frac{15}{1/3} = 45 \ \Omega \textbf{ Ans.}$$

Example 16. *The 6V zener diode shown in Fig. P (3.17) has zero zener resistance and Knee current of 5 mA. The minimum value of R so that voltage across it does not fall below 6V.*

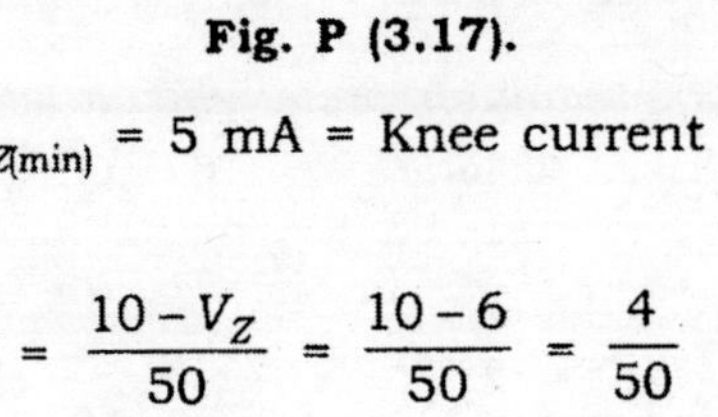

Fig. P (3.17).

Solution : Given that $I_{Z(min)}$ = 5 mA = Knee current from Fig. P (3.17) current in 50Ω resistance, say I

$$I = \frac{10 - V_Z}{50} = \frac{10 - 6}{50} = \frac{4}{50}$$

$$I = 80 \text{ mA}$$

$$\therefore \quad I = I_Z + I_L$$

$$I_{R(max)} = I - I_{Z(min)} = 80 - 5 = 75 \text{ mA}$$

Now, $$R = \frac{V_Z}{I_{R(max)}} = \frac{6\text{V}}{75\text{mA}} = 80 \ \Omega. \textbf{ Ans.}$$

Example 17. *In the voltage regulator shown in Fig. P (3.18). The power dissipation in the zener diode will be*

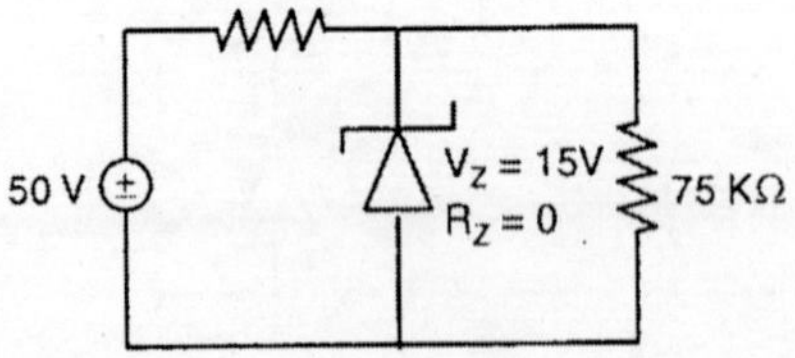

Fig. P (3.18).

Solution : From Fig. P (3.18), we have $R_Z = 0$. So, in order to calculate the power dissipation across the zener diode. We must calculate the current flowing in zener diode *i.e.,* I_Z.

And I_Z is given by Relation.

$$I_Z = \frac{V_{th}}{R_{th}} \quad \text{where, } V_{th} = \text{therenin voltage;}$$

$$R_{th} = \text{therenin resistance}$$

From Fig. P (3.18)

$$V_{th} = 50 \times \frac{75\text{K}\Omega}{(75+150)\text{K}\Omega} = \frac{50}{3}\text{V}$$

$$R_{th} = (150 \| 75)\text{K}\Omega = 50\ \text{K}\Omega$$

$$I_Z = \frac{50/3\text{V}}{50\text{K}\Omega} = 0.33\ \text{mA.}$$

Now,

Power dissipation, $P_Z = V_Z \cdot I_Z$

or $P_Z = 15 \times 0.33 = 0.5$ mW **Ans.**

Example 18. *In the voltage regulator circuit shown below in Fig. P (3.19). The power rating of zener diode is 400 mW. Calculate the value of* R_L *that will establish maximum power in zener diode.*

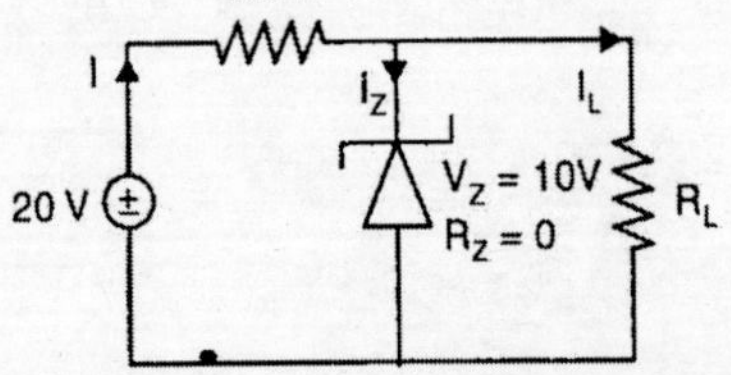

Fig. P (3.19).

Solution : From Fig. P (3.19)

$$I_{Z(max)} = \frac{P}{V_Z}$$

or $$I_{Z(max)} = \frac{400mW}{10V} = 40 \text{ mA}$$

$$\therefore \quad I = I_Z + I_L = \frac{20 - V_Z}{222} = \frac{20 - 10}{222} = 45 \text{ mA}$$

$$I_{L(min)} = 45 - I_{Z(max)}$$
$$= 45 - 40 = 5 \text{ mA}$$

Now,

$$R_L = \frac{V_Z}{I_{L(min)}} = \frac{10V}{5mA} = 2k\Omega.$$

Example 19. *In the voltage regulator circuit shown below in Fig. P (3.20). The zener diode current is to be limited to the range $5 \le I_Z \le 100$ mA.*

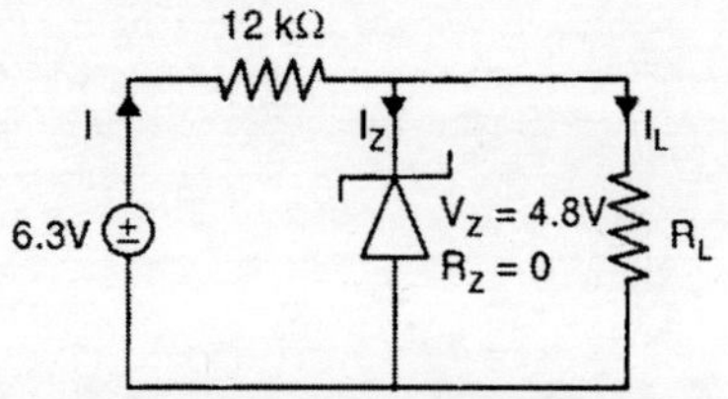

Fig. P (3.20).

Calculate (i) The range of possible load currnet.

(ii) The range of possible load resistance.

(iii) The power rating required for load resistance.

Solution : (i) From Fig. P (3.20)

$$I = \frac{6.3 - V_Z}{12 \text{ k}\Omega} = \frac{6.3 - 4.8}{12 \text{ k}\Omega} = \frac{1.5 \text{ V}}{12 \text{ k}\Omega} = \frac{0.5}{4} \text{mA} = 125 \text{ mA}$$

The range of I_L :

when, $I_Z = 5$ mA

$I_L = I - I_Z = 125 - 5 = 120$ mA

when, $I_Z = 100$ mA

$I_L = 125 - 100 = 25$ mA

Hence the range of I_L is

$25 \text{ mA} \le I_L \le 120$ mA **Ans.**

(*ii*) The range of possible load resistance

$$R_L = I_L \cdot V_Z$$

$$R_L = 25 \times 10^{-3} \times 4.8 = .120\ \Omega \qquad \text{(when } I_L = 25 \text{ mA)}$$

$$R_L = 120 \times 10^{-3} \times 4.8 = .576\ \Omega \qquad \text{(when } I_L = 120 \text{ mA)}$$

$.120\ \Omega \leq R_L \leq .576\ \Omega$ **Ans.**

(*iii*) Power rating, $P_{Z(\max)} = I_{L(\max)} \times V_Z$

$$= 120 \times 10^{-3} \times 4.8$$

$= 5.76$ mW. **Ans.**

□□□

UNIT—II

(Chapters 4 and 5)

At the end of this unit you will able to learn about the :

- Different types of transistor configurations
- Concept of D.C. and A.C. load line
- Biasing
- *h*-parameters
- Voltage gain, current gain
- Concept of feedback
- Oscillators

4

BIPOLAR JUNCTION TRANSISTOR

4.1. INTRODUCTION

The transistor was invented in 1948 by John Bardeen, Walter Brattain and William Shockley at Bell Laboratory in America. The invention of transistor completely revolutionized the electronic industry. Since then, there has been a rapidly expanding effort to utilize and develop many types of semiconductor devices such as FET, MOSFET, UJT, SCR etc. transistors have replaced bulky vacuum tubes in performing many jobs.

When a third doped element is added to crystal diode in such a way that two PN-junctions are formed, the resulting device is known as *transistor.* The transistor—an entirely new type of electronic device is capable of achieving amplification of weak signals.

4.2. TRANSISTOR

A transistor consists of two *p-n* junctions formed by sand-wiching either *p*-type or *n*-type semiconductor between a pair of opposite types. There are two types of transistors possible namely :

(*i*) *n-p-n* transistor

(*ii*) *p-n-p* transistor.

An *n-p-n* transistor is composed of two *n*-type semiconductors separated by a thin section of *p*-type as shown in Fig. 4.1(*a*). However, a *p-n-p* transistor is formed by two *p*-sections separated by a thin section of *n*-type as shown in Fig. 4.1(*b*).

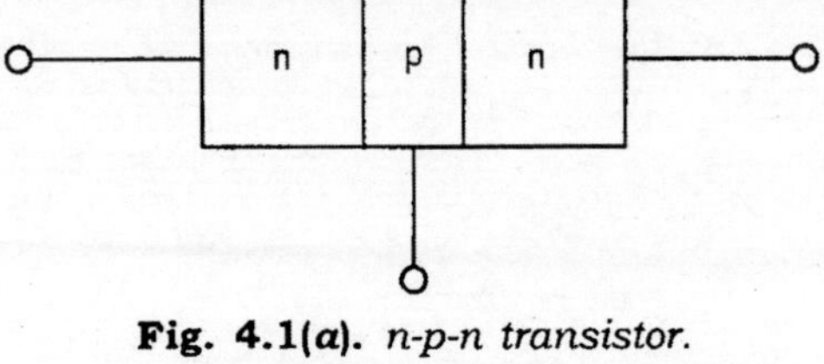

Fig. 4.1(*a*). *n-p-n transistor.*

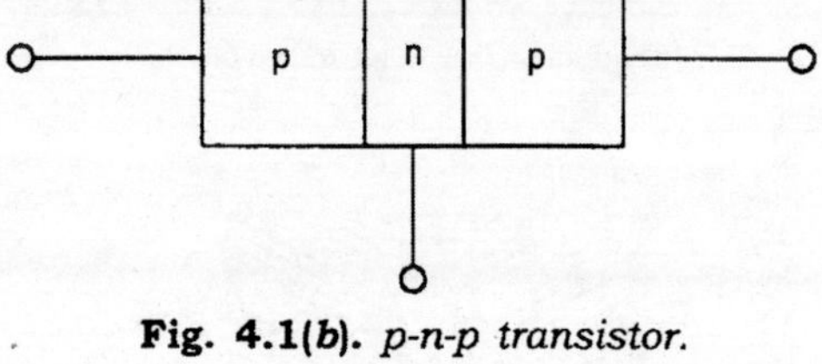

Fig. 4.1(*b*). *p-n-p transistor.*

Some points regarding transistor :

(1) There are two PN-junctions. Therefore a transistor may be regarded as a combination of two diodes connected back to back.

(2) The middle section is very thin layer. This is the most important factor in the function of a transistor.

Meaning of the Transistor

As the transistor has two PN-junction. As discussed latter, one junction is forward biased and the other junction is reversed biased. The forward biased junction has a low resistance path whereas a reverse biased junction has a high resistance path. The weak signal is introduced in the low resistance, circuit and output is taken from the high resistance circuit. Therefore, a transistor *transfers* a signal from a low resistance to high resistance. The prefix 'trans' means the signal transfer property of device while 'istor' classifies it as a solid element in the same general family with resistors.

i.e., Transfer + Resistor = Transistor

4.3. BASIC CONSTRUCTION OF A JUNCTION TRANSISTOR

A transistor is basically a silicon or germanium crystal containing three separate regions. Transistor may be either *npn*-type or *pnp*-type. Fig. 4.2(*a*) shows an *npn*-transistor. It has three region namely :

(1) Emitter

(2) Base

(3) Collector

The section on one side is the *Emitter* and the section on the opposite side is the *collector.* The middle section is called the *base* and forms two junctions between the emitter and the collector. One junction is between the emitter and the base, and is called emitter-base junction or simply the emitter junction. The other junction is between the base and the collector, and is called the collector base junction, or simply collector junction. Thus, a transistor is like two PN-junction diode connected back-to-back.

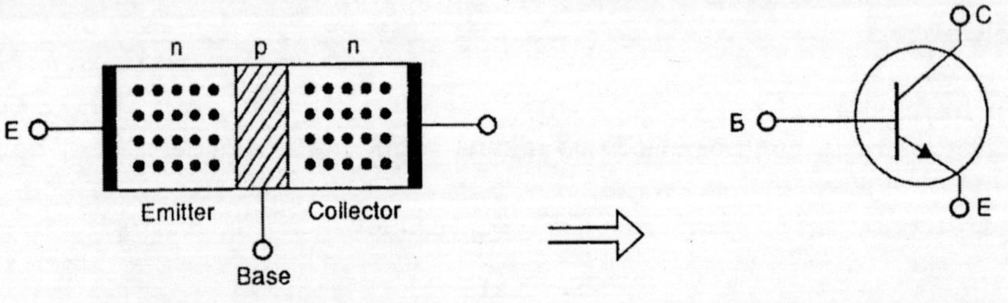

(*a*) Junction transistor *npn*-type.　　(*b*) Symbol of *npn*-transistor.

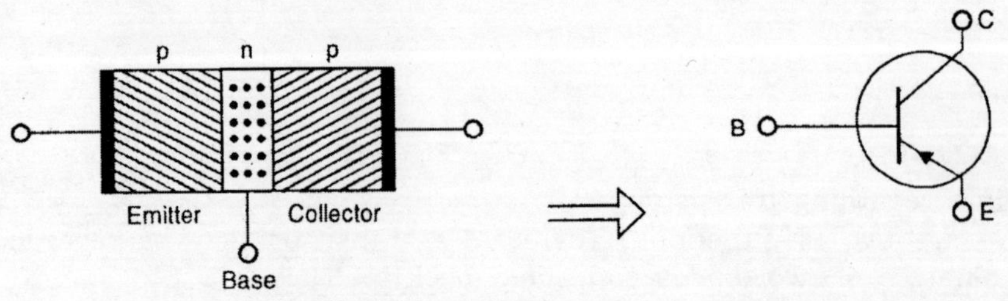

(*c*) Junction transistor *pnp*-type.　　(*d*) Symbol of *pnp*-transistor.

Fig. 4.2. *Junction transistor of npn- and pnp-type.*

Now we will discuss in brief about :

(*a*) **Emitter.** The section on one side of the junction that supplies charge carriers (*i.e.*, electrons or holes) is called the *emitter.* The emitter is always forward biased with respect to base so as to supply a large number of majority carriers. Fig. 4.3 shows the arrangement of forward biased for *npn* as well *pnp*-transistor. In the case of *npn*-transistor forward biased, emitter supplies electrons to the junction whereas in the case of *pnp*-transistor forward biased emitter supplies holes to the junction. It may be noted that emitter is a heavily doped region.

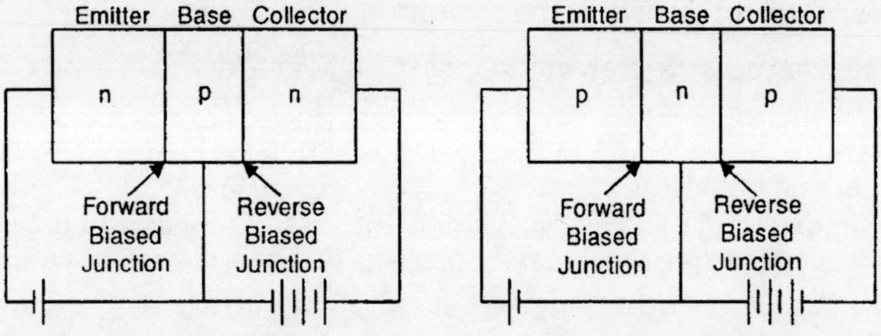

Fig. 4.3. *Forward and reverse bias in different types of transistor.*

(*b*) **Base.** The middle section which forms two *pn*-junctions between the emitter and collector is called the *base.* The base-emitter junction is forward biased, allowing low resistance for the emitter circuit. The base-collector junction is reverse biased and provides high resistance in the collector circuit. The base is very lightly doped and very thin so that all charge carriers easily passed through the junctions.

(*c*) **Collector.** The section on the one side of junction that do the job of collecting or gathering these charge carriers (electrons or holes) from the base is called collector. The function of collector is to remove charges from its junction with the base. The collector is always reversed biased with respect to base. The doping of the collector is between the heavy doping of the emitter and light doping of the base. It may be noted that the collector of a transistor is always larger than the emitter and base of a transistor.

Reason why collector region is larger than that of the emitter and base :

As we know that the main function of collector is to collect the charge carriers in the collector region. Because due to the interaction of the charge carrier, the collector region becomes heat up, as all the heat dissipates through this region and the transistor may burn out. So, in order to save the transistor from the burning the collector is always larger than that of the emitter and base region.

4.4. TRANSISTOR ACTION (OPERATION OF TRANSISTOR)

As we have already discussed that a transistor has two junctions emitter-base junction and collector-base junction. There are four possible ways of biasing these two junctions. Table 4.1 shows all four combination with their region (will be discussed later) of operation.

Table 4.1

S.No.	Emitter-Junction	Collector-Junction	Region of Operation
1.	Forward-biased	Reversed-biased	Active
2.	Forward-biased	Forward-biased	Saturation
3.	Reverse-biased	Reverse-biased	Cutoff
4.	Reverse-biased	Forward-biased	Inverted

NOTE : The modes of operation are the same for both NPN and PNP transistors. However, in PNP transistor biasing (which is discussed later), the polarity of battery voltages are opposite to that of NPN transistor.

Operation (Working) of an NPN Transistor

Fig. 4.4 shows an NPN transistor biased in forward-active mode *i.e.*, the emitter base junction of a transistor is forward biased while the collector-base junction of a transistor is reverse biased. The emitter-base junction is forward biased only if V_{EB} is greater than barrier potential or cut in voltage which 0.7 V for silicon and 0.3 V for Germanium transistors. The forward bias on the emitter-base junction causes the free electrons in the N-type emitter to move towards the base region. This constitutes the emitter current (I_E). Due to the forward biased of the junction, the barrier potential is reduced. Holes move from base to the emitter region. Thus the resulting current consists of electrons and holes. However, it may be noted that only the electron current is useful in the action of the transistor as the doping of the base region is more lightly than emitter region which gives very low quantity of hole current.

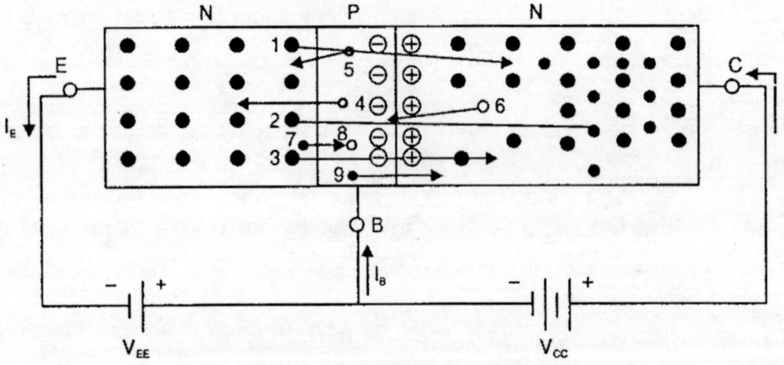

Fig. 4.4. *Operation of an npn-transistor.*

From Fig. 4.4 we see that electrons 1, 2, 3 crossing from the emitter to the base, and hole 4, 5 from the base to the emitter. The total sum of these charge-carrier movements constitutes the emitter current I_E electrons 1, 2, 3 are injected by the emitter into base. The ratio of the electron current to the total emitter current known as *emitter injection ratio* as the *emitter efficiency*. Typically this ratio is equal to 0.995. Once the electrons are injected by the emitter into the base, now these electrons become the minority carriers in the base region. These electrons do not have separate identities from those which are thermally generated in the base region itself. As the base region is very narrow and very lightly doped, because of this, most of the minority carriers (electrons) travelling from the emitter end of the base region to its collector end do not recombine with holes in this journey. Only a few electrons (like 7) may recombine with holes (like 8). The ratio the number of electrons arriving at collector to the number of emitted electrons is known as the *base transportation factor.* Typically its value is 0.995.

Referring to Fig. 4.4 movement of hole (like 6) from the collector region and electron 9 from the base region constitutes leakage current, I_{CBO}. Actually, the number of electrons and holes crossing the emitter-base junction is much more than the number of electrons and holes crossing the collector-base junction. The difference of these two currents in the base region makes the base current I_B.

The collector current is less than the emitter current, there are two reasons for this. Firstly a part of emitter current consists of holes that do not contribute to the collector current. Secondly, not all the electrons injected into the base are successful in reaching the collector. The ratio of collector current to the emitter current is called the d.c. alpha (α_{dc}) of the transistor.

Relation between different currents in a transistor

The emitter current is equal to the sum of the collector and base currents, *i.e.*,

$$\boxed{I_E = I_B + I_C} \quad \text{... (4.1)}$$

NOTE : (*i*) The emitter current of a transistor consists of two components namely base current and collector current. The base current is about 2% of the emitter current, while collector current is about 98% of the emitter current.

(*ii*) The collector current is mainly due to the electrons injected from the emitter. However, there is another small component of collector current due to thermally generated carriers. This small value of collector current is called leakage current or reverse saturation current.

4.5. DIFFERENT TYPES OF TRANSISTOR CONFIGURATION

There are three types of transistor configuration possible, namely :

(*i*) Common base (CB) configuration

(*ii*) Common emitter (CE) configuration

(*iii*) Common collector (CC) configuration

Here it is interesting to note that transistor may be connected in any one of the three configuration stated above. Now, we will discuss these three configuration briefly one by one.

(*i*) **Common Base (CB) Configuration**

The circuit diagram of *npn*-transistor and *pnp*-transistor for common base configuration is shown below in Fig. 4.5 (*a*) and (*b*) respectively. In this case the input is connected between emitter and base. However output is taken across collector and base. Thus the base of transistor is common to both input and output circuits and hence the name common base configuration.

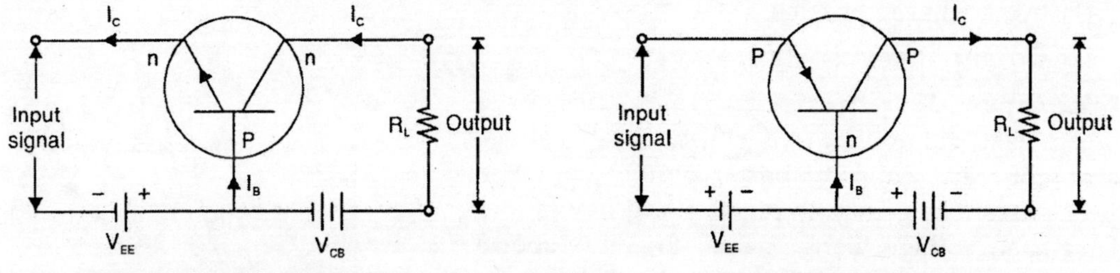

(*a*) Common base configuration with *npn* transistor

(*b*) Common base configuration with *pnp* transistor

Fig. 4.5.

Emitter current amplification factor (α)

Current amplification factor (α) may be defined as the ratio of output current to the input current.

Or in other words *the ratio of change in collector current to emitter current at constant collector base voltage (V_{CB}) is known as current amplification factor of transistor in common base configuration. It is generally represented by Greek letter.*

i.e., $$\boxed{\alpha = \frac{\Delta I_C}{\Delta I_E}} \quad \text{at constant } V_{CB}$$

We know that $I_E = I_C + I_B$ or $\Delta I_E = \Delta I_C + \Delta I_B$, then from the above relation it is clear that value of current amplification factor is always less than or equal to unity. The value of α approaches to unity if value of I_B reduces to zero. This can be achieved by doping the base lightly and making very thin.

Total collector current of a common base (CB) configuration

During the recombination process very small percentage of electrons combine with holes in the base region. Due to this, whole of the emitter current could not reach the collector. On the other hand, the collector current is slightly increased because of leakage current that flows due to minority carriers as the collector-base junction is reversed biased. It means the total collector current consists of :

(*i*) A large value of emitter current that reaches the collector terminals *i.e.*, $\alpha\, I_E$.

(*ii*) The leakage current (I_{CBO}) means current between collector and base at open emitter terminal. This current is due to movement of minority carriers across the collector-base junction as the junction is heavily reverse biased. This value is much smaller than αI_E.

Thus, total collector current

$$I_C = \alpha\, I_E + I_{\text{leakage}}$$

or $$I_C = \alpha\, I_E + I_{CBO}$$

Here, I_{CBO} (leakage current) is due to minority charge carriers *i.e.*, collector base junction open emitter (or $I_E = 0$).

Characteristic of common base (CB)

There are two important characteristics of CB. Configuration namely :

(1) Input characteristics.

(2) Output characteristics.

1. Input Characteristics. It is the curve between base current I_B and emitter-base voltage V_{BE} at constant collector-base voltage V_{CB}. The emitter current is generally taken along *y*-axis and emitter-base voltage along *x*-axis. Fig. 4.6 shows the input characteristics of a typical transistor in CB arrangement. The following points may be noted from these characteristics :

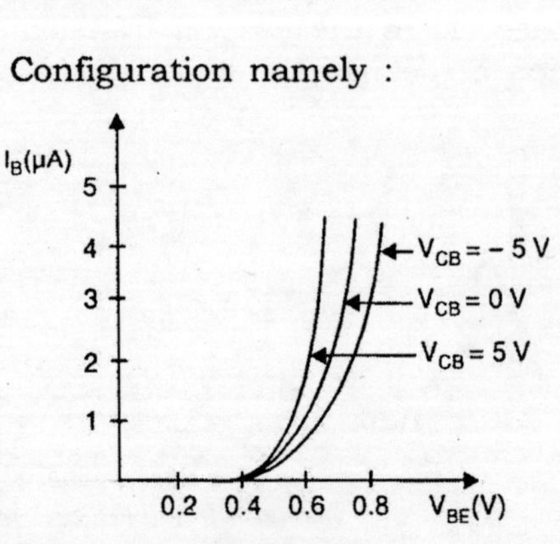

Fig. 4.6.

(*i*) The emitter current I_E increases rapidly with small increase in emitter-base voltage V_{EB}. It means that input resistance is very small.

(*ii*) The emitter current is almost independent of collector-base voltage V_{CB}. This leads to the conclusion that emitter current (and hence collector current) is almost independent of collector voltage.

NOTE : In fact, input resistance is the opposition offered to the signal current. As a very small V_{EB} is sufficient to produce a large flow of emitter current I_E, therefore, input resistance is quite small, of the order of a few ohms.

2. Output Characteristic. It is the curve between collector current I_C and collector-base voltage V_{CB} at constant emitter current I_E. Generally, collector current is taken along *y*-axis and collector-base voltage along *x*-axis. Fig. 4.7 (*a*) shows the output characteristics of a typical transistor in CB arrangement.

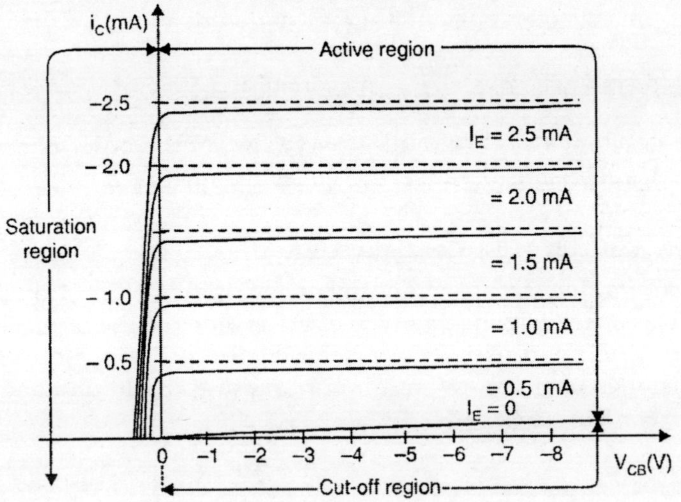

Fig. 4.7 (*a*).

The following points may be noted from the characteristics :

(*a*) The collector current I_C varies with V_{CB} only at very low voltages (< 1V). The transistor is *never* operated in this region.

(*b*) When the value of V_{CB} is raised above 1 – 2 V, the collector current becomes constant as indicated by straight horizontal curves. It means that now I_C is independent of V_{CB} and depends upon I_E only. This is consistent with the theory that the emitter current flows *almost* entirely to the collector terminal. The transistor is *always* operated in this region.

(*c*) The collector current is not zero when I_E = 0. It has a very small value. This is the reverse leakage current I_{CO}. The conditions that exist when I_E = 0 for CB configuration as shown in Fig. 4.7(*b*). The notation must frequently used for I_{CO} is I_{CBO}.

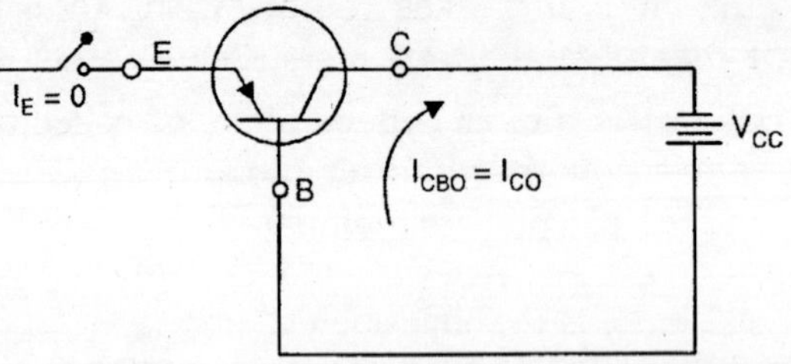

Fig. 4.7 (b). *Reverse leakage current in CB configuration.*

The meaning of I_{CBO} is that the current due is collector and base when emitter is open-circuited.

(*d*) A very large change in collector-base voltage produces only a tiny change in collector current. This means that output resistance is very high.

Output resistance. It is the ratio of change in collector-base voltage (ΔV_{CB}) to the resulting change in collector current (ΔI_C) at constant emitter current *i.e.*,

$$\text{Output resistance, } r_o = \frac{\Delta V_{CB}}{\Delta I_C} \text{ at constant } I_E$$

The output resistance of *CB* circuit is very high, of the order of several tens of kilo-ohms. This is not surprising because the collector current changes very slightly with the change in V_{CB}.

(*ii*) **Common Emitter (CE) Configuration**

The common emitter circuit arrangement for *npn* and *pnp* transistor is shown in Fig. 4.8(*a*) and (*b*) respectively. In this type of arrangement, the input is applied between base and emitter while output is taken between the collector and emitter. Thus, the emitter of transistor is common to both input and output circuits and hence the name common emitter configuration.

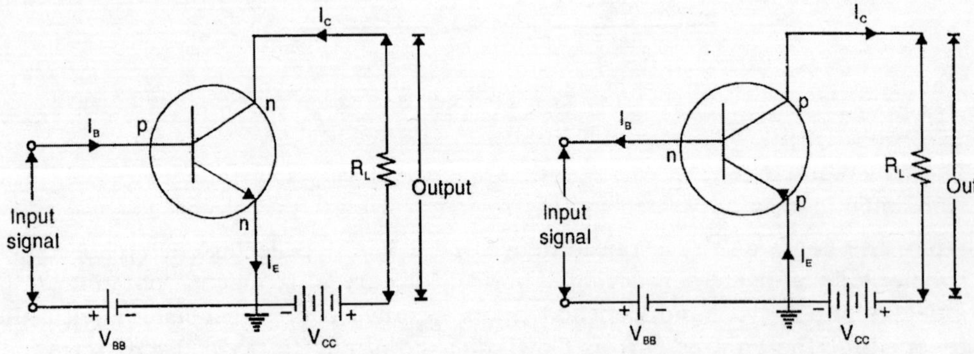

(*a*) Common emitter circuit of *npn* transistor

(*b*) Common emitter circuit of *pnp* transistor

Fig. 4.8.

Base current amplification factor (β)

The ratio of output current (I_C) to the input current (I_B) in common emitter configuration is known as base current amplification factor.

Thus, *the ratio of change in collector current to the change in base current is known as base current amplification factor. It is represented by Greek letter* β.

i.e.,

$$\boxed{\beta = \frac{\Delta I_C}{\Delta I_B} = \frac{\text{Variation in collector current}}{\text{Variation in base current}}}$$

From this relation it is clear that base current amplification factor β is much higher than α, as ΔI_B is very small as compared to ΔI_C.

Relation between α and β

We know that

$$\alpha = \frac{\Delta I_C}{\Delta I_E} \quad \text{... (4.2)}$$

and

$$\beta = \frac{\Delta I_C}{\Delta I_B} \quad \text{... (4.3)}$$

Also $\Delta I_E = \Delta I_C + \Delta I_B$ ... (4.4)

Eqn. (4.4) divided by ΔI_C

$$\frac{\Delta I_E}{\Delta I_C} = \frac{\Delta I_C}{\Delta I_C} + \frac{\Delta I_B}{\Delta I_C}$$

$$\frac{1}{\alpha} = 1 + \frac{1}{\beta}$$

$$\boxed{\beta = \frac{\alpha}{1 - \alpha}}$$

Therefore, above relation shows that as α approaches to unity, β approaches to infinity. In other words, the current gain in common emitter configuration is very high.

Total collector current in common emitter configuration

As we know that in CE configuration, the input current is I_B and the output current is I_C, so from the relation :

$$I_E = I_C + I_B \quad \text{... (4.5)}$$

and

$$I_C = \alpha\, I_E + I_{CBO} \quad \text{... (4.6)}$$

Apply (4.5) in (4.6) $I_C = \alpha\, (I_C + I_B) + I_{CBO}$

$$I_C = \alpha\, I_C + \alpha\, I_B + I_{CBO}$$

$$I_C\,(1 - \alpha) = \alpha\, I_B + I_{CBO}$$

$$\boxed{I_C = \frac{\alpha}{1 - \alpha} \cdot I_B + \frac{1}{1 - \alpha} \cdot I_{CBO}}$$

or

$$\boxed{I_C = \beta\, I_B + (1 + \beta)\, I_{CBO}}$$

$$\left\{ \begin{array}{l} \therefore \beta = \dfrac{\alpha}{1 - \alpha} \\ \\ \therefore 1 + \beta = \dfrac{1}{1 - \alpha} \end{array} \right\}$$

In CE configuration, if the base circuit is open (*i.e.*, $I_B = 0$) the collector current will be the current to the emitter, denoted by I_{CEO} (means leakage current of collector-emitter, junction at open base)

$$I_{CEO} = \frac{1}{1-\alpha} \cdot I_{CBO}$$

So relation $I_C = \beta\, I_B + (1 + \beta)\, I_{CBO}$ may be written as

$$\boxed{I_C = \beta\, I_B + I_{CEO}}$$

⇒ ***Reason why leakage current I_{CEO} is very much larger than the leakage current I_{CBO} ($I_{CEO} > I_{CBO}$)***

Ans. There are two factors responsible for this reason :

(*i*) There exists a leakage current which flows, not through the function, but around it and across the surface the leakage current is proportional to the voltage across the junction.

(*ii*) New carriers may be generated by collision in the collector junction region, leading to avalanche multiplication of current. It is clear from the relation : $I_{CEO} = (1 + \beta) \cdot I_{CBO}$

Characteristics of Common-Emitter (CE)

There are two important characteristics of CE configuration namely :

(1) Input characteristics.

(2) Output characteristics.

(1) Input Characteristics. It is the curve between the base current I_B and base-emitter voltage V_{BE} at constant collector emitter voltage V_{CE}.

The input characteristics of a CE connection can be determined by the circuit shown in Fig. 4.8 keeping V_{CE} constant (say at –10 V), note the base current I_B for various values of V_{BE}. Then plot the reading obtained on the graph, taking I_B along *y*-axis and V_{BE} along *x*-axis. This gives the input characteristic at V_{CE} = –10 V as shown in Fig. 4.9. Following a similar procedure, a family of input characteristics can be drawn. The following points may be noted from the characteristics :

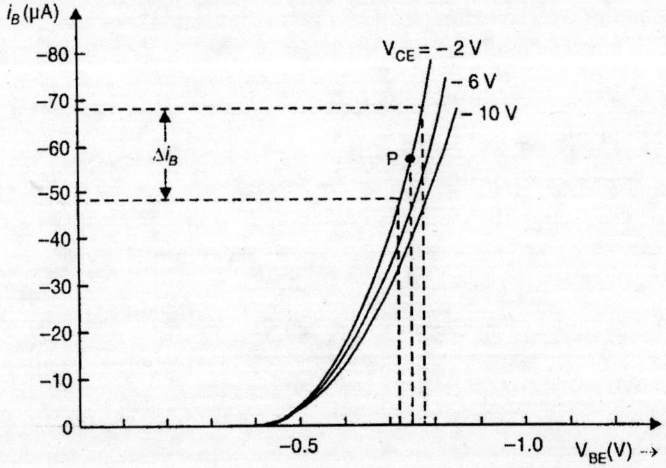

Fig. 4.9. *Common-emitter input characteristics of a PNP transistor.*

(*i*) The characteristic resembles that of a forward biased diode curve. This is expected since the base-emitter junction of transistor is a diode and it is forward biased.

(*ii*) As compared to CB arrangement, I_B increases less rapidly with V_{BE}. Therefore, input resistance of CE circuit is higher than that of CB.

Input Resistance. It is the ratio of change in base-emitter voltage (ΔV_{BE}) to the change in base current (ΔI_B) at constant V_{CE} *i.e.*,

$$\text{Input resistance,} \qquad r_i = \frac{\Delta V_{BE}}{\Delta I_B} \text{ at constant } V_{CE}$$

Input resistance for a *CE* circuit is of the order of a few hundred ohms.

(2) Output Characteristics. It is the curve between collector current I_C and collector-emitter voltage V_{CE} at constant base current I_B.

The output characteristics of a *CE* circuit is shown in Fig. 4.10.

The following point may be noted from the characteristics :

(*i*) The collector current I_C varies with V_{CE} for V_{CE} between 0 and 1 V only. After this collector current becomes almost constant and independent of V_{CE}. This value of V_{CE} upto which transistor current I_C changes with V_{CE} is called the *knee voltage* (V_{knee}). *The transistors are always operated in the region above knee voltage.*

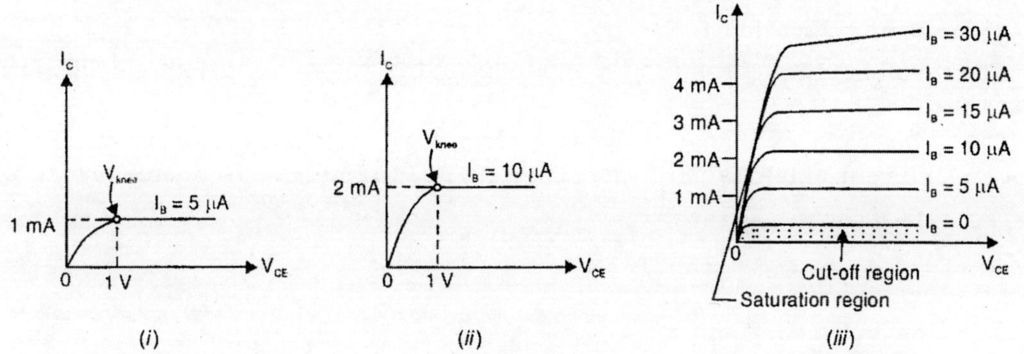

Fig. 4.10.

(*ii*) Above knee voltage, I_C is almost constant. However, a *small increase in I_C with increasing V_{CE} is caused by the collector depletion layer getting wider and capturing a few more majority carriers before electron-hole combinations occur in the base area.*

(*iii*) For any value of V_{CE} above knee voltage, the collector current I_C is approximately equal to $\beta \times I_B$.

Output resistance. It is the ratio of change in collector-emitter voltage (ΔV_{CE}) to the change in collector current (ΔI_C) at constant I_B *i.e.*,

$$\text{Output resistance, } r_o = \frac{\Delta V_{CE}}{\Delta I_C} \text{ at constant } I_B$$

It may be noted that whereas the output characteristics of CB circuit are horizontal, they have noticeable slope for the CE circuit. Therefore, the *output resistance of a* CE *circuit is less than that of* CB *circuit.* Its value is of the order of 50 kΩ.

(*iii*) Common Collector (CC) Configuration

The common collector circuit arrangement for *npn* and *pnp* transistor is shown in Fig. 4.11 (*a*) and (*b*) respectively. In this arrangement, the input is connected between base and collector while output is taken across the emitter and collector. Thus, the collector of a transistor is common to both input and output circuits and hence the name common collector configuration.

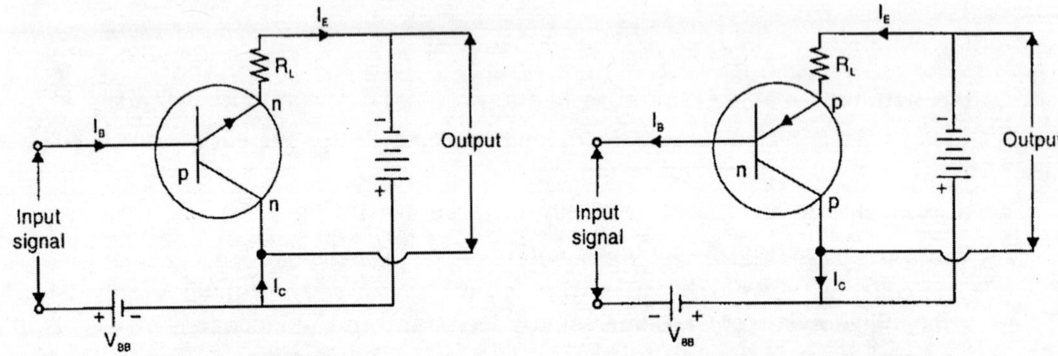

(*a*) Common collector circuit of *npn* transistor

(*b*) Common collector circuit of *pnp* transistor

Fig. 4.11.

Current amplification factor (γ)

The ratio of output current to the input current is known as current amplification factor.

Thus, the ratio of change in emitter current to the change in base current is known as current amplification factor. It is generally represented by Greek letter γ.

i.e.,
$$\boxed{\gamma = \frac{\Delta I_E}{\Delta I_B}}$$

Relation between γ and α

We know that

$$\alpha = \frac{\Delta I_C}{\Delta I_E} \quad \text{... (4.7)}$$

$$\gamma = \frac{\Delta I_E}{\Delta I_E} \quad \text{... (4.8)}$$

$$\Delta I_E = \Delta I_C + \Delta I_B \quad \text{... (4.9)}$$

From eqn. (4.7), (4.8) and (4.9)

$$\boxed{\gamma = \frac{1}{1 - \alpha}}$$

The above relation shows that value of γ is nearly equal to β. However, the common collector circuit is seldom used for amplification because in this arrangement input resistance is high (above 700 Ω) and output resistance is very low (about 25 Ω). Due to this reason voltage gain of common collector configuration is nearly equal to unity. However current gain is very large (about 100 to 150).

4.5.1. Comparison in Different Configurations

We have seen that a transistor can be connected in any one of the three configuration discussed earlier. Transistor behaves differently in different configuration. In which configuration should we connect a transistor ? This depends upon the particular application as we desire. A configuration may be suitable for some application, whereas it may not be suitable for the other. What are the important parameters that govern the suitability of the configuration ? We should know the input dynamic resistance, output dynamic resistance, d.c. current gain, a.c. current gain, a.c. voltage gain and leakage current of the transistor in a given configuration. Table shown below describes the comparison of CB, CE and CC configurations.

Table 4.2

Parameters	*C-B configurations*	*C-E configuration*	*C-configuration*
1. Input dynamic resistance	Very low (20 Ω)	Low (1 KΩ)	Very high (1 MΩ)
2. Output dynamic resistance	Very high (1 MΩ)	High (10 KΩ)	Very low (25 Ω)
3. Current gain	Less than unity	Large (50 to 100)	Very high (150)
4. Voltage gain	Very high (150)	Large (100 to 125)	Less than unity.

From the Table 4.2 as shown we see that :

NOTE : CB-configuration circuit gain A_1 is less than unity, voltage gain A_V is high (approximately equal to that of CE stage), input resistance, R_i is lowest and output resistance R_o is highest of the three configuration. The CB stage has few applications. It is sometimes used to match a very low impedance source, to drive a high-impedance load. It is also used as a constant-current source.

NOTE : In the case of CE-configuration we see that only the common-emitter stage is capable of both a voltage gain and current gain greater than unity. This configuration is the most widely used of the three configurations. We also observe from the above table that the magnitudes of R_i and R_o lie between those for the CB and CC configurations. It may be remember that CE-configuration is usually used in the intermediate stages.

NOTE : In the case of CC-configuration, we observe that voltage gain is approximately unity while the current gain A_1 is very high. Input impedance is the highest and R_o output impedance R_o is the lowest of the three configurations. That's why this circuit finds wide application as a buffer stage between a high-impedance source and a low-impedance load.

4.5.2. Reason Behind the Amplification Action of a Transistor

Although a transistor can perform a number of other functions, its main use lies in amplifying electrical signals. Here the transistor (*npn*) is connected in common-emitter configuration. The transistor is biased (biasing and their types will be discussed latter) to operate in the active region. In the given C-E configuration B-E. (Base-emitter) junction is forward biased however (C-E (collector-emitter) junction is reversed biased.

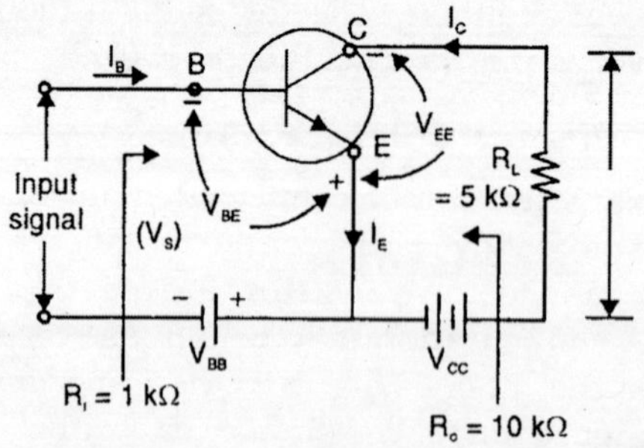

Fig. 4.12. *A basic transistor amplifier in common-base configuration.*

A signal source V_S is connected in the input circuit. A load resistance of 5 kΩ is connected across the output terminal. Assumed that, the input impedance of the transistor is approximately 1 kΩ while output impedance is approximately 10 kΩ. An output voltage V_0 is developed across this resistor when the signal V_S is superimposed on the d.c. voltage V_{BB}, the base voltage V_{BE} varies with time. As a result, the base current I_B also varies with time as a result collector as well as emitter current varies with time. This varying current when passes through the load resistor R_L, a varying voltage is developed across it. This varying voltage is the output voltage V_0.

Here, an interesting question arises, how the signal voltage is magnified (or amplified) ? to understand this, let us consider how the transistor responds to the a.c. signal. Since the base-emitter junction is forward biased, it offers very low resistance to the signal source V_S. However, the collector-emitter (CE) junction due to reverse biased, offers high resistance. Typically, the input and output resistance for a CE transistor vary from 500 Ω to 2 kΩ and 10 kΩ to 50 kΩ respectively. Assume that the input signal voltage is 20 mV (rms or effective value).

Using an average value of 1 kΩ for the input resistance we get the effective value of the emitter-current variation as :

$$I_B = \frac{20 \times 10^{-3}}{1 \times 10^3} = 20\ \mu A$$

Assumed d.c. current gain (β) of CE transistor is 100. So, the effective value of collector current variation is :

$$I_C = \beta\ I_B = 100 \times 20\ \mu A = 2\ mA$$

Now, the output resistance of the transistor is very high. So output voltage is approximately :

$$V_0 = I_C\ R_L = 2\ mA \times 5\ k\Omega = 10\ V$$

The ratio of output voltage (V_0) to the input voltage (V_S) is known as voltage amplification or voltage given (A_V) of amplifier. So,

$$A_V = \frac{V_0}{V_S} = \frac{10}{20\ mV} = 500$$

Thus we say that the transistor's amplifying action is basically due to its capability of transferring its signal current from a low resistance circuit to high resistance circuit or we can say that

i.e., | Transfer + Resistor = Transistor |

⇒ ***Why CE configuration is widely used in amplifier circuits.***

The main utility of a transistor lies in its ability to amplify weak signals. The transistor alone cannot perform this function. We have to connect some passive components (such as resistors and capacitors) and a biasing battery, such a circuit is then called an amplifier. Thus an amplifier is an electronic circuit which is capable of amplifying or increasing the level of signal.

It may be noted that, a single transistor amplifying stage is not sufficient. In almost all applications we use a number of amplifier stages, connected in cascaded (in which output of one stage is connected to the input of next stage) form. Such a connection of amplifier stages is known as cascaded amplifier. Voltage gain and current gain of CE configuration is high. That's why CE configuration is mostly used for *intermediate stage* in cascaded configuration.

In other words reasons why CE-configuration is mostly used due to :

1. high voltage gain
2. high current gain and
3. high power gain.

4.6. REGIONS OF A TRANSISTOR

Generally regions of a transistor is divided into three parts :

1. Active region
2. Saturation region
3. Cut-off region.

1. **Active Region.** In this region the collector junction is biased in the reverse direction while the emitter junction is biased in the forward direction. If we want to use transistor as a amplifier, we will biased it in such a manner its operating point or quiescent point (will be discussed later) remains in the active region after applying the signal.

2. **Saturation Region.** In this region both the input output terminals are biased in forward direction. If we wants to transistor as a switch in one condition, we will biased transistor in this region.

Condition for saturation :

$$\boxed{I_B \geq (I_B)_{min}}$$

where $(I_B)_{min} = \dfrac{I_C}{h_{FE}}$, h_{FE} = d.c. current gain

I_B = Base current

I_C = Collector current

If this condition is satisfied then the transistor will in the saturation region.

3. **Cut off Region.** In this region both the input and output terminals are biased in reversed direction. If we wants to transistors as a switch in off condition, we will biased transistor in this region. In simple words OFF means there is no current *i.e.*, open circuit. Which is possible when both the junctions are reverse biased. Because it offers very high impedance *i.e.*, open circuit.

In short, cut off means that $I_E = 0$, $I_C = I_{CO}$, $I_B = - I_C = - I_{CO}$, and V_{BE} is a reverse voltage whose magnitude v of the order of 0.1 V for germanium and 0.5 V for silicon transistor.

Required condition of cut off region is given by the relation :

$$\boxed{V_{BE} = - V_{BB} + R_B I_{CO} \leq - 0.1 \text{ V}} \quad \text{for Ge transistor.}$$

4.7. STABILIZATION

As we have discussed earlier that quiescent point or operating point is nothing but zero signal values of collector current (I_C) and collector emitter voltage (V_{CE}). If either of the two changes, the operating point is shifted.

Cause of unstabilization ?

A transistor is said to be unstabilized when its operating point is shifted. Usually unstabilization in a transistor is occurred because of the change in collector current. The collector current in a transistor changes rapidly due to following two reasons namely :

1. If the transistor is replaced by another one of same type. Because no two transistor can have same transistor parameters (*i.e.*, β)

2. The temperature changes affects I_C. It can be more clearly understood by the equation $I_C = \beta I_B + (1 + \beta) I_{CO}$. We know that leakage current (I_{CBO} or I_{CO}) increase due to increase in temperature, results increase in collector current I_C. The rise in collector current further increases the temperature resulting in further increase in I_{CBO}. Such a cumulative effect leads to thermal runaway (which will be discussed later.)

Effect of inherent variations of transistor parameter

Inherent variations of a transistor parameter is nothing but a replacement of one transistor by another of the same type, that may have different parameters. The inherent variations of transistor parameters shifts the operating point when it is not properly stabilised.

Here it is important to note that in order to avoid thermal runaway or the self-destruction of transistor because of rise in temperature and to overcome the effect of inherent variations of transistor parameters, it is very necessary to stabilize the operating point or to fix the operating point at all connection, we require *stabilisation.*

Thus stabilisation may be defined as the process of making operating point independent of temperature changes or inherent variations in transistor parameters is called stabilisations.

4.7.1. Stability Factor

We have been already discussed that stability of the operating point means the value of V_{CE} and I_C remains constant even though there is variation in leakage current I_{CBO} (or I_{CO}) because of rise in temperature.

Stability factor may be defined as the rate of change of collector current I_C with respect to rate of change of collector leakage current. I_{CBO} at constant I_B and β is called stability factor.

Mathematically, it may be written as

Stability factor, $$S = \frac{d\,I_C}{d\,I_{CBO}} \text{ at constant } I_B \text{ and } \beta.$$

Note : A circuit having stability factor is equal to 5 and another circuit having stability factor is equal to 10 which one is the better circuit and why ? Explain it with reference to stability factor.

The circuit having lower the value of stability factor, greater is the thermal stability of the transistor. Because lower the value of stability factor means that there is large change (variation) in leakage current (which is temperature dependent) there is less change in the collector current (I_C). Thus we can say that the circuit having stability factor equal to 5 is better circuit than the circuit having stability factor equal to 10.

Another form of stability factor S′ and S″

Stability factor (S′). The stability factor S' is defined as the rate of change of I_C with V_{BE} keeping I_{CBO} and β constant.

Mathematically,

Stability factor $$S' = \frac{d\,I_C}{d\,V_{BE}} \text{ at constant } I_{CBO} \text{ and } \beta.$$

Stability factor (S″). The term stability factor S'' may be defined as the rate of change of I_C with respect to β, keeping I_{CBO} and V_{BE} constant.

Mathematically,

Stability factor, $$S'' = \frac{dI_C}{d\beta} \text{ at constant } I_{CBO} \text{ and } V_{BE}.$$

Steps involved during the calculation of stability factor for any configuration

Step 1. Apply the KVL the input of the given circuit, write the expression and let eqn. (1).

Step 2. Use

$$I_C = \beta I_B + (1 + \beta)\, I_{CO}$$

Step 3. Calculate the value of I_B from eqn. (1) in terms of I_C and other parameters and substitute in eqn. (2) which gives eqn. (3).

Step 4. Differentiate eqn. (3) with respect to I_C and put

$$S = \frac{d\,I_C}{d\,I_{CBO}} \text{ and then solve for } S.$$

These four steps find use in determining the stability of any transistor-amplifier.

4.8. OPERATING POINT

The D.C. operating point and load line

We have discussed earlier that of proper operation of a transistor, in any application, we set a fixed levels of certain currents and voltages in a transistor. These values of currents and voltages define the point at which transistor operates. This point is known as operating point or quiescent point or simply *Q*-point. Since the level of the currents and voltage are fixed therefore the operating point is also called d.c. operating point.

Or in other words, zero signal values of V_{CE} and I_C are called the operating point or Q-points. Here zero signal means in the absence of signal (i.e., V_S).

In order that the circuit amplifies the signal properly, a judicious selection of the operating point is very necessary. Operating point can be better understood by considering the circuit as shown in Fig. 4.13.

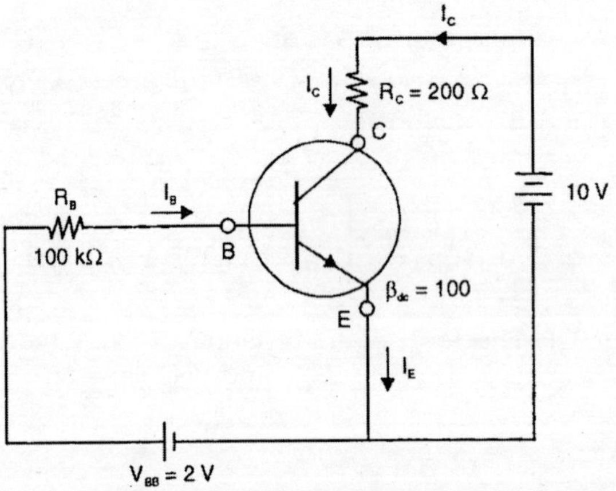

Fig. 4.13. *CE-Configuration circuit.*

Neglecting the cut-off voltage V_{BE} applying KVL to the input section.

$$V_{BB} = R_B I_B$$

or

$$I_B = \frac{V_{BB}}{R_B} = \frac{2}{10} = .2 \text{ mA}$$

$$I_C = \beta I_b = 100 \times .2 = 20 \text{ mA}$$

Again applying KVL to the output section.

$$V_{CE} + I_C R_C = 10$$

$$V_{CE} = 10 - 20 \times 200 \times 10^{-3} = 6 \text{ V}$$

Here the values $I_C = 20$ mA and $V_{CE} = 6$ V define a *Q*-point.

Transistor load line analysis

Transistor load line or d.c. load line is a graphical representation of quiescent point.

Before applying a transistor, the transistor circuit is generally required to be analysed, this can be obtained by plotting the output characteristics and then determining the collector current at a derived collector-emitter voltage.

This (drawing of d.c. load line) can better understand by considering the output characteristics of a common emitter transistor shown in Fig. 4.14.

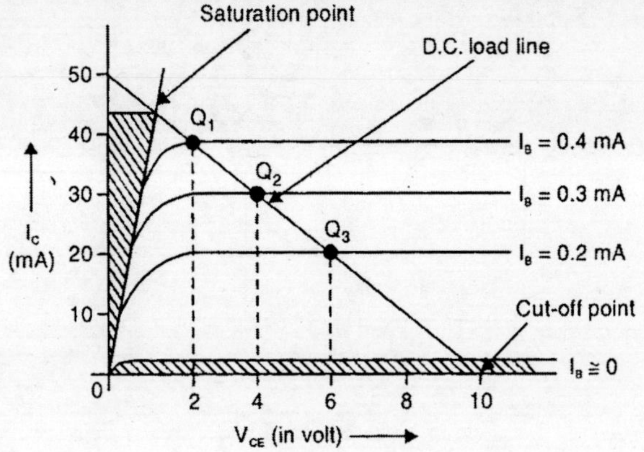

Fig. 4.14.

In the given figure there are three operating point Q_1 (4 V, 30 mA), Q_2 (2 V, 40 mA) and Q_3 (6 V, 20 mA) on both sides, we shall obtain a straight line as shown in Fig. 4.14.

The values of base current, collector current and collector-to-emitter voltage can be read at any point along this line. The d.c. load line gives us the information about the following two important points.

1. The load line intersects the horizontal axis at a point marked V_{CE}, which is equal to 10 V. This point is called *transistor cut-off point or lower end of the load line.* At this point the values of base current and collector current are zero (ideally). But actually, there is a small value of reverse saturation current (I_{CBO}) at cut-off point as indicated in the figure. Due to this reason, the V_{CE} is slightly less than V_{CC}.
2. The load line intersects the vertical axis at a point marked I_C equal to 50 mA. This point is called *transistor saturation point or upper end of the load line.* At this point, the collector current in maximum and collector-to-emitter voltage is very small equal to $V_{CE(Sat)}$.

It may be noted that if the upper and lower ends of the load line are known, the load line can be obtained by joining the upper and lower ends.

Simpler way of drawing a d.c. load line

Steps involved in drawing a d.c. load line :

Step 1. Applying KVL in he output of the circuit.

Step 2. On putting the value V_{CE} = 0 we get I_C in terms of V_{CC} and R_C.

Step 3. On putting the value of I_C = 0 we get value of V_{CE}.

Step 4. Join the values obtained in step 2 and step 3 to understand the above steps more clearly. Consider a circuit having R_C = 10 kΩ, V_{CC} = 20 V, R_B = 47 kΩ, V_{BB} = 10 V and β = 100 as shown in Fig. 4.15.

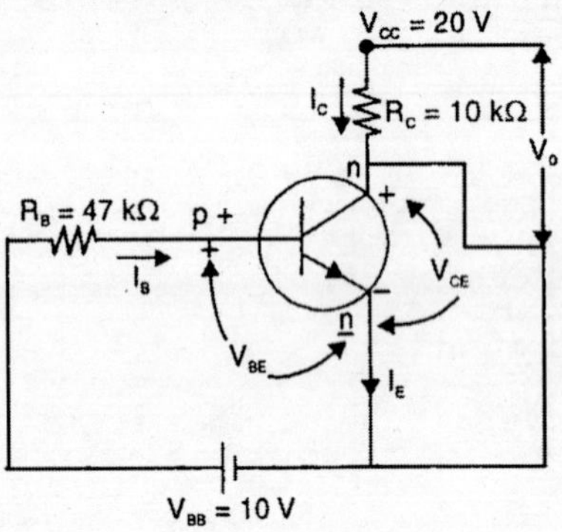

Fig. 4.15.

Step 1. Applying KVL in the output of circuit.

$$V_{CC} - I_C R_C - V_{CE} = 0$$

or
$$V_{CC} = I_C R_C + V_{CE} \quad \text{... (4.10)}$$

Step 2. On putting $V_{CE} = 0$ in eqn. (4.10), we get

$$I_C = \frac{V_{CC}}{R_C} = \frac{20}{10 \text{ k}\Omega} = 2 \text{ mA.}$$

Step 3. On putting $I_C = 0$ in eqn. (4.10) we get,

$$V_{CE} = V_{CC} = 20 \text{ V}$$

Step 4. Join these values (*i.e.*, I_C = 2 mA and V_{CE} = 20 V) as shown in Fig. 4.16.

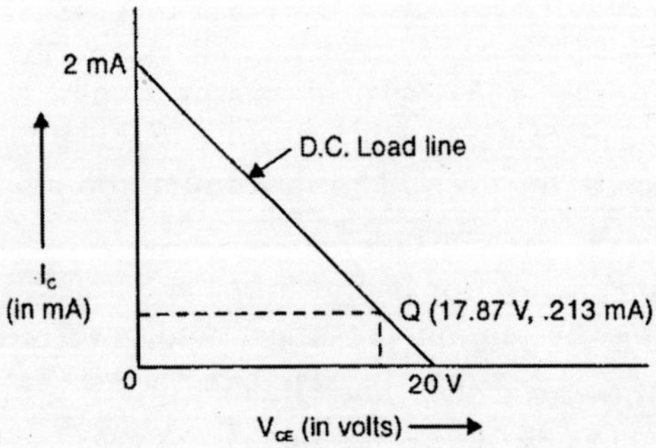

Fig. 4.16.

If we want to calculate the *Q*-point for the same problem then from eqn. (4.10)

$$V_{CC} = I_C R_C + V_{CE}$$

Again applying KVL at the input section (neglecting V_{CE} *i.e.*, cut in voltage)

$$V_{BB} = I_B R_B$$

$$I_B = \frac{V_{BB}}{R_B} = \frac{10}{47 \text{ k}\Omega} = .213 \text{ mA}$$

$$I_C = \beta \, I_B = 100 \times .213 = 21.3 \text{ mA}$$

Again for eqn. (4.10)

$$V_{CE} = V_{CC} - I_C R_C$$

or $$V_{CE} = 20 - .213 \times 10^{-3} \times 10 \times 10^3$$

or $$V_{CE} = 17.87 \text{ V}.$$

The quiescent point (17.87 V, .213 mA) is shown in Fig. 4.16.

4.9. TRANSISTOR BIASING

Biasing is nothing but a selection of operating point (*i.e.*, quiescent point) to operate a transistor in a desired region. Here an interesting question arises, what is need of biasing ? The solution is that when the signal is applied across the input, if the transistor is not properly biased then there is more chances to clip the input signal or in other words we can say that we get distorted output. To avoid this difficulty we need biasing.

The circuitary which provides the necessary conditions of transistor biasing is known as *biasing circuit.*

Selection of operating point

As we have discussed earlier biasing is nothing but a selection of operating point. Let us assume that we want to transistor work as an amplifier, so we will bias a transistor in such manner that its quiescent point (θ) always remain in the active region after applying the input signal. Here an important question arises that in which place in the active region we bias the transistor. For simple understanding there are three possibilities *i.e.*, D, C, E to bias a transistor which will work as an amplifier. The points D, C, E are shown in Fig. 4.17. From the figure it is clear that there are three possibilities to bias a transistor in active region to work as an amplifier (namely D, C and E points).

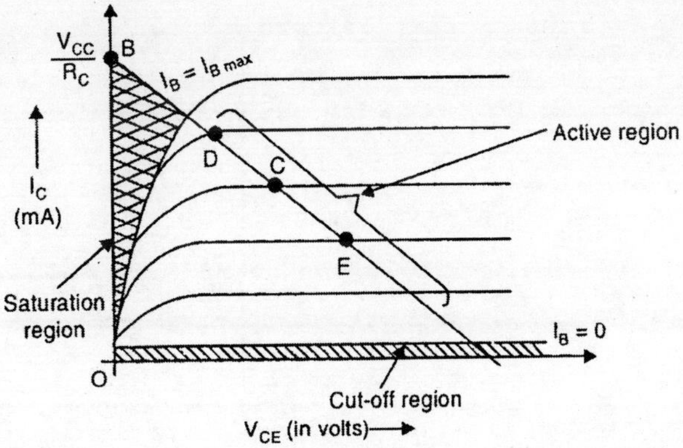

Fig. 4.17.

If point D is selected as the operating point, the upper portion of the positive half will be clipped off as this point lies very near to the *saturation region* [see in Fig. 4.18(*a*)].

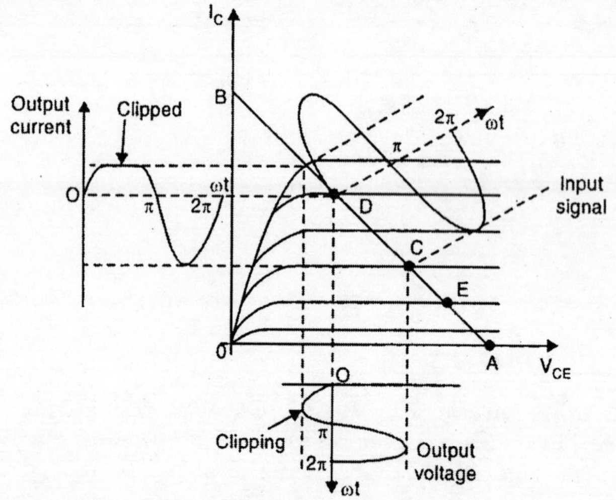

Fig. 4.18(*a*). *Output characteristics if D is selected as the operating point.*

On the other hand, if point E is selected as the operating point, the peak of the negative half will be clipped off as this point lies very near to the *cut-off region* [see in Fig. 4.18(*b*)].

Thus, in both the cases distorted signal is obtained at the output.

However, if point C is selected as the operating point, full cycle of the signal is obtained in the amplified form at the output [see Fig. 4.18(*c*)]. In this case, signal is not distorted at all.

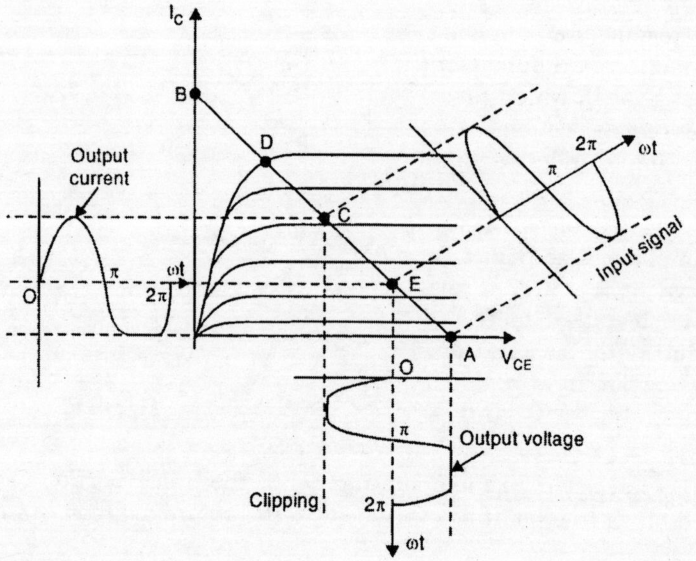

Fig. 4.18(*b*). *Output characteristics if E is selected as the operating point.*

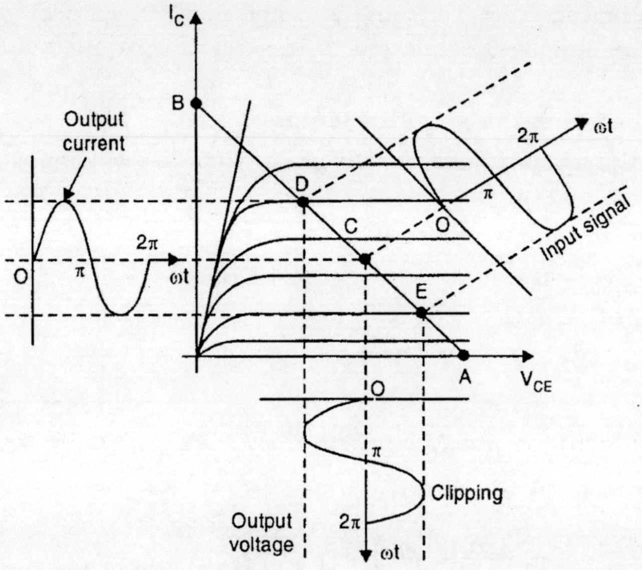

Fig. 4.18(c). *Output characteristics if C is selected as the operating point.*

4.9.1. Methods of Transistor Biasing

The basic principles of the biasing circuitary is to obtain the required value of I_C and V_{CE} from the given circuit in the zero signal conditions. From the above discussion, it clear that transistor biasing is required for *faithful amplification* (when an amplifier does not distort the original shape of input signal after amplification process except increased magnitude, the phenomenon is referred to as faithful amplification). The biasing circuit should have following :

(*a*) It should ensure proper zero signal collector current and V_{CE} *i.e.*, it should establish the operating point in the centre of the active region of the characteristics, so the signal may not cut-off at any part.

(*b*) It should make the operating point independent of transistor parameters.

(*c*) It should ensure that V_{CE} does not fall below the required limit (*i.e.*, 1.0 V for Si and 0.5 V for Ge transistors)

The following methods are used for obtaining transistor biasing :

(*i*) Fixed biasing or base resistor biasing

(*ii*) Collector-to-base biasing or voltage feedback biasing

(*iii*) Emitter resistor biasing

(*iv*) Potential divider or self biasing.

(*i*) Fixed Biasing or Base Resistor Biasing

The circuit arrangement of fixed biasing for an *npn* transistor is shown in Fig. 4.19. In this biasing a high valued resistance is connected between supply and base of the transistor. The required zero signal base current I_B is provided by V_{CE}. The supply also keeps the base positive with respect to emitter and hence makes the base emitter junction forward biased. However, on the other hand in case of *pnp* transistor, the negative end of the supply is connected to the collector through R_C and base resistor R_B.

Circuit analysis for the stability factor

To calculate the stability factor for the given circuit we will apply the steps discussed earlier.

Apply KVL to the input section

$$V_{CC} - I_B R_B - V_{BE} = 0 \quad \text{...(4.11)}$$

also we know that

$$I_C = \beta\, I_B + (1 + \beta)\, I_{CO} \quad \text{...(4.12)}$$

From eqn. (4.11) $I_B = \dfrac{V_{CC} - V_{BE}}{R_B}$, on putting the value of I_B in eqn. (4.12) we get,

$$I_C = \beta\left[\frac{V_{CC} - V_{BE}}{R_B}\right] + (1 + \beta)\, I_{CO} \quad \text{...(4.13)}$$

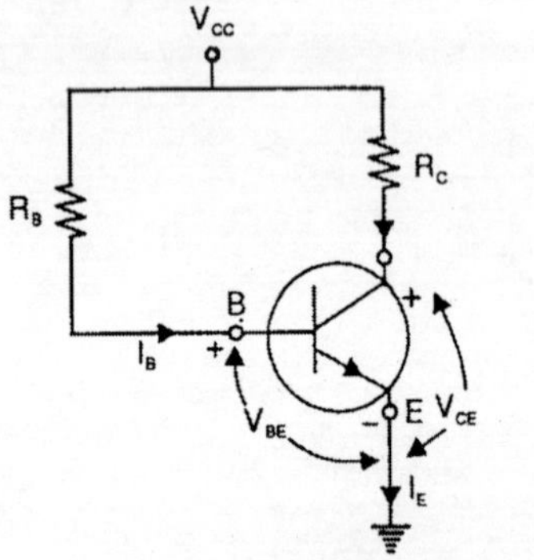

Fig. 4.19. *Fixed biasing circuit for npn transistor.*

Now differentiating eqn. (4.13) w.r.t. I_C, we get

$$\frac{d}{dI_C} . I_C = \frac{d}{dI_C}\left[\beta\left(\frac{V_{CC} - V_{BE}}{R_B}\right) + (1 + \beta)\, I_{CO}\right]$$

or $$1 = 0 + (1 + \beta)\frac{dI_{CO}}{dI_C}$$

or $$1 = 0 + (1 + \beta).\frac{1}{S} \qquad \left\{\therefore S = \frac{dI_C}{dI_{CO}}\right\}$$

$$\boxed{S = 1 + \beta}$$

Thus, from the above relation it is clear that the value of stability factor in fixed biasing method is very high which shows its poor thermal stability.

Advantages

(*i*) In this method, calculations are simple.

(*ii*) Simple in construction as only one resistor R_B is required to set the conditions.

(*iii*) There is no loading of the source as no resistor is employed across base emitter junction.

Disadvantages

(*i*) It provided high stability factor or in other words poor thermal stability and therefore chances of thermal run way.

(*ii*) This method provides poor stability against inherent variations of transistor parameters.

Due to the disadvantages, this method is rarely employed in practical circuits.

(*ii*) Collector-to-base Biasing or Voltage Feedback Biasing

The biasing arrangement for an *npn* transistor is shown in Fig. 4.20. In this biasing a very high value resistance R_B is connected between collector and base and resistance R_C is connected between supply voltage and collector of the transistor. Here the required zero signal base current is determined by collector-base voltage V_{CB} and not by V_{CC}. Hence, the base-emitter junction is forward biased and base current I_B flows through R_B.

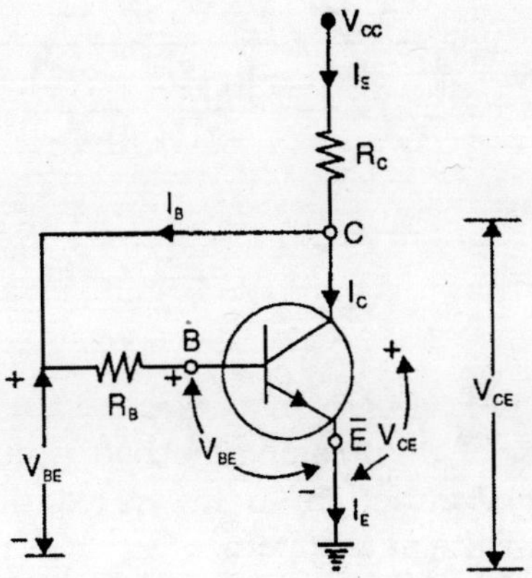

Fig. 4.20. *Collector-to-base biasing for an npn transistor.*

It may be noted that the collector C is connected to base (input) through R_B. Through this resistance (R_C) current I_B feedbacks from the output. In other words base current depends upon the collector voltage. This is why this circuit is also called *a voltage feedback biasing circuit.*

Circuit analysis for stability factor (S)

Taking into account the steps discussed earlier, applying KVL in the input *i.e.*,

$$V_{CC} - I_E R_C - I_B R_B - V_{BE} = 0 \quad \text{...(4.14)}$$

Also we know that $\quad I_C = \beta\, I_B + (1 + \beta)\, I_{CBO} \quad$ ...(4.15)

From eqn. (4.14),

$$V_{CC} - (I_B + I_C)R_C - I_B R_B - V_{BE} = 0$$

or

$$I_B = \frac{V_{CC} - V_{BE} - I_C\, R_C}{(R_C + R_B)} \quad \text{... (4.16)} \ \{\because I_E = I_B + I_C\}$$

On putting the values of I_B in eqn. (4.15), we get

$$I_C = \beta\left[\frac{V_{CC} - V_{BE} - I_C\, R_C}{R_C + R_B}\right] + (1 + \beta)\, I_{CBO} \quad \text{... (4.17)}$$

Now, differentiating eqn. (4.17) w.r.t. the I_C, we get

$$\frac{d}{dI_C} . I_C = \frac{d}{dI_C}\left\{\beta\left[\frac{V_{CC} - V_{BE} - I_C R_C}{R_C + R_B}\right] + (1 + \beta)\, I_{CBO}\right\}$$

$$1 = \frac{-\beta R_C}{(R_C + R_B)} + (1 + \beta).\frac{d\,I_{CBO}}{d\,I_C} \qquad \left\{\because S = \frac{d\,I_C}{d\,I_{CBO}}\right\}$$

$$1 + \frac{\beta R_C}{R_C + R_B} = (1 + \beta) . \frac{1}{S}$$

$$R_C (1 + \beta) + R_B = (1 + \beta)\frac{1}{S}$$

$$\boxed{S = \frac{(1 + \beta)(R_C + R_B)}{R_B + (1 + \beta) R_C} = \frac{(R_C + R_B)}{R_C + \dfrac{R_B}{(1 + \beta)}}}$$

The main advantage is that it is a simple biasing arrangement as only one resistance R_B is required.

Disadvantages

(*i*) The stability factor of this biasing method is quite high. Hence, the frequency point does change, although to lower extent, due to temperature rise or inherent variations in parameters.

(*ii*) The resistor R_B not only provides d.c. feedback for stabilization of operating point, but it also causes a.c. feedback. This reduces the voltage gain of the amplifier which is not desirable. Because of this drawback, this circuit is seldom used in operational amplifiers.

(*iii*) Emitter Resistance Biasing

Circuit arrangement for emitter resistance biasing for an *npn* transistor is shown in Fig. 4.21. This arrangement is just a modification to the collector-to-base biasing circuit. In this biasing an additional resistance R_E is connected in the emitter. Thus we can say that this circuit contains three resistors R_B, R_E and R_C.

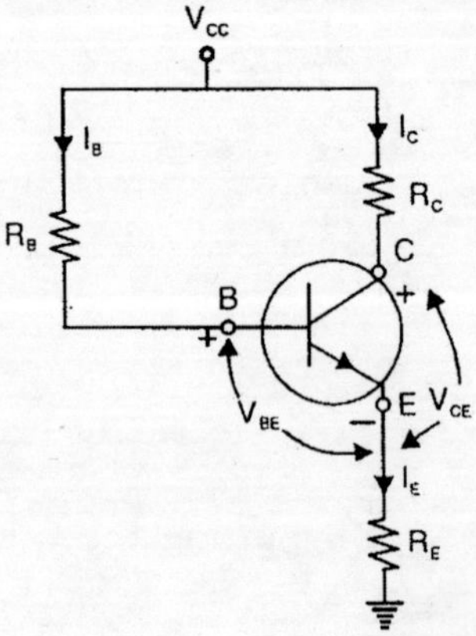

Fig. 4.21. *Emitter biasing circuit for an npn transistor.*

Circuit analysis for stability factor

Taking into account the steps discussed earlier applying KVL in the input section

$$V_{CC} - I_B R_B - V_{BE} - I_E R_E = 0 \qquad ...(4.18)$$

Also we know that

$$I_C = \beta I_B + (1 + \beta) I_{CBO} \qquad ...(4.19)$$

From eqn. (4.18)

$$V_{CC} - I_B R_B - V_{BE} - (I_C + I_B) R_E = 0$$

$$I_B = \frac{V_{CC} - V_{BE} - I_C R_E}{R_E + R_B} \qquad ...(4.20)$$

On putting the value of I_B in equation of

$$I_C = \beta \left[\frac{V_{CC} - V_{BE} - I_C R_E}{R_E + R_B} \right] + (1 + \beta) I_{CBO} \qquad ...(4.21)$$

Now, differentiating the eqn. (4.21) w.r.t. I_C

$$\frac{d}{dI_C} I_C = \frac{d}{dI_C} \left\{ \beta \left(\frac{V_{CC} - V_{BE} - I_C R_E}{R_E + R_B} \right) + (1 + \beta) I_{CBO} \right\}$$

$$1 = -\frac{\beta R_E}{(R_E + R_B)} + (1 + \beta) \frac{d I_{CBO}}{dI_C}$$

$$1 + \frac{\beta R_E}{(R_E + R_B)} = (1 + \beta) \frac{1}{S}$$

$$\frac{R_E + R_B + \beta R_E}{(R_E + R_B)} = (1 + \beta) \frac{1}{S}$$

or

$$\boxed{S = \frac{(R_E + R_B) \cdot (1 + \beta)}{R_E (1 + \beta) + R_B}}$$

From the above expression it is clear that the stabilization is improved by inserting R_E in the emitter circuit. The emitter resistance is present in outside as well as input side. Therefore, a feedback occurs through this resistor. The feedback voltage is proportional to the emitter current I_E. Hence, the circuit is also called current feedback biasing circuit.

This arrangement provides better stabilization than other biasing circuits.

Disadvantages

(*i*) The stability factor of the biasing is quite high. Hence the operating point does change, although to lower extent, due to temperature rise or variations in parameters.

(*ii*) The d.c. feedback helps in the stabilization of operating point but at the same time a.c. feedback reduces the voltage gain of the amplifier. However, this drawback can be removed by putting a capacitor C_E across the resistor R_E. This is possible due to the fact that C_E offers very low impedance to the a.c. signal (current) and does not allow it to pass through R_E and hence a.c. feedback is restricted. However for d.c. supply it (capacitor) offers very high impedance (ideally open circuit) this property of capacitor provides better d.c. stabilization. Thus the process of amplification remains unaffected.

(*iv*) Potential Divider Biasing or Self Biasing

The potential divider or self biasing for an *npn* transistor of common emitter configuration is shown in Fig. 4.22. This biasing circuit consist of two resistors R_1 and R_2 connected across the supply voltage V_{CC} and provide the necessary biasing. Whereas, a resistor connected in the emitter circuit R_E provides thermal stability and improves stability factor. Since the resistor R_1 and R_2 form the potential divider hence the name *potential divider biasing.* This biasing is the most widely used method of providing biasing and stabilisation to transistor since the operating point, in this case can be made almost independent of β.

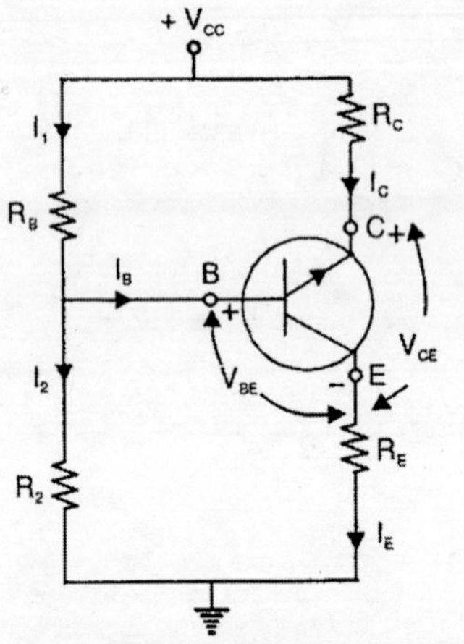

Fig. 4.22. *Potential divider biasing for an npn transistor.*

Circuit analysis of stability factor

Let us consider the base-emitter loop of the voltage divider bias circuit. A basic assumption is that the resistance looking into the base is much longer than that of the resistor R_2. If this is so, then, the current through resistor R_1 flows almost completely into resistor R_2 and the two resistors may be considered effectively.

Before applying the steps discussed earlier, first of all we will draw Thevenin equivalent model (*i.e.*, in which an equivalent voltage source (V_{th}) followed by equivalent series resistance (R_{th})).

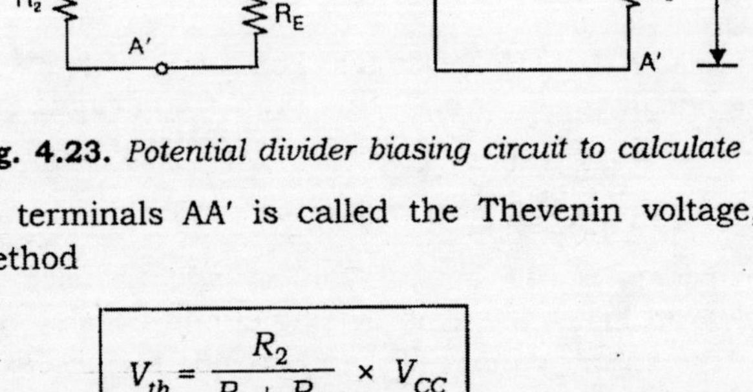

Fig. 4.23. *Potential divider biasing circuit to calculate V_{th}.*

Voltage across terminals AA' is called the Thevenin voltage, given by applying potential divider method

i.e.,
$$V_{th} = \frac{R_2}{R_1 + R_2} \times V_{CC}$$

To determine Thevenin resistance R_{th}, reduce the circuit as shown in Fig. 4.24.

As we know that R_{th} can be calculated by taking all the energy sources replaced by their *internal resistance* (*i.e.*, voltage source as a short circuit and current source as a open circuit)

so,
$$\boxed{R_{th} = R_1 \parallel R_2 = \frac{R_1 R_2}{R_1 + R_2}}$$

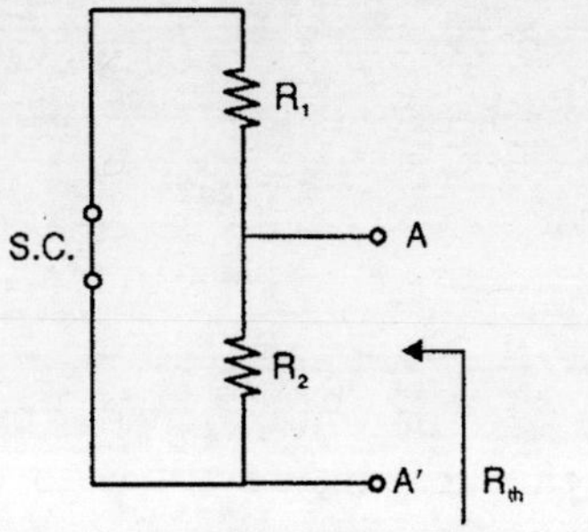

Fig. 4.24.

The resultant thevenin equivalent model circuit is shown in Fig. 4.25.

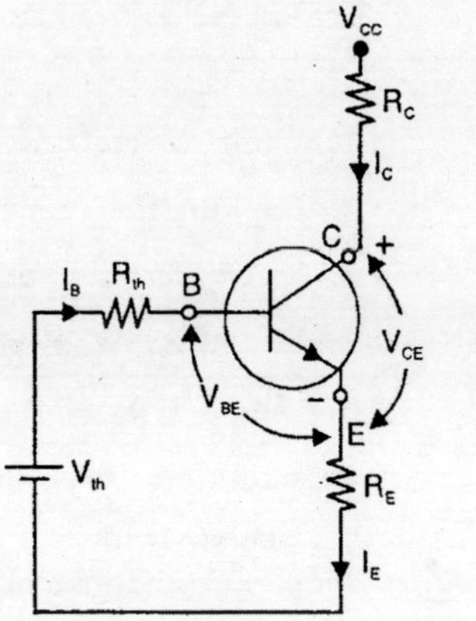

Fig. 4.25.

Applying KVL to the input side.

$$V_{th} - I_B R_{th} - V_{BE} - I_E R_E = 0 \qquad \text{... (4.22)}$$

We know that

$$I_C = \beta\, I_B + (1 + \beta)\, I_{CBO} \qquad \text{... (4.23)}$$

From eqn. (4.22)

$$I_B = \frac{V_{th} - V_{BE} - I_C R_E}{(R_E + R_{th})} \qquad \text{... (4.24)}$$

$$\{\because I_E = I_B + I_C,\ I_{CO} \equiv I_{CBO}\}$$

On putting the value of I_B in eqn. (4.23), we get

$$I_C = \beta \left[\frac{V_{th} - V_{BE} - I_C R_E}{R_E + R_{th}}\right] + (1 + \beta) I_{CO} \qquad \text{... (4.25)}$$

Differentiating eqn. (4.25) w.r.t. I_C, we get

$$\frac{dI_C}{dI_C} = \frac{d}{dI_C}\left[\beta\left(\frac{V_{th} - V_{BE} - I_C R_E}{(R_E + R_{th})}\right) + (1 + \beta)\, I_{CO}\right]$$

$$1 = \frac{-R_E \cdot \beta}{(R_E + R_{th})} + (1 + \beta) \cdot \frac{dI_{CO}}{dI_C}$$

$$1 + \frac{\beta \cdot R_E}{R_E + R_{th}} = (1 + \beta) \cdot \frac{1}{S}$$

$$\frac{R_E + R_{th} + \beta\, R_E}{R_E + R_{th}} = (1 + \beta) \cdot \frac{1}{S}$$

$$\frac{R_E\,(1 + \beta) + R_{th}}{R_E + R_{th}} = (1 + \beta) \cdot \frac{1}{S}$$

or

$$S = (1+\beta)\left[\frac{1+\dfrac{R_{th}}{R_E}}{1+\beta+\dfrac{R_{th}}{R_E}}\right] \quad \text{or} \quad S = (1+\beta)\frac{R_E + R_{th}}{R_{th} + R_E(1+\beta)}$$

The voltage divider or self bias circuit provides magnificent stabilization, therefore this biasing circuit is invariably employed for transistor biasing.

SELF-TEST (TRANSISTOR)

1. What do you mean by transistor ?
2. Explain the basic construction of a transistor.
3. List different types operation of a transistor.
4. Explain why collector region is larger than that of the emitter and base.
5. Which type of transistor is widely used *npn* or *pnp* and why ?
6. What are the different types of transistor configuration ?
7. Explain with the help of input characteristics of CB-configuration the input resistance is very small while output resistance is very high.
8. Explain why leakage current I_{CEO} is very much larger than the leakage current I_{CBO} (*i.e.*, $I_{CEO} > I_{CBO}$).
9. Explain the reason behind the amplification action of a transistor.
10. Why CE configuration is widely used in amplifier circuits ?
11. Explain the required condition for the transistor to be in saturation region.
12. Explain why CE configuration is generally used in intermediate stage while CB-configuration in input section and CC-configuration in the output section.
13. For a common emitter transistor, define β_1 $B_{dc} = h_{FE}$ and $\beta_1'' = h_{FC}$. Derive the relationship between h_{FE} and h_{fe}. For what condition is $\beta' \cong h_{FE}$.
14. Define *h*-parameters of a common emitter amplifier. How can these parameters be determined from the characteristic curve ?
15. (*i*) Determine common-emitter hybrid equivalent model and

    (*ii*) Determine common-base hybrid equivalent model.
16. Draw and describe *h*-parameter equivalent circuit for a common collector transistor amplifier. Show how the values of h_{ic}, h_{rc}, h_{fc} and h_{oc} are obtained from the common collector input characteristic curves and CC output characteristic curves.
17. Define the active, saturation and cut-off regions in a transistor. Sketch a family of CB output characteristics for a transistor. Indicate the active, cut-off and saturation regions.
18. Draw output characteristics for an *n-p-n* transistor used in common base configuration and label clearly the cut-off, active and saturation regions on the diagram.

19. List three factors that contribute to thermal instability in transistor amplifiers. Also, draw a circuit for CE amplifier with fixed bias and show that stability factor 'S' for the circuit is (β + 1).

20. If a transistor has its collector-base junction forward bias and the other junctions reverse biased, will it work ? Explain.

SOLVED EXAMPLES

Example 1. *In CB-configuration, the value of a = .98, a voltage drop of 4.9 V is obtained across of 5 kW when connected in collector circuit. Find the base current.*

Solution : α = .98, V_0 = 4.9 V, R_L = 5 kΩ.

So,
$$I_C = \frac{V_0}{R_L} = \frac{4.9 \text{ V}}{5 \text{ k}\Omega} = .98 \text{ mA.}$$

also
$$\alpha = \frac{I_C}{I_E} = .98$$

$$I_E = \frac{.98}{.98} = 1 \text{ mA}$$

We know that
$$I_B = I_E - I_C = (1 - .98) \text{ mA} = .02 \text{ mA.} \quad \textbf{Ans.}$$

Example 2. *The emitter current I_E in a transistor is 3 mA. If the leakage current I_{CBO} is 5 μA and α = .95, calculate the collector and base current.*

Solution : We know that

$$I_C = \alpha\, I_E + I_{CBO}$$

Given
$$I_E = 3 \text{ mA}$$
$$I_{CBO} = 5\ \mu\text{A}$$
$$\alpha = .95$$

So,
$$I_C = .95 \times 3 \times 10^{-3} + 5 \times 10^{-6} = 2.855 \text{ mA.} \quad \textbf{Ans.}$$
$$I_B = I_E - I_C = (3 - 2.855) \text{ mA}$$

or
$$I_B = .145 \text{ mA.} \quad \textbf{Ans.}$$

Example 3. *Find the value of β if* (*i*) α = 0.9

(*ii*) α = .99

(*iii*) α = .98

Solution : We know that the relation between the α and β is given by the relation

$$\beta = \frac{\alpha}{1 - \alpha}$$

(*i*)
$$\beta = \frac{.9}{1 - .9} = 9 \quad \textbf{Ans.}$$

(ii) $$\beta = \frac{.99}{1 - .99} = 99 \quad \textbf{Ans.}$$

(iii) $$\beta = \frac{.98}{1 - .98} = 49 \quad \textbf{Ans.}$$

Example 4. *For a transistor, β = 50 and voltage dross across 2 kΩ which is connected in the collector circuit is 2 V. Find the base current for common emitter connection.*

Solution : Fig. P (4.1) shows the required common emitter connection.

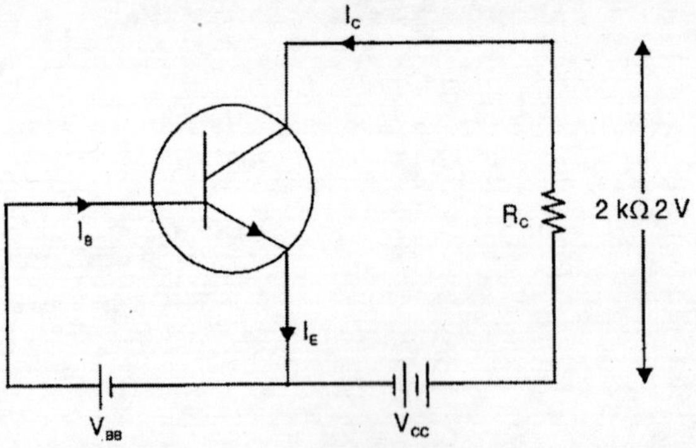

Fig. P (4.1).

The voltage drop across R_C (= 2 kΩ) is 2 volt

$$I_C = \frac{2\ \text{V}}{2\ \text{k}\Omega} = 1\ \text{mA.}$$

$$I_B = \frac{I_C}{\beta} = \frac{1}{50}\ \text{mA} = 20\ \mu\text{A.} \quad \textbf{Ans.}$$

Example 5. *Find α rating of the transistor shown in Fig. P (4.2). Hence determine the value of I_C using both α and β rating of the transistor.*

Solution : We know that

$$\alpha = \frac{\beta}{1 + \beta}$$

or $$\alpha = \frac{49}{1 + 49} = .98$$

The value of I_C can be found by using either α or β rating, as under :

$$I_C = \alpha\ I_E = .98\ (12\ \text{mA}) = 11.76\ \text{mA}$$

$$I_C = \beta\ I_B = 49\ (240\ \mu\text{A}) = 11.76\ \text{mA.} \quad \textbf{Ans.}$$

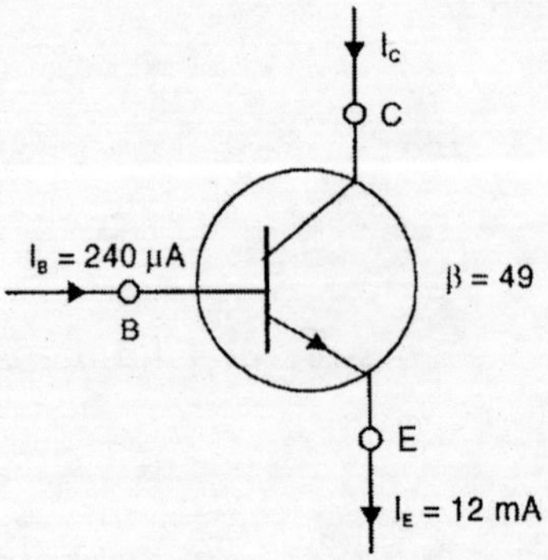

Fig. P (4.2).

Example 6. (*a*) *A transistor has an α = .975. What is the value of β,* (*b*) *What is the value of α, if β = 200.*

Solution : (*a*) Where α = .975

From relation, $$\beta = \frac{\alpha}{1 - \alpha}$$

or $$\beta = \frac{.975}{1 - .975}$$

or $$\beta = 39$$

(*b*) Where $\beta = 200$.

From relation, $$\alpha = \frac{\beta}{1 + \beta}$$

or $$\alpha = \frac{200}{1 + 200}$$

or $\alpha = .995$. **Ans.**

Example 7. *A certain transis has $\alpha = .98$, $I_{CO} = 5\ \mu A$, $I_B = 100\ \mu A$. Find the values of collector current and emitter current.*

Solution : Given $\alpha = .98$

$I_{CO} = 5\ \mu A$

$I_B = 100\ \mu A$

We know that the collector current given by the relation,

$I_C = \beta\ I_B + (1 + \beta)\ I_{CO}$ and

From relation, $$\beta = \frac{\alpha}{1 - \alpha} = \frac{.98}{1 - .98} = 49.$$

Then $I_C = 49.100 \times 10^{-6} + (1 + 49)\ 5 \times 10^{-6}$

$= 4900 \times 10^{-6} + 250 \times 10^{-6}$ A

$= 5150 \times 10^{-6}$ A

$= 5.150$ mA **Ans.**

Also we know that the emitter current

$I_E = I_C + I_B$

or $I_E = 5.150$ mA $+ 100\ \mu A$

or $I_E = 5.25$ mA. **Ans.**

Example 8. *A germanium transistor used in a complementary symmetry amplifier has a collector cut-off current I_{CBO} (i.e. I_{CO}) equals to 10 μA at a temperature of 20°C and $h_{FE} = 50$.*

(*a*) *What is the collector current, when the base current is 0.25 mA ?*

(*b*) *Assuming h_{FE} does not increase with temperature, what would be the value of new collector current, if the transistor's temperature rise to 50°C.*

Solution : Given $I_{CBO} = 10\ \mu A$

$h_{FE} = \beta = 50$

(*a*) We know that collector current,

$I_C = \beta\ I_B + (1 + \beta)\ I_{CBO}$

or $I_C = 50 \times 0.25 \times 10^{-3} + (1 + 50) \times 10 \times 10^{-6}$

or $I_C = 13.01$ mA. **Ans.**

(*b*) Value of new collector current if temperature rises to 50°C.

Also we know that I_{CBO} doubles for every 10°C rise in temperature. Thus I_{CBO} at 50°C

$$I_{CBO \text{ at } 50°C} = I_{CBO \text{ at } 25°C} \times 2^{\frac{(T_2 - T_1)}{10}}$$

or $$I_{CBO \text{ at } 50°C} = 10 \times 2^{2.5}\ \mu A = 56.56\ \mu A.$$

∴ Collector current at 50°C

$$I_C = \beta \,.\, I_B + (1 + \beta)\ I_{CBO} \text{ (at 50°C)}$$

or $$I_C = 50 \times 0.25 \times 10^{-3} + (1 + 50)\ 56.56 \times 10^{-6}$$

or $$I_C = 12.5 \times 10^{-3} + 2884.56 \times 10^{-6} \qquad \left\{\begin{array}{l}\text{here, } T_2 = 50°C \\ T_1 = 25°C\end{array}\right\}$$

or $$I_C = 15.384 \text{ mA.} \quad \textbf{Ans.}$$

? SELF-TEST (BIASING)

1. What do you mean by stabilization ?
2. List the causes of unstabilization.
3. What do you mean by stability factor ?
4. Define the stability factor S, S' and S''.
5. What are the operating points or *Q*-points of a transistor ?
6. List the steps involved in drawing a d.c. load line.
7. What do you mean by biasing ?
8. List the different types of biasing.
9. What is the need of biasing ?
10. What do you mean by fixed or Base Resistor Biasing ?
11. What are the advantages and disadvantages of Base Resistor biasing ?
12. Describe emitter biasing and list their advantages and disadvantages.
13. Describe potential divider biasing and explain why it is most widely used among the biasing.
14. Compare merits and demerits of different types of biasing.
15. Define α and β of a transistor and derive the relationship between them.
16. Why is biasing needed for a transistor to work as an amplifier ? Explain with the help of a neat diagram.
17. Differentiate between J_{CO} and I_{CBO}. What is the effect of temperature on I_{CBO} ?
18. Sketch the circuit of a common collector configuration and also the input and output curves.
19. Draw *npn* and *pnp* transistors. Label all the currents and show the direction of flow. How are all the currents of a transistor related ?
20. The common base configuration factor α < 1, why this is called amplification factor ?
21. Compare CE, CB and CC transistor amplifiers for their parameters *e.g.*, A_V, A_I, R_0, R_i.

SOLVED EXAMPLES

Example 9. *A base resistor biasing circuit is shown in Fig. P (4.3a). Determine (assume V_{BE} = 0.6 V and β = 60).*

(*i*) *The collector current I_C.*

(*ii*) *Collector emitter voltage V_{CE}*

Solution : Given

$$V_{BE} = 0.6 \text{ V}$$
$$V_{CC} = 10 \text{ V}$$
$$\beta = 60$$
$$R_B = 20 \text{ k}\Omega$$
$$R_C = 300 \text{ }\Omega$$

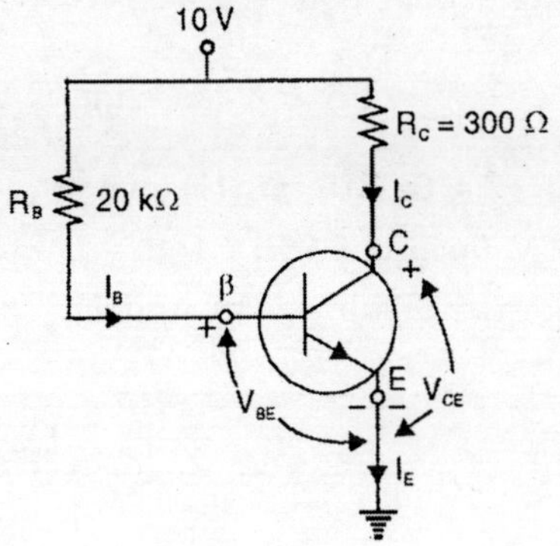

Fig. P (4.3*a*).

Applying KVL in the input section. For easy to understanding the circuit diagram of input section is shown in Fig. P (4.3*b*).

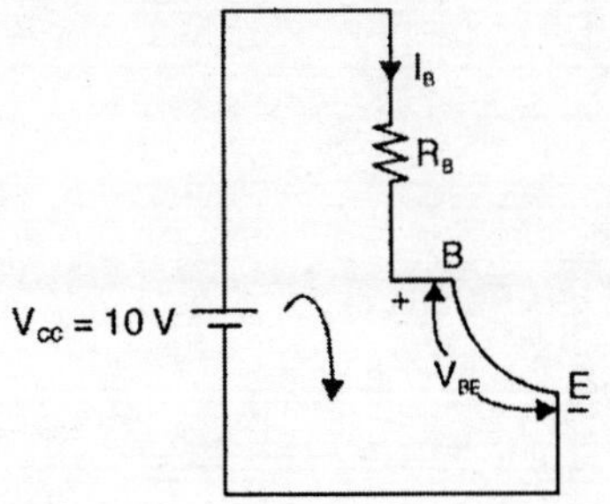

Fig. P (4.3*b*).

$$10 - I_B R_B - V_{BE} = 0.$$

$$I_B = \frac{10 - V_{BE}}{R_B} = \frac{10 - .6}{20 \text{ k}\Omega} = \frac{9.4 \text{ V}}{20 \text{ k}\Omega} = .47 \text{ mA.}$$

(*i*) $I_C = \beta\, I_B = 60 \times .47 = 28.2$ mA. **Ans.**

(*ii*) Again, applying the KVL in the output section

$$10 - I_C R_C - V_{CE} = 0.$$

$$V_{CE} = 10 - I_C R_C$$

$$= 10 - 28.2 \times 300 \times 10^{-3}$$

$$= 1.54 \text{ V.} \quad \textbf{Ans.}$$

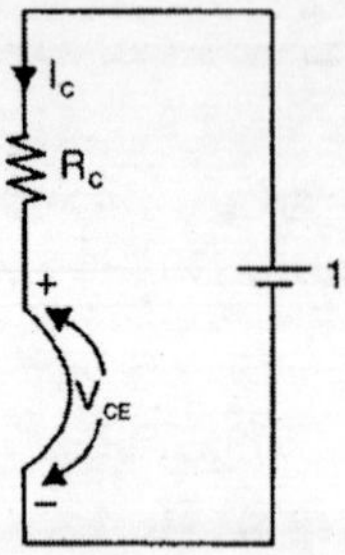

Fig. P (4.3c).

Example 10. *Draw the dc load line and locate the operating point for the fixed biasing transistor circuit shown in Fig. P (4.4a) (assume V_{BE} = .6 V)*

Solution : For drawing load line taken into account the steps discussed earlier.

Applying KVL to the output the circuit.

$$V_{CC} - I_C R_C - V_{CE} = 0 \qquad ...(1)$$

On putting the value V_{CE} = 0 in eqn. (1), we get

$$I_C = \frac{V_{CC}}{R_C} = \frac{9}{2 \text{ k}\Omega} = 4.5 \text{ mA (assume OA)}$$

On putting the value of I_C = 0. In eqn. (1), we get,

$$V_{CE} = V_{CC} = 9 \text{ V (assume OB)}$$

To draw the load line join the values OA and OB.

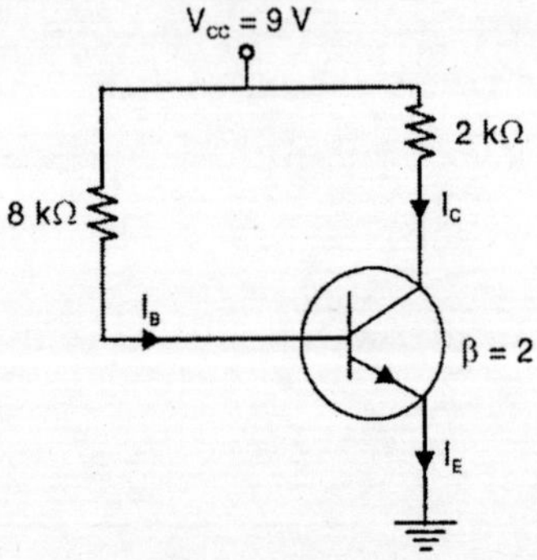

Fig. P (4.4a).

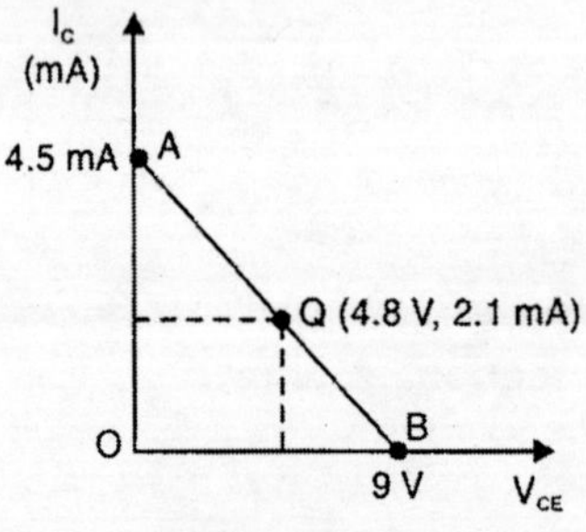

Fig. P (4.4b).

Now, $$I_B = \frac{V_{CC} - V_{BE}}{R_B} = \frac{9 - .6}{8 \text{ k}\Omega} = \frac{8.4}{8} = 1.05 \text{ mA}$$

$$V_{CE} = V_{CC} - I_C R_C = 9 - \beta\, I_B R_C$$

$$= 9 - 2 \times 1.05 \times 2 = 9 - 4.2 = 4.8 \text{ V}$$

$$I_C = \beta\, I_B = 2 \times 1.05 = 2.1 \text{ mA}$$

Hence, operating point $Q\,(V_{CE}, I_C)$ = 4.8 V, 2.1 mA. **Ans.**

Example 11. *For the circuit determine the emitter current the collector voltage V_C. Comment on the result. Neglecting V_{BE}.*

Solution : Here one thing is interesting to note that collector voltage V_C is nothing but V_{CE}.

Since emitter is grounded so

$$V_{CE} = V_C \quad \therefore \quad \left.\begin{aligned} V_{CE} &= V_C - V_E \\ &= V_C - 0 \\ &= V_C \end{aligned}\right\}$$

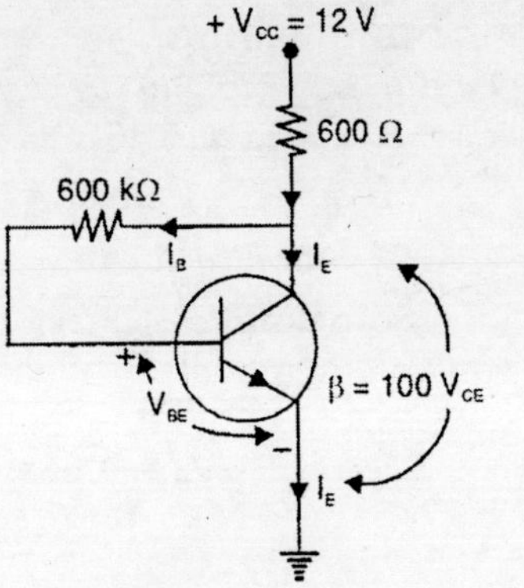

Fig. P (4.5).

Now applying KVL to the input side

$$V_{CC} - I_E\,(600) - I_B \times 600 \times 10^3 - V_{BE} = 0$$

or $$V_{CC} - (I_B + I_C)\,600 - I_B \times 600 \times 10^3 = 0 \; \{\because\; V_{BE} = 0\}$$

$$V_{CC} - (\beta\, I_B + I_B.\beta)\,600 - I_B \times 600 \times 10^3 = 0$$

$$I_B = \frac{V_{CC}}{600 \times 10^3 + 100 \times 600}$$

$$= \frac{12}{600 \times 10^3 + 100 \times 600} = 18.18\ \mu\text{A}$$

$$I_C = \beta\, I_B = 100 \times 18.18\ \text{mA} = 1.818\ \text{mA}.$$

$$I_E = I_C + I_B = 1.818\ \text{mA} + 18.18\ \mu\text{A} \cong 1.818\ \text{mA}.$$

($\because$ 18.18 μA is nearly equal to zero)

Now, applying KVL to the output circuit.

$$V_{CE} = V_C = V_{CC} - I_E R_C = 12 - 1.818 \times 10^{-3} \times 600$$

$$= 10.91\ \text{V}.$$

Comments. It may be noted that transistor V_{CE} = 10.91 V which is slightly less than V_{CC}. Hence the operating point is very near to the cut-off region. Or in other words we can say that given transistor is used as a switch in the off condition.

Example 12. *Calculate the value of R_B in the biasing circuit shown in Fig. P (4.6) so that the operating point becomes I_C = 6.4 mA and V_{CE} = 8.4, β = 80. Neglecting V_{BE}.*

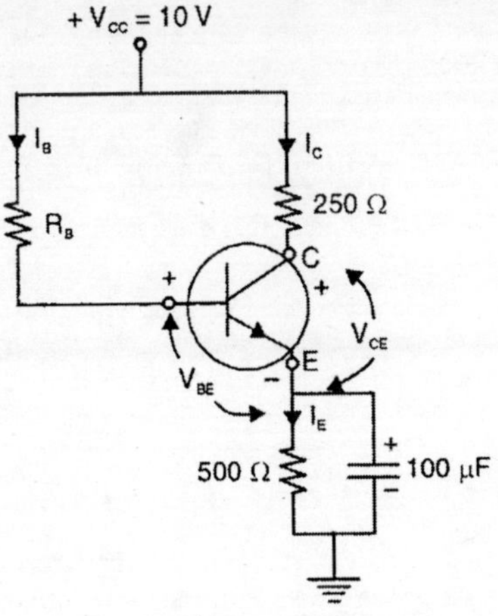

Fig. P (4.6).

Solution : Here it is important to note that for DC analysis capacitor acts as an open circuit. Applying KVL in the input side,

$$V_{CC} - I_B R_B - V_{BE} - I_E\,(500) = 0$$

$$V_{CC} - I_B R_B - 0 - (I_B + I_C)\,(500) = 0$$

$$R_B = \frac{V_{CC} - (\beta + 1)\,500\,.\,I_B}{I_B}$$

$$R_B = \frac{10 - (80 + 1)\,500 \times 80 \times 10^{-6}}{80 \times 10^{-6}}$$

$$\left\{ I_B = \frac{I_C}{\beta} = \frac{6.4\ \text{mA}}{80} = 80\ \mu\text{A},\; V_{BE} = 0\ \text{given} \right\}$$

or $$R_B = 84.5\ \text{k}\Omega.$$ **Ans.**

Example 13. *For the circuit given in Fig. P (4.7a) is used in self-biasing arrangement with* $\beta = 44$, $R_L = 1.51\ k\Omega$, $R_E = 270\ k\Omega$, *and* $V_{CC} = 4.5\ V$. *(Assume* $V_{BE} = 0$*) Estimate*

(a) quiescent point

(b) Stability factor.

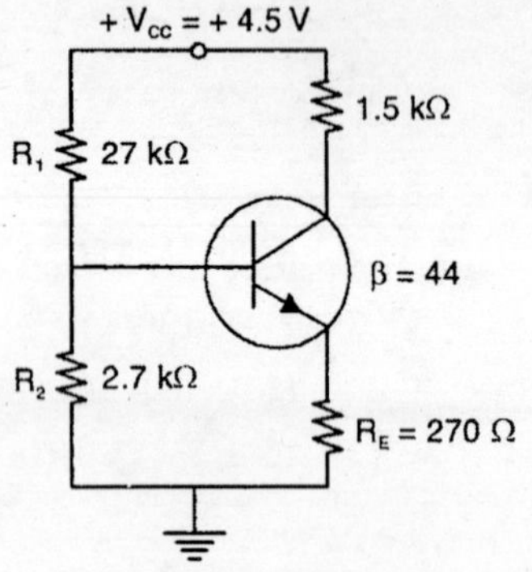

Fig. P (4.7*a*).

Solution : (*a*) First of all we will take the Thevenin equivalent model.

$$V_{th} = 4.5 \times \frac{2.7}{2.7 + 27} = .40\ V$$

$$R_{th} = \frac{R_1 R_2}{R_1 + R_2} = \frac{27 \times (2.7)}{27 + 27} = 2.45\ k\Omega.$$

Now applying KVL to the input circuit,

$$V_{th} - I_B R_{th} - V_{BE} - I_E R_E = 0$$

$$V_{th} - I_B R_{th} - V_{BE} - (I_B + I_C)\, R_E = 0$$

or $\quad V_{th} - I_B R_{th} - V_{BE} - (I_B + \beta\, I_B)\, R_E = 0 \quad (\because I_C = \beta\, I_B)$

$$I_B = \frac{V_{th}\ .\ \beta}{R_{th} + \beta R_E} \qquad (\therefore V_{BE} = 0)$$

$$I_C = \beta\ I_B = \frac{44\ (.40)}{2.45\ k\Omega + 44 \times .270\ k\Omega}$$

or $\quad I_C = 1.228$ mA

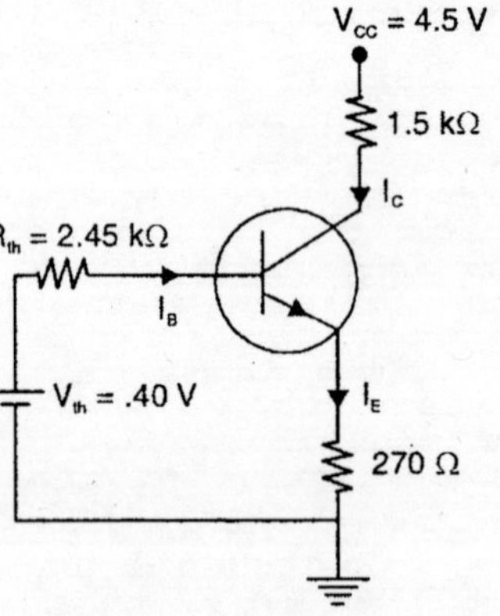

Fig. P (4.7*b*).

And, applying KVL to the output section,

$$V_{CC} - I_C R_C - V_{CE} - I_E R_E = 0$$

$$V_{CE} = V_{CC} - I_C\ (R_C + R_E) \qquad (\therefore \text{neglecting } I_B)$$

$$= 4.5 - 1.228\ (1.5 + .270)$$

$$= 4.5 - 2.173 = 2.326\ V$$

Stability factor for potential divider biasing

$$S = (\beta + 1)\left[\frac{1 + \dfrac{R_{th}}{R_E}}{1 + \beta + \dfrac{R_{th}}{R_E}}\right] \qquad \text{(As we derived earlier)}$$

$$= (44 + 1)\left[\frac{1 + \dfrac{2.45}{.27}}{1 + 44 + \dfrac{2.45}{.27}}\right]$$

$$= 45\left[\frac{1 + 9.07}{1 + 44 + 9.07}\right] = 8.38 \quad \textbf{Ans.}$$

(*b*) The quiescent points are

$$I_C \text{ and } V_{CE} = (1.28\ \text{mA},\ 4.01\ \text{V}) \quad \textbf{Ans.}$$

Example 14. *For the two battery transistor circuit shown in Fig. P (4.8 a), prove that the stabilization factor S is given by*

$$S = \frac{1+\beta}{1+\beta+\dfrac{R_C}{R_C+R_B}}$$

Solution : Taken into account the steps discussed earlier. Applying KVL to the input circuit. Draw its equivalent circuit.

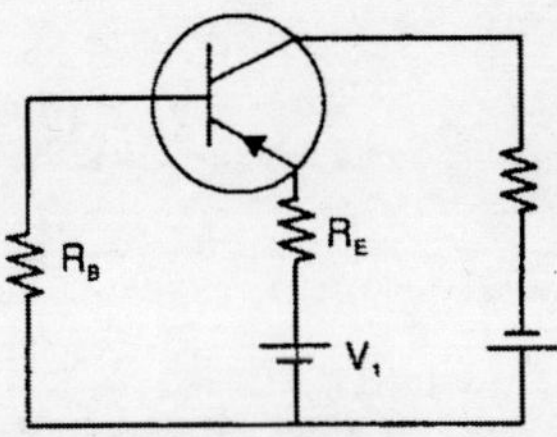

Fig. P (4.8*a*).

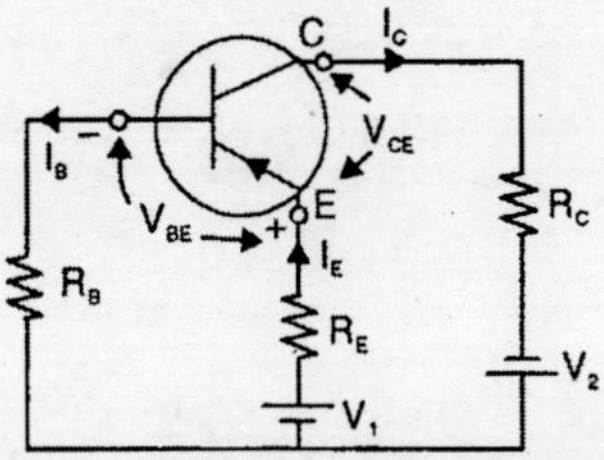

Fig. P (4.8*b*).

$$V_I - I_E R_C + V_{EB} - I_B R_B = 0 \qquad \text{...(1)}$$

We know that

$$I_C = \beta\, I_B + (1+\beta)\, I_{CBO} \qquad \text{...(2)}$$

From eqn. (1)

$$I_B = \frac{V_I - I_C R_C - V_{BE}}{(R_B + R_C)} \qquad \text{...(3)}$$

On putting the value of I_B in eqn. (2), we get

$$I_C = \beta\left[\frac{V_1 - I_C R_C - V_{BE}}{(R_B + R_C)}\right] + (1+\beta)\, I_{CO} \qquad \text{...(4)}$$

Now, differentiating eqn. (4) w.r.t., I_C

$$\frac{d\,I_C}{d\,I_C} = \frac{d}{d\,I_C}\left[\beta\left(\frac{V_I - I_C R_C - V_{BE}}{R_B + R_C}\right) + (1+\beta)\, I_{CO}\right]$$

$$1 = -\frac{\beta . R_C}{R_B + R_C} + (1+\beta) . \frac{d\,I_{CO}}{d\,I_C}$$

$$1 + \beta\frac{R_C}{(R_B + R_C)} = (1+\beta) . \frac{1}{S}$$

or

$$S = \frac{(1+\beta)}{1+\dfrac{\beta\, R_C}{(R_B + R_C)}} \quad \text{Hence proved.}$$

Example 15. *A germanium transistor is used in the self biasing arrangement with V_{CC} = 16 V and R_C = 1.5 kΩ. The quiescent point is chosen to be V_{CE} = 8 V and I_C = 4 mA. A stability factor S = 12 is desired if β = 50 find R_1, R_2 and R_E.*

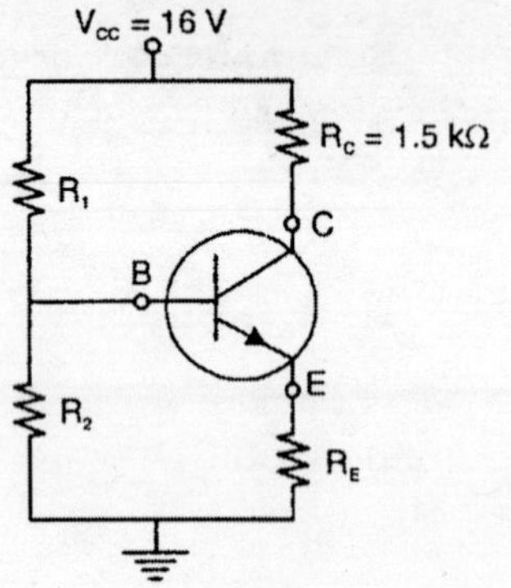

Fig. P (4.9*a*).

Solution : The circuit arrangement of self biasing as shown in Fig. P (4.9*a*).

Given

$$S = 12$$
$$V_{CC} = 16\text{ V}$$
$$R_C = 1.5\text{ K}$$
$$V_{CE} = 8\text{ V}$$
$$I_C = 4\text{ mA}$$
$$\beta = 50$$
$$R_1,\ R_2,\ R_E = ?$$

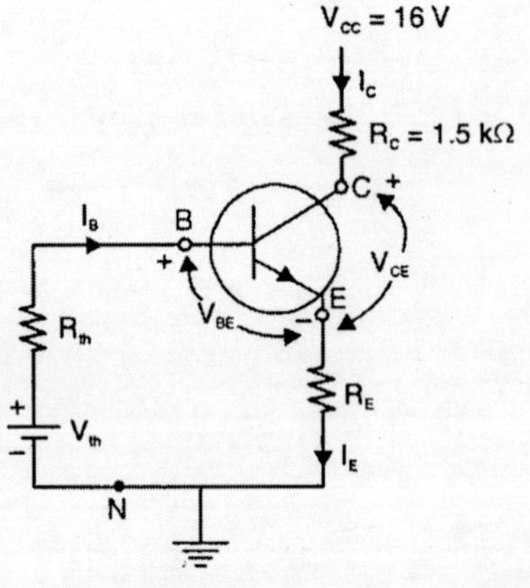

Fig. P (4.9*b*).

Taking Thevenin's equivalent circuit

$$V_{th} = V_{CC} \cdot \frac{R_2}{R_1 + R_2}$$

$$R_{th} = \frac{R_1 R_2}{R_1 + R_2}$$

Applying KVL to the input side

$$V_{th} - I_B R_{th} - V_{BE} - I_E R_E = 0 \qquad ...(1)$$

$$I_B = \frac{I_C}{\beta} = \frac{4\text{ mA}}{50} = 80\ \mu\text{A}.$$

and
$$I_E = I_B + I_C = I_B (1 + \beta) = 4.08\text{ mA}$$

Applying KVL to the output section

$$R_E = \frac{V_{CC} - I_C R_C - V_{CE}}{(I_B + I_C)}$$

$$= \frac{16 - 4 \times 10^{-3} \times 1.5 \times 10^3 - 8}{4.08\text{ mA}} = 0.49\text{ k}\Omega$$

Given that

$$S = \frac{(R_{th} + R_3)(1 + \beta)}{R_{th} + (1 + \beta) R_E} = 12 \qquad ...(2)$$

$$(R_{th} + R_E)(1 + \beta) = 12 \{R_{th} + (1 + \beta) R_E\}$$

$$\left(\frac{R_{th}}{R_E} + 1\right)(1 + \beta) = 12\left\{\frac{R_{th}}{R_E} + (1 + \beta)\right\}$$

$$\left(\frac{R_{th}}{R_E} + 1\right)(1 + 50) = 12\left\{\frac{R_{th}}{R_E} + 51\right\}$$

$$\frac{R_{th}}{R_E} + 50\frac{R_{th}}{R_E} + 1 + 50 = 12\frac{R_{th}}{R_E} + 12 \times 51$$

or $$\frac{R_{th}}{R_E} = 14.4$$

or $$R_{th} = .49 \times 14.4 = 7.05 \text{ k}\Omega$$

and $$R_{th} = \frac{R_1 R_2}{R_1 + R_2} = 7.05 \text{ k}\Omega$$

$$V_{BN} = V_{BE} + I_E R_E = 0.2 + 4.08 \times .49 = 2.2 \text{ V} \quad ...(3)$$

{for Ge transistor $V_{BE} = 0.2$ V}

$$V_{th} = V_{BN} + I_B R_{th} = 2.2 + 0.08 \times 7.05 = 2.764 \text{ V} \quad ...(4)$$

From eqn. (4)

$$V_{th} = \frac{V_{CC} \cdot R_2}{R_1 + R_2} = 16\frac{R_2}{R_1 + R_2} \quad ...(5)$$

$$2.764 = 16 \cdot \frac{R_2}{R_1 + R_2} \cdot R_1 \quad \text{(Multiplying and dividing by } R_1)$$

$$2.764 = 16 \cdot \frac{R_{th}}{R_1}$$

$$2.764 = 16 \cdot \frac{(7.05)}{R_1}$$

$$R_1 = 41 \text{ k}\Omega$$

and $$R_2 = 8.56 \text{ k}\Omega$$

$$\left.\begin{aligned} R_1 &= 41 \text{ k}\Omega \\ R_2 &= 8.56 \text{ k}\Omega \\ R_E &= 0.49 \text{ k}\Omega. \end{aligned}\right\} \textbf{Ans.}$$

Example 16. *A potential divider biasing circuit is shown in Fig. P (4.10). What will happen when*

(*i*) *Biasing resistor R_2 is shorted.*

(*ii*) *Biasing resistor R_1 is opened.*

(*iii*) *Biasing resistor R_2 is opened.*

(*iv*) *Biasing resistor R_1 is shorted.*

Solution : (*i*) When the biasing resistor R_2 is shorted, the base will be grounded. The emitter-base junction will no more be forward biased and the transistor will be cut-off. Hence the output will be reduced to zero.

(*ii*) When the biasing resistor R_1 is opened, the emitter-base junction will no more remain in the forward biased and the transistor will be cut-off. Hence, the output will be reduced to zero.

(*iii*) When the biasing resistor R_2 is opened, the emitter base junction will be highly forward biased through R_1. This will rise the collector current I_C to large value and the transistor will be saturated. Or in other words we can say that V_{CE} (collector to emitter voltage) will be reduced to zero.

(*iv*) When the biasing resistor R_1 is shorted, the emitter base-junction will be highly forward biased. Since the voltage at the base is V_{CE}. This will raise the collector current I_C to large value and the transistor will be saturated. Hence, the collector-emitter voltage V_{CE} will be reduced to almost zero.

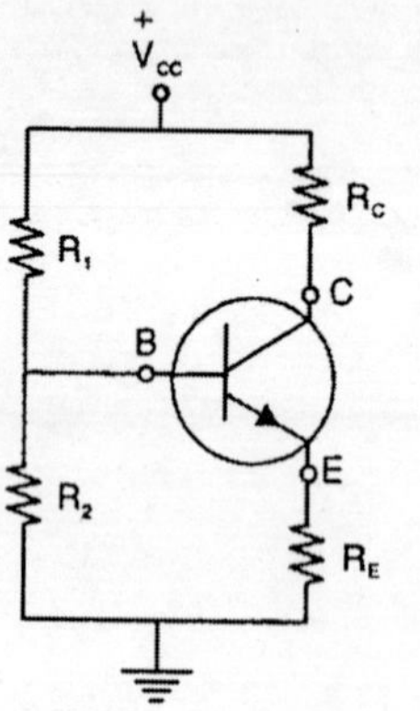

Fig. P (4.10).

Example 17. *Determine the stability factor S for the circuit shown in Fig. P (4.11a).*

Solution : To determine the stability factor (*S*) first of all we will make its equivalent (*i.e.*, Thevenin equivalent) circuit.

$$R_{th} = \frac{R_1 \, R_2}{R_1 + R_2} \quad \text{...(1)}$$

$$V_{th} = V_A \cdot \frac{R_2}{R_1 + R_2} \quad \text{...(2)}$$

where $V_A = (V_{CC} - I_E R_C)$

Applying KVL to the input section,

$$V_{th} - R_{th} I_B - V_{BE} - I_E R_E = 0$$

or $$V_{CC} = I_E R_C - R_{th}\, I_B - V_{BE} - I_C R_C = 0 \quad \text{...(3)}$$

Also we know that

$$I_C = \beta\, I_B + (1 + \beta)\, I_{CO} \quad \text{...(4)}$$

From eqn. (3)

$$I_B = \frac{V_{CC} - V_{BE} - I_C\,(R_C + R_E)}{(R_{th} + R_C + R_E)} \quad \text{...(5)}$$

On putting the value of I_B in eqn. (4), we get

$$I_C = \beta\left[\frac{V_{CC} - V_{BE} - I_C\,(R_C + R_E)}{(R_{th} + R_C + R_E)}\right] + (1 + \beta)\, I_{CO} \quad \text{...(6)}$$

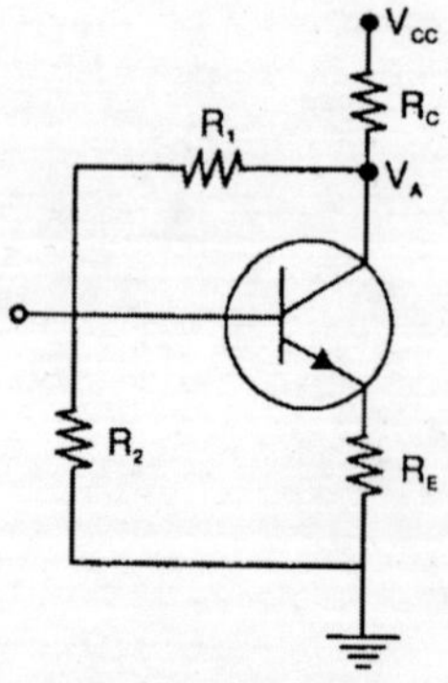

Fig. P (4.11a).

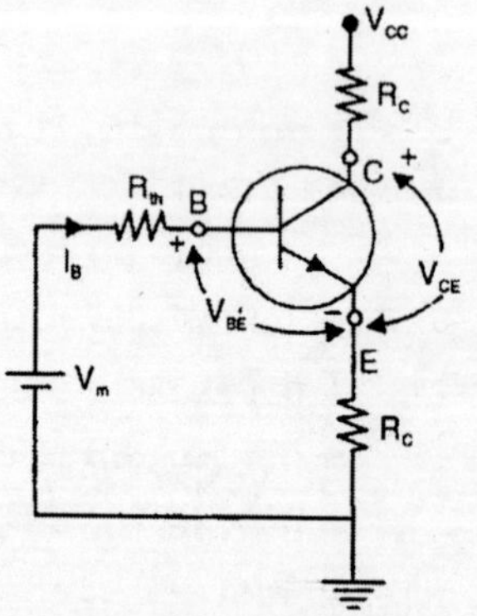

Fig. P (4.11b).

Now, differentiating eqn. (6) w.r.t. I_C

$$\frac{d\,I_C}{d\,I_C} = \frac{d}{d\,I_C}\left\{\beta\left[\frac{V_{CC} - V_{BE} - I_C\,(R_C + R_E)}{(R_{th} + R_C + R_E)}\right] + (1+\beta)\,I_{CO}\right\}$$

$$1 = \frac{-\beta\,(R_C + R_E)}{(R_{th} + R_C + R_E)} + (1+\beta)\frac{d\,I_{CO}}{d\,I_C}$$

or $$1 + \beta\frac{(R_C + R_E)}{(R_{th} + R_C + R_E)} = (1+\beta)\frac{1}{S}$$

or $$S = \frac{(1+\beta)\,(R_{th} + R_C + R_E)}{R_{th} + (R_C + R_E)\,(1+\beta)} \quad \text{where } R_{th} = \frac{R_1\,R_2}{R_1 + R_2}.$$

Example 18. *Prove that for the circuit shown the stability factor S' is given by*

$$S' = \frac{-S}{R_B + R_E} \times \frac{\beta}{\beta + 1}$$

Solution : To calculate stability factor S' $\left(i.e.,\ S' = \frac{\partial I_C}{\partial V_{BE}}\right)$

First of all we must calculate the value of S $\left(i.e.,\ S = \frac{d\,I_C}{d\,I_{CO}}\right)$

We know that

$$S = \frac{(1+\beta).(R_E + R_B)}{R_B + R_E\,(1+\beta)} \quad ...(1)$$

Now, applying KVL to the input section

$V - I_B R_B - V_{BE} - I_E R_E = 0$ ...(2)

We know that $I_C = \beta\,I_B + (1+\beta)\,I_{CO}$ ...(3)

From eqn. (2), the base current

$$I_B = \frac{V - V_{BE} - I_C\,R_E}{(R_B + R_E)} \quad ...(4)$$

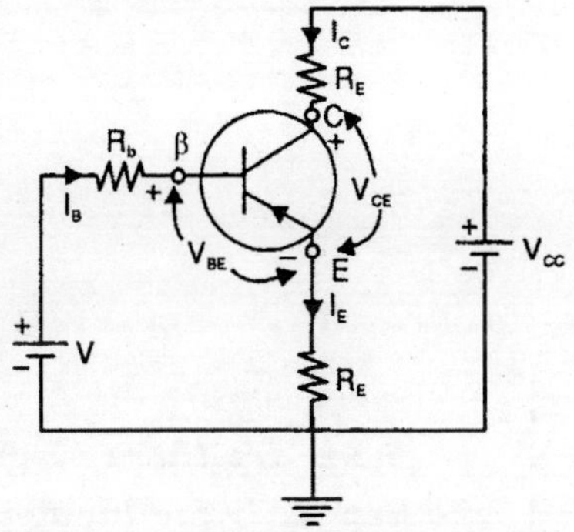

Fig. P (4.12).

On substituting the value of I_b in eqn. (2)

$$I_C = \beta\left[\frac{V - V_{BE} - I_C R_E}{(R_B + R_E)}\right] + (1+\beta)\,I_{CO} \quad ...(5)$$

Now, differentiating eqn. (5) w.r.t. V_{BE}

$$\frac{\partial I_C}{\partial V_{BE}} = \frac{\partial}{\partial V_{BE}}\left\{\beta\left(\frac{V - V_{BE} - I_C\,R_E}{R_B + R_E}\right) + (1+\beta)\,I_{CO}\right\}$$

$$S' = -\frac{\beta}{R_B + R_E} - S'\frac{R_E}{R_B + R_E}$$

$$S'\left\{1+\beta\frac{R_E}{R_B+R_E}\right\}=\frac{-\beta}{R_B+R_E}$$

$$S'=-\frac{\beta}{(R_B+R_E)}\cdot\frac{(R_B+R_E)}{R_E(1+\beta)+R_B} \quad \text{...(6)}$$

From eqn. (1) $$S=\frac{(1+\beta)(R_E+R_B)}{R_B+R_E(1+\beta)} \quad \text{...(7)}$$

or $$R_b+R_c(1+\beta)=\frac{(1+\beta).(R_E+R_B)}{S}$$

Now $$S'=\frac{-\beta}{\beta+1}\cdot\frac{S}{R_B+R_E}$$ Hence proved.

Example 19. *If for the given circuit h_{FE} = 100, determine the region of transistor.*

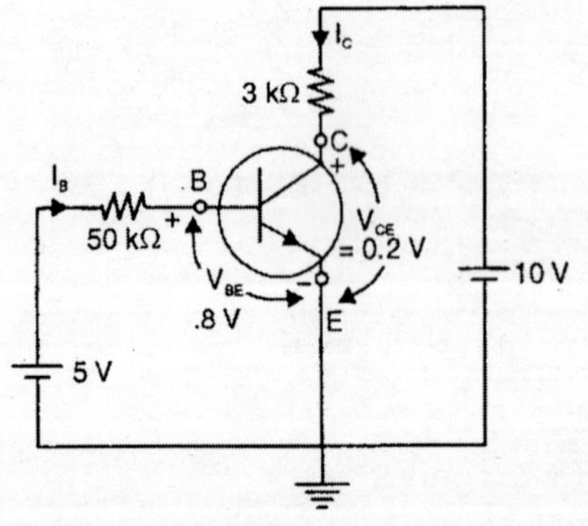

Fig. P (4.13).

Before solving the problem, it is necessary to discuss following three steps.

There are three steps which will decide the region of transistor.

Step 1. If the input (*i.e*, B-E junction) is in forward biased and output (*i.e.*, C-E junction) in the reverse biased, it means transistor will be in the *active region.*

Step 2. If I_B (base current) greater than or equal to minimum base current (I_B min) then transistor will be in *saturation region* or in other words

$$\boxed{I_B \geq (I_B)_{\min}} \qquad \text{where } (I_B)_{\min}=\frac{I_C}{h_{FE}}$$

Step 3. If both input and output section are in reversed biased the transistor will be in *cut-off region.*

Solution : From Fig. P (4.13) it is clear that input side (*i.e.*, B-E junction) is in forward biased so transistor will never be in the cut-off region.

If the condition $(I_B) \geq I_{B\ \min}$ is satisfied the transistor will be in *saturation region* otherwise in *active region.*

Now, applying KVL in input side,

$$5 - 50 \times 10^3 I_B - 0.8 = 0$$

or $$I_B = .084 \text{ mA.}$$

Again KVL at output side, we have

$$10 - 3 \times 10^3 I_C - 0.2 = 0$$

$$I_C = \frac{9.8}{3 \times 10^3} = 3.266 \text{ mA.}$$

Now, $$(I_B)_{min} = \frac{I_C}{h_{FE}} = \frac{3.266}{100} = 0.03266 \text{ mA}$$

Since $I_B \geq I_{B\ min}$ is satisfied. Hence the given transistor will be in *saturation region.*

Example 20. *For the circuit shown, assume* $\beta = h_{FE} = 100$.

(a) Find if the silicon transistor is in cut-off, saturation or in active region.

(b) Find V_O.

(c) Find the minimum value for the emitter resistor R_C *for which the transistor operates in the active region.*

[Assume $(V_{BE})_{sat} = 0.8$ V *and* $(V_{CE})_{sat} = 0.2$ V*]*

Solution : (*a*) Since the input side is in forward biased so, it will never in cut-off region, *i.e.*, either it will be in saturation or in active region.

Now, applying KVL in the input side, we have

$$3 - (I_B + I_C)\, 500 - 0.8 - 7 \times 10^3 I_B = 0$$

or $$2.2 - (500 + 7000)\, I_B - 500\, I_C = 0 \quad \text{...(1)}$$

Again applying KVL to the output side, we have,

$$-10 + 3 \times 10^3 I_C + 0.2 \text{ V} + 500\,(I_B + I_C) = 0$$

$$9.8 = 3500\, I_C + 500\, I_B \quad \text{...(2)}$$

From eqns. (1) and (2)

$$I_C = \frac{144.8}{52000} = 2.78 \text{ mA}$$

$$I_B = \frac{9.8 - 3500 \times 2.78 \times 10^{-3}}{500}$$

or $$I_B = \frac{9.8 - 9.746}{500}$$

or $$I_B = 1.4 \times 10^{-4}$$

or $$I_B = .14 \text{ mA}$$

$$(I_B)_{min} = \frac{I_C}{h_{FE}} = \frac{2.278}{100} = 0.027 \text{ mA.}$$

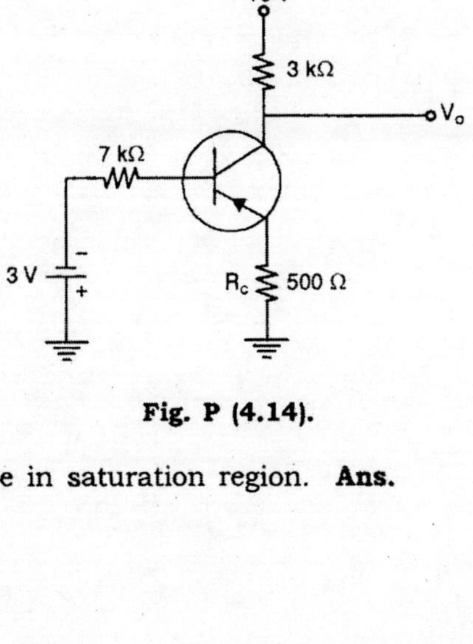

Fig. P (4.14).

Now, since $I_B > (I_B)_{min}$. So the transistor will be in saturation region. **Ans.**

(*b*) We have,

$$V_O = -V_{CE} - (I_B + I_C)\,500$$
$$= -0.2 - (2.99 \times 10^{-3})\,500$$
$$= -1.69 \text{ V. } \textbf{Ans.}$$

(*c*) According to the question for the transistor to be in active region.

$$V_{BC} \geq -0.5 \text{ V}$$

(Note that but for *npn* transistor $V_{BC} > 0.5$ V)

0.5 V is the emitter voltage of the collector junction

$$-3 = I_B \times 7 + (I_B + I_C)\,R_C - 0.7$$

Put $\quad I_C = \beta\, I_B \qquad (\because V_{BE\,(active)} = .7 \text{ V})$

$$-2.3 = I_B\,(7 + 101\,R_C)$$

(Note that all the resistors are taken in kΩ)

$$I_B = \frac{-2.3}{(7 + 101\,R_C)} \text{ mA}$$

$$I_C = \beta\, I_B = \frac{-230}{(7 + 101\,R_C)} \text{ mA}$$

Now $\quad V_{BN} = -0.7 - \dfrac{232.30\,R_C}{(7 + 101\,R_C)}$

$$V_{CN} = -10 + \frac{3 \times 230}{7 + 101\,R_C}$$

and $\quad V_{BC} = V_{BN} - V_{CN} \geq -0.5$

$$\frac{232.30\,R_C}{7 + 101\,R_C} - \frac{3 \times 230}{7 + 101\,R_C} \geq -0.5$$

After calculating we get

$$R_C \geq 819\ \Omega \quad \textbf{Ans.}$$

Example 21. *Calculate the Q-point values for the d.c. bias circuit shown in Fig. P (4.15).*

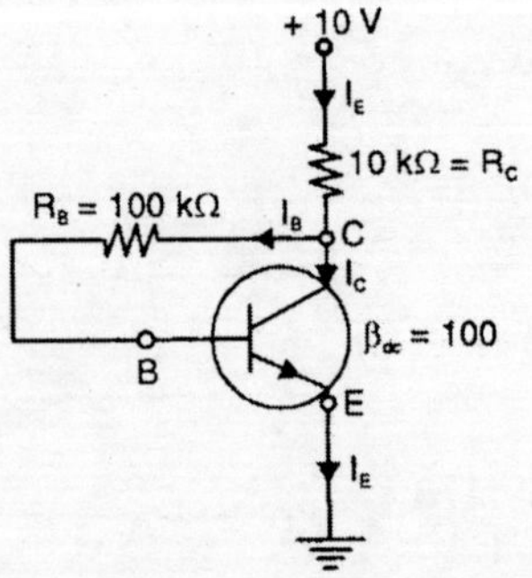

Fig. P (4.15).

Also draw the load line and locate Q-point on it. Assume V_{BE} = 0.7 V.

Solution : Given

$$V_{CC} = 10 \text{ V}$$
$$R_C = 10 \text{ k}\Omega$$
$$R_B = 100 \text{ k}\Omega$$
$$\beta_{dc} = 100$$
$$V_{BE} = 0.7 \text{ V.}$$

Q-point.

Applying KVL to the input section,

$$V_{CC} = I_E R_E + I_B R_B + V_{BE}$$

$$I_C = \frac{V_{CC} - V_{BE}}{R_C + \dfrac{R_B}{\beta}}$$

or
$$I_C = \frac{10 - 0.7}{10 \times 10^3 + \dfrac{100 \times 10^3}{100}}$$

or
$$I_C = .845 \times 10^{-3} \text{ A}$$

or
$$I_C = .845 \text{ mA.} \quad \textbf{Ans.}$$

and
$$V_{CE} = V_{CC} - I_C R_C$$

or
$$V_{CE} = 10 - .845 \times 10^{-3} \times 10 \times 10^3$$

or
$$V_{CE} = 10 - 8.45$$

$$V_{CE} = 1.55 \text{ V.} \quad \textbf{Ans.}$$

$$Q\text{-points} = (1.55 \text{ V}, .845 \text{ mA})$$

D.C. load line is shown in Fig. P (4.16).

$$I_{C(sat)} = \frac{V_{CC}}{R_C} = \frac{10}{10 \times 10^3} = 1 \text{ mA}$$

$$V_{CE(cut\text{-}off)} = V_{CC} = 10 \text{ V.}$$

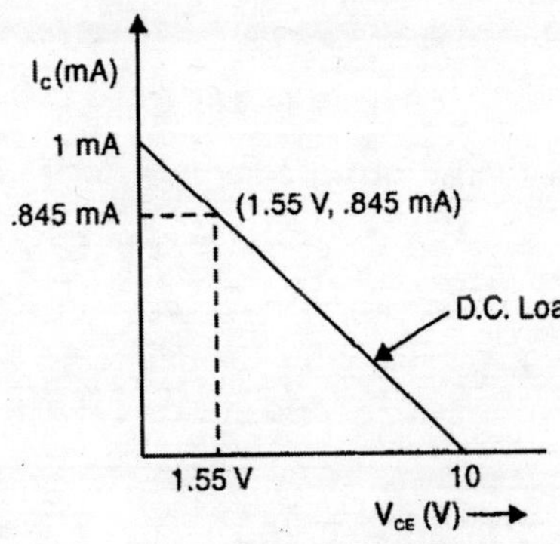

Fig. P (4.16).

Example 22. *Determine the d.c. bias current and voltage for the d.c. bias current as shown in Fig. P (4.17).*

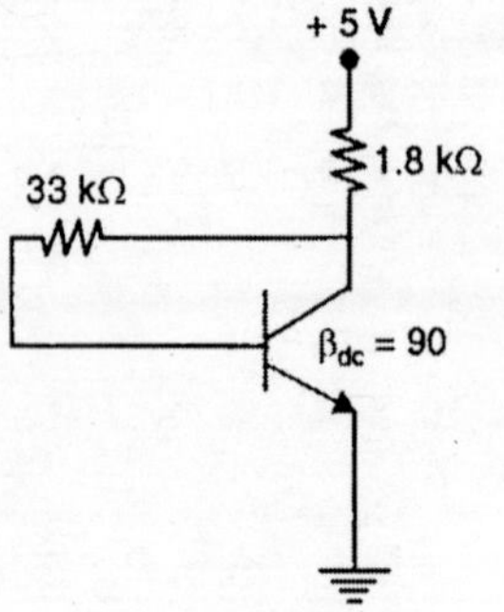

Fig. P (4.17).

and also find the stability.

Solution :

$$V_{CE} = 1.1 \text{ V}$$
$$I_C = 1.06 \text{ mA}$$
$$\text{Stability, } S = 16.1$$

Example 23. *A silicon transistor with $\beta = 100$ is to be used in the self-biasing circuit shown in Fig. P (4.18) such that the quiescent point corresponds to $V_{CE} = 12$ V and $I_C = 2$ mA and R_E if $V_{CC} = 24$ V and $R_C = 5$ kΩ.*

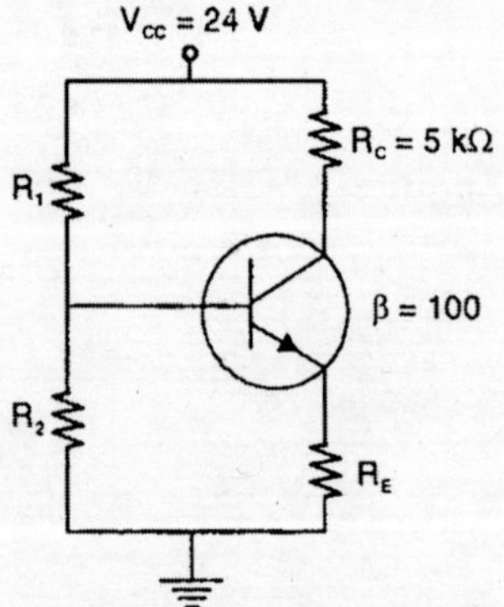

Fig. P (4.18).

Solution : Applying KVL to the output section,

$$V_{CC} = I_C R_C + V_{CE} + I_E R_E$$

or
$$V_{CC} = I_C R_C + V_{CE} + (I_C + I_B) R_E$$

or
$$V_{CC} = I_C R_C + V_{CE} + I_C R_E + \frac{I_C}{\beta} . R_E$$

or
$$V_{CC} = I_C \left(R_C + R_E + \frac{R_E}{\beta}\right) + V_{CE}$$

or $$\frac{V_{CC} - V_{CE}}{I_C} = R_C + R_E\left(1 + \frac{1}{\beta}\right)$$

or $$\frac{24 - 12}{2 \times 10^{-3}} - R_C = R_E\left(1 + \frac{1}{100}\right)$$

or $$\frac{12}{2 \times 10^{-3}} - 5 \times 10^3 = R_E\left(\frac{101}{100}\right)$$

or $$R_E = \frac{2}{2.02 \times 10^{-3}} = .99 \times 10^3 = 990\ \Omega.$$ **Ans.**

Example 24. *If in the case of example 15 it is desired to have $S \le 3$. Find the limiting values of R_1 and R_2, assuming V_{BE} (active) = 0.65 V.*

Solution : We know that stability factor in case of potential divider biasing is given by the relation,

$$S = \frac{(\beta + 1)(R_{th} + R_E)}{R_{th} + R_E(1 + \beta)}$$

$$3 = \frac{(100 + 1)(R_{th} + 990)}{R_{th} + 990(1 + 100)} \quad \left\{\text{Where } R_{th} = \frac{R_1 R_2}{R_1 + R_2}\right\}$$

$$3R_{th} + 299970 = 101\ R_{th} + 9990$$

or $$R_{th} = 2040.6 \text{ ohm}$$

$R_{th} < 2.041$ kΩ for $S \le 3$

Also, $$R_{th} = \frac{R_1 R_2}{R_1 + R_2} \qquad ...(1)$$

$$V_{th} = V_{CC} \cdot \frac{R_2}{R_1 + R_2} \qquad ...(2)$$

$$V_{th} = I_B R_{th} + V_{BE} + I_E R_E \qquad ...(3)$$

From eqns. (1), (2) and (3)

$$R_1 = \frac{V_{CC}}{V_{th}} \cdot \left(\frac{R_1 R_2}{R_1 + R_2}\right) = \frac{V_{CC}}{V_{th}} R_{th}$$

$$R_2 = \frac{V_{th} \cdot R_1}{V_{CC} - V_{th}}$$

Now $$V_{th} = \frac{I_C}{\beta} \cdot R_{th} + V_{BE} + (I_B + I_C)\ R_E$$

or $$V_{th} = \frac{2 \times 10^{-3}}{100} \times 2.041 \times 10^3 + .7 + \left(\frac{2 \times 10^{-3}}{100} + 2 \times 10^{-3}\right) 990$$

or $$V_{th} = 2.691 \text{ V.}$$

After manipulation we get,

$$R_1 = \frac{24 \times 2.041 \times 10^3}{2.691} = 18.203 \text{ k}\Omega$$

$$R_2 = \frac{18.023 \times 10^3 \times 2.691}{24 - 2.691} = 2.299 \text{ k}\Omega. \quad \textbf{Ans.}$$

Example 25. *Determine the values of the resistors in figure such that $I_C = 5$ mA, $V_{CE} = 8$ V, $V_E = 6$ V, $S = 10$, $h_{fc} = 200$ and $V_{CC} = 20$ V, $V_{BE} = .6$ V.*

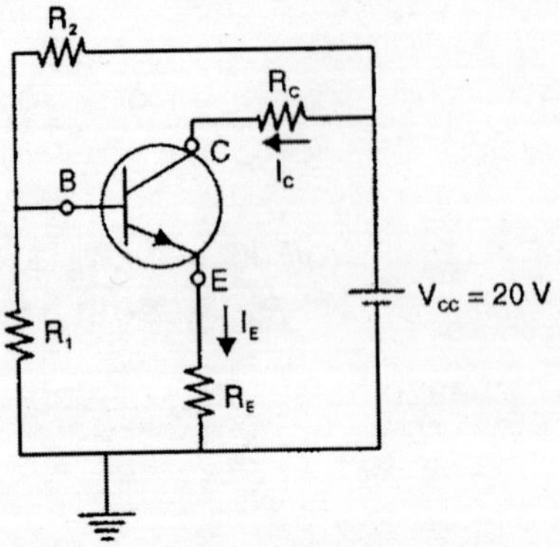

Fig. P (4.19).

Solution : From Fig. P (4.19).

Voltage at point C, $$I_C R_C = V_{CC} - V_{CE} - V_E$$

or $$I_C R_C = 20 - 8 - 6 = 6 \text{ V}$$

or $$I_C R_C = 6$$

or $$R_C = \frac{6}{I_C} = \frac{6}{5 \times 10^{-3}} = 1.2 \text{ k}\Omega. \quad \textbf{Ans.}$$

We know that

$$I_B = \frac{I_C}{\beta} = \frac{5 \times 10^{-3}}{200} = 0.025 \text{ mA.}$$

Voltage at point E

$$V_E = I_E R_E$$
$$6 = (I_C + I_B) R_E$$
$$6 = (5 \times 10^{-3} + 0.025 \times 10^{-3}) R_E$$
$$R_E = \frac{6}{5.025 \times 10^{-3}} = 1.194 \text{ k}\Omega. \quad \textbf{Ans.}$$

Given $$S = 10$$

$$10 = \frac{(\beta + 1)(R_B + R_E)}{R_B + R_E(1 + \beta)} \qquad \left\{\text{where } R_B = \frac{R_1 R_2}{R_1 + R_2}\right\}$$

$$10 = \frac{(200 + 1)(R_B + 1.194 \text{ k}\Omega)}{R_B + 1.194 \text{ k}\Omega (1 + 200)}$$

$$10R_B + 10(2399.940 \text{ k}\Omega) = 201(R_B + 1.194 \text{ k}\Omega)$$

or $$R_B = 11308.62\ \Omega$$

$$V_{th} = V_{CC}\left(\frac{R_2}{R_1 + R_2}\right)$$

or $$V_{th} = V_{CC}\frac{R_B}{R_1} \qquad \text{...(1)}$$

Applying KVL we get,

$$V_{th} = I_B R_B + V_{BE} + V_E$$

or $$V_{th} = 0.025 \times 10^{-3} \times 11308.62 + 0.6 + 6$$

or $$V_{th} = 6.883 \text{ V.}$$

From eqn. (1) $$R_1 = \frac{V_{CC}}{V_{th}} \cdot R_B$$

or $$R_1 = \frac{20}{6.883} \times 11308.62$$

$$R_1 = 32.861 \text{ k}\Omega. \quad \textbf{Ans.}$$

$$R_2 = \frac{R_1 \cdot V_{th}}{V_{CC} - V_{th}} = \frac{32.861 \times 6.883}{20 - 6.883}$$

$$= 17.243 \text{ k}\Omega. \quad \textbf{Ans.}$$

Example 26. *In the circuit shown in Fig. P (4.20), determine V_{BB} to saturate the transistor. Given $V_{CE\,sat} = 0.1$ V, $V_{BE\,sat} = 0.6$ V, $h_{fc} = 50$ (i.e., $\beta = 50$).*

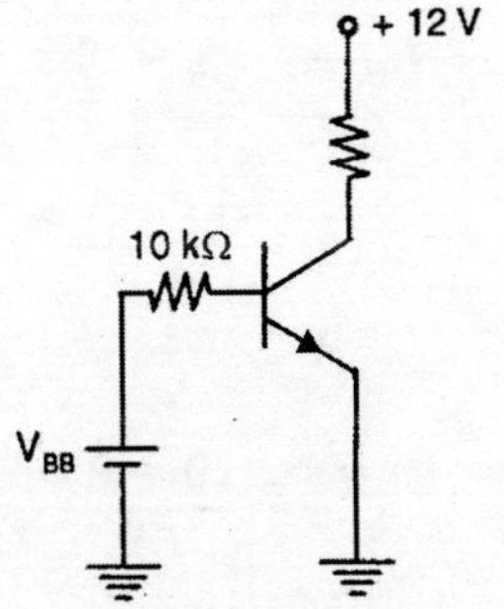

Fig. P (4.20).

Solution : From Fig. P (4.20).

$$I_C = \frac{V_{CC} - V_{CE\,(Sat)}}{R_C}$$

or
$$I_C = \frac{12 - 0.1}{1\ \text{k}\Omega} = 11.9\ \text{mA}$$

$$I_B = \frac{I_C}{\beta} = \frac{11.9}{50} = 0.238\ \text{mA}$$

so,
$$V_{BB} = I_B R_B + V_{BE\ \text{sat}}$$

or
$$V_{BB} = 0.238 \times 10^{-3} \times 10 \times 10^3 + 0.6$$

or
$$V_{BB} = 2.38 + .6$$

or
$$V_{BB} = 2.98\ \text{V}.\ \ \textbf{Ans.}$$

Example 27. *For the circuit shown in Fig. P (4.21), given $I_{CBO} = 10\ \mu A$, $h_{FE} = 100$. Calculate the value of I_C and V_O.*

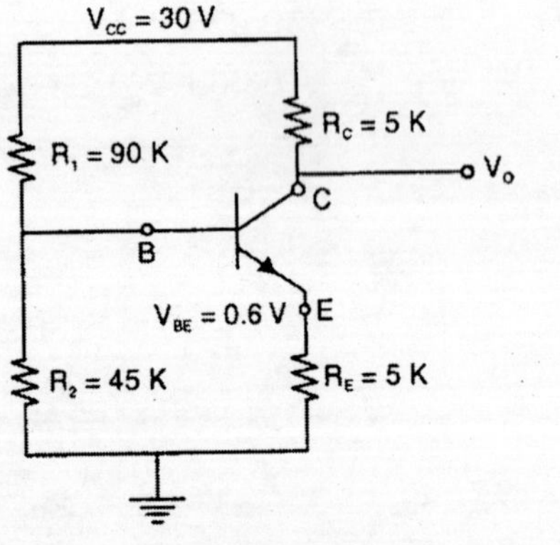

Fig. P (4.21).

Solution : Given $h_{FE} = 100$, $I_{CBO} = 10\ \mu\text{A}$.

From Fig. P (4.21),
$$V_{th} = V_{BB} = 30 \times \frac{45}{45 + 90} = 10\ \text{V}$$

$$R_{th} = R_{BB} = \frac{R_1 R_2}{R_1 + R_2} = \frac{90 \times 45}{90 + 45} = 30\ \text{K}.$$

$$I_C = \frac{V_{CC} - V_{BE}}{\dfrac{R_{th}}{\beta} + R_E}$$

or
$$I_C = \frac{(20 - 0.6)\ \text{V}}{\left(\dfrac{30}{100} + 5\right)\text{K}}$$

or $$I_C = 1.757 \text{ mA}.$$

$$I_{CEO} = (\beta + 1)\, I_{CO} = 101\ (10 \times 10^{-6}) = 1.01 \text{ mA}.$$

Total $$I_C = 1.757 \text{ mA} + 1.01 \text{ mA} = 2.767 \text{ mA}. \quad \textbf{Ans.}$$

$$V_O = V_{CC} - I_C R_C$$

or $$V_O = 30 - 2.767 \times 10^{-3} \times 5 \times 10^3 = 16.165 \text{ V}. \quad \textbf{Ans.}$$

Example 28. *Find the I_{CO} and V_{CEQ} for Fig. P (4.22), given $\beta_{dc} = 130$. (Assume $V_{BE} = 0.7$ V)*

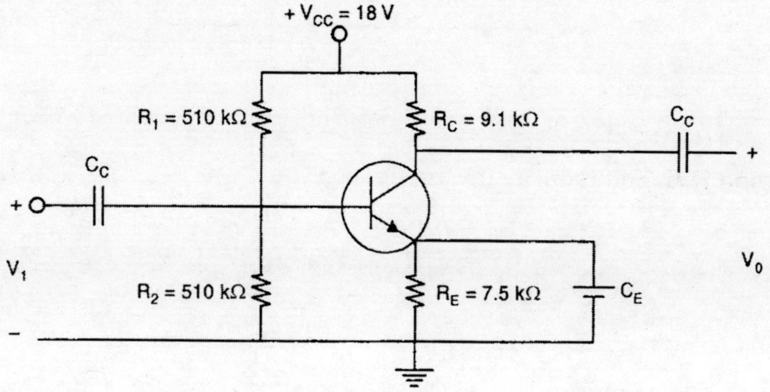

Fig. P (4.22).

Solution : First of all we will draw DC equivalent circuit by open circuiting the capacitors as shown in Fig. P (4.23).

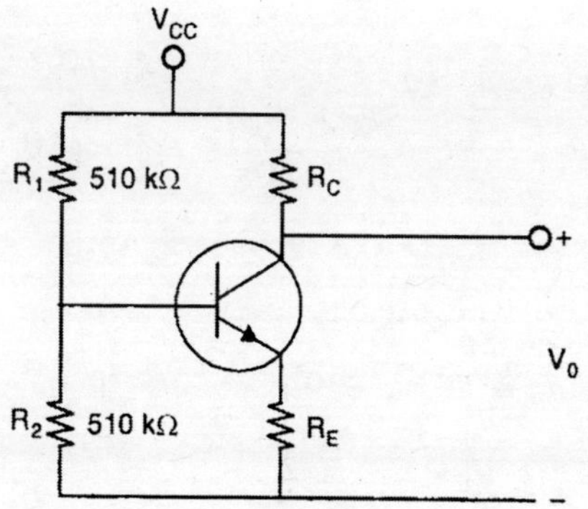

Fig. P (4.23). *DC equivalent circuit of the given circuit.*

This circuit can be redrawn using Thevenin's equivalent in the base circuit, as in Fig. P (4.24).

Here, $$V_{th} = V_{CC} \cdot \frac{R_2}{R_1 + R_2} = 18 \times \frac{510}{510 + 150} = 9 \text{ V}$$

$$R_{th} = \frac{R_1 R_2}{R_1 + R_2} = \frac{510 \times 510}{510 + 510} = 255 \text{ k}\Omega$$

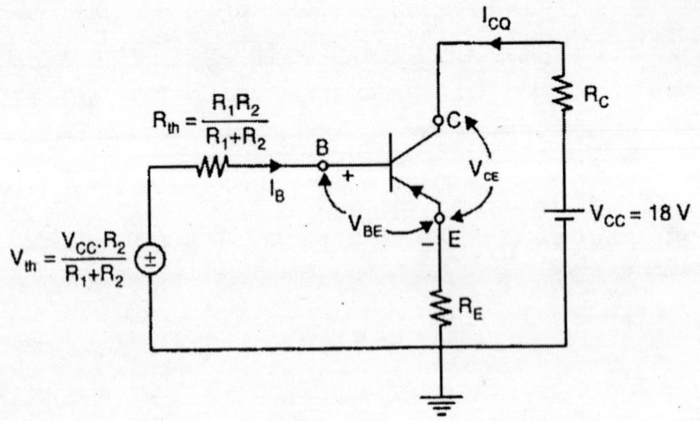

Fig. P (4.24). *Thevenin equivalent circuit of the above circuit.*

Applying KVL equation in the input loop

$$V_{th} = I_B R_{th} + V_{BE} + I_E R_E$$

$$V_{th} = I_B R_{th} + V_{BE} + (I_C + I_B) R_E \qquad (\because I_E = I_C + I_B)$$

$$V_{th} = \frac{I_C}{\beta} R_{th} + V_{BE} + \left(I_B + \frac{I_C}{\beta} \right) R_E$$

or $$I_C \cong I_{CQ} = \frac{V_{th} - V_{BE}}{\left(\frac{R_{th}}{\beta} + R_E + \frac{R_E}{\beta} \right)} \qquad (\because I_C = \beta I_B)$$

or $$I_{CQ} = \frac{9 - 0.7}{\frac{255 \times 10^3}{130} + 7.5 \times 10^3 + \frac{7.5 \times 10^3}{130}}$$

$$I_{CQ} = 0.872 \times 10^{-3} \text{ Amp} = .872 \text{ mA.} \quad \textbf{Ans.}$$

Again applying KVL at the output loop, we get,

$$V_{CC} = I_C R_C + V_{CEQ} + I_E R_E$$

or $$V_{CEQ} = V_{CC} - I_C R_C + I_E R_E$$

or $$V_{CEQ} = V_{CC} - I_C \left(R_C + R_E + \frac{R_E}{\beta} \right)$$

or $$V_{CEQ} = 18 - .872 \times 10^{-3} \left(9.1 \times 10^3 + 7.5 \times 10^3 + \frac{7.5 \times 10^3}{130} \right)$$

or $$V_{CEQ} = 3.475 \text{ V.} \quad \textbf{Ans.}$$

Hence the required values are

$$I_{CQ} = 0.872 \text{ mA}, \; V_{CEQ} = 3.475 \text{ V.}$$

Example 29. *What will happen to V_{CE} if R_2 in open circuited in the previous example.*

Solution : The circuit becomes like afer opening the R_2 is given below :

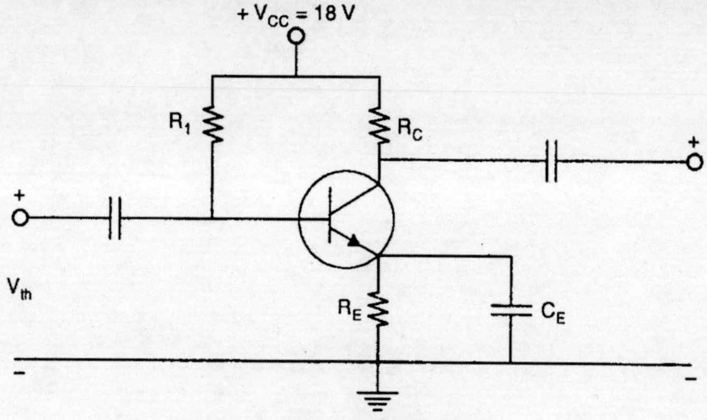

Fig. P (4.25).

DC equivalent for this circuit will be as below :

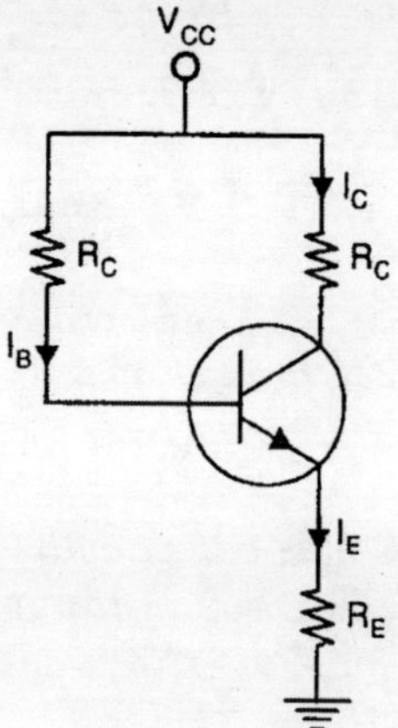

Fig. P (4.26).

For better understanding this circuit is replaced by a redrawn circuit having two supplies in place of one supply.

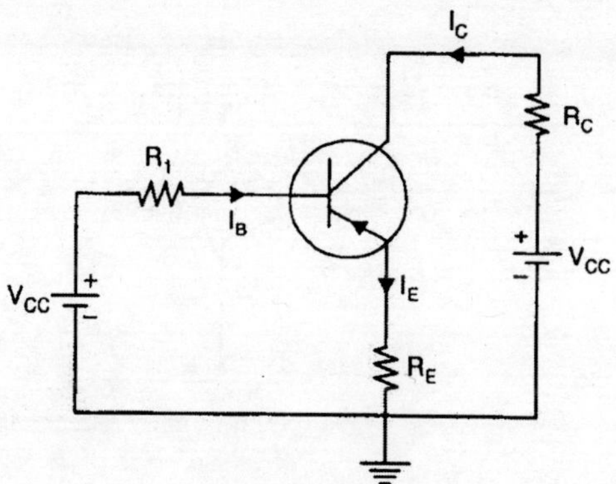

Fig. P (4.27).

This circuit is similar to the Thevenin equivalent circuit shown in the previous problem. The only difference is V_{th} is replaced by V_{CC} and R_{th} is replaced by R_1. By similarty,

$$I_{CQ} = \frac{V_{CC} - V_{BE}}{\frac{R_1}{\beta} + R_E + \frac{R_E}{\beta}}$$

or
$$I_{CQ} = \frac{18 - 0.7}{\left(\frac{510 \times 10^3}{130} + 7.5 \times 10^3 + \frac{7.5 \times 10^3}{130}\right)}$$

or
$$I_{CQ} = 1.507 \text{ mA. } \textbf{Ans.}$$

And
$$V_{CEQ} = V_{CC} - I_{CQ}\left(R_C + R_E + \frac{R_E}{\beta}\right)$$

or
$$V_{CEQ} = 18 - 1.507 \times 10^{-3}\left(9.1 \times 10^3 + 7.5 \times 10^3 + \frac{7.5 \times 10^3}{130}\right)$$

or
$$V_{CEQ} = -7.103 \text{ V. } \textbf{Ans.}$$

NOTE : Here it may be noted that, V_{CEQ} can not become negative it means base current is too much so that transistor gets saturated and β reduces and will not remain 130.

$$V_{CEQ} \cong V_{CE\,(sat)} \cong 0.2 \text{ V.}$$

Example 30. *Consider a DC bias circuit with voltage feedback, given below. Determine the quiescent points I_{CQ} and V_{CEQ}. The β of transistor is 90, and cut in voltage is 0.7 V. What will be the effect on I_{CQ} and V_{CEQ} if β is increased by 50% ?*

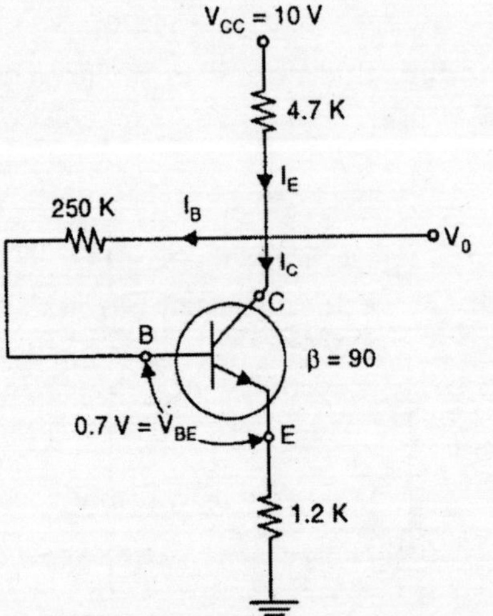

Fig. P (4.28).

Solution : Apply KVL for the input loop :

$$V_{CC} = 10 = 4.7 \times 10^3 I_E + 250 \times 10^3 I_B + V_{BE} + I_E \times 1.2 \times 10^3$$

$$10 - 0.7 = 4.7 \times 10^3 \left(I_C + \frac{I_C}{\beta}\right) + 250 \times 10^3 \frac{I_C}{\beta} + \left(I_C + \frac{I_C}{\beta}\right) 1.2 \times 10^3$$

or $$I_C = I_{CQ} = \frac{\beta (10 - 0.7)}{(1 + \beta)(4.7 \times 10^3 + 1.2 \times 10^3) + 250 \times 10^3}$$

or $$I_{CQ} = \frac{90 (9.3)}{(1 + 90)(5.9 \times 10^3) + 250 \times 10^3}$$

or $$I_{CQ} = 1.06 \text{ mA.} \quad \textbf{Ans.}$$

Again applying KVL for the output loop,

$$V_{CC} = (I_C + I_B)\, 4.7 \times 10^3 + V_{CEQ} + (I_C + I_B)\, 1.2 \times 10^3$$

$$V_{CEQ} = 10 - \left(1.06 \times 10^{-3} + \frac{1.06 \times 10^{-3}}{90}\right)\left[4.7 \times 10^3 + 1.2 \times 10^3\right]$$

or $$V_{CEQ} = 3.68 \text{ V.} \quad \textbf{Ans.}$$

If β in increased by 50% *i.e.*, $\beta' = 90 + 45 = 135$.

Now, $$I'_{CQ} = \frac{\beta' (V_{CC} - V_{BE})}{(\beta' + 1)(4.7 \times 10^3 + 1.2 \times 10^3) + 250 \times 10^3}$$

$$I'_{CQ} = \frac{135 (10 - 0.7)}{(135 + 1)(5.9 \times 10^3) + 250 \times 10^3} = 1.19 \text{ mA.} \quad \textbf{Ans.}$$

$$V'_{CEQ} = 10 - \left(1.06 \times 10^{-3} + \frac{1.06 \times 10^{-3}}{135}\right)\left(4.7 \times 10^3 + 1.2 \times 10^3\right)$$

or $$V'_{CEQ} = 2.93 \text{ V.} \quad \textbf{Ans.}$$

Example 31. *Given I_E = 1.2 mA, β = 120, and r_o = 40 K ohm. Sketch the*

(i) Common emitter hybrid equivalent model.

(ii) Common base hybrid equivalent model.

Solution : Try yourself.

Example 32. *Sketch the circuit of a common collector configuration and also the input and output curves.*

Solution : Try yourself.

Example 33. *Draw npn and pnp transistors. Label all the currents and show the direction of flow. How are all the currents of a transistor related ?*

Solution : Try yourself.

Example 34. *Find the value of* β, V_{CC} *and* R_B *of the following circuit.*

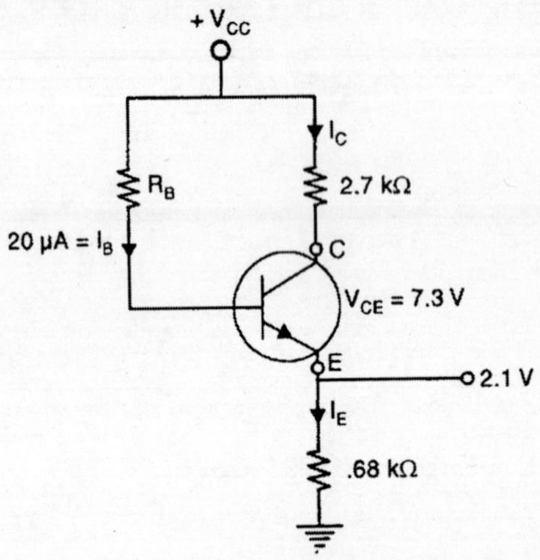

Fig. P (4.29).

Solution : $$I_E = \frac{V_E}{R_E}$$

or $$I_E = \frac{2.1\text{ V}}{0.68\text{ k}\Omega} = 3.088\text{ mA}$$

$$I_E = I_C + I_B$$

or $$I_C = I_E - I_B$$

or $$I_C = 3.088 - 20 \times 10^{-3}\text{ mA} \quad (\because I_B = 20\ \mu\text{A} = 20 \times 10^{-6}\text{ Amp.})$$

or $$I_C = 3.068\text{ mA.}$$

we know, $$\beta = \frac{I_C}{I_B} = \frac{3.068 \times 10^{-3}}{20 \times 10^{-6}} = 153.4. \quad \textbf{Ans.}$$

Apply KVL to the output loop, we get,

$$V_{CC} = I_C R_C + V_{CE} + I_E R_E$$

or $$V_{CC} = 3.068 \times 10^{-3} \times 2.7 \times 10^3 + 7.3 + 3.088 \times 10^{-3} \times 0.68 \times 10^3$$

or $$V_{CC} = 17.684\text{ volts.} \quad \textbf{Ans.}$$

Again apply KVL to the input loop, we get

$$V_{CC} = I_B R_B + V_{BE} + I_E R_E$$

$$R_B = \frac{V_{CC} - V_{BE} - I_E R_E}{I_B}$$

or $$R_B = \frac{17.684 - 0.7 - 3.088 \times 10^{-3}}{20 \times 10^{-6}} \times .68 \times 10^3$$

or $$R_B = 744.2\text{ k}\Omega. \quad \textbf{Ans.}$$

Example 35. *Find A_v, R_i and R_0 of the following circuit.*

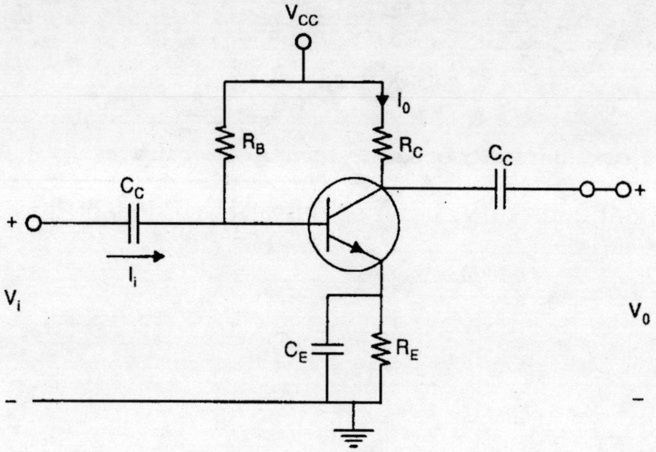

Fig. P (4.30).

Solution : Try yourself.

Example 36. *Draw the load line for the following Fig. P (4.31), what is I_C at saturation point ? Find V_{CE} at cut off point.*

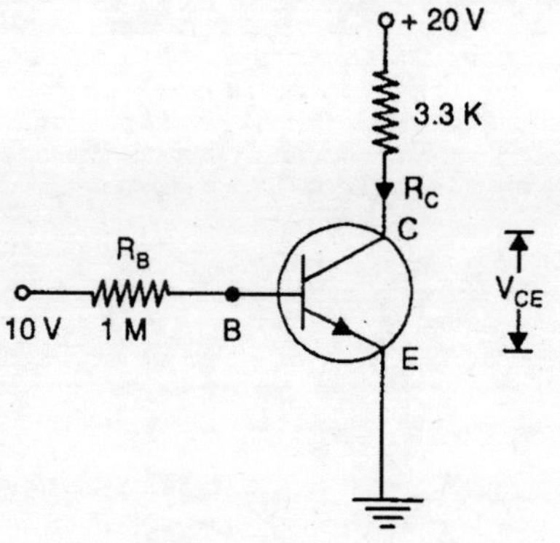

Fig. P (4.31).

Solution : Equation for DC load line will be,

$$I_C = -\frac{1}{R_C}V_{CE} + \frac{V_{CC}}{R_C}$$

It cuts the X-axis on output characteristics at,

$$V_{CE} = V_{CC} \qquad (\because V_{CE} \text{ at cut-off})$$
$$= 20 \text{ volts.}$$

It cuts the Y-axis on output characteristics at,

$$I_C = \frac{V_{CC}}{R_C} = \frac{20}{3.3\text{ K}} \qquad (I_C \text{ is at saturation})$$

$$= 6.06 \text{ mA.} \quad \textbf{Ans.}$$

Load line in draw joining two points having co-ordinates $\left(0 \cdot \frac{V_{CC}}{R_C}\right)$ and $(V_{CC}, 0)$ *i.e.*, (0, 6.06 mA) and (20 V, 0) as shown in Fig. P (4.32).

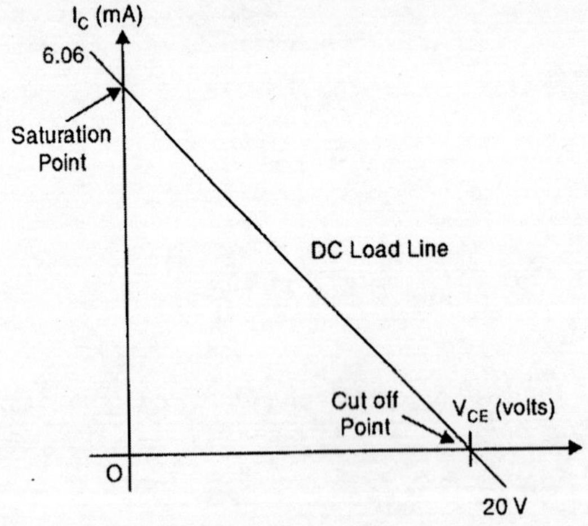

Fig. P (4.32).

Example 37. *Find V_{CE} and I_E in Fig. P (4.33).*

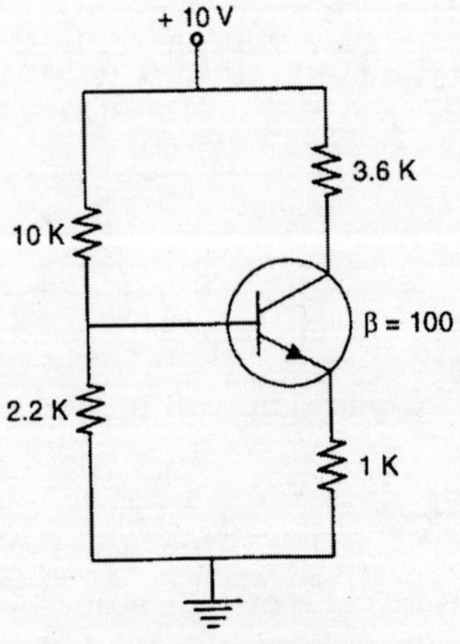

Fig. P (4.33).

Solution : Thevenin's equivalent circuit is drawn in Fig. P (4.34) for the given circuit.

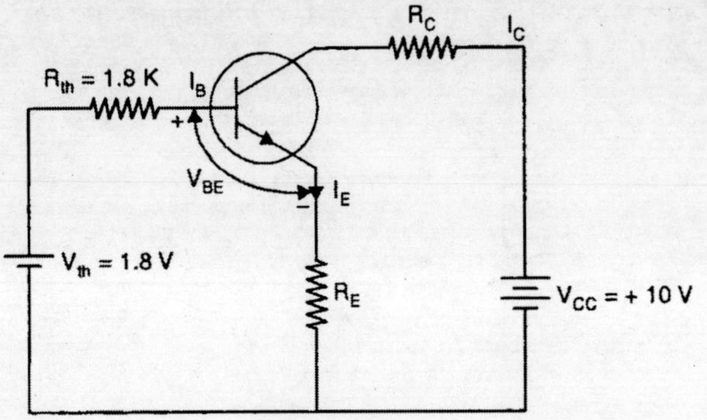

Fig. P (4.34).

Here, $$V_{th} = \frac{R_2}{R_1 + R_2} \times V_{CC} = \frac{2.2}{10 + 2.2} \times 10 = 1.8 \text{ volts}$$

$$R_{th} = \frac{R_1 R_2}{R_1 + R_2} = \frac{2.2 \times 10}{10 + 2.2} = 1.8 \text{ K}$$

KVL is input loop

$$V_{th} = I_B R_{th} + V_{BE} + I_E R_E$$

We know that $$I_E = (\beta + 1)\, I_B$$

So $$V_{th} = \frac{I_E}{\beta + 1} R_{th} + V_{BE} + I_E R_E$$

or $$I_E = \frac{V_{th} - V_{BE}}{R_E + \frac{R_{th}}{(\beta + 1)}}$$

here
$$V_{th} = 1.8 \text{ V}$$
$$V_{BE} = 0.7 \text{ V}$$
$$R_E = 1 \text{ K}$$
$$R_{th} = 1.8 \text{ K}$$
$$\beta = 100$$

So $$I_E = \frac{1.8 - 0.7}{1000 + \frac{1800}{101}} = 1.081 \text{ mA.}$$ **Ans.**

$$I_B = \frac{I_E}{\beta + 1} = \frac{1.081}{101} = 10.7\ \mu\text{A} = 1.07 \text{ mA.}$$

KVL in output loop

$$V_{CC} = I_C R_C + V_{CE} + I_E R_E$$
$$V_{CE} = V_{CC} - I_C R_C - I_E R_E$$
$$= 10 - 1.07 \times 10^{-3} - 1.081 \times 10^{-3} \times 10^3$$
$$= 5.067 \text{ volts.}$$ **Ans.**

Example 38. *Determine Z_i, Z_0 and A_v for the following network in Fig. P (4.35). Given h_{fc} = 110, h_{ie} = 1.6 K, h_{re} = 2 × 10^{-4} and h_{oe} = 20 μA/V.*

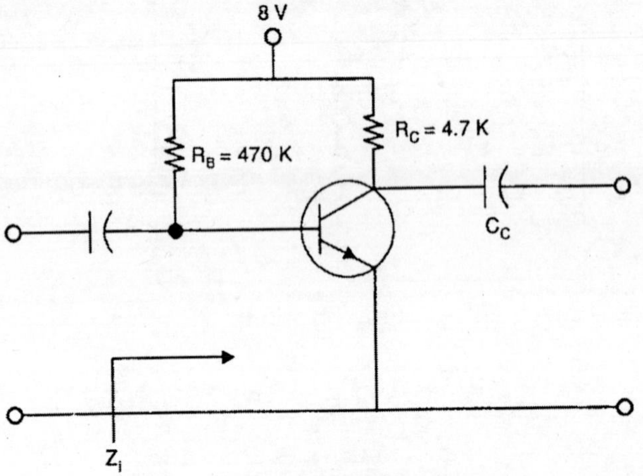

Fig. P (4.35).

Solution : AC equivalent circuit using C-E, *h*-parameter model of transistor is drawn in Fig. P (4.36).

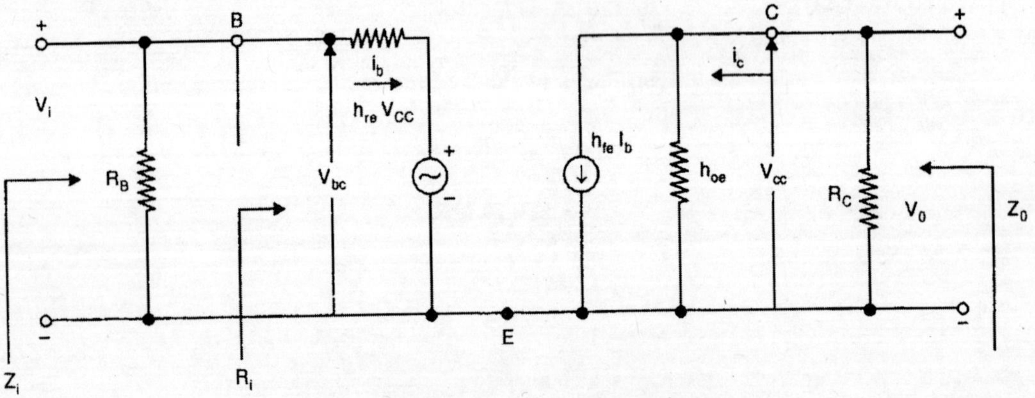

Fig. P (4.36). *CE, h-parameter model.*

Here, $$R_{in} = h_{ie} - \frac{h_{fe}\, h_{re}\, R_C}{1 + h_{oe}\, R_C} = 1.6 \times 10^3 - \frac{110 \times 2 \times 10^{-4} \times 4.7 \times 10^3}{1 + 20 \times 10^{-6} + 4.7 \times 10^3}$$

$$= 1.51 \text{ K ohm}$$

$$Z_i = R_B \parallel R_i = \frac{R_B R_i}{R_B + R_i} = \frac{470 \times 10^3 \times 1.51 \times 10^3}{470 \times 10^3 + 1.51 \times 10^3}$$

$$= 1.50 \text{ K ohm.}$$ **Ans.**

We know $$A_V = \frac{-h_{fe}}{Z_{in}\left(h_{oe} + \frac{1}{R_L}\right)}$$

here $R_L = R_C$

So
$$A_V = \frac{-110}{1.5\times10^3\left(20\times10^{-6}+\dfrac{1}{4.7\times10^3}\right)} = -313. \textbf{ Ans.}$$

We know
$$Z_0 = \frac{1}{h_{oe} - \dfrac{h_{fe}\,h_{re}}{Z_{in}+R_s}}$$

here
$$R_s = 0$$

So
$$Z_0 = \frac{1}{20\times10^{-6} - \dfrac{110\times2\times14^{-4}}{1.51\times10^3}} = \frac{10^6}{20-14.57}$$
$$= 184 \text{ K ohm. } \textbf{Ans.}$$

Example 39. *Determine the dc level of I_B, I_C and V_C for the following circuit Fig. P (4.37).*

Solution : The level of I_B, I_C, V_C may be obtained as follow:

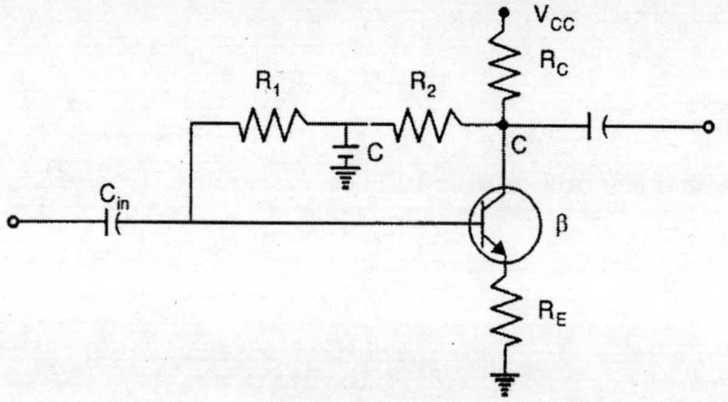

Fig. P (4.37).

We obtain the I_B considering the current eqn. of I_B as follows.

Applying KVL in the loop of R_1 and R_2

$$V_{CC} - I_C'R_C + (R_1 + R_2)\,I_B - V_{BE} - I_ER_E = 0$$
$$I_C' = I_C + I_B = I_E$$

and we get,
$$I_C = \beta I_B$$
$$I_E = (\beta + 1)\,I_B$$

Substituting the values, we get,

or
$$V_{CC} - I_ER_C - (R_1 + R_2)\,I_B - V_{BE} - I_ER_E = 0$$

or
$$V_{CC} - (\beta + 1)\,I_BR_C + (R_1 + R_2)\,I_B - V_{BE} - (\beta + 1)\,I_BR_E = 0$$

or
$$I_B[(\beta + 1)\,R_C + R_1 + R_2 + (\beta + 1)\,R_E] = V_{CC} - V_{BE}$$

$$I_B = \frac{V_{CC}-V_{BE}}{(\beta+1)(R_C+R_E)+R_1+R_2}$$

$$\beta + 1 \cong \beta$$

or
$$I_B = \frac{V_{CC} - V_{BE}}{R_1 + R_2 + \beta(R_C + R_E)}$$
$$I_C = \beta I_B$$
$$V_C = V_{CC} - I_C' R_C$$

Example 40. *Draw the dc load line on the BJT characteristics for the following circuit Fig. P (4.38).*

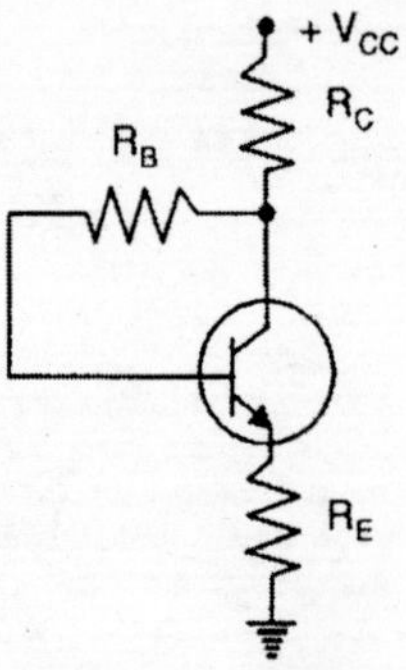

Fig. P (4.38).

Solution : DC load line can be drawn by finding the value of I_{CQ} and V_{CEQ} applying KVL we get (from the previous problem)

$$I_B = \frac{V_{CC} - V_{BE}}{R_B + \beta(R_E + R_C)}$$

$$I_C = \beta I_B = \frac{(V_{CC} - V_{BE})\beta}{R_B + \beta(R_E + R_C)} = I_{CQ}$$

and V_{CE} is given by

$$V_{CE} = V_{CC} - I_C(R_E + R_C) = V_{CQ}$$

thus by given values of R_C, R_E, V_{CC} and β we can draw load lines of Q point.

□□□

5

TRANSISTOR AMPLIFIER

5.1. INTRODUCTION

Sometimes when we are not able to hear a stereo system, we increase the volume, when the picture on our television is too dark, we increase the brightness control. In both of these cases, we are talking a relatively weak signal and making it stronger by changing the d.c. power into a.c. power. Thus, the process of increasing the power of an a.c. signal is called amplification.

One things should be always kept in mind that an amplifier never changes the shape of the actual signal *i.e.,* input signal. However, it increases the level of the actual signal or in other words an amplifier may also be defined as a device, which produces a larger electrical output of a similar characteristics than that of the input parameters. It increases the magnitude of current or voltage input by means of energy drawn from an external source *i.e.,* Battery.

As discussed above, an amplifier needs a source of energy for amplification. This may be provided by a battery or a d.c. source resulting from a rectifier and filter. An amplifier also contains at least one active device. This may be an electron tube, BJT or FET transistor, which can provide a control function. Basically the active device converts the energy from the d.c. source into energy at the output of the amplifier, which is proportional to the input signal.

The amplifier, in which the instantaneous output signal is directly proportional to the corresponding input signal, is called a linear amplifier. On the other hand, an amplifier, in which the output is not directly proportional to the input, is called a non-linear amplifier.

5.2. CLASSIFICATION OF AMPLIFIERS

The linear amplifiers may be classified according to their mode of operation :

As based on the input → Small-signal amplifier
→ Large-signal amplifier

As based on transistor configuration → Common-emitter (CE) amplifier
→ Common-base (CB) amplifier
→ Common-collector (CC) amplifier

As based on biasing conditions →
- Class-A amplifier
- Class-B amplifier
- Class-AB amplifier
- Class-C amplifier

As based on the number of stages →
- Single-stage amplifier
- Multi-stage amplifier

The amplifier can be analysed for its performance by the following two methods :

(*i*) Graphical method.

(*ii*) Equivalent circuit method.

Graphical method. For analysing an amplifier by the graphical method, we need the output characteristics of the transistor. When the a.c. voltage (*i.e.*, signal) is applied to the input, the base current varies. The corresponding variations in collector current and collector voltage can be seen on the characteristics. As the graphical method involves no approximations, therefore the results obtained by this method are more accurate than the equivalent circuit method.

5.3. GRAPHICAL ANALYSIS OF CE AMPLIFIER

Fig. 5.1(*a*) shows a basic circuit of an CE amplifier, which uses potential divider biasing here potential divider biasing is used as its circuit is simple and provides good stabilization of the operating point. If this circuit is to amplify a.c. voltages, some more components must be added to it. The result is shown in Fig. 5.1(*b*), here we have added three capacitors.

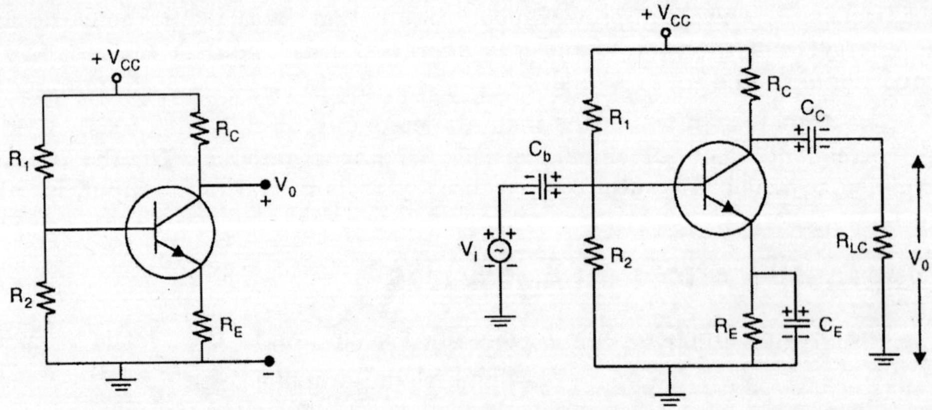

(*a*) Potential divider biasing circuit (*b*) Basic common emitter amplifier circuit

Fig. 5.1.

As we have discussed that here circuit uses three capacitors namely :

(*i*) C_b called blocking capacitor.

(*ii*) C_C called coupling capacitor.

(*iii*) C_E called bypass capacitor.

A coupling capacitor passes an a.c. signal from one side to the other, at the same time, it does not allow the d.c. voltage to pass through. Hence it is also called blocking capacitor, usually the main function of coupling capacitor in the cascading of the amplifier (*i.e.*, multistage amplifier). Thus we can say that coupling capacitor and blocking capacitor do the same job, however its working time is different. In other words, we can say that blocking capacitor used to provide good stability to the circuit as it blocks the a.c. signal at the time of biasing of the circuit.

The main function of bypass capacitor (C_E) to bypass all the a.c. currents from the emitter to the ground. If C_E is not put in the circuit, the a.c. voltage developed across R_E will affect the input a.c. voltage. Such a feedback of a.c. signal is reduced by putting the capacitor C_E, resulting increases the gain of the amplifier *i.e.*, transistor. As we know that capacitance is given by the relation $X_C = \frac{1}{2\pi fC}$ which depends upon the value of capacitor and the frequency of the signal. For a particular circuit arrangement value of capacitor is fixed, now we have full control on the frequency of the signal. For a high frequency signal capacitance X_C becomes lowest as compared to R_E. As a practical guide, we make the reactance of the capacitor C_E at the lowest frequency, not more than one-tenth the value of R_E *i.e.*,

$$\boxed{X_{CE} \leq \frac{R_E}{10}} \qquad ...(5.1)$$

The resistor R_0 represents the resistance of whatever is connected at the output.

To what extent an amplifier enlarges signal is expressed in terms of its voltage gain. The voltage gain of an amplifier is given as :

$$A_v = \frac{\text{Output a.c. voltage}}{\text{Input a.c. voltage}} = \frac{V_o}{V_i} \qquad ...(5.2)$$

One things should be always kept in mind that amplifier never changes the frequency of the input signal. However, it increases the amplitude of the input signal.

5.3.1. Calculation of Voltage and Current Gain by Graphical Method

In order to calculate the current gain and voltage gain by the graphical method, we consider a typical amplifier circuit as shown in Fig. 5.1(*b*). In this circuit let the values of R_1 = 33 kΩ, R_2 = 3.3 kΩ, V_i = 5 mV, R_C = 1 kΩ, V_{CC} = 9 V, R_E = 100 Ω, and C_E = 250 μF, R_0 = 470 Ω. The output characteristics of this circuit is shown in Fig. 5.2.

The simplest way to analyse the circuit is to split the analysis into the following two parts.

(1) d.c. analysis.

(2) a.c. analysis.

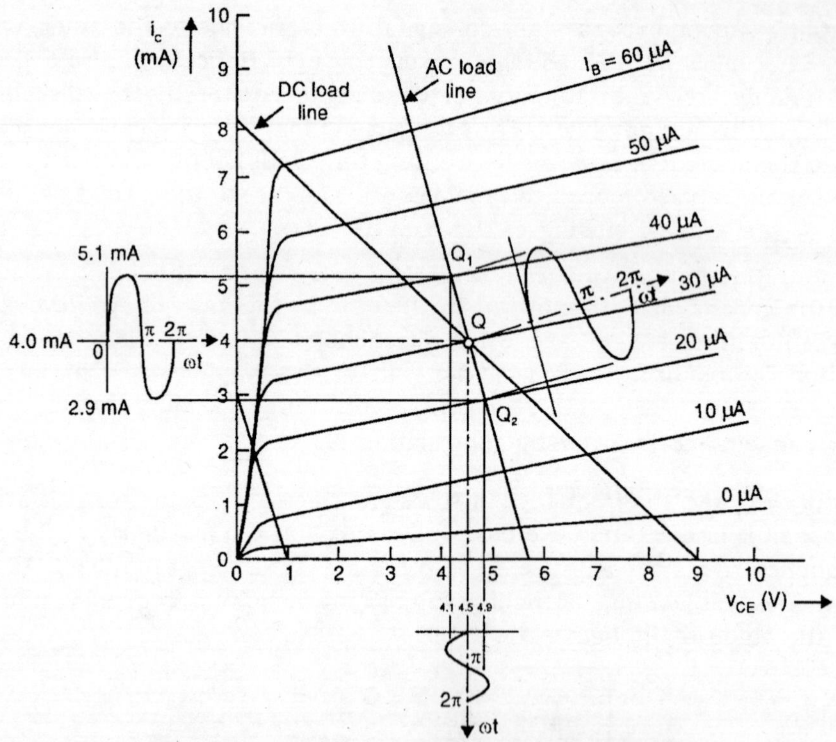

Fig. 5.2. *Graphical method for calculating gain.*

The following procedure is adapted for the d.c. analysis :

1. First of all, reduces the a.c. sources to zero. It means the voltage source is replaced by a short circuit and the current source is replaced by an open-circuit.
2. Now open all the capacitors, as they block direct current.

The circuit, which remains after applying the above two steps, is called a d.c. equivalent circuit. The d.c. equivalent circuit for Fig. 5.1(*b*) is shown in Fig. 5.3.

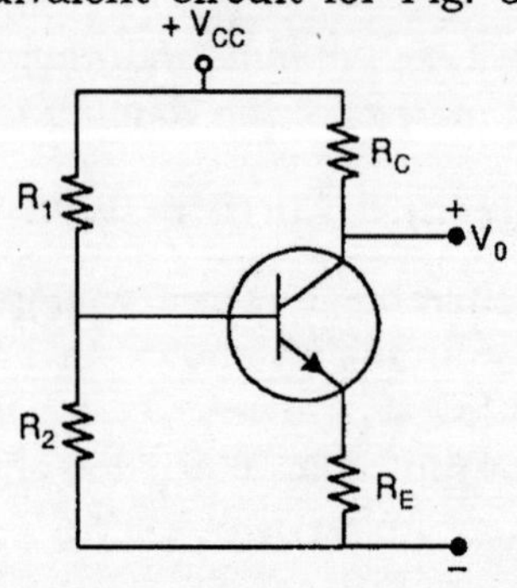

Fig. 5.3. *D.C. equivalent circuit of Fig. 5.1(b).*

The following procedure is adapted for the a.c. analysis :

(*i*) First of all, reduces the d.c. sources to zero. It means that a voltage source is replaced by a short-circuit and a current source by an open circuit.

(*ii*) Now short all the capacitors as they pass alternating current.

(*iii*) Then replace the transistor with its small signal (which will discussed later) equivalent circuit.

The circuit, which remains after applying the above three steps, is called an a.c. equivalent circuit. The circuit can be used to determine the a.c. currents and voltages, which are required for analysis. The a.c. equivalent circuit for Fig. 5.1(*b*) is shown in Fig. 5.4.

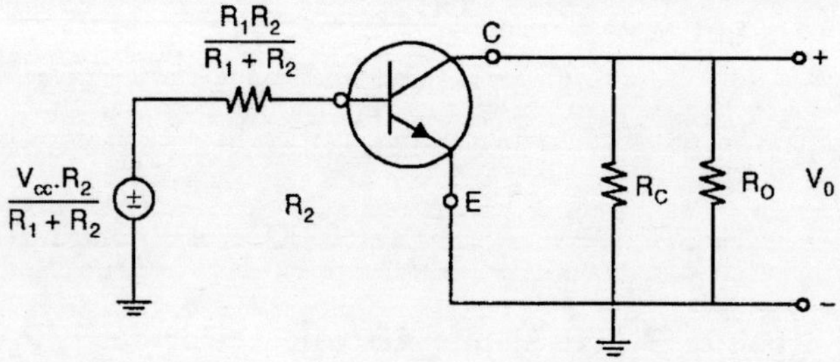

Fig. 5.4. *a.c. equivalent circuit of Fig. 5.1(b).*

Now, we first plot the d.c. load line on the output characteristics (as shown in Fig. 5.2). The equation of this d.c. load line is given by the relations :

$$V_{CC} = I_C R_C + V_{CE} + I_E R_E$$

or $$V_{CC} = V_{CE} + I_C (R_C + R_E) \qquad \{\because I_C \simeq I_E\}$$

or $$I_C = \frac{V_{CC}}{(R_C + R_E)} - \frac{V_{CE}}{(R_C + R_E)}$$

The line is drawn by simply joining the points. (V_{CC}, 0) and (0, $\frac{V_{CC}}{R_{dc}}$). This may be seen that the slope of the d.c. load line is $-\frac{1}{R_{dc}}$.

Here $R_{dc} = R_C + R_E$. Thus the two points for plotting the d.c. load line are (9 V, 0) and (0, 8.2 mA).

Let us assume that the biasing arrangement is such that the d.c. base current is 30 μA. The quiescent operating point *Q* is given by the intersection of the d.c. load line and the output characteristics corresponding to I_B = 30 μA. The d.c. collector current and collector-to-emitter voltage of the *Q*-point may be seen to be 4 mA and 4.5 V respectively.

When we apply an a.c. input V_i, the circuit behaves like the one shown in Fig. 5.1(*b*). It is clear that $R_{ac} = R_C \parallel R_O$ (*i.e.*, here 1 kΩ ∥ 470 Ω = 320 Ω). The a.c. load line will have a slope of $-\frac{1}{R_{ac}}$. Since the *Q*-point describes the zero-signal conditions of the circuit, the a.c. load line should also pass through *Q*-point. To draw such a line we first draw any line AB with the given slope. We can then draw the a.c. load line parallel to this line and passing through the *Q*-point.

Fig. 5.5 explains how a line with the given slope $-\frac{1}{R_{ac}}$ is drawn.

$$\text{Slope} = -\frac{1}{R_{ac}} = \tan\theta = \tan(180-\alpha) = -\tan\alpha$$

$$\therefore \quad \tan\alpha = \frac{1}{R_{ac}} = \frac{OB}{OA}$$

Let us take OA = 1 V. Then

$$OB = \frac{OA}{R_{ac}} = \frac{1}{320} = 0.0031 \text{ A} = 3.1 \text{ mA}.$$

After locating the points B and A, line AB whose slope = $1/R_{ac}$ can be drawn.

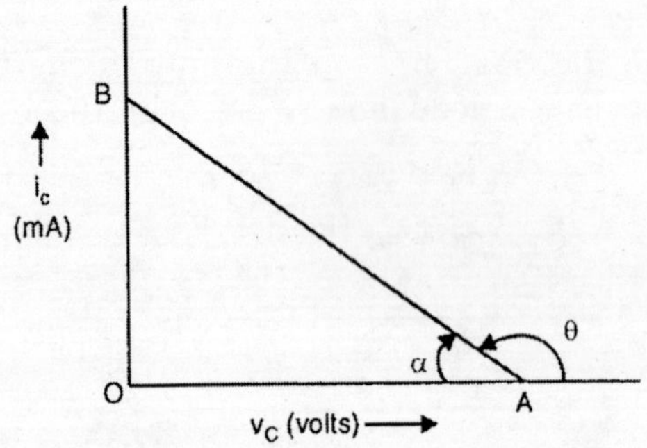

Fig. 5.5. *To draw a line with a slope of* $-1/R_{ac}$.

Suppose an a.c. voltage of 5 mV is applied at the input. This corresponds to a peak-to-peak variation of $5 \times \sqrt{2} \times 2 = 14.14$ mV.

Assume that the input characteristics of the transistor are such, as to produce a 20 μA peak-to-peak variation in the base current corresponding to this input a.c. voltage. When the base current varies within these limits (from 20 μA to 40 μA), the instantaneous operating point moves along the a.c. load line between points Q_1 and Q_2. The corresponding variations in collector current and collector-to-emitter voltage are shown in Fig. 5.2. The collector current varies between the limits 2.9 mA to 5.1 mA. The collector-to-emitter voltage varies between the limits 4.1 V to 4.9 V. The current gain and the voltage gain of the amplifier are given as :

$$\text{Current gain} = \frac{I_{C(\max)} - I_{C(\min)}}{I_{B(\max)} - I_{B(\min)}} = \frac{(5.1-2.9)\text{ mA}}{(40-20)\text{ μA}} = \mathbf{110}$$

$$\text{Voltage gain} = \frac{V_{CE\,(\max)} - V_{CE\,(\min)}}{V_{i\,(\max)} - V_{i\,(\min)}} = \frac{(4.9-4.1)\text{ V}}{14.14\text{ mV}} = \mathbf{56.58}$$

5.4. *h*-PARAMETER MODEL FOR LOW FREQUENCY

5.4.1. Introduction

Very often transistor is used in one or another form in almost all amplifier circuits. Biasing arrangement, resistances used and capacitances of junctions altogether force to analyze the behaviour of BJT amplifier in different configurations and working conditions. As we know that transistor parameters like β, I_{CBO} etc. vary considerably by change in temperature, they can't be uniquely used to describe efficiently the performance of the BJT in the amplifier circuits. So, in order to analyze the various amplifier circuits using BJT, we will use *h*-parameters. For low frequency applications, various parameters will be defined and interpreted physically in forthgoing topics. The equivalent model for BJT based upon these parameters will be drawn. A very clear procedure for a.c. and d.c. analysis for BJT amplifier and evaluation of performance parameters (voltage gain, current gain, input impedance and output impedance) of transistor are the basic ideas to be discussed in this unit. Furthermore, we will discuss about hybrid π model (for high frequency applications) and representation of tuned amplifiers, cascade amplifier will also be discussed.

5.4.2. Hybrid Parameters of a Generalized Electrical Network

Here, first of all we learn about the representation of electrical networks, in general, using *h*-parameters and then the model so obtained is implemented for the transistor amplifier.

For this purpose, let us consider the two port network given in Fig. 5.6 as below :

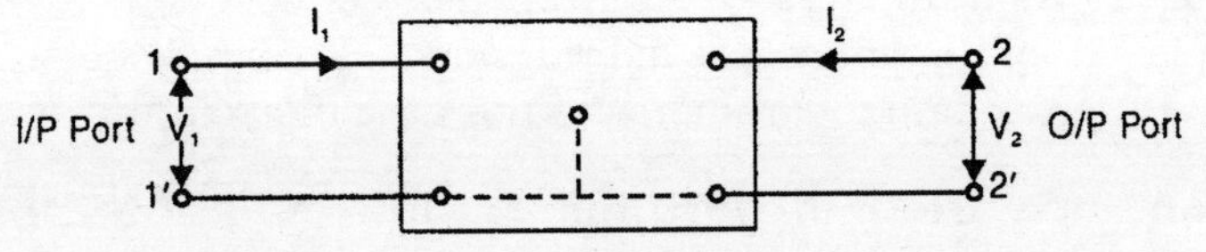

Fig. 5.6. *Representation of general two port (four terminal) electrical network.*

h-parameter representation is often used in modelling of electronic components and circuits, particularly transistors. The name *hybrid parameter* comes out from the fact that here we use both types of terminal conditions *i.e.*, open circuit and short circuit for the purpose of evaluating the unknown parameters. In this form of representation, the voltage of the input port and the current of the output port are expressed in terms of the current of the input port and the voltage of the output port.

Thus, mathematically we can say that,

$$(V_1, I_2) = f(I_1, V_2) \qquad ...(5.3)$$

The above function f shows that in the *h*-parameter representation, the input voltage and output current are to be expressed as input current and output voltage. The function defined by eqn. (5.3) can be represented in the form of matrix as under :

$$\begin{bmatrix} V_1 \\ I_2 \end{bmatrix} = \begin{bmatrix} h_{11} & h_{12} \\ h_{21} & h_{22} \end{bmatrix} \begin{bmatrix} I_1 \\ V_2 \end{bmatrix} \qquad ...(5.4)$$

The matrix representation of eqn. (2) yields following two equations :

$$V_1 = h_{11} I_1 + h_{12} V_2 \quad \text{...(5.5)}$$

$$I_2 = h_{21} I_1 + h_{22} V_2 \quad \text{...(5.6)}$$

Here, in eqn. (5.4); the matrix other than $[I_1, V_2]$ represents the h-parameter matrix for the electrical network depicted in Fig. 5.6.

Thus, the parameters h_{11}, h_{12}, h_{21} and h_{22} which relate the input and output variables of the two port network are called h-parameters.

Now, we can draw easily the equivalent circuit of Fig. 5.7 using eqns. (5.5) and (5.6) as under :

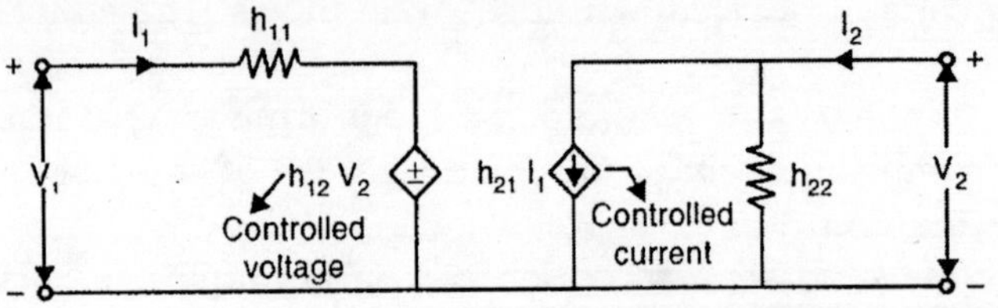

Fig. 5.7. *h-parameter equivalent circuit.*

5.4.3. Representation of a Linear Two Port Network using Hybrid Parameters

As the name suggests, a two port network consists of two ports *i.e.*, input port and output port. We can say that in reference of amplifier circuits working linear (active) region ; at input port, the signal is applied and at the output port we get the amplified or replicated version of the input signal.

With the help of h-parameters given in previous subsection, the two port network can be easily replaced by its equivalent hybrid model. The analysis of this hybrid model yields the required performance parameter of the amplifier circuit under consideration.

5.4.4. Definitions of various *h*-parameters and their Interpretation

Considering again the equations given by (5.5) and (5.6)

$$\boxed{\begin{aligned} V_1 &= h_{11} I_1 + h_{12} V_2 \\ I_2 &= h_{21} I_1 + h_{22} V_2 \end{aligned}}$$

As mentioned earlier, the representation of h-parameters utilizes both terminal conditions.

Now, if we arbitrarily set $V_2 = 0$ *i.e.*, short circuiting the output terminals and solve for h_{11}, the following result is obtained.

$$\boxed{h_{11} = \left.\frac{V_1}{I_1}\right|_{V_2 = 0}} \quad \Rightarrow V_1 = h_{11} I_1 + h_{12} V_2 \text{ at } V_2 = 0$$

Clearly, the unit of h_{11} is ohms, because the ratio indicates that the parameter h_{11} is an impedance parameter. Since, it is the ratio of the input voltage to the input current with the output terminals shorted, it is called the *short-circuit input impedance parameter.*

The subscript 11 of h_{11} defines the fact that the parameter is determined by a ratio of quantities measured at the input terminals.

Now, if I_1 in eqn. (5.5) is set to zero by opening the input port, the following relation is obtained :

$$\boxed{h_{12} = \left.\frac{V_1}{V_2}\right|_{I_1 = 0}} \qquad \Rightarrow V_1 = h_{11}\ I_1 + h_{12}\ V_2 \text{ at } I_1 = 0$$

The parameter h_{12}, being the ratio of the input voltage to the output voltage is a unit less quantity.

Clearly, it is defined and obtained under open circuit condition, so it is called the *open-circuit reverse transfer voltage ratio parameter.*

The subscript 12 of the h_{12} reveals that the parameter is a transfer quantity determined by a ratio of input to output measurements.

Now considering eqn. (5.6) as under :

$$I_2 = h_{21}\ I_1 + h_{22}\ V_2$$

If in above equation, V_2 is set to zero by again shorting the output terminals the following result is obtained for h_{21}

$$\boxed{h_{21} = \left.\frac{I_2}{I_1}\right|_{V_2 = 0}} \qquad \Rightarrow I_2 = h_{21}\ I_1 + h_{22}\ V_2 \text{ at } V_2 = 0$$

It is to be noted that we now have the ratio of an output quantity to an input quantity. The term forward will now be used rather than reverse as indicated for h_{12}.

The parameter h_{21} is the ratio of the output current to the input current with the output terminals shorted. It is normally called as the *short-circuit transfer current ratio parameter.*

The last parameter, h_{22}, can be found by again opening the input leads to set $I_1 = 0$ and solving for h_{22} eqn. (5.6) :

$$\boxed{h_{22} = \left.\frac{I_2}{V_2}\right|_{I_1 = 0}} \qquad \Rightarrow I_2 = h_{21}\ I_1 + h_{22}\ V_2 \text{ at } I_1 = 0$$

Since, it is the ratio of the output current to the output voltage, it is the *output conductance parameter.* The subscript 22 in h_{22} reveals that it is determined by a ratio of output quantities. The units of h_{22} is of conductance *i.e.*, ohm^{-1} or mho or siemens. Instead of using digits 1 and 2 for *h*-parameter designations, we now will use following relations in all forthgoing subtopics :

$h_{11} \longrightarrow$ input resistance $\longrightarrow h_i$
$h_{12} \longrightarrow$ reverse transfer voltage ratio $\longrightarrow h_r$
$h_{21} \longrightarrow$ forward transfer current ratio $\longrightarrow h_f$
$h_{22} \longrightarrow$ output conductance or admittance $\longrightarrow h_o$

5.4.5. Hybrid Model for different Configurations of Transistor

Depending upon the transformed notation of h-parameters *i.e.*, h_i, h_r, h_f, h_o, the complete hybrid model for a transistor can be generalized as under in Fig. 5.8.

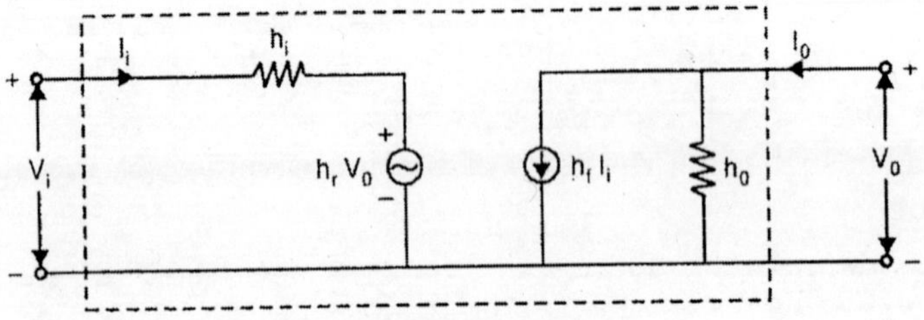

Fig. 5.8. *Complete hybrid-equivalent circuit for transistor.*

Here,

$V_i \longrightarrow$ input voltage

$I_i \longrightarrow$ input current

$V_o \longrightarrow$ output voltage

$I_o \longrightarrow$ output current

Now, we shall discuss the equivalent hybrid models for each of the configurations of transistor *i.e.*, CE, CC or CB.

The circuit of Fig. 5.9 is applicable to any linear three-terminal device with no internal independent sources. The transistor is a three terminal device *i.e.*, emitter, base and collector.

For the purpose of converting a three terminal electronic device into an equivalent two port electrical network one terminal must be made common to both input and output terminals. Thus, we obtain three configurations in which a transistor can be placed in an amplifier circuit. These are as under :

1. Common emitter (CE) configuration.
2. Common collector (CC) configuration.
3. Common base (CB) configuration.

The h-parameters, however, will change with each configuration. To distinguish which configuration has been used or which is available, a second subscript has been added to the h-parameter notation.

Table (5.1) better describes the arrangement for configurational distinction :

* The first letter of subscript defined in h-parameters takes the position in Numerator and the second in Denominator.
If Numerator has I subscript we use reverse otherwise we use forward.

Approximate conversion formulas for hybrid parameters

Common collector to common emitter

$$h_{ic} = h_{ie}$$
$$h_{rc} = 1$$
$$h_{fc} = -(1 + h_{fe})$$
$$h_{oc} = h_{oc}$$

Common base to common emitter

$$h_{ib} = \frac{h_{ie}}{1 + h_{fe}}$$

$$h_{rb} = \frac{h_{ie}\ h_{oe}}{1 + h_{fe}} - h_{fe}$$

$$h_{fb} = \frac{-h_{fe}}{1 + h_{fc}}$$

$$h_{ob} = \frac{h_{oe}}{1 + h_{fe}}$$

Table 5.1

S. No.	*Particular parameter*	*CE*	*CC*	*CB*
1.	h_i	h_{ie}	h_{ic}	h_{ib}
2.	h_r	h_{re}	h_{rc}	h_{rb}
3.	h_f	h_{fe}	h_{fc}	h_{fb}
4.	h_o	h_{oe}	h_{oc}	h_{ob}

Common emitter configuration

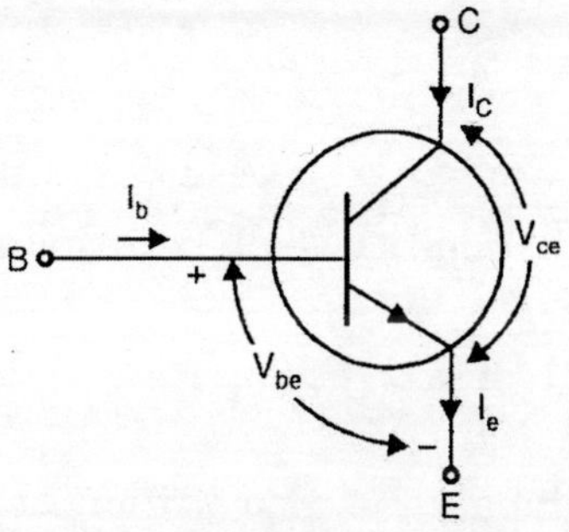

(*a*) Graphic symbol

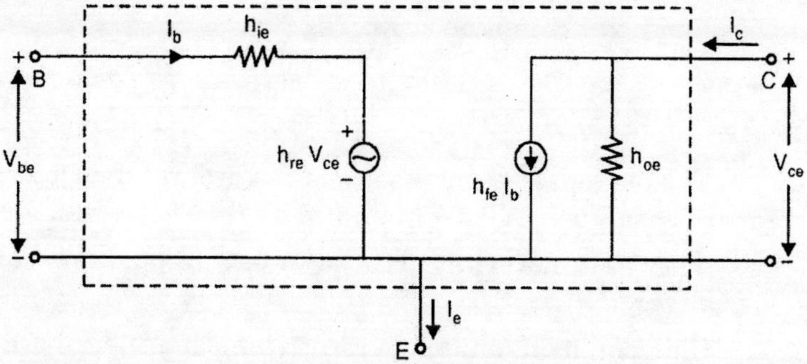

(*b*) Hybrid equivalent circuit

Fig. 5.9.

Common base configuration

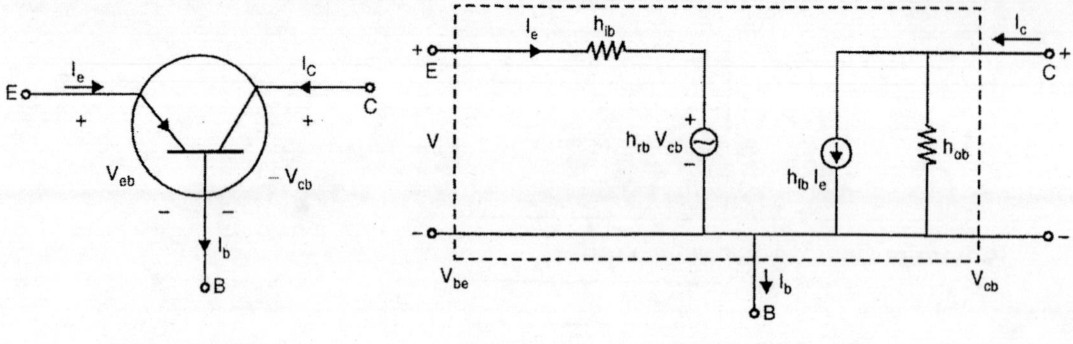

(*a*) Graphic symbol (*b*) Hybrid equivalent circuit

Fig. 5.10.

> Similarly, we can draw the hybrid equivalent circuit for CC configuration where subscript *c* will be used for second designation.

The parameter h_{fe}

For a common emitter connection the output characteristics are shown in Fig. 5.11.

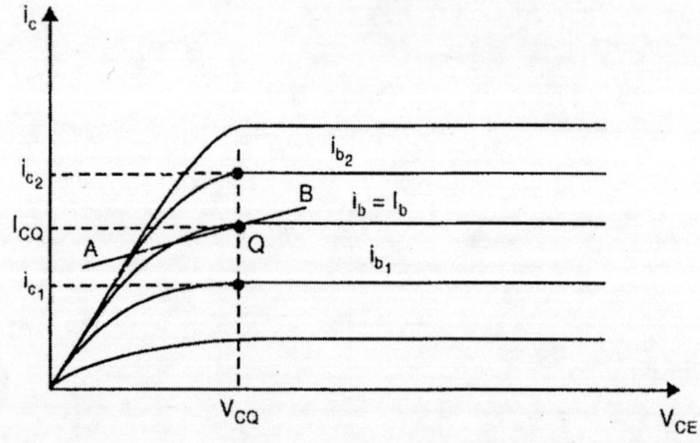

Fig. 5.11. *Determination of h_{f_e} and h_{o_e} from the CE output characteristics.*

From the definition of h_{fe} :

$$h_{fe} = \frac{\partial i_C}{\partial i_B} \approx \frac{\Delta i_C}{\Delta i_B} = \frac{i_{C_2} - i_{C_1}}{i_{B_2} - i_{B_1}}$$

The current increments are taken around the Quiescent point Q which corresponds to the base current $i_B = I_B$ and to the collector voltage $V_{CE} = V_C$.

The parameter h_{fe} is the most important small signal parameter of the transistor.

This common-emitter current transfer ratio, or CE alpha (α) is also written as α_e or β', and called the small-signal beta (β) of the transistor. The relationship between $\beta' = h_{fe}$ and $\beta \approx h_{FE}$ (large signal beta).

The parameter h_{oe}

By the definition of output admittance discussed earlier, we have,

$$h_{oe} = \frac{\partial i_C}{\partial V_C} \approx \left.\frac{\Delta i_C}{\Delta V_C}\right|_{I_B \to \text{constt.}}$$

The value of h_{oe} at the Quiescent-point Q is given by the slope of the output characteristics curve at that point. This slope can be evaluated by drawing the line AB in Fig. 5.11 tangent to the characteristics curve at the point Q.

Similarly, we can find out the parameters h_{ie} and h_{re} graphically from the input CE characteristics, in a manner analogous to that illustrated above for h_{fe} and h_{oe}.

Hybrid-parameter variations

Generally it is assumed that once the transistor has been manufactured, the specified *h*-parameters for the transistor do not change. But, in actual practice, it is not so. The *h*-parameters vary greatly as the Quiescent point shifts up and down. Generally, it is seen that the *h*-parameters are affected by the change in following quantities :

1. I_B
2. V_{CE}
3. Junction temperature

Table 5.2 depicts the values of *h*-parameters for three different transistor configurations of a typical BJT :

Table 5.2. Typical *h*-parameter values at I_E = 1.3 mA

Parameter	*CE*	*CC*	*CB*
$h_{11} = h_i$	1,100 Ω	1,100 Ω	21.6 Ω
$h_{12} = h_r$	2.5×10^{-4}	– 1	2.9×10^{-4}
$h_{21} = h_f$	50	– 51	– 0.98
$h_{22} = h_o$	24 μA/V	25 μA/V	0.49 μA/V
$\frac{1}{h_0}$	40 Ω	40 kΩ	2.04 MΩ

5.4.6. Procedural Analysis for A.C. and D.C. Equivalent of different Transistor Circuits and Analysis of Transistor Amplifiers using *h*-Parameters

A transistor amplifier typically consists of a transistor, external load and the source of signal as depicted in Fig. 5.12.

Proper biasing arrangement must be prevailed before applying the signal because in absence of it there is no meaning of amplification. The transistor appearing in above network can have any of the three configurations. This transistor, in general, can be replaced by its small signal hybrid model. Fig. 5.13 is valid for any type of the load, whether it be pure resistance, an impedance, or even another transistor. This is true because transistor hybrid-model has been derived irrespective of the external circuit in which the transistor is employed.

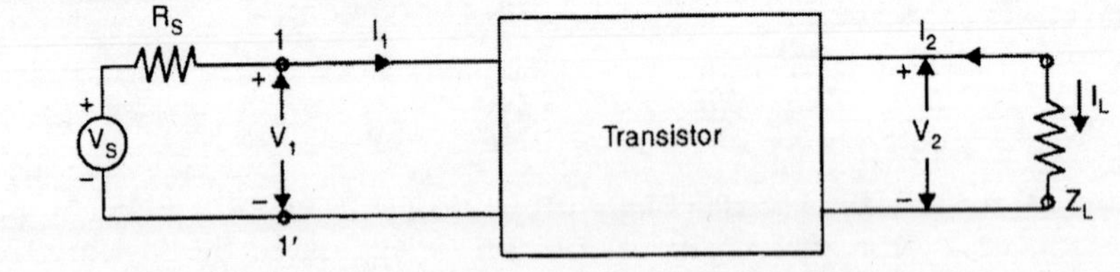

Fig. 5.12. *A basic amplifier circuit.*

Here it is noteworthy that the h-parameters of transistor don't get changed by the presence of external circuit.

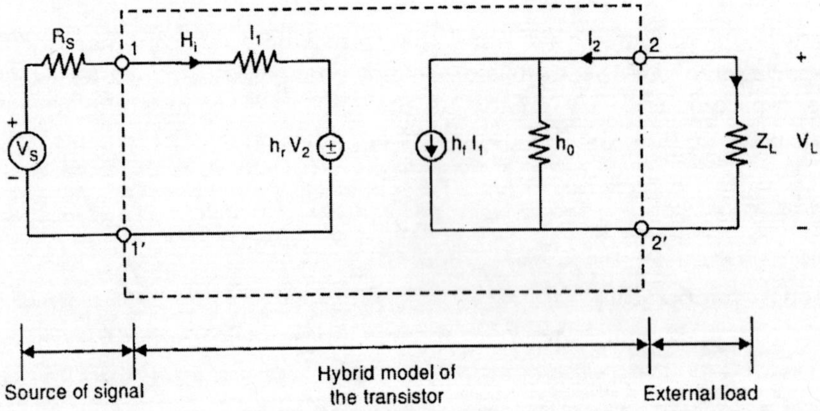

Fig. 5.13. *The transistor in Fig. 5.12 replaced by its h-parameter model.*

Following quantities are of utmost interest while analyzing the amplifier circuit :

(*a*) Current gain

(*b*) The input impedance

(*c*) The voltage gain

(*d*) The output impedance.

Now, we shall determine each *performance parameter* in succession. Care must be taken that all these expressions so obtained are for transistor, not for the amplifier circuit as a whole.

The current gain or current amplification A_I

The current gain of a transistor amplifier, A_I is defined as the ratio of output to input currents

or $$A_I = -\frac{I_L}{I_1} = -\frac{I_2}{I_1}$$ ($\because$ I_L and I_2 are in opposite direction)

Referring to Fig. 5.13, we have

$$I_2 = h_f I_1 + h_o V_2$$ (Using KCL)

Substituting $V_2 = -I_2 Z_L$ in above expression, we obtain

$$I_2 = h_f I_1 + h_o(-I_2 Z_L)$$

or $$I_2(1 + h_o Z_L) = h_f I_1$$

or $$\frac{I_2}{I_1} = \frac{h_f}{1 + h_o Z_L}$$

or $$\boxed{A_1 = \frac{I_2}{I_1} = \frac{-h_f}{1 + h_o Z_L}}$$

The input impedance Z_i

The resistance R_S appearing in Fig. 5.12 and Fig. 5.13 represents the signal-source resistance.

The amplifier *input impedance* is obtained by looking into the amplifier input terminals (1, 1′) and is expressed as :

$$Z_i = \frac{V_1}{I_1}$$

Applying KVL in input side of Fig. 5.13 we have

$$V_1 = h_i I_1 + h_r V_2$$

Hence, $$Z_i = \frac{h_i I_1 + h_r V_2}{I_1} = h_i + h_r \frac{V_2}{I_1}$$

Again, substituting above expression

$$V_2 = -I_2 Z_L$$
$$= A_I I_1 Z_L$$
$$Z_i = h_i + h_r A_I Z_L$$

$$\boxed{Z_i = h_i - \frac{h_f h_r}{\gamma_L + h_o}}$$

Here A_I is replaced by $\frac{-h_f}{1 + h_o Z_L}$ and Z_L is replaced by $\frac{1}{\gamma_L}$.

From above it is clear that the input impedance is a function of the load impedance.

Voltage amplification or voltage gain, A_V

The ratio of output voltage V_2 to input voltage V_1 gives the voltage gain of the transistor or equivalently.

$$A_V = \frac{V_2}{V_1}$$

We already have,

$$V_2 = -I_2 Z_L = A_I I_1 Z_L$$

$$\therefore \quad A_V = \frac{A_I I_1 Z_L}{V_1} = \frac{A_I Z_L}{\left(\frac{V_1}{I_1}\right)} = \frac{A_I Z_L}{Z_i}$$

*** Cor. The voltage amplification, A_{V_S}, taking into account the resistance R_S of the source.**

This overall voltage gain A_{V_S} is defined by

$$A_{V_S} = \frac{V_2}{V_S} = \frac{V_2}{V_S}\frac{V_1}{V_S} = A_V \frac{V_1}{V_S}$$

The value of V_1 can be calculated using following input circuit of transistor amplifier :

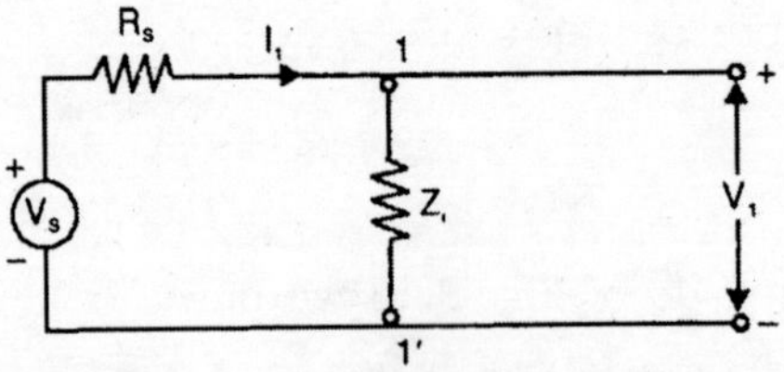

Fig. 5.14. *Input circuit of a transistor amplifier using Thevenin's equivalent.*

Now, referring to Fig. 5.14

$$V_1 = \frac{V_S Z_i}{V_i + R_S} \qquad \text{\{Using voltage-divider rule\}}$$

Then,
$$A_{V_S} = \frac{A_V Z_i}{Z_i + R_S} = \frac{A_I Z_L}{Z_i + R_S}$$

It is to be notable here that if $R_S = 0$ the A_{V_S} reduces to A_V.

Hence, Δ_V *is the voltage gain for an ideal voltage source.*

The current amplification A_{I_S}, taking into account the source resistance R_S

If the input source is a current generator with a resistance R_S in parallel as in Fig. 5.15, then this overall current gain A_{I_S} is defined by

$$A_{I_S} = \frac{-I_2}{I_S} = \frac{-I_2 I_1}{I_1 I_S} = A_I \frac{I_1}{I_S}$$

where A_{I_S} also called current gain for the circuit

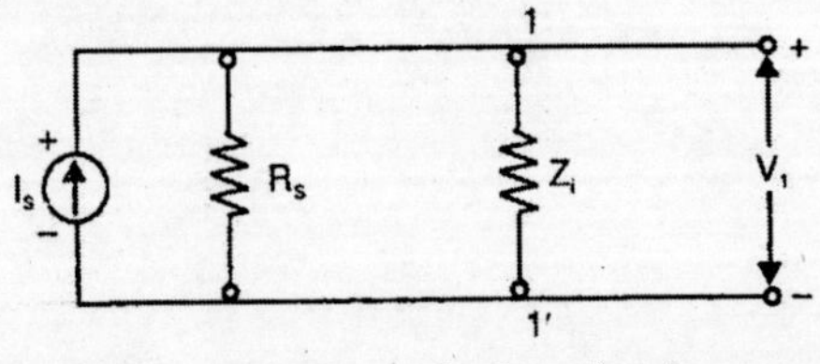

Fig. 5.15. *Norton's equivalent for the source.*

Referring to Fig. 5.15,

$$I_1 = \frac{I_S R_S}{Z_i + R_S} \qquad \text{\{Using current-division rule\}}$$

and hence,

$$\boxed{A_{I_S} = \frac{A_I R_S}{Z_i + R_S}}$$

It is to be that noted here that if $A_{I_S} \longrightarrow A_I$.

Hence, A_I *is the current gain for an ideal current source (one with infinite source resistance).*

Imp.

$$\boxed{A_{V_S} = A_{I_S} \frac{Z_L}{R_S}}$$

This relationship is obtained by making use of expressions for A_{I_S} and A_{V_S}.

The output admittance

By definition, the output impedance $Z_o = \dfrac{1}{\gamma_0}$ is obtained by setting the source voltage V_S to zero and the load impedance Z_L to infinity and by driving the output terminals from a generator V_2.

Let us assume that the current drawn form V_2 is I_2, then

$$\gamma_0 = \frac{I_2}{V_2} \text{ with } V_S = 0 \text{ and } R_2 = \infty$$

We have already,

Due to Norton's equation, $\gamma_0 = h_f \dfrac{I_2}{V_2} + h_o$ and

Due to Thevenin's equation,

$$R_S I_1 + h_i I_1 + h_r V_2 = 0$$

or

$$\frac{I_1}{V_2} = -\frac{h_r}{h_i + R_S}$$

Thus, from initial expression

$$\boxed{\gamma_o = h_o - \frac{h_f h_r}{h_i + R_S}}$$

It is interesting to note that the output admittance is a function of the source resistance.

If the source impedance is resistive, as we have assumed, then γ_o is a real (*i.e.*, conductance).

For quick review, Table 5.3 must be memorized :

Table 5.3. Small signal analysis of a transistor amplifier

$A_I = -\dfrac{h_f}{1 + h_o Z_L}$	$Y_o = h_o - \dfrac{h_f h_r}{h_i + R_S} = \dfrac{1}{Z_o}$
$Z_i = h_i + h_r A_I Z_L$	$A_{V_S} = \dfrac{A_V Z_i}{Z_i + R_S} = \dfrac{A_I Z_L}{Z_i + R_S}$
$A_V = \dfrac{A_I Z_L}{Z_i}$	$A_{I_S} = \dfrac{A_I R_S}{Z_i + R_S} = A_{V_S} \dfrac{R_S}{Z_L}$

Procedural analysis of a transistor circuit

Upto now we have studied about the various transistor configurations, various biasing techniques and a sophisticated tool (*h*-parameter) for analysis of transistor circuits.

This section basically deals with the analysis of transistor circuit in a very lucid manner.

There are many transistor circuits which do not employ simply CE, CB or CC configurations. Every configuration has its own advantages and disadvantages. There may be requirement of feedback from output to input in any manner or there may be more than two transistors connected in cascade. So, an analytic determination of the small-signal behaviour of complicated amplifier circuits may be made by following these simple rules :

1. Draw the actual wiring diagram of the circuit.
2. Designate the points, B, C and E for base, collector and emitter respectively.
3. Locate these points as the start of the equivalent circuit. Maintain the same relative positions as in the original circuit.
4. Replace each transistor by its *h*-parameter model.
5. Transfer all circuit elements (resistors, sources, capacitors, external load etc.) from the actual circuit to the equivalent circuit of the amplifier.
6. Replace each independent d.c. source by its internal resistance. The ideal voltage source is replaced by short circuit and the ideal current source by an open circuit.
7. Solve the resultant linear circuit for mesh or branch currents and node voltages by applying Kirchhoff's current and voltage laws.

5.5. COMPUTATION OF A_I, A_V, Z_i, Z_o OF SINGLE STAGE TRANSISTOR CE AND CC AMPLIFIER CONFIGURATION

5.5.1. Common-Collector Configuration

The circuit diagram of common collector transistor amplifier is shown in Fig. 5.16. This configuration is also called the *emitter follower,* because its voltage gain is close to unity, and hence a change in base voltage appears as an equal voltage, change across the load at the emitter. In other words, the emitter follows the input signal.

It is shown below that the input impedance R_i of an emitter follower is very high and the output impedance R_o is very low. Hence the most common use for the common collector (CC) configuration is as a Buffer stage which performs the function of resistance transformation (from high to low resistance) over a wide range of frequencies, with voltage gain close to unity. In addition, the emitter follower increases the power level of the signal.

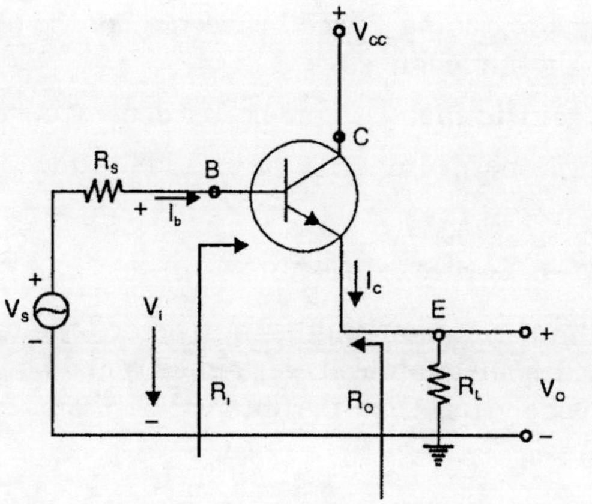

Fig. 5.16. *A common-collector or emitter follower.*

5.5.2. Analysis of Emitter Follower or CC-Configuration

Current gain $$A_I = \frac{-I_C}{I_b} = -\frac{h_{fc}}{1 + h_{oc} R_L} = \frac{1 + h_{fc}}{1 + h_{oc} R_L}$$

{from approximate conversion formula}

Input impedance $$R_i = \frac{V_i}{I_b} = h_{ic} + h_{rc} A_I R_L = h_{ic} + A_I R_L \qquad (\because h_{rc} = 1)$$

Voltage gain $$A_V = \frac{V_o}{V_i} = \frac{A_I R_L}{R_i} = 1 - \frac{h_{ic}}{R_i}$$

(slightly less than one and no phase reversal between input and output voltages).

Output admittance, $$Y_o = h_{oc} - \frac{h_{rc} . h_{fc}}{h_{ic} + R_S}$$

or $$Y_o = h_{oc} + \frac{1 + h_{fe}}{h_{ic} + R_S} \approx \frac{1 + h_{fe}}{h_{ie} + R_S}$$

5.5.3. Properties of Common Collector or Emitter Follower

(1) Voltage gain ≈ 1, no phase reversal.

(2) Very high input impedance.

(3) Very low output impedance.

(4) Uses negative voltage series feedback which increases good stability.

5.5.4. Applications of Emitter Follower

(1) As an impedance matching device between a source of high internal impedance and a load of low impedance.

(2) Since current gain is high. It can be used as a current amplifier.

(3) As a buffer stages between oscillator and load to avoid the effect of loading on the oscillator.

5.5.5. Common-Emitter Configuration

The circuit diagram of common-emitter transistor amplifier is shown in Fig. 5.17. This configuration provides adequate voltage gain as well as current gain that's why this configuration is most widely used for the intermediate stages in cascaded amplifier.

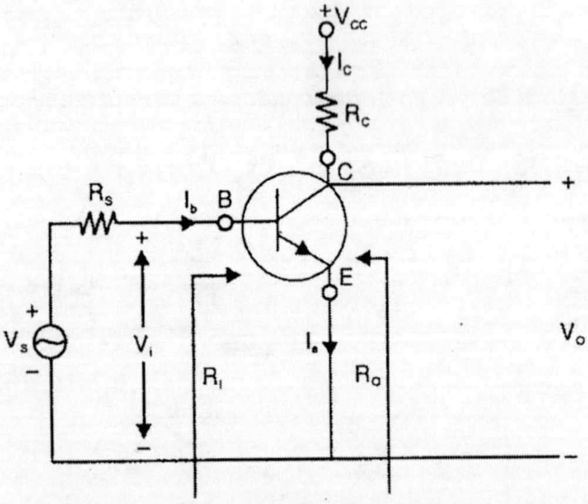

Fig. 5.17. *Common-emitter configuration.*

Now, in order to describe voltage gain, current gain, input impedance and output impedance, the above circuit is replaced by its *h*-parameter model.

(*i*) **Current gain or current amplification.** A_I for the transistor amplifier stage, A_I is defined as the ratio of output to input currents, or

$$A_I = \frac{I_C}{I_b} = \frac{-h_{fe}}{1 + h_{oe} R_L} \qquad \{\text{where } R_L = R_C \parallel R_0\}$$

(*ii*) **Input impedance, (R_i).** The resistance R_S in Fig. 5.17 represents the signal-source resistance. The impedance we see looking into the amplifier input terminals is the amplifier input impedance R_i, or

$$R_i = \frac{V_I}{I_b}$$

or
$$\boxed{R_i = h_{ie} + h_{re}\, A_I \,.\, R_L}$$

(*iii*) **Voltage gain, or voltage amplification, (A_V).** The ratio of output voltage to the input voltage gives the voltage gain of the transistor, or

$$A_V = \frac{V_o}{V_i}$$

or
$$\boxed{A_V = \frac{A_I \,.\, R_L}{R_i}}$$

(*iv*) **Voltage amplification A_{V_S},** taking into account the resistance R_S of the source the overall voltage gain A_{V_S} is defined by

$$A_{V_S} = \frac{V_o}{V_S} = A_V \cdot \frac{V_i}{V_S}$$

or
$$\boxed{A_{V_S} = A_V \cdot \frac{R_i}{R_i + R_S} = \frac{A_I \,.R_L}{R_i + R_S}}$$

(*v*) **The current amplification, A_{I_S}** taking into account the source resistance R_S

$$A_{I_S} = -\frac{I_C}{I_S} = A_I \cdot \frac{I_b}{I_s}$$

or
$$\boxed{A_{I_S} = A_I \cdot \frac{R_S}{R_S + R_i}}$$

(*vi*) **The output admittance, (γ_o).** By definition the output impedance $Z_o = \dfrac{1}{\gamma_o}$ is obtained by setting the source voltage V_S to zero and the local impedance to infinity.

$$\boxed{\gamma_o = h_{oe} - \frac{h_{fe} \cdot h_{re}}{h_{ie} + R_S}}$$

Note that the output admittance is a function of the source resistance. If the source impedance is resistive, then γ_o is real (a conductance).

5.6. INTRODUCTION TO FEEDBACK IN AMPLIFIER

The important characteristics of an amplifier are its voltage gain, bandwidth, input and output impedances. The parameters are more or less constant for an amplifier. The value of these parameters are required to change. The designer's does not control the value of these parameters. This problem can be solved with number of ways. For example, if the gain could be reduced by voltage divider circuit in the input or in the ouptut circuit of an amplifier, the input impedance of an amplifier could be increased if required by connecting a series resistance in the circuit. But these methods results in the loss of useful signals. A new technique is introduced called as feedback in the amplifiers.

The feedback is a process of injecting some energy (*i.e.*, the form of voltage or current) from the output and then return it back to the input when the fraction or a part of output is feedback to the input, the process is known as *Feedback*. When amplifier circuit uses the feedback, then it is called feedback amplifier.

5.7. BASIC CONCEPT OF FEEDBACK

As Fig. 5.18 (*a*) shows a block diagram of a basis amplifier. Here, V_i is the input signal and V_o is the output signal. If A is the voltage gain of the amplifier, the output V_o is related to the input V_i by,

$$A = \frac{V_o}{V_i}$$

In this amplifier, the input does not know what is happening at the output. If due to some reason, the output changes the net input remains unaffected. Such a system is called open-loop or non-feedback system.

However, Fig. 5.18 (*b*) shows a block diagram of feedback amplifier network. This feedback network is called a β network or a feedback network. A fraction βV_o of output voltage is going back to the input. This changes the net input voltage to the amplifier. The input knows at every instant what the output is. Such a system is called a closed-loop or feedback system.

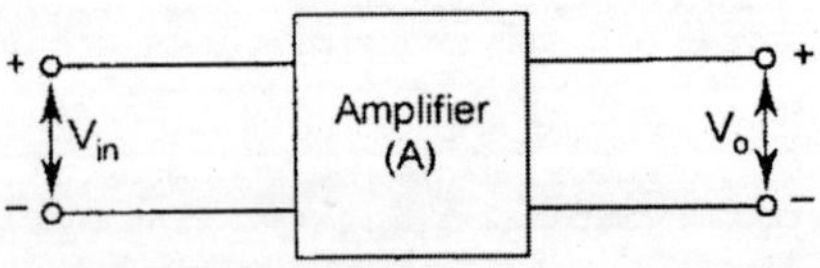

(*a*) Block diagram of basic amplifier.

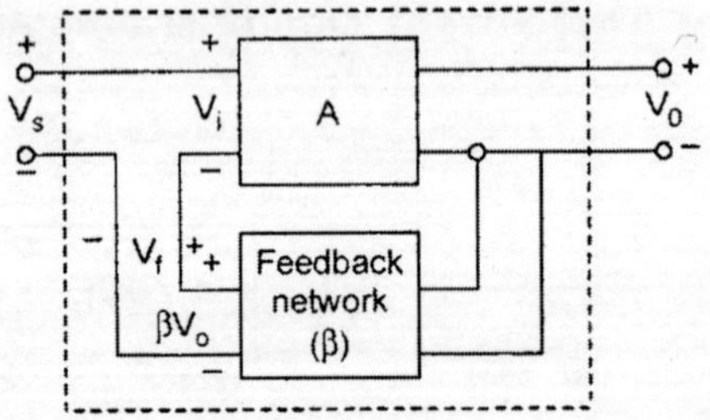

(*b*) Feedback introduced in the amplifier.

Fig. 5.18.

The voltage gain of the feedback amplifier is then

$$A_f = \frac{V_o}{V_s} \quad \text{...(5.7)}$$

The more general structure of the feedback amplifier is given in Fig. 5.18 (*c*). Here this is a signal flow diagram, and the quantities *x* represent either voltage or currency signal.

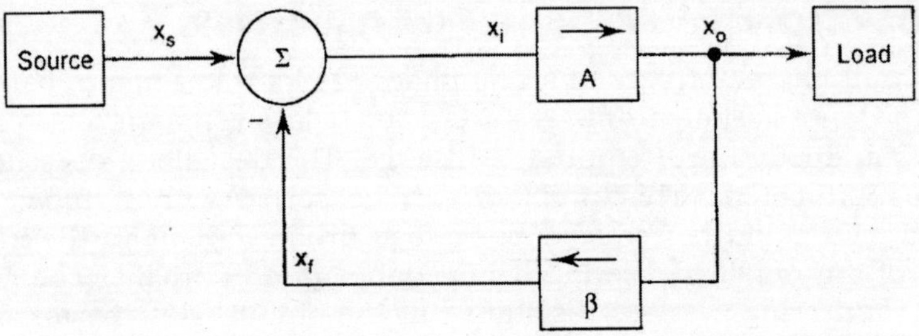

Fig. 5.18 (c). *General structure of the feedback amplifier.*

$$x_0 = Ax_i \quad ...(i)$$

These are general equation of a feedback amplifier

$$x_f = \beta x_o \quad ...(ii)$$

$$x_i = x_s - x_f \quad ...(iii)$$

$$A_f = \frac{x_o}{x_s} = \frac{A}{1 + A\beta} \quad ...(iv)$$

But from Fig. 5.18 (*b*) (*i.e.*, on applying KVL), we get

$$V_s = V_i + V_f \quad ...(5.8)$$

where V_f = feedback voltage. The output voltage (V_o) and feedback voltage (V_f) are related with feedback network (β) as

$$\beta = \frac{V_f}{V_o}$$

or

$$V_f = \beta V_o \quad ...(5.9\ (a))$$

For the basic amplifier, the input is V_i and the output V_o. Hence its voltage gain A (called internal gain) is given as

$$A = \frac{V_o}{V_i} \quad ...(5.10)$$

We shall now derive the expression of the gain A_i in term of A (internal gain) and β (feedback network)

Now,

$$A_f = \frac{V_o}{V_s}$$

or

$$A_f = \frac{V_o}{V_i + V_f} \quad ...(5.11)$$

or

$$A_f = \frac{V_o}{\frac{V_o}{A} + \beta V_o} \quad \text{(from equations (5.10) and (5.11))}$$

$$A_f = \frac{A}{1 + A\beta} \quad ...(5.12)$$

Thus, it is clear from the equation (5.12) that the gain of amplifier decreases when we apply negative feedback. However, gain of an amplifier when we apply positive feedback is given by the relation.

$$A_f = \frac{A}{1 - A\beta} \quad ...(5.13)$$

Thus, positive feedback increases the gain.

Thus, finally $A_f = \frac{A}{1 + A\beta}$ – for Negative feedback

and $A_f = \frac{A}{1 - A\beta}$ – for Positive feedback

The term 'β' is called **'feedback factor'**. Whereas β is known as **feedback ratio**. The expression $(1 \pm A\beta)$ is called **loop gain**).

Amplifiers are not the only things where the feedback is used. We use the idea of feedback in our daily life too. You may not have realised it, but even we use feedback in the process of learning. When a child is asked to write a letter A, he will probably write it as shown in Fig. 5.18 (*d*). When he finds that the stroke is not going in the correct direction, the information goes to his brain through the eyes. The brain immediately orders the hand to correct the direction of the stroke. With much effort and with constant feedback, the child writes the letter 'A' as shown in Fig. 5.18 (*e*).

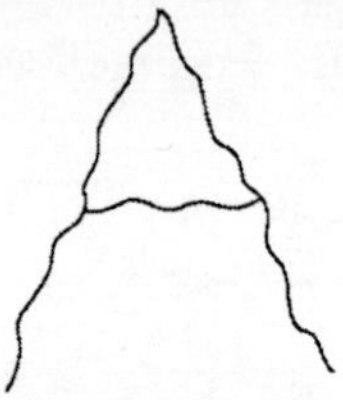

Fig. 5.18 (*d*)

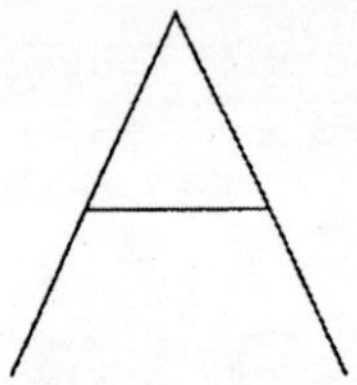

Fig. 5.18 (*e*)

From the equations (5.12) and (5.13), we have the following three cases :

(*i*) $A_f > A$; for positive feedback.

(*ii*) $A_f < A$; for negative feedback.

(*iii*) In the case of positive feedback, if $A\beta = 1$ and $A_f = \infty$. It is possible only in the case : when input is zero. But the amplifier is capable of producing output at input zero. This stage of an amplifier works as an oscillator.

5.8. TYPES OF FEEDBACK

As we have discussed a feedback amplifier is consist of two parts namely amplifier circuits and feedback circuit. Depending upon whether the feedback signal increase or decreases the input signal there are two types of feedback in amplifier.

(*a*) **Positive Feedback.** If the feedback signal (voltage or current) is applied in such a way that it is in phase with the input signal and thus increases it. Then it is called positive feedback. It is also known as regenerative feedback or direct feedback.

Advantages

(*i*) It increases the gain of the amplifier

(*ii*) If positive feedback is sufficiently large it leads to oscillations. So it used in oscillators.

Disadvantages

It increases the distortion and instability.

(*b*) **Negative feedback.** If the feedback signal (*i.e.*, voltages or current) is applied in such a way that it is out of phase with the input signal and thus decrease it. Then it is called negative feedback. It is also known as degenerative feedback.

Advantages

(*i*) It stabilize the gain of the amplifier.

(*ii*) It reduces the distortion and noise.

(*iii*) It reduces the output impedance.

(*iv*) It increase the input impedance.

(*v*) It increase the range of uniform amplification or bandwidth.

Disadvantages

It reduces the gain of amplifier. But due to the large number of advantages of negative feedback is frequently employed in the amplifiers. So lets discuss the negative feedback in more detail.

5.8.1. Classification of Negative Feedback

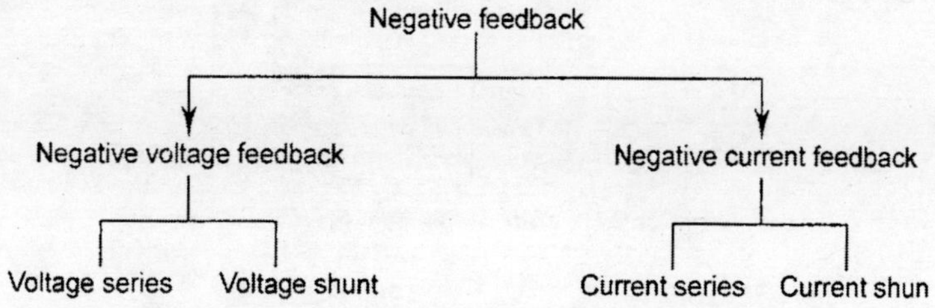

Fig. 5.19. *Classification of –ve feedback.*

There are two types of negative feedback circuits *i.e.*,

(*i*) **Negative voltage feedback.** The voltage is fedback to the input of amplifier is proportional to the output voltage irrespective to the load. It is further classified into two categories *i.e.*,

(*a*) Voltage series feedback

(*b*) Voltage shunt feedback

(*ii*) **Negative current feedback.** The voltage is fedback to the input of the amplifier is proportional to the output current, irrespective to the load. It is further classified into two catagories :

(*a*) Current-series feedback

(*b*) Current-shunt feedback

Let us discuss and analyze the effect of these four types of Negative feedback on the input and output impedance.

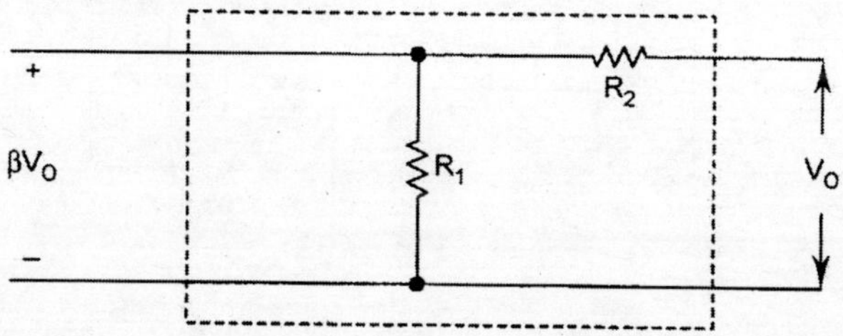

Fig. 5.20. *General representation of feedback circuit with resistors.*

$$\beta\, V_o = \frac{R_1}{R_1 + R_2} \cdot V_o \qquad \text{(By applying potential divider rule)}$$

or
$$\beta = \frac{R_1}{R_1 + R_2}$$

(*a*) **Voltage series feedback :** In the voltage series feedback circuit there is amplification of voltage into a voltage (*i.e.*, voltage $\xrightarrow{\text{amplified}}$ voltage) and a fraction of output voltage is fedback in series with the input voltage through the feedback network. The Basic circuit or a block diagram representation of voltage series feedback is shown in Fig. 5.21.

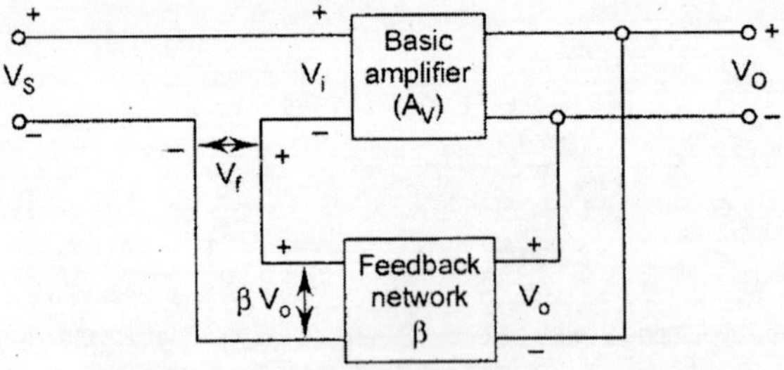

Fig. 5.21. *Voltage series feedback circuit.*

Derivation for input impedance with feedback

From the circuit shown input impedance with feedback is given as :

$$Z_{if} = \frac{V_s}{V_i}$$

$$Z_{if} = \frac{V_f + V_i}{I_i} \qquad (\because V_s = V_i + V_f)$$

$$Z_{if} = \frac{\beta V_o + V_i}{I_i}$$

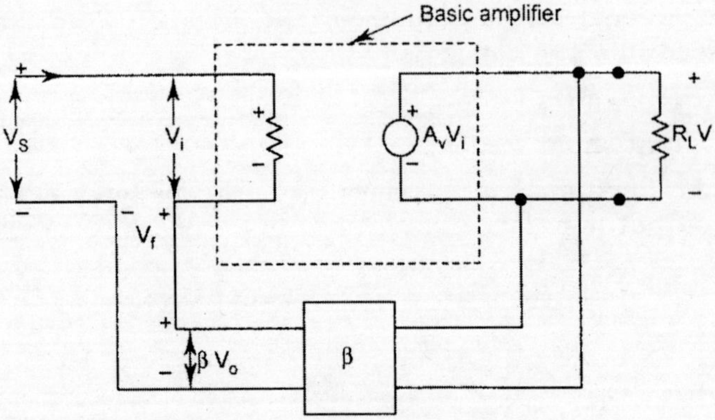

Fig. 5.22

NOTE : During the calculation of input impedance for any circuit output is open circuited or $$Z_{if} = \frac{\beta . A . V_i + V_i}{Z_i} \quad \{\because V_o = A_v V_i \text{ then output is open}\}.$$

or $$Z_{if} = (A\beta + 1)\frac{V_i}{I_i} \quad \left\{\because \frac{V_i}{I_1} = Z_i, \text{ input impedance of the amplifier}\right\}$$

$$Z_{if} = Z_i (1 + A\beta)$$

Thus, the impedance increases in voltage series feedback.

Derivation of output impedance with feedback

NOTE : To calculate the output impedance for any circuit all the independent sources are replaced by the internal resistance (*i.e.*, voltage source will be treated as short circuited and current source as open circuited). Take imaginary voltage source (V) at the output of the terminal and also assume that current (I) is flowing through this source then output impedance with feedback will be equal to the ratio of imaginary voltage source (V) to the current (I)

i.e., $$Z_{of} = \frac{V}{I}$$

This procedure will apply in all the feedback circuits. The circuit diagram of voltage series configuration for calculating the output impedance is shown here. Source voltage V_s is the independent source replaced by short-circuit (S.C.)

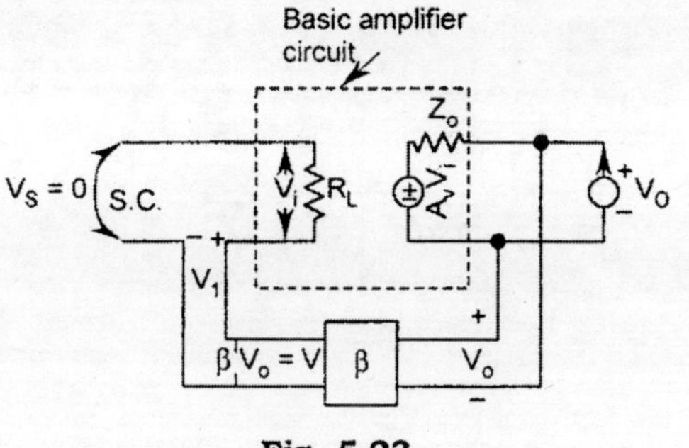

Fig. 5.23

Now,

Applying KVL at the output

$$V = I Z_o + A_v V_i \qquad (\because V_s = V_i + V_f$$
$$V = I Z_o + A_v (- V_f) \qquad 0 = V_i + V_f$$
$$V = I Z_o - A_v . \beta V_o \qquad \text{or } V_i = - V_f)$$
$$V = I Z_o - A_v \beta . V \qquad (\because V_o = V \text{ here})$$
$$V + V A_v \beta = IZ_o$$

or $$\frac{V}{I} = \frac{Z_o}{1 + A_v}$$

or $$Z_{of} = \frac{Z_o}{1 + A_v}$$

Thus, we see that in voltage series feedback output impedance decreases.

(*b*) **Voltage shunt feedback :** In the voltage shunt feedback there is amplification of current into a voltage (*i.e.*, current amplified voltage) and a fraction of output voltage is fedback in shunt with the input current through a fedback network. Basic circuit for voltage shunt feedback is shown in Fig. 5.24.

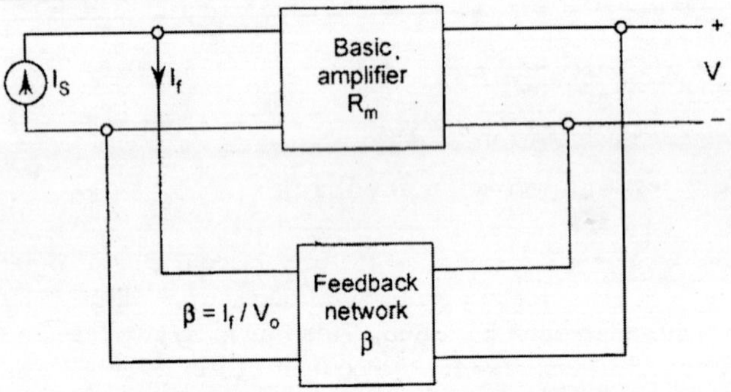

Fig. 5.24. *Voltage shunt feedback circuit.*

Derivation for input impedance

From Fig. 5.25,

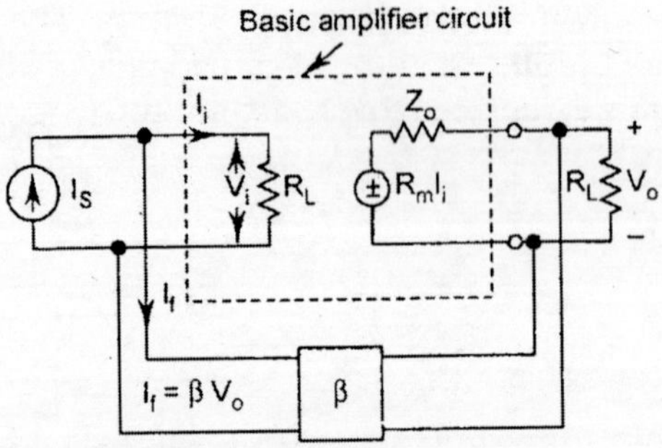

Fig. 5.25

$$Z_{if} = \frac{V_i}{I_s}$$

or $$Z_{if} = \frac{V_i}{I_f + I_i} \qquad (\because I_s = I_i + I_f)$$

or $$Z_{if} = \frac{V_i}{\beta V_o + I_i} \qquad (\because I_f = \beta V_o)$$

or $$Z_{if} = \frac{V_i}{\beta . R_m I_i + I_i} \qquad (\because V_o = R_m I_i)$$

or $$Z_{if} = \frac{V_i}{I_i (R_m \beta + 1)}$$

or $$Z_{if} = \frac{Z_i}{1 + R_m \beta}$$

Thus, input impedance in the case of voltage shunt feedback is decreases.

Derivation for output impedance

According to the procedure for calculating output impedance the circuit diagram shown as in Fig. 5.26.

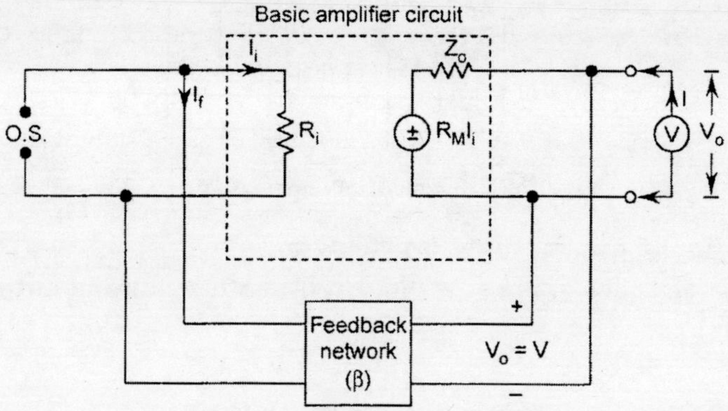

Fig. 5.26

$$Z_{of} = \frac{V}{I}$$

Now, $V = I\,Z_o + R_m\,I_i$

or $V = I\,Z_o + R_m\,(-\,I_f)$ $\qquad (\because I_s = I_i + I_f)$

or $V = I\,Z_o - R_m\,.\,\beta V$ $\qquad 0 = I_i + I_f$

$\qquad I_i = -\,I_f$

or $V\,(1 + R_m\,\beta) = I\,Z_o,$ $\qquad \{I_f = \beta\,V_o\}\ V = V_o$

or $$\frac{V}{I} = \frac{Z_o}{1 + R_m\,\beta}$$

or $$Z_{of} = \frac{Z_o}{1 + R_m\,\beta}$$

Thus we see that in voltage shunt feedback output impedance decreases with feedback.

(*a*) **Current series feedback :** In the current series feedback circuit there is amplification of voltage into current (voltage $\frac{V_o}{V_i + V_f}$ current) and a fraction of output current is fedback in series with the voltage through a feedback network. Basic circuit for current series feedback is as shown in Fig. 5.27).

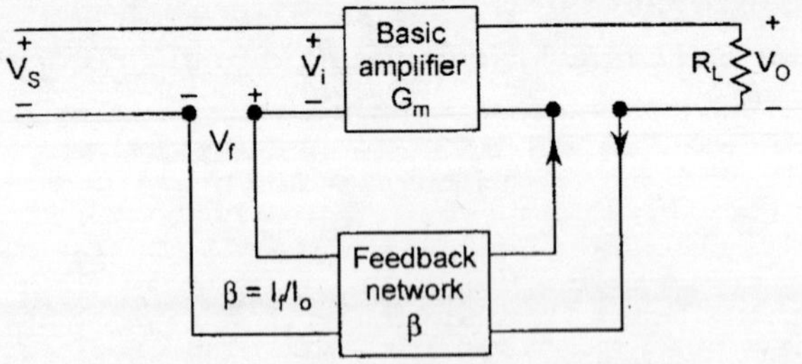

Fig. 5.27. *Current series feedback circuit.*

Derivation for input and output impedance

It is an exercise for students. In this circuit both input and output impedance after feedback are :

$$Z_{if} = (1 + G_m \beta) Z_i$$

and $$Z_{of} = (1 + G_m \beta) Z_o$$

(*b*) **Current shunt feedback :** In the current shunt feedback circuit there is amplification of current into a current (*i.e.*, current $\xrightarrow{\text{amplified}}$ current) and a fraction of output current is fedback in shunt with the input current, through a feedback network. Basic circuit for current shunt feedback is shown in Fig. 5.28.

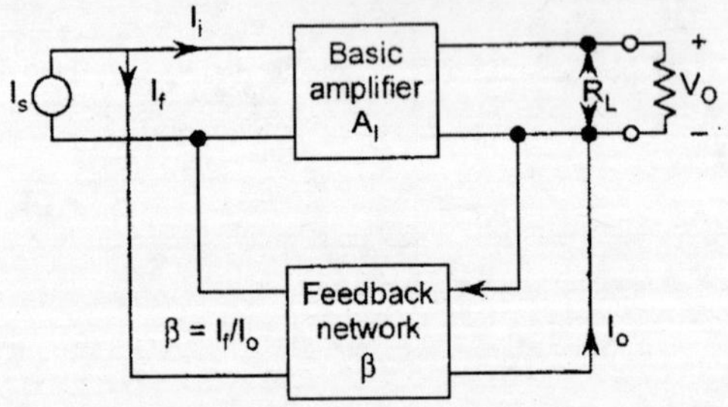

Fig. 5.28. *Current shunt feedback circuit.*

Derivation for input impedance

$$I_s = I + I_f = I_i + \beta I_o$$

$$= I_o (1 + \beta A)$$

$$Z_{if} = \frac{V_I}{I_s} = \frac{V_i}{I_i(1+\beta A)}$$

$$= \frac{V_I/I_i}{1+A\beta} = \frac{Z_i}{1+A\beta}$$

So, the input impedance of amplifier by a factor (1 + Aβ).

Derivation for output impedance

To calculate the change in output impedance, the output resistance R_L is disconnected and V_o is set to zero. External voltage V_o is applied across the output terminals and output current is measured.

$$Z_{if} = \frac{V_o}{I_o}$$

$$(I_o + \beta A I_o) = \frac{V_o}{Z_o}$$

$$Z_o (1 + \beta A) I_o = V_o$$

$$Z_{if} = \frac{V_o}{I_o} = Z_o (1 + A\beta)$$

Now, Table 5.4 shows the effects of feedback connection on input and output impedance after feedback.

Table 5.4

Type/Parameter	*Voltage series*	*Current series*	*Voltage shunt*	*Current shunt*
Z_{if}	$Z_i (1 + A\beta)$ increases	$Z_i (1 + A\beta)$ increases	$\frac{Z_i}{1 + A\beta}$ decreases	$\frac{Z_i}{1 + A\beta}$ decreases
Z_{of}	$\frac{Z_o}{1 + A\beta}$ decreases	$Z_o (1 + A\beta)$ increases	$\frac{Z_o}{1 + A\beta}$ decreases	$Z_o (1 + A\beta)$ increases

Table 5.5 shows some parameters and their relationship under different feedback circuits.

Table 5.5

Type/Parameter	*Voltage series*	*Voltage shunt*	*Current series*	*Current shunt*
Gain without feedback (A)	$A = \frac{V_o}{V_i}$	$A = \frac{V_o}{I_i}$	$A = \frac{I_o}{V_i}$	$A = \frac{I_o}{I_i}$
Feedback ratio (β)	$\beta = \frac{V_f}{V_o}$	$\beta = \frac{I_f}{V_o}$	$\beta = \frac{V_f}{I_o}$	$\beta = \frac{I_f}{I_o}$
Gain with feedback (A_f)	$A_f = \frac{V_o}{V_s}$	$A_f = \frac{V_o}{I_s}$	$A_f = \frac{I_o}{V_s}$	$A_f = \frac{I_o}{I_s}$

5.9. PRINCIPLE OF FEEDBACK IN AMPLIFIERS

Before proceeding with the concept of feedback, it is very necessary to study the classification of amplifiers. Generally amplifiers can be broadly classified in four categories, as given below :

(*i*) Voltage amplifier

(*ii*) Current amplifier

(*iii*) Transconductance amplifier

(*iv*) Transresistance amplifier

This classification is based on the magnitudes of the input and output impedances of an amplifier relative to source and load impedance respectively.

(*i*) **Voltage amplifier :** The general representation of voltage amplifier is shown in Fig. 5.29 (*a*). If the amplifier input resistance R_i is large compared with the source resistance R_s then $V_i = V_s$. If the external load source resistance R_L is large compared with the output resistance R_o of the amplifier will provide a output voltage proportional to input voltage and the proportionality factor is independent of the magnitudes of the source and load resistance. Such a circuit is called a voltage amplifier.

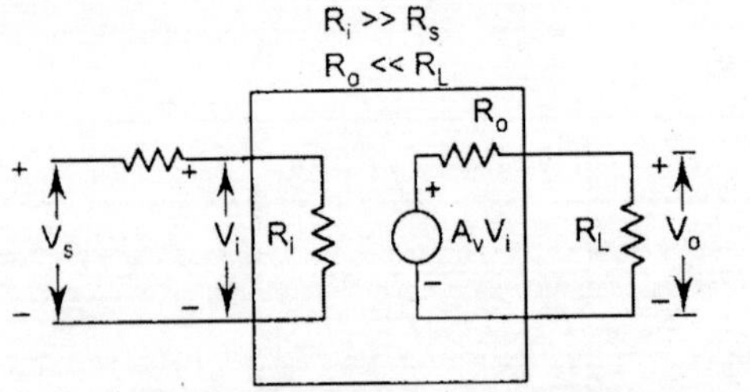

Fig. 5.29 (*a*). *Voltage amplifier.*

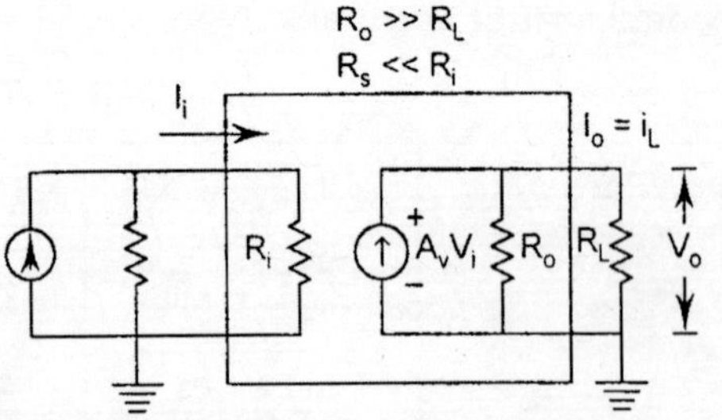

Fig. 5.29 (*b*). *Current amplifier.*

(*ii*) **Current amplifier :** The general representation of current amplifier is shown in Fig. 5.29 (*b*). If the amplifier R_i is small compared with the source resistance R_s and output resistance R_o is very large compared with the load resistance then $I_L = A_i\, I_i$.

$$I_L = A_i\, I_i.$$

This amplifier will provide a current proportional to the signal current and the proportionality factor is independent of the magnitudes of the source and load resistance, such a circuit is called current amplifier.

(*iii*) **Transconductance amplifier :** The general representation of transconductance amplifier is shown in Fig. 5.30. The ideal transconductance amplifier supplies an output current which is proportional to the signal voltage, independent of the magnitudes of source resistance and load resistance. To achieve this source resistance should be small compared to the input resistance and output resistance of amplifier should be large compared to the load resistance.

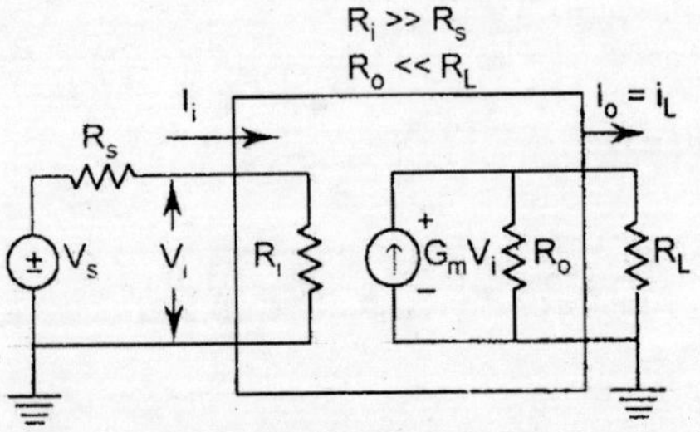

Fig. 5.30. *Transconductance amplifier.*

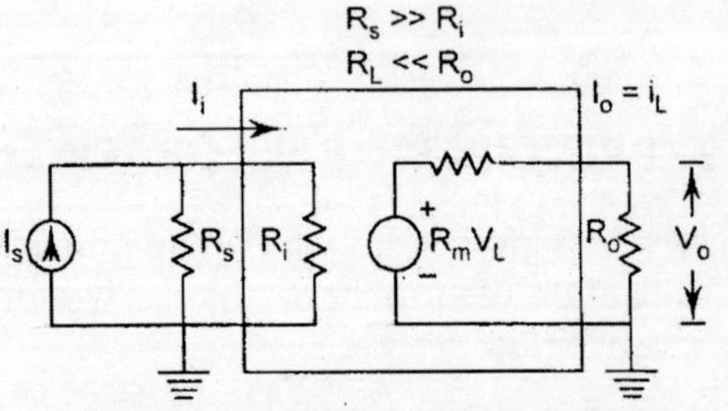

Fig. 5.31. *Transresistance amplifier.*

(*iv*) **Transresistance amplifier :** General representation of transresistance amplifier is shown in Fig. 5.31. For a practical transresistance amplifier we must have $R_s >> R_i$ and $R_L >> R_o$; when both input and output resistance of a amplifier are lower as compared to source and load resistance respectively. Since if $R_i << R_s$, $I_i \approx I_s$ and if $R_L >> R_o$, $V_o \approx R_m I_i \approx R_m I_s$. This type of amplifier is called *transresistance amplifier.*

Now, Table 5.6 shows the ideal requirement of amplifier characteristics like input and output resistance.

Table 5.6

Parameter	*Amplifier Type*			
	Voltage	*Current*	*Transconductance*	*Transresistance*
R_i	∞	0	∞	0
R_o	0	∞	∞	0
Transfer characteristics	$V_o = A_v V_s$	$I_L = A_i I_s$	$I_L = G_m V_s$	$V_o = R_m I_s$

5.10. SINGLE LOOP FEEDBACK AMPLIFIER (PRINCIPLE)

Now for better understanding we will draw a schematic representation of a single loop feedback amplifier. The transfer gain A may be A_v, A_i, G_m, R_m as the case may be.

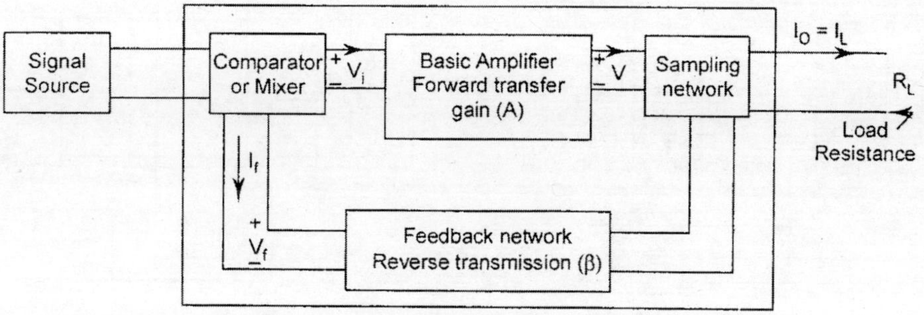

Fig. 5.32. *Simple loop feedback amplifier.*

The function of each block is given below :

(*a*) **Signal Source :** This block is either a signal voltage V_s in series with resistance (R_s) or a current source (I_s) in parallel with a resistance R_s.

(*b*) **Feedback Network :** This block is usually a passive two part network which may contain resistors, capacitors and inductors. Most often it is simply a resistive configuration.

(*c*) **Sampling Network :** In the sampling network the output voltage is sampled by connecting the feedback network in shunt across the output. Another feedback connection which samples the output current where the feedback network is connected in series with the output. This type of connection is referred to as current or loop sampling.

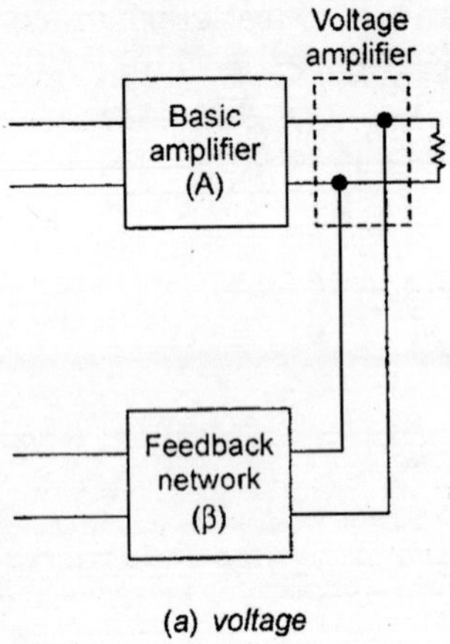

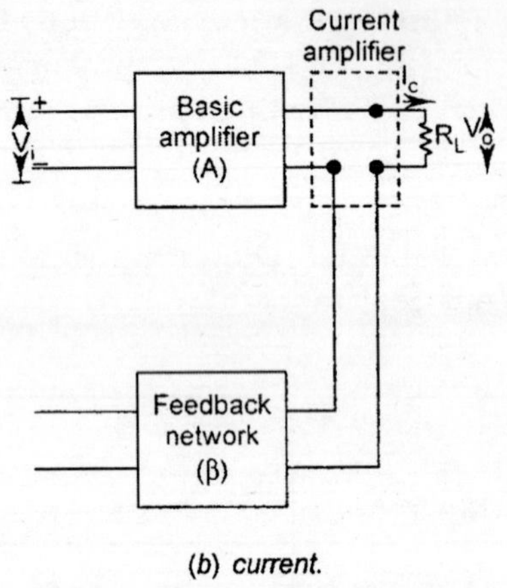

Fig. 5.33. *Feedback connection of the output of basic amplifier, sampling the output.*

(*d*) **Comparator or mixer network :** Two mixing blocks are very common, series (loop) input and shunt (node) input connection respectively. A differential amplifier is often used as a mixer. Such an amplifier has two inputs and gives an output proportional to the difference between the signal at the two inputs.

Now, we will discuss the feedback connections at the input of a basic amplifier as shown in Fig. 5.34.

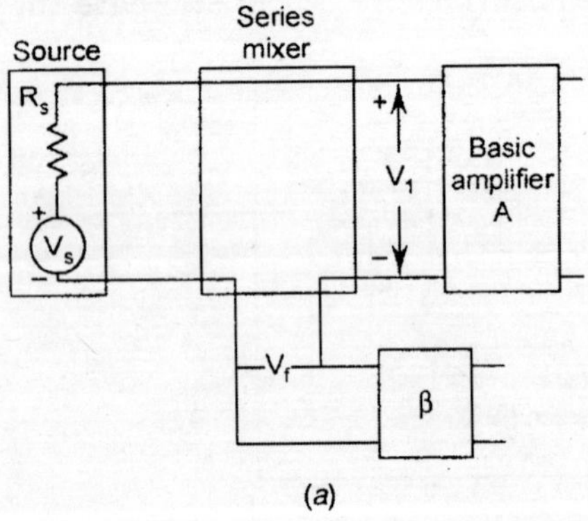

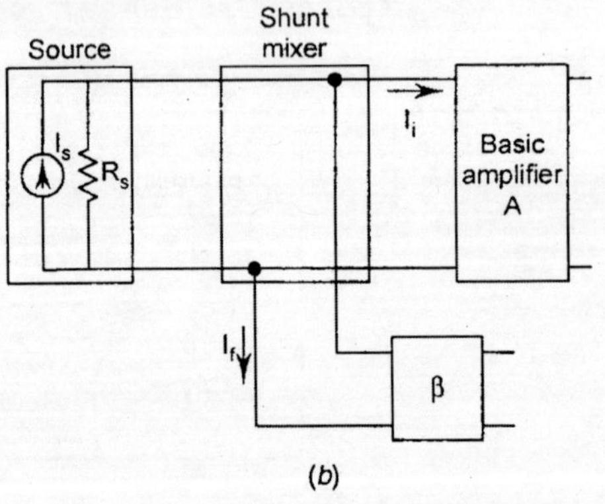

Fig. 5.34. *Schematic diagram of feedback connections at the input of a basic amplifier (a) series mixer (b) shunt mixing.*

Transfer ratio or gain : Transfer ratio or gain of amplifier is the ratio of output signal to the input signal. The transfer ratio $\frac{V_o}{V_i}$ is the voltage amplification or voltage gain. In the same fashion on transfer ratio $\frac{I_o}{I_i}$ represents current amplification or current gain. However, the transfer function $\frac{I_o}{V_i}$ represents the transconductance g_m of amplifier and the transfer ratio $\frac{V_o}{I_i}$ represents transresistance R_M. From the above discussions it is clear that the symbol g_m and R_M does not represent an amplification in the usual sense. Nevertheless, it is convenient to refer to each of the four quantities A_v, A_i, g_m and R_M as a transfer gain of basic amplifier without feedback and to use the symbol A to represent any one of these quantities.

The symbol A_f is defined as the ratio of the output signal to the input signal of the amplifier configuration and is called the transfer gain of the amplifier with feedback. Hence A_f is used to represent any one of the four ratios $\frac{V_o}{I_s} \cong A_{vf}$, $\frac{I_o}{I_s} = I_{if}$, $\frac{I_o}{V_s} = g_{mf}$, and $\frac{V_o}{I_s} = R_{Mf}$

Fig. 5.35 shows the schematic representation of a single loop feedback amplifier.

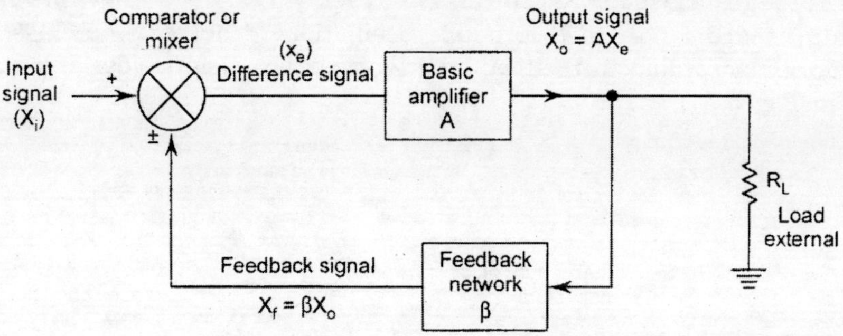

Fig. 5.35

5.11. INTRODUCTION TO OSCILLATORS

Wave generators plays a prominent role in the field of electronics. The generate signals from few hertz to several giga hertz (10^9 Hz). Modern wave generators use many different circuits and generate such outputs as, sinusoidal, square, rectangular, saw tooh and trapezoial waveshapes. These waveshapes serves many useful purposes in the electronics circuit you will studying. For example, they are used extensively throughout the television receiver to reproduce both picture and sound.

One type of wave generator is known as an oscillator. The oscillator can be regarded as an amplifier which provides its own input signal. They are used to generate high frequency wave (carrier wave) in the tunning stages. Oscillators are widely use in radar, computer and digital electronic circuits.

As we know that an amplifier produces an output signal whose waveform is similar to the input signal but whose power level is different. This extra or additional power is supplied by external *dc* source. There is no doubt in saying that an amplifier is essentially an energy *i.e.*, it takes energy from the *dc* power source and converts it into as energy at signal frequency. It is interesting here to note that if there is no input signal. There is no energy conversion and hence there is no output signal.

However, an oscillators differs from an amplifier in the way that the oscillators does not require an input source to start or maintain energy conversion process. It keeps producing an output signal so long as the *dc* power source is connected.

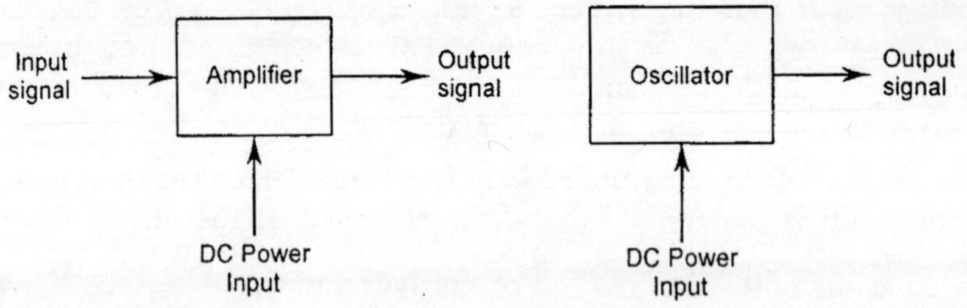

Fig. 5.36. *Illustrating the comparison between an amplifier and an oscillator.*

Classification of Oscillators

Oscillators are classified according to the waveshapes they produce and the requirements needed for them to produce oscillations. They can be classified into two broad categories according to their output. Waveshapes ; sinusoidal and non-sinusoidal (as shown in Fig. 5.37).

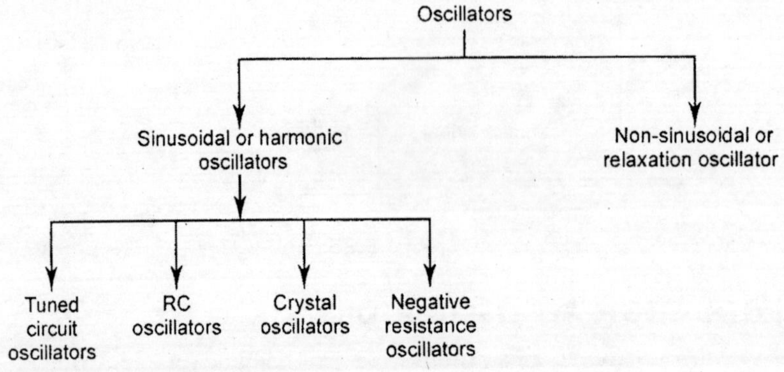

Fig. 5.37. *Classification of oscillator.*

(1) **Sinusoidal oscillators.** A sinusoidal oscillator produces a sin-wave output signal. Ideally, the output signal is of constant amplitude with no variation in frequency. Actually, something less than this is usually obtained. The degree to which the ideal is approached depends upon such factors as class of amplifier operation, amplifier characteristics, frequency stability and amplitude stability. Sine-wave oscillators produce signal ranging from low audio frequencies to ultra high radio and microwave frequencies (from 20 Hz to 1 MHz). The sinusoidal oscillators may be further sub-divided into following type :

(*a*) **Tuned circuit oscillator :** These oscillators use a tuned circuit consisting inductors (L) and capacitors (C) and are used to generate high frequency signals. Thus, they are also known as radio frequency (RF) oscillators. Such oscillators are Hartley, Colpitts, Clapp oscillators etc.

(*b*) **RC oscillators :** These oscillators use resistors and capacitors and are used to generate low or radio frequency signals. Thus, they are also known as audio frequency (AF) oscillators. Such oscillator are Phase shift oscillator and Wien bridge oscillator.

(c) **Crystal oscillators :** These oscillators use quartz crystal and are used to generate highly stabilized output signal with frequency upto 10 MHz. For example, peizo oscillators.

(d) **Negative resistance oscillator :** These oscillators use negative resistance characteristics of the device such as tunnel diodes. For example, tuned diode oscillator.

(2) **Non-sinusoidal oscillators :** They generates complex waveforms such as square, rectangular, trigger, sawtooth, or trapezoial. Because their outputs are generally characterized by a sudden change, or relaxation, they are often referred as relaxation oscillators . The signal frequency of these oscillators is usually governed by the charge or discharge time of a capacitor in series with resistor. Some oscillators, both RC and LC Networks are used for determining the frequency of oscillation. Within this category of non-sinusoidal oscillators are multivibrators, blocking oscillators, sawtooth generators and trapezoidal generators.

Advantages of Oscillators

Following are the advantages of oscillators :

- An oscillator uses non-rotating device. Consequently, there is little wear and tear hence longer life.
- Due to absence of moving parts, the operation of an oscillator is quite silent.
- The frequency of an oscillator can be easily changed when desired.
- It has a good frequency stability *i.e.,* frequency once set remains constant for a considerable period of time.
- It has a very high efficiency.

Nature of sinusoidal oscillations : The name of oscillations, produced by sinusoidal oscillators may be of two types namely (*i*) damped oscillations and (*ii*) undamped oscillations.

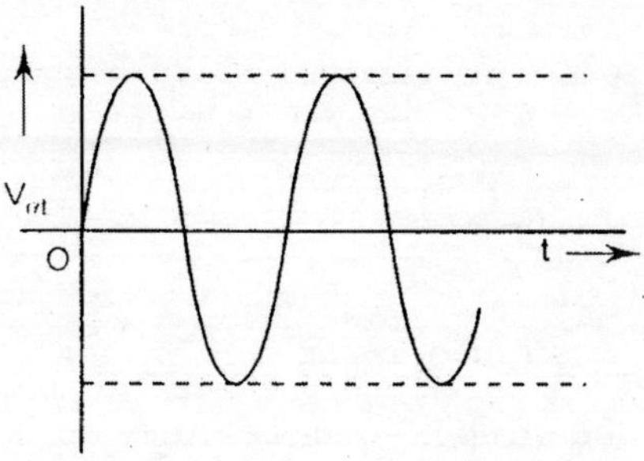

Fig. 5.38. *Damped oscillations.*

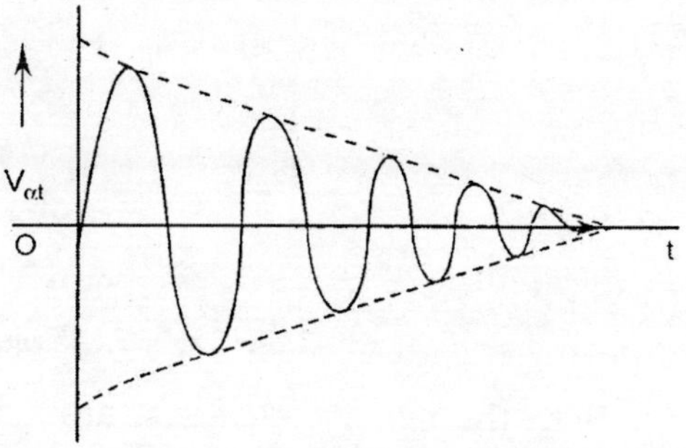

Fig. 5.39. *Undamped oscillations.*

(*i*) **Damped oscillations :** The electrical oscillations, whose magnitude goes on decreasing with time are called damped oscillations, as shown in Fig. 5.38. The damped oscillations are produced by those oscillator circuits in which power losses place continuously during each oscillation, however, it may be noted that frequency changes with time, which is undesirable for practical purpose.

(*ii*) **Undamped oscillations :** The electrical oscillations whose amplitude remain constant with time, are called undamped oscillations as shown in Fig. 5.39. The undamped oscillations are produced by those circuits in which either there are no power losses or if they have any power loss they have provision for compensating those losses.

5.12. OSCILLATORY CIRCUIT

A circuit which produces electrical oscillations of any desired frequency is known as oscillatory circuit or tank circuit. A simple oscillatory circuit consists of a capacitance (C) and inductance (L) in parallel. Both the reactive devices are capable of storing electrical energy.

Basic principle of sinusoidal oscillator : It is a well known fact that an inductor stores energy in its magnetic field whenever a current flows through it. Similarly a capacitor stores energy in its electric field, whenever a voltage is applied across its plates.

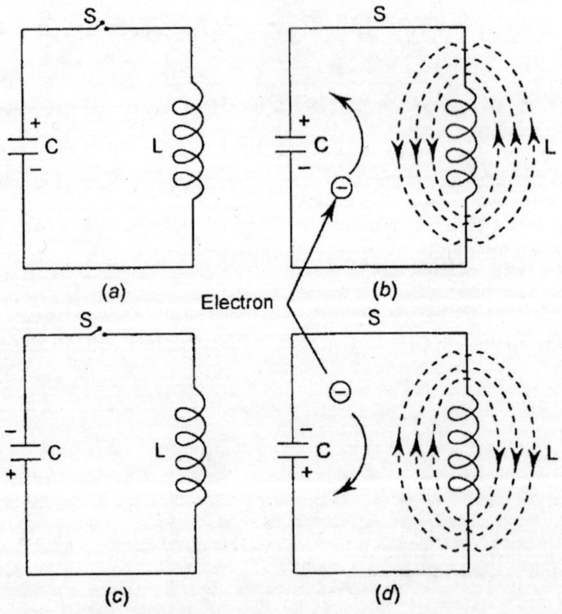

Fig. 5.40. *LC circuit acting as oscillatory circuit.*

Both the reactive elements are assumed to be an ideal one (*i.e.,* they do not have any power losses). Now we shall discuss as to how an LC circuit generates oscillations. Suppose a capacitor has been charged *dc* voltage source with a polarity as shown in Fig. 5.40. As the switch (S) is opended it cannot discharge through the coil. When the switch (S) is closed as shown in Fig. 5.40, the capacitor, discharges through the coil (L). This current flow set up magnetic field around the coil. This magnetic field stores the energy released by electric field. Because of the inductive effect, the current grows up slowly towards the maximum value. This situation occurs, when the capacitor is fully discharged. At this instant, the electrical energy stored in the capacitor becomes zero and the energy stored in magnetic field is maximum energy around the coil.

Once the capacitor is discharged completely, the magnetic field around the coil begins to collapse and produces a counter (or back) emf according to Lenz's law, the counter emf, keeps electrons moving in the same direction as shown in Fig. 5.40 (*b*). This again charges the capacitor, though in the opposite direction, as shown in Fig. 5.40 (*c*). When the capacitor is charged completely in opposite direction, the magnetic field around the coil is also collapsed completely. At this instant, the energy previously stored in the magnetic field, has now converted into the energy stored in the electric field of the capacitor.

The above mentional sequence of charging and discharging of a capacitor results in an alternating motion of electrons or an alternating current. As a result of this the energy is alternatively stored in the electric field of the capacitor and magnetic field of the coil. This interchange of energy between the capacitor and coil continues to be repeated, and results in the production of electrical oscillations.

If there is no power losses in the tank circuit, the waveform of electrical oscillations will be undamped (*i.e.,* constant amplitude). However, in a practical tank circuit, the waveform of the oscillations is damped (*i.e.,* of decreasing amplitude). It is because of the energy lost by L and C element, which may take place because of the following two reasons :

1. Some energy is lost in the form of heat produced in the resistance of the coil, connecting wires and within the dielectric material of the capacitor.
2. Some energy is lost from the circuit due to radiation in the form of electromagnetic waves. As a result of these losses, the amplitude of oscillating current in tank circuit decreases gradually and eventually it becomes zero.

So, for achieving the undamped oscillation the following condition must be fulfilled :

- The amount of energy supplied should be of such a value so as to meet the losses in the tank circuit and *ac* energy removed from the circuit by the load for example, if losses in LC circuit amount to 5 mW and the *ac* output being taken is 100 mW, when power of 105 mW should be continuously applied to the circuit.
- The applied energy should have the same frequency as that of oscillations in the tank circuit.
- The applied energy should be in the same phase with the oscillations set up in the tank circuit.

If these conditions are fulfilled the circuit will produce continuous undamped oscillation. Finally we conclude that the essentials of a transistor oscillator is oscillator circuit and transistor amplifier. For better understanding Fig. 5.41 shows the block diagram of transistor amplifier.

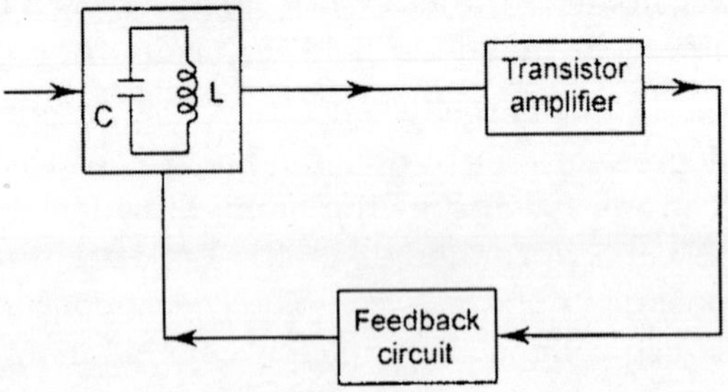

Fig. 5.41. *Block diagram of transistor oscillator.*

5.13. CONDITIONS FOR OSCILLATIONS (BARK HAUSEN CRITERION)

We know that overall voltage gain (*i.e.*, gain with feedback for a positive feedback is given by the expression).

$$A_f = \frac{A}{1 - \beta A}$$

where,

A = Gain of amplifier without feedback or open loop gain.

A_f = Gain of amplifier with feedback or closed loop gain.

βA = loop gain.

It may be noted that if $\beta A = 1$ then the $A_f \rightarrow \infty$ but in actual practice it is impossible. Therefore the condition $\beta A = 1$ means the circuits has stopped amplifying and started oscillating. Also it may be noted that undamped oscillation will not be maintained if the value of βA is less than unity.

We know that generally an amplifier causes a phase shift of 180° between the input and output signals. In order to provide positive feedback, the feedback network must provide a phase shift of 180° to provide a signal with a phase shift of 360° or 0° at the amplifier input.

Thus, the conditions for undamped oscillation is that there should be a positive feedback with loop gain ($\beta A = 1$) equal to unity, and the net phase shift around the loop is equal to 360° or 0° are called the condition for oscillation. Thus, if βA is a complex quantity then its real part must be equal to unity and imaginary part equal to zero *i.e.*,

$$\beta A = 1 + j\,0$$

However, an oscillator in which βA is exactly unity, is realizable in practice. In every practical oscillator βA is slightly larger than unity to avoid the losses discussed earlier. Finally, we concluded that an oscillator circuit must satisfy the two conditions mentioned below to produce sustained undamped oscillation. The practical conditions are

(*i*) Loop gain $\beta A \geq 1$

(*ii*) Phase shift between the input and output signal must be 360° or 0°.

Different conditions for oscillations are :

(*i*) If Aβ is less than unity, the output will die out and known as damped oscillations as shown in Fig. 5.42 (*a*).

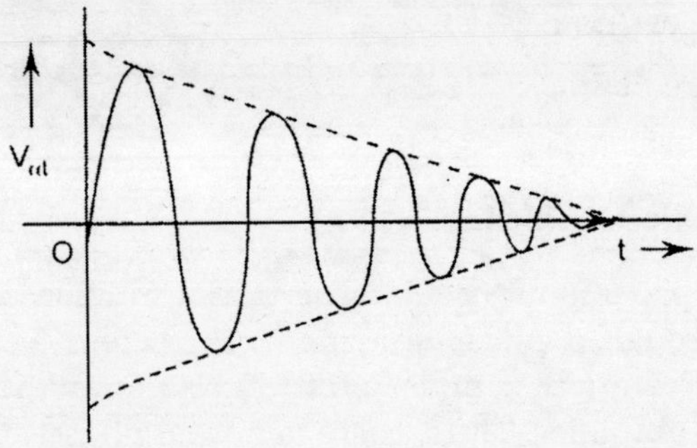

Fig. 5.42 (*a*). *Damped oscillation when Aβ < 1.*

(*ii*) If Aβ is greater than unity, the output signal will build up is known as growing oscillations as shown in Fig. 5.42 (*b*).

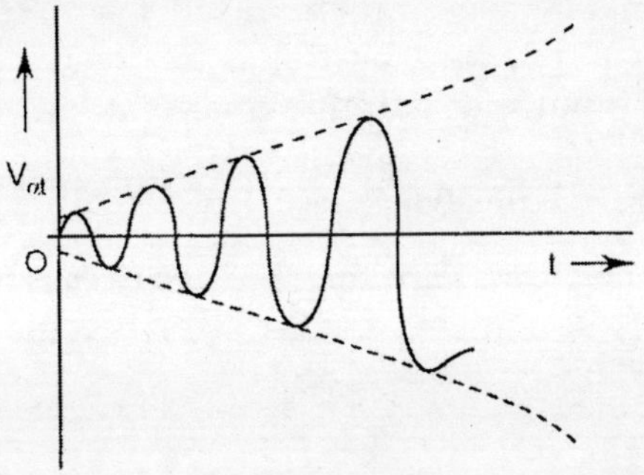

Fig. 5.42 (*b*). *Growing oscillation when Aβ > 1.*

(*iii*) If Aβ is equal to unity, no change occurs in the output and we get an output with constant magnitude is known as undamped oscillations as shown in Fig. 5.42 (*c*).

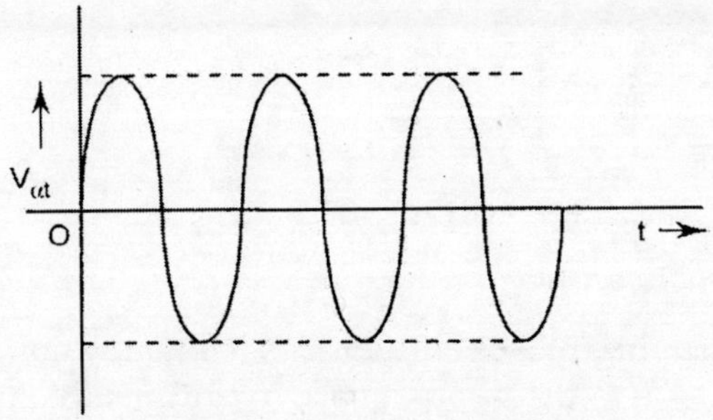

Fig. 5.42 (c). *Undamped oscillations when Aβ = 1.*

5.14. RC OSCILLATORS

These oscillators use resistors and capacitors are used to generate low or audio frequency signals. Thus, they are also known as audio frequency (AF) oscillators, such oscillators are given below namely :

(*i*) Phase shift oscillator

(*ii*) Wein bridge oscillator

5.14.1. RC Phase Shift Oscillator

Basic principle of phase shift oscillator is that a fraction of the output single-stage amplifier is passed through a phase-shift network, before feeding back to input. The phase-shift network gives another phase-shift of 180° in addition to the phase-shift of 180° introduced by the amplifier. Thus, there is a total phase-shift of 360°, which is also equal to 0°. The RC oscillator, utilizing this principle is known as phase-shift oscillator.

Circuit arrangement of phase-shift oscillator is shown in Fig. 5.43. Resistance R_1, R_2 and R_3 are the biasing parameter and R_c is the load resistance. There are three R-C combinations forming feedback network. Each R-C combination provides a phase shift of 60°. Hence the net phase shift produced by three RC networks is 180° and another 180° phase shift is provided by transistor itself. Thus, total 360° phase shift is produced between the input and output signal. The phase shift ϕ, given by each RC section is $\phi = \tan^{-1}\dfrac{1}{\omega CR}$ if R is made zero then ϕ will be become 90°. But making R = 0 is impractical because if R is zero then voltage across it will become zero, therefore in practice the value of R is adjusted such that ϕ become 60°.

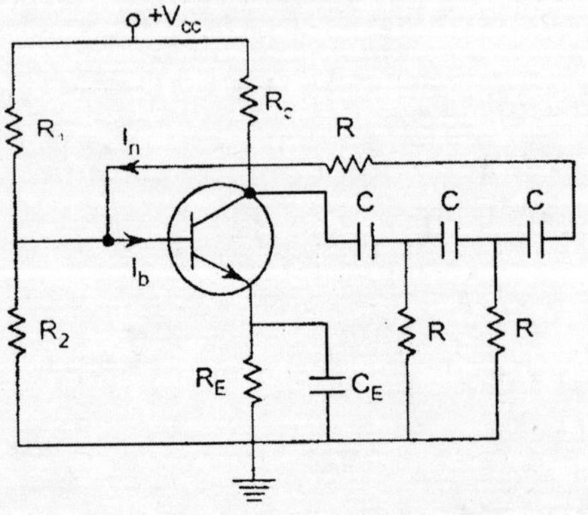

Fig. 5.43. *Phase shift oscillator.*

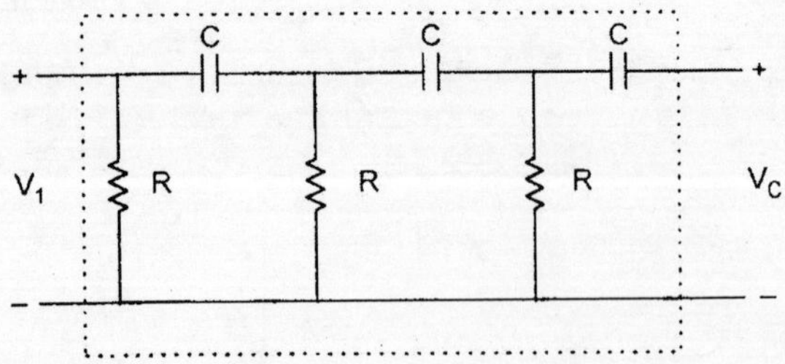

Fig. 5.44. *Feedback network.*

The transitor used in this circuit is C-E configuration. Let for this circuit $h_{oe} R_C < 0.1$ so that we may use the approximate hybrid model (as discussed earlier). Also, it is assumed that biasing resistor R_1, R_2, and R_E have no effect on the signal operation and are neglected in the analysis. The circuit for the feedback network is shown in Fig. 5.44. Since it is the case of voltage shunt feedback hence input and output can be separated without feedback by using $V_0 = 0$ and $I_1 = 0$ respectively. The frequency of oscillation is given by

$$f_r = \frac{1}{2\pi RC\sqrt{6+4K}}$$

At this frequency, it is found that the feedback factor of the network is $|\beta| = \frac{1}{29}$. In order that $|A\beta|$ shall not be less than unity. It is required that the amplifier gain $|A|$ must be more than 29 for oscillator operation.

Derivation of Frequency Oscillation

First of all draw its approximate hybrid model as shown in Fig. 5.45 (*a*).

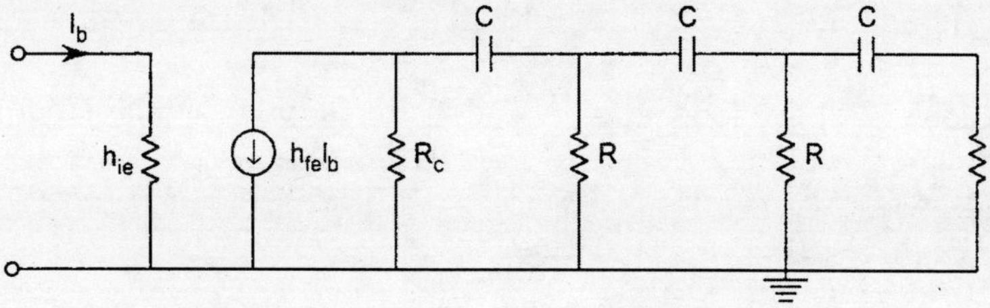

Fig. 5.45 (*a*). *Approximate hybrid model for h_{oe} R_C > 0.1.*

From Fig. 5.45 (*a*), loop gain $= \frac{I_3}{I_b}$

Now, to make calculation easy, apply the source transformation technique in Fig. 5.45 (*a*) and the equivalent circuit is shown in Fig. 5.45 (*b*).

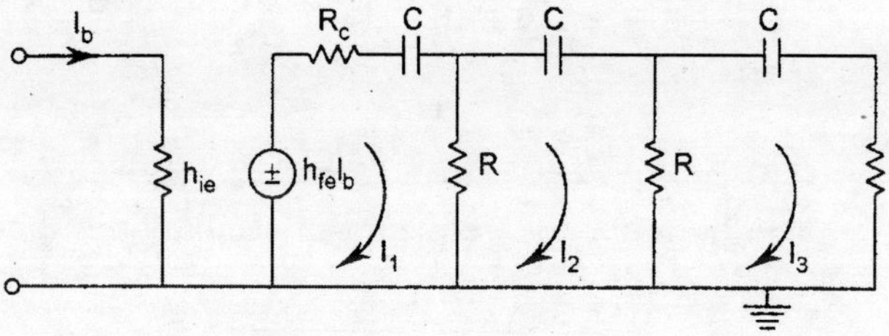

Fig. 5.45 (*b*). *The equivalent circuit for which to calculate the loop gain.*

Now, applying KVL in the loop (1), (2) and (3) we get

$$-h_{fe} I_b R_C - I_1 [(R_C - JX_C) + R] + I_2 R = 0 \quad ...(5.14)$$

$$(2R - JX_C) I_2 - I_1 R - I_3 R = 0 \quad ...(5.15)$$

$$(2R - JX_C) I_3 - I_2 R = 0 \quad ...(5.16)$$

Let $K = \frac{R_C}{R}$ and $\alpha = \frac{X_C}{R} = \frac{1}{\omega CR}$, on putting these value equations (5.14), (5.15) and (5.16) may be rewritten as

$$-h_{fe} I_b R_C - I_1 \{K + 1 - J\alpha\} + I_2 = 0 \quad ...(5.17)$$

$$I_2 (2 - J\alpha) - I_1 - I_3 = 0 \quad ...(5.18)$$

$$(2 - J\alpha) I_3 - I_2 = 0 \quad ...(5.19)$$

From equations (5.18) and (5.19), we get

$$(2 - J\alpha)\ I_3\ (2 - J\alpha) = I_1 + I_3 \quad ...(5.20)$$

or $$I_3\ (2 - J\alpha)\ (2 - J\alpha) = I_1 + I_3$$

or $$I_3\ \{(2 - J\alpha)\ (2 - J\alpha) - 1\} = I_1$$

or $$I_3\ \{3 - \alpha^2 - 4J\alpha\} = I_1 \quad ...(5.21)$$

On substituting the values of I_1 and I_2 in equation (5.17). In order to calculate ratio $\frac{I_3}{I_b}$, we get

$$-h_{fe}\ I_b\ K = I_3\ \{(3 - \alpha^2 - 4J\alpha)\ (k + 1 - J\alpha)\} - (2 - J\alpha)\ I_3$$

$$-h_{fe}\ I_b\ K = I_3\ \{3K + 3 - 3J\alpha - K\alpha^2 - \alpha^2 + J\alpha^3 - J4K\alpha - J4\alpha - 4\alpha^2 - 2 + J\alpha\}$$

$$-h_{fe}\ I_b\ K = I_3\ \{1 + 3K - (5 + K)\ \alpha^2 - J\ (6 + 4K)\ \alpha - \alpha^3\} \quad ...(5.22)$$

From equation (5.22) the loop gain $\frac{I_3}{I_b}$ and if this is to be real co-efficient of J, must be equal to zero *i.e.*,

$$(6 + 4K)\ \alpha - \alpha^3 = 0$$

or $$\alpha^2 = 6 + 4K$$

$$\alpha = \sqrt{6 + 4K} \quad \left\{\therefore \alpha = \frac{1}{\omega CR}\right\}$$

$$\frac{1}{\omega CR} = \sqrt{6 + 4K}$$

$$\frac{1}{2\pi f CR} = \sqrt{6 + 4K} \quad \{\therefore \omega = 2\pi f\}$$

$$f = \frac{1}{2\pi CR\sqrt{6 + 4K}} \quad ...(5.23)$$

This is the required condition for the frequency of oscillation in phase shift oscillator.

Now, $$\frac{I_3}{I_b} = \frac{-h_{fe}\,K}{1 + 3K - (5 + K)\,\alpha^2}$$

or $$\frac{I_3}{I_b} = \frac{-h_{fe}\,K}{1 + 3K - (5 + K)(6 + 4K)} \quad \{\therefore \alpha^2 = 6 + 4K\}$$

or $$\frac{I_3}{I_b} = \frac{-h_{fe}\,K}{1 + 3K - 30 - 24K - 4K^2}$$

or $$\frac{I_3}{I_b} = \frac{-h_{fe}\,K}{4K^2 + 23K + 29}$$

But loop gain ≥ 1

or $$\frac{I_3}{I_b} \geq 1$$

or $$\frac{-h_{fe}K}{4K^2+23K+29} \geq 1$$

$$h_{fe} \geq 4K + 23 + \frac{29}{K} \quad \text{...(5.24)}$$

Now, we will calculate the minimum or maximum value of h_{fe} (*i.e.*, gain of the transistor). To calculate it we must differentiate equation (5.24) with respect to K, we get

$$\frac{\partial h_{fe}}{\partial K} \geq 4 - \frac{29}{K^2}$$

For maximum or minimum value, $\frac{\partial h_{fe}}{\partial K} = 0$

$$0 = 4 - \frac{29}{K^2}$$

$$K = \sqrt{\frac{29}{4}} \cong 2.75$$

On putting the value of K in equation (5.24), we get

$$h_{fe} \geq 44.5 \quad \text{...(5.25)}$$

Thus, the value of K which gives the minimum h_{fe} turns out to be approximate 2.75 and for this value h_{fe} = 44.5, *i.e.*, a transistor with a small signal common emitter short circuit current gain less than 44.5 cannot be used in the phase shift oscillator.

Advantages of Phase Shift Oscillator

Following are the advantages of phase shift oscillater :

- It does not require transformers or inductors, that's why it is less bulky.
- Cheap and simple circuit as it contains resistors and capacitor only.
- Waveform is exceptionally pure and sinusoidal since the core saturation effect and harmonic distortion are absent as no transformer is used.

Disadvantages of Phase Shift Oscillator

Following are the disadvantages of phase shift oscillater :

- The main disadvantage of this circuit is the high gain requirement (approximately > 44.5) which is practically impossible.
- It gives only small output due to smaller feedback.
- Feedback is less and it is difficult for the circuit to start oscillations. This is because of high reactance of R and C.
- It requires high supply voltage *i.e.*, V_{CC} > 12 V.

Example 1. *Select the value of capacitor C and transistor h_{fe} to provide an oscillator frequency f = 2 KHz in phase shift oscillator*

Given $R_C = 10\ K\Omega,\ R = 8\ K\Omega$

Solution : Given that

$$f = 2\ \text{KHz} = 2 \times 10^3\ \text{Hz}$$

$$R_C = 10\ k\Omega = 10 \times 10^3\ \Omega$$

$$R = 8\ K\Omega = 8 \times 10^3\ \Omega$$

We know that the frequency of oscillation in the case of phase shift oscillation is given by relation (equation (5.23)).

$$f = \frac{1}{2\pi RC\sqrt{6 + \frac{4R_C}{R}}}$$

$$2 \times 10^3 = \frac{1}{2\pi \times 8 \times 10^3 \times C\sqrt{6 + 4 . \frac{10 \times 10^3}{8 \times 10^3}}}$$

or $$2 \times 10^3 = \frac{1}{2\pi \times 8 \times 10^3 \times C\sqrt{6+5}}$$

or $$2 \times 10^3 = \frac{1}{2\pi \times 8 \times 10^3 \times C \times 3.32}$$

or $$C = \frac{1}{16\pi \times 10^3 \times 2 \times 10^3 \times 3.32}$$

or $$C = 3.0 \times 10^{-9}\ \text{F} = 0.003\ \mu\text{F}$$

Value of transistor gain h_{fe} is given by relation

$$h_{fe} \geq 23 + 29\frac{R}{R_C} + 4\frac{R_C}{R}$$

or $$h_{fe} \geq 23 + 29\left(\frac{8 \times 10^3}{10 \times 10^3}\right) + 4 . \left(\frac{10 \times 10^3}{8 \times 10^3}\right)$$

or $$h_{fe} \geq 23 + 23.2 + 5$$

or $$h_{fe} \geq 51.2$$

or $$h_{fe} = 51.2.\ \textbf{Ans.}$$

5.14.2. Wein Bridge Oscillator

In the circuit for wein bridge oscillator as shown in Fig. 5.46. A second-stage of amplifier is used for producing another 180° phase shift in addition to the phase shift of 180° produced by the first stage. Thus, there is total phase shift of 360°, which is the basic requirement. A fraction of output from the second stage is feedback to the input of the first stage without producing any further phase-shift. The RC oscillator utilizing this principle, is known as wien bridge oscillator.

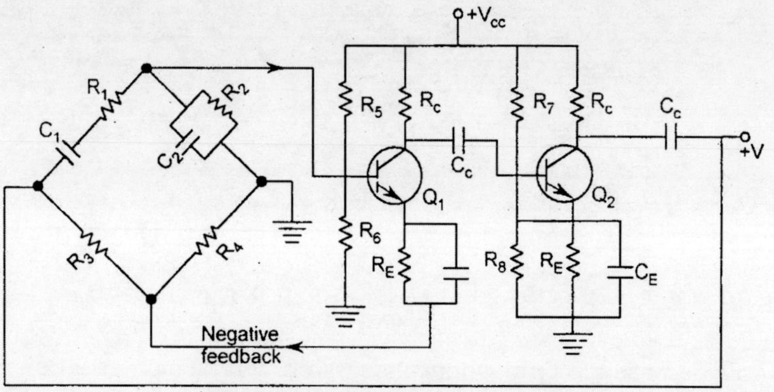

Fig. 5.46. *Wein bridge oscillator.*

The wein bridge oscillator consists of two transistor (CE configuration) which provides an approximately 360° or 0° phase shift so the feedback network has no need to introduce any additional phase shift where R_5, R_6, R_7 and R_8 are biasing resistors.

The feedback network consists of C_1–R_1 and C_2–R_2 (called a lead-lag network) and R_3–R_4 (called a voltage divider). The lead-lag network provides a positive feedback to the input of the first stage and the voltage divider provides the negative feedback to the emitter of the transistor Q_1.

Derivation for frequency of oscillation

For calculating the frequency of oscillation the feedback network of wein bridge oscillator is shown in Fig. 5.47.

We know that at balanced condition phase shift must be zero.

According to wheat stone bridge,

$$\frac{P}{Q} = \frac{R}{S}$$

So,

$$\frac{R_3}{R_1 + \frac{1}{j\omega C_1}} = \frac{R_4}{\frac{R_2 \frac{1}{j\omega C_2}}{R_2 + \frac{1}{j\omega C_2}}}$$

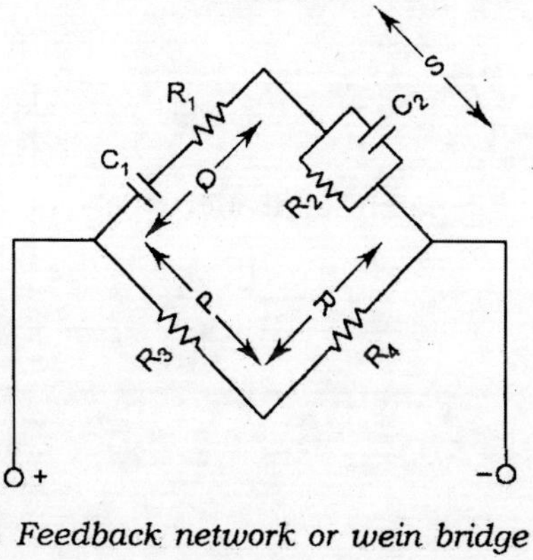

Fig. 5.47. *Feedback network or wein bridge oscillator.*

$$\frac{R_3 j\omega C_1}{jR_1\omega C_1 + 1} = \frac{R_4}{R_2}(j\omega C_2R_2 + 1)$$

$$R_2R_3j\omega C_1 = R_4R_1j^2\omega^2C_1C_2R_2 + R_4R_1j\omega C_1 + R_4 + R_4j\omega C_2R_2$$

$$j\left(R_2R_3 - R_4\frac{C_2}{C_1}R_2\right)\omega C_1 = R_4 - R_1R_2R_4C_1C_2\omega^2$$

$$-\left(R_2R_3 - R_4R_1 - R_4\frac{C_2}{C_1}R_2\right) = j\left\{\frac{R_4}{\omega C_1} - R_4R_1R_2C_1C_2\right\} \qquad \text{...(5.26)}$$

the oscillation will be possible only when phase shift zero. The oscillation frequency is given by

$$\frac{R_4}{\omega C_1} - \omega C_2R_4R_2R_1 = 0$$

or

$$\omega^2 = \frac{1}{C_1C_2R_1R_2} \qquad \{\because \omega = 2\pi f\}$$

$$f = \frac{1}{2\pi\sqrt{R_1R_2C_1C_2}} \qquad \text{...(5.27)}$$

Condition for oscillation :

$$R_2R_3 - R_1R_4 - R_2R_4\frac{C_2}{C_1} = 0$$

$$\frac{R_3}{R_4} - \frac{R_1}{R_2} = \frac{C_2}{C_1} \qquad \text{...(5.28)}$$

This is the required condition for oscillation.

If, $R_1 = R_2 = R$ and $C_1 = C_2 = C$, then from equation (5.28).

$$\frac{R_3}{R_4} - 1 = 1$$

$$\frac{R_3}{R_4} = 2$$

We know that $\beta = \frac{V_f}{V_o}$

But from Fig. 5.48 applying potential divider rule

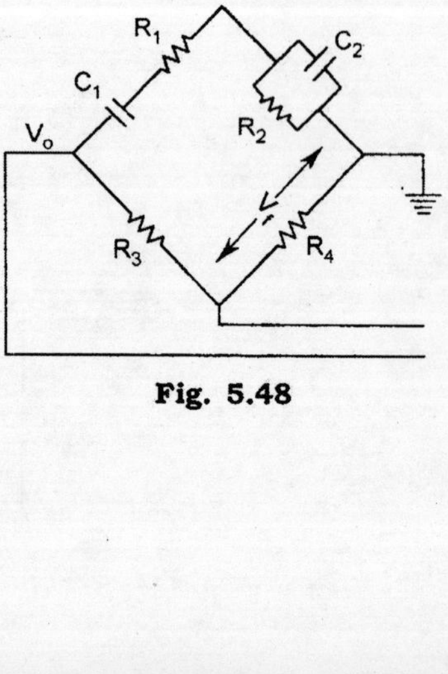

Fig. 5.48

$$V_f = V_o \cdot \frac{R_4}{R_3 + R_4}$$

or

$$\frac{V_f}{V_o} = \frac{R_4}{R_3 + R_4}$$

or $$\beta = \frac{R_4}{2R_4 + R_4} \qquad ...(5.29)$$

or $$\beta = \frac{1}{3}$$

Also since as we know that

$$A\beta \geq 1$$

$$A \times \frac{1}{3} \geq 1$$

$$A \geq 3 \qquad ...(5.30)$$

Thus, for oscillation produced gain must be equal to or greater than 3.

Advantages of Wein Bridge Oscillator

Following are the advantages of wein bridge oscillator :

- It has better stability.
- Output is constant.
- Its working quite simple and easy.
- Overall gain in high as two transistors are used.
- Frequency of oscillations can be easily adjusted by varying gang capacitors C_1 and C_2.

Disadvantages of Wein Bridge Oscillator

Following are the disadvantages of wein bridge oscillator :

- Costlier as more components are used.
- It cannot be used to generate very high frequency (> 1 MHz) this is the main draw back of this circuit.

NOTE : Why wein bridge oscillator used for audio frequency oscillator ?

Since the frequency of oscillation for the wein bridge oscillator is $f = \frac{1}{\sqrt{R_1R_2C_1C_2}}$ or $f = \frac{1}{2\pi RC}$, when $R_1 = R_2$ and $C_1 = C_2$ and the practical feasible value of RC gives the frequency of audio range that's why this oscillator is mainly used as audio frequency oscillator.

Example 2. *A wein bridge oscillator is used for operation at f = 10 KHz. If the value of R is 100 KΩ. Find the value of the capacitor C.*

Solution : Given :

$$f = 10 \text{ KHz} = 10 \times 10^3 \text{ Hz}$$

$$R = 100 \text{ K}\Omega = 100 \times 10^3 \ \Omega$$

We know that frequency of oscillation in a wein bridge oscillation is given by relation.

$$f = \frac{1}{2\pi RC}$$

$$10 \times 10^3 = \frac{1}{2\pi \times 100 \times 10^3 \times C}$$

or
$$C = \frac{1}{2\pi \times (100 \times 10^3) \times 10 \times 10^3}$$

or
$$C = 159 \times 10^{-12} \text{ F}$$
$$C = 159 \text{ pF. } \textbf{Ans.}$$

5.15. HARTLEY AND COLPITTS OSCILLATOR

These oscillators called tuned circuit oscillator are also known as LC oscillators resonant circuit oscillators or tank circuit oscillators. These oscillator are used to produce an output with frequencies ranging from 1 MHz to 500 MHz. Hence they are also known as radio frequency (RF) oscillators.

5.15.1. Hartley Oscillator

Fig. 5.49 shows the circuit of a Hartley oscillator. Tank circuit consists of two coils L_1 and L_2 and a capacitor C. The coil L_1 is inductively coupled to the coil L_2 and the combination works as an auto transformer. A coil called Radio Frequency Choke (RFC) is connected between the oscillator and V_{CC} supply. The feedback between the output and input circuit is accomplished through auto transformer action, which also introduceds a phase shift of 180°. The phase reversal between the output and input voltage occurs because they are taken from the opposite ends of the coils (L_1 and L_2) with respect to the tap, which is grounded.

Since the transistor also introduces a phase shift of 180°, therefore, the total phase shift is 360° and hence feedback is positive. Ignoring the loading effects of the base, the feedback fraction is given by the relation.

$$\beta = \frac{L_2}{L_1}$$

Also since
$$A\beta \geq 1$$

$$A \cdot \frac{L_2}{L_1} \geq 1$$

$$A \geq \frac{L_1}{L_2}$$

It means voltage gain must be equal to $\frac{L_1}{L_2}$

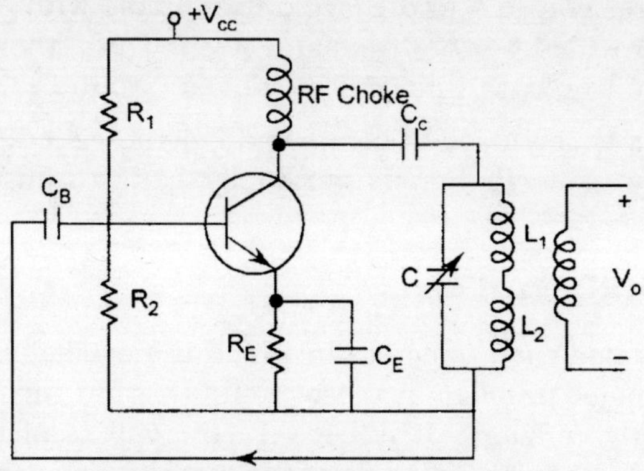

Fig. 5.49. *Hartley oscillator.*

The resistor R_1, R_2 and R_E are used to provide *dc* bias to the transistor. When the circuit is energised, switching on the supply, the collector current flows. The oscillations are produced because of positive feedback from the tank circuit.

5.15.2. Colpitts Oscillator

Colpitts oscillator is similar to Hartley oscillator. The only difference is that in case of colpitts oscillator, coupling is capacitive instead of being inductive. Fig. 5.50 shows the colpitts oscillator circuit. The tank circuit is made up of two capacitors C_1 and C_2 connected in series with each other across a fixed inductance (L). The resistors R_1, R_2, R_E and RF choke have the same function as mentioned in Hartley oscillator.

The feedback between the output and input circuit is accomplished by the voltage developed across the capacitor C_2. Ignoring the loading effect of the base, the feedback fraction

$$\beta = \frac{C_1}{C_2}$$

But, since $$A\beta \geq 1$$

or $$A\frac{C_1}{C_2} \geq 1$$

or $$A \geq \frac{C_2}{C_1}$$

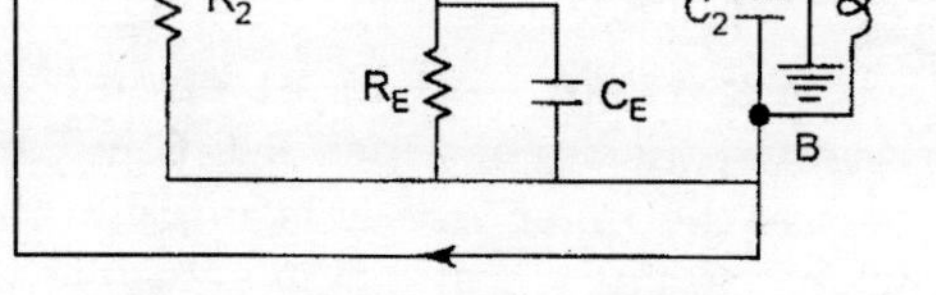

Fig. 5.50. *Colpitts oscillator.*

Thus, to start the oscillation, the voltage gain (A_V) must be greater than $\left(\frac{C_2}{C_1}\right)$. The frequency of oscillation (neglecting mutual inductance) is given by the relation.

$$f = \frac{1}{2\pi\sqrt{LC}}$$

where $$C = \frac{C_1 C_2}{C_1 + C_2}$$ (C is the effective value of capacitance)

It may be noted that in a Colpitts oscillator C_1 and C_2 act as a simple alternating voltage divider. Therefore, points A and B are out of phase with each other and another phase shift of 180° is provided by the transistor itself. Thus, there is a total phase shift of 360° between the emitter-base and collector-base circuits.

Colpitts oscillator is widely used in commercial signal generator ranging from 1 MHz to 500 MHz. Frequency of oscillation is varied by gang-tuning the two capacitor C_1 and C_2.

5.15.3. Limitation of RC Oscillators

So far, we have studied the oscillators in which the oscillations are produced by the oscillatory circuit. The major problem in these circuits is that their frequency of operation does not remain perfectly constant. It is because the values of resistors and inductors change with temperature. However, in some of the applications it is necessary to maintain constant frequency with an extremely low tolerance.

The solution to this problem is use of crystal oscillators. In crystal oscillators, piezoelectric crystal are employed in place of RL or RC circuit. The frequency of crystal oscillators remains more or perfectly constant even if temperature changes.

5.16. CRYSTAL OSCILLATOR

The crystal is usually made of quartz material and provides a high degree of frequency stability and accuracy or we can say that crystal oscillator is basically a tuned oscillator. It uses a piezoelectric crystal (when an *ac* voltage is applied across a crystal it starts vibrating at the frequency of supply voltage, the effect is known as piezoelectric effect and the crystal which exhibit this effect is known as **piezoelectric crystals**, conversely, when these crystals are placed under mechanical strain to vibrate, they produce an *ac* voltage. For example, rochelle salt, quartz and tourmaline are known as piezoelectric materials. Out of the three, rochelle salts exhibit the greatest piezoelectric activity) as a reasonant tank circuit. The crystal is usually, made of quartz material and provides high degree of frequency stability and accuracy. Therefore, the crystal oscillators are very useful in those applications where frequency stability is very essential. The crystal oscillators are widely used in communication transmiters, digital watches and clocks etc.

Electrical equivalent circuit of a crystal : The electrical equivalent circuit of a crystal consists of series R-L-C in parallel with capacitor C_m when the crystal is applied across the *ac* source it is not vibrating, it is equivalent to the capacitor C_m. However, when crystal vibrating, it acts like tuned R-L-C circuit.

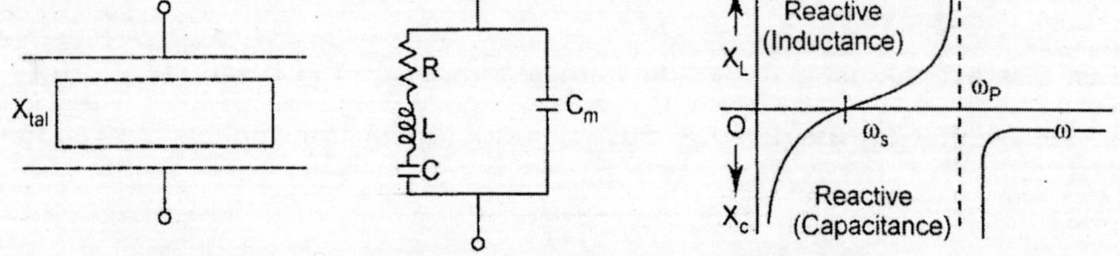

(*a*) Symbol (*b*) Equivalent cirecuit (c) Reactance function (if R = 0)

Fig. 5.51

Fig. 5.51 (*a*) shows the symbol of piezoelectric crystal while Fig. 5.51 (*b*) shows its equivalent circuit. However, Fig. 5.51 (*c*) shows the graph between reactance versus frequency. If this device is properly mounted deformation takes place within the crystal, and an electromechanical system is formed which will vibrate when properly excited. The resonant frequency and the quality factor (Q) depend upon the crystal dimensions.

The crystal has two resonant frequencies viz series resonant frequency and parallel resonant frequency. In this case when the impedance of the circuit is equal to the resistance R

$$X_L = X_C.$$

Thus,

$$\omega_S L = \frac{1}{\omega_S} C$$

or

$$\omega_S = \frac{1}{\sqrt{LC}}$$

Then frequency of oscillations is given by

$$f_S = \frac{1}{2\pi\sqrt{LC}} \qquad \text{...(5.31)}$$

this frequency of oscillation is called series resonant frequency.

This is occur when resistance of series arm equals the reactance of C_m.

$$\omega_p L - \frac{1}{\omega_p C} = \frac{1}{\omega_p C_m}$$

or

$$\omega_p L = \frac{1}{\omega_p C_m} + \frac{1}{\omega_p C}$$

or

$$\omega_p^2 L = \frac{1}{C_m} + \frac{1}{C} = \frac{C + C_m}{C \cdot C_m}$$

$$\omega_p = \sqrt{\frac{C + C_m}{LCC_m}}$$

$$f_p = \frac{1}{2\pi\sqrt{LC_{eq}}} \qquad \text{...(5.32)}$$

where

$$C_{eq} = \frac{C \cdot C_m}{C + C_m}$$

usually the value of C_m is much larger than C therefore the frequencies F_p and F_s are very close to each other otherwise F*p* is always more than F*s*. However, when crystal is used as oscillator, the oscillation frequency always lie between F*s* and F*p*.

The impedance versus frequency graph of crystal is shown in Fig. 5.52.

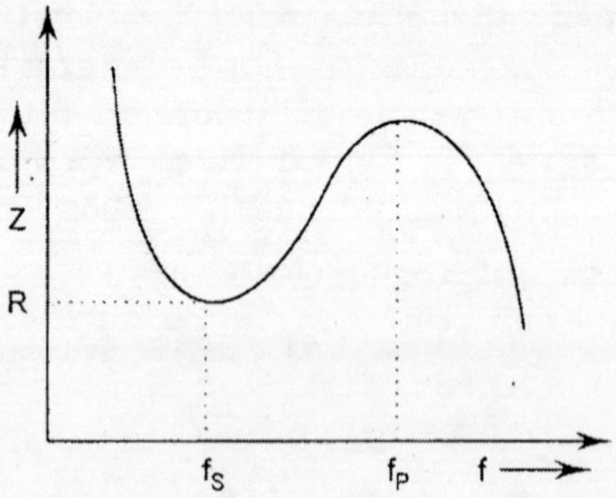

Fig. 5.52. *Impedance versus frequency graph of the crystal.*

Transistor crystal oscillator : The circuit arrangement for transistor crystal oscillator is shown in Fig. 5.53. This arrangement is used in the place when a constant high frequency (25 kHz – 5 MHz) is required, a transistor crystal oscillator is always preferred.

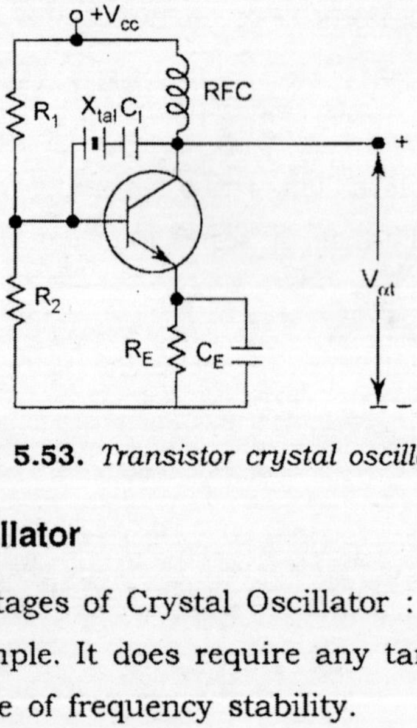

Fig. 5.53. *Transistor crystal oscillator.*

Advantages of Crystal Oscillator

Following are the advantages of Crystal Oscillator :

- The circuit is very simple. It does require any tank circuit other crystal itself.
- It provides high degree of frequency stability.
- It possess very high quality factor.
- Different oscillation frequencies can be achieved by simply replacing one crystal with another.

Disadvantages Crystal Oscillator

Following are the disadvantages of Crystal Oscillator :

- These oscillators are used are not fit for frequencies less than 100 kHz.
- The crystal oscillators have very limited tunning range.
- The crystal oscillators are fragile and therefore, can be used in low power circuits.

Example 3. *Calculate the parallel resonant frequency, series resonant frequency and Q factor of the crystal oscillator, if the crystal of the oscillator has the following parameters : C = 0.06 μF ; L = 0.5 H ; C_m = 1 pF and R = 10 KΩ.*

Solution : Given :

$$C = 0.06 \text{ pF}$$
$$L = 0.5 \text{ H}$$
$$R = 10 \text{ K}\Omega$$
$$C_m = 1 \text{ pF}$$

Parallel resonant frequency of crystal is given by

$$f_p = \frac{1}{2\pi}\sqrt{\frac{C + C_m}{LCC_m}}$$

$$= \frac{1}{2\pi}\sqrt{\frac{(0.006+1)\times 10^{-12}}{(0.5\times 0.06\times 10^{-12})(1\times 10^{-12})}} = 946 \text{ kHz.}$$

Series resonant frequency of crystal is given by

$$f_s = \frac{1}{2\pi\sqrt{LC}} = \frac{1}{2\pi\sqrt{(0.5)(0.06\times 10^{-12})}} = 918.9 \text{ kHz.}$$

Q factor of the crystal is given by

$$Q = \frac{\omega_s L}{R} = \frac{2\pi f_s L}{R}$$

$$= \frac{2\pi\times(918.9\times 10^3)\times 0.5}{10\times 10^3}$$

$$= 288.54.$$

5.17. FREQUENCY STABILITY OF OSCILLATOR

The frequency stability of an oscillator measures its ability to maintain a constant frequency over a long time interval. However it has been found that if an oscillator is set at some particular frequency, it does not maintain for a longer period. In other words the frequency of oscillator changes slowly (or drifts away) from the initially set values. But at some times, it may be changing quite erratically. The change in oscillation frequency my arises due to the following factors.

1. **Supply voltage :** The change in dc supply voltage applied to the active device, shifts the oscillator frequency. This problem can be avoided by using highly regulated power supply.
2. **Operating point of the active device :** The operating point of the active device (*i.e.,* bipolar transistor or FET) is selected in such a way that its operation in non-linear region, changes the values of device parameters which, in turn affects the frequency stability of the oscillator.
3. **Circuit components :** The values of circuits components (*i.e.,* resistor, inductors and capacitors) change with the variation in temperature. Such changes take place slowly, they also cause a drift in oscillator frequency.

4. **Output load :** A change in the output load may cause a change in the Q-factor of the tank circuit, thereby causing a change in oscillator output frequency.
5. **Stray capacitances and inter element capacitances :** Any change in the inter element capacitances of a transistor causes changes in the oscillator output frequency and thus affects the frequency stability. Similarly, the stray capacitance also affects the frequency stability of an oscillator. The effect of change in inter element. However it is very difficult to avoid the effect of stray capacitance.

? SELF-TEST (AMPLIFIERS)

1. What are the different classification of amplifiers ?
2. List different classification of amplifier on the basis of biasing.
3. What is the basic difference between the graphical method and the equivalent circuit method ?
4. What is the main function of blocking capacitor, coupling capacitor and bypass capacitor ?
5. What do you mean by gain ?
6. Explain the procedure adapted for the a.c. analysis and d.c. analysis.
7. Explain why common emitter configuration is most widely used as an amplifier.

? SELF-TEST (*h*-PARAMETERS)

1. What do you mean by hybrid ?
2. What are the different hybrid parameters ?
3. Draw the equivalent circuit of *h*-parameter model.
4. List conversion formulas for hybrid parameters under different configurations.
5. List procedural analysis for a.c. and d.c. equivalent of different transistor configuration. Using *h*-parameters.

SOLVED EXAMPLES

Example 4. *Determine the value of h-parameters for the circuit shown in Fig. P (5.1).*

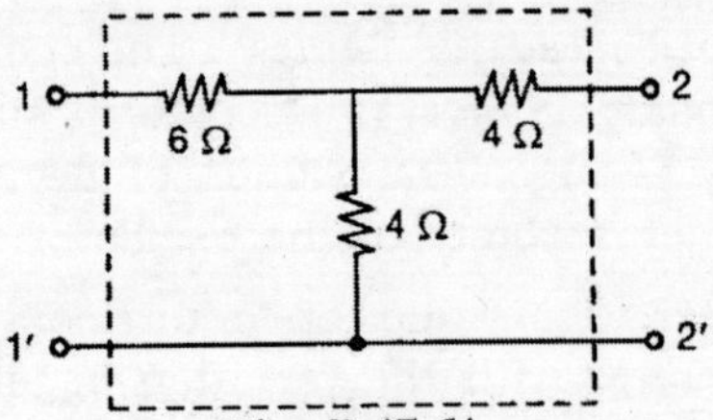

Fig. P (5.1).

Solution : *Parameter h_{11} and h_{21}*

We know that the values of h_{11} and h_{21} can be determined by putting a short circuit across the output terminals of the given circuit.

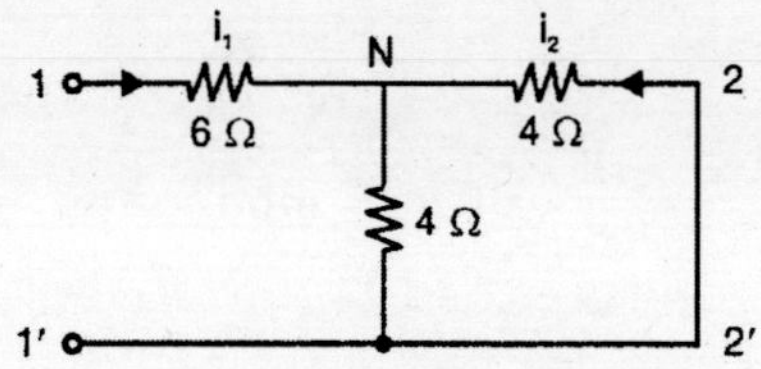

Fig. P (5.2).

$$h_{11} = 6 + (4 \parallel 4) \text{ \{Input resistance looking into the terminals 11'\}}$$

$$= 6 + \frac{4 \times 4}{4 + 4} = 8\ \Omega. \quad \textbf{Ans.}$$

It may be noted that the input current is divided into two equal parts at point N. A current of $i_1/2$ flows outwards and $i_1/2$ flows downwards. The output current

$$i_2 = -\,i_1/2$$

$$\therefore \qquad h_{21} = \frac{i_2}{i_1} = \frac{-\,i_1/2}{i_1} = -\frac{1}{2} = -\,0.5. \quad \textbf{Ans.}$$

Parameters h_{12} and h_{22}

We know that the values of h_{12} and h_{22} may be determined by keeping the input terminals open and applying a voltage source V_2 across the output terminals as follows :

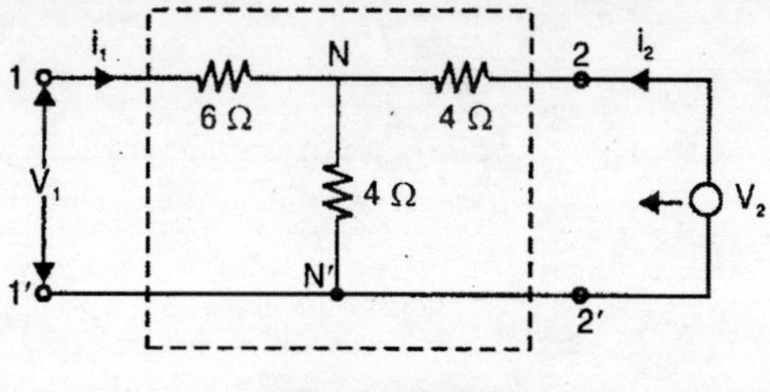

Fig. P (5.3).

Due to the open input $i_1 = 0$. The current flowing through 6 Ω resistor is zero, and hence there is no voltage drop across it. The voltage (V_2) produces a current (i_2) in the circuit. This current will produce equal voltage drops across the two 4 Ω resistors. The voltage drop across the 4 Ω resistor connected between N and N' appears as the V_1.

Thus, $$V_1 = \frac{V_2}{2}$$

$$\therefore \qquad h_{12} = \frac{V_1}{V_2} = \frac{V_2/2}{V_2} = 0.5. \quad \textbf{Ans.}$$

The output resistance (*i.e.*, the resistance looking from the output terminals 2, 2′ with input open)

$$= 4 + 4 = 8\ \Omega$$

∴ h_{22}, output conductance

$$= \frac{1}{Z_0} = \frac{1}{8} = 0.125 \text{ mho.} \quad \textbf{Ans.}$$

Alternative method

Assuming V_1 and V_2 as input and output voltages respectively. Also, I_1 and I_2 are input and output currents respectively. Applying KVL in input side.

$$V_1 = 6\ I_1 + 4\ (I_1 + I_2)$$

$$V_1 = 10\ I_1 + 4\ I_2 \qquad \text{...(1)}$$

Similarly applying KVL in output side

$$V_2 = 4\ I_2 + 4\ (I_1 + I_2)$$

$$V_2 = 4\ I_2 + 4\ I_1 + 4\ I_2$$

$$V_2 = 4\ I_1 + 8\ I_2 \qquad \text{...(2)}$$

From eqn. (2), $I_2 = \dfrac{V_2 - 4I_1}{8}$

Substituting this result in eqn. (1)

$$V_1 = 10I_1 + 4\frac{(V_2 - 4I_1)}{8}$$

$$= 10I_1 + \frac{V_2}{2} - \frac{4I_1}{2}$$

$$V_1 = 8\ I_1 + \frac{V_2}{2}$$

Now, comparing above with

$$V_1 = h_{11}\ I_1 + h_{12}\ V_2$$

We have, $h_{11} = 8$

$h_{12} = 0.5$

Similarly, $I_2 = \dfrac{V_2}{8} - \dfrac{4I_1}{8}$

∴ $h_{21} = -\ 1.25$

$h_{22} = 0.125$

Thus, we obtain same results.

Example 5. *Determine the h-parameter for the circuit shown below in Fig. P (5.4).*

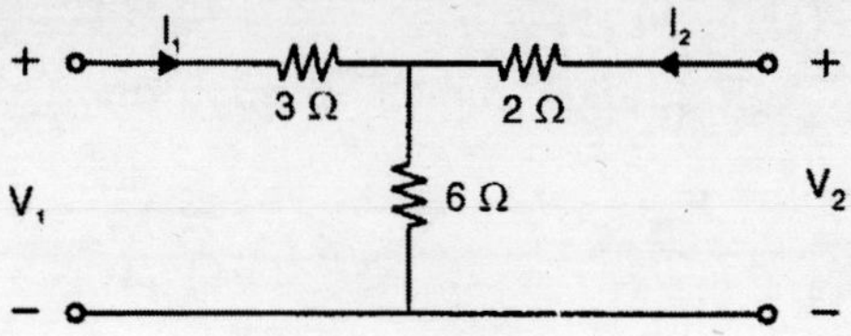

Fig. P (5.4).

Solution : Applying KVL in the input and output terminal.

$$V_1 = 3\ I_1 + 6\ (I_1 + I_2) = 9\ I_1 + 6\ I_2 \qquad ...(1)$$

$$V_2 = 2\ I_2 + 6\ (I_1 + I_2) = 6\ I_1 + 8\ I_2 \qquad ...(2)$$

On comparing eq. (1) and (2) with standard *h*-parameter equation

$$V_1 = h_{11}\ I_1 + h_{12}\ V_2 \text{ and } V_2 = h_{21}\ I_1 + h_{22}\ I_2 \text{ we get}$$

$$h_{11} = 9, \qquad h_{12} = 6$$

$$h_{21} = 6, \qquad h_{22} = 8 \quad \textbf{Ans.}$$

Example 6. *Find the CC h-parameters in terms of the CE h-parameters.*

Solution : In CE configuration,

$$V_{be} = h_{ie}\ i_b + h_{re}\ V_{ce} \text{ and} \qquad ...(1)$$

$$i_c = h_{fe}\ i_b + h_{oe}\ V_{ce} \qquad ...(2)$$

By KCL for any configuration

$$i_b + i_c + i_e = 0 \quad \text{or} \quad i_c = -\ (i_b + i_e) \qquad ...(3)$$

For CE circuit,

$$V_{be} = V_{bc} + V_{ce} \qquad ...(4)$$

$$= V_{bc} - V_{ec}$$

Putting these values in (1)

$$V_{bc} - V_{ec} = h_{ie}\ i_b + (-\ h_{re}\ V_{ec})$$

$$\therefore \quad V_{bc} = h_{ie}\ i_b + (1 - h_{re})\ V_{ec}$$

$$-\ (i_b + i_e) = h_{fe}\ i_b + h_{oe}\ (-\ V_{ec})$$

$$\therefore \quad i_e = -\ (1 + h_{fe})\ i_b + h_{oe}\ V_{ec}$$

In CC configuration we know that

$$V_{bc} = h_{ie}\ i_b + h_{rc}\ V_{ec} \text{ and}$$

$$i_e = h_{fe}\ i_b + h_{oc}\ V_{ec}$$

Comparing eqns. (2), (3) and (4), we get

$$h_{ic} = h_{ie}$$

$$h_{rc} = 1 - h_{re} \approx 1$$

$$h_{fc} = -\ (1 + h_{fe})$$

$$h_{oc} = h_{oe}. \quad \textbf{Ans.}$$

Example 7. *In a single stage amplifier the h-parameter*

$$h_{ic} = 1.1\text{ k},\ h_{rc} = 1,\ h_{fc} = -51,\ h_{oc} = 25\ \mu A/V,$$

Calculate : A_I, A_V, A_{VS}, R_i and R_o for CC configuration with $R_S = R_L = 10$ kΩ.

Solution :

$$A_I = \frac{-h_{fc}}{1 + h_{oc} R_L}$$

$$= \frac{51}{1 + (25 \times 10^{-6} \times 10 \times 10^3)} = 40.8. \quad \textbf{Ans.}$$

$$R_i = h_{ic} + h_{rc} A_I R_L = 409.1 \text{ k}. \quad \textbf{Ans.}$$

$$A_V = \frac{A_I R_L}{R_i} = \frac{40.8 \times 10^4}{409.1 \times 10^3} = 0.997. \quad \textbf{Ans.}$$

$$\gamma_o = h_{oc} - \frac{h_{fc} h_{rc}}{h_{ic} + R_S}$$

$$= 25 \times 10^{-6} + \frac{(51)(1)}{1100 + 10{,}000}$$

$$= 4.62 \times 10^{-3} \text{ mho}. \quad \textbf{Ans.}$$

Thus,
$$R_o = \frac{1}{\gamma_o} = 216\ \Omega$$

$$A_{VS} = \frac{A_V R_i}{R_i + R_S}$$

$$= \frac{0.998 + 409.1}{409.1 + 10}$$

$$= 0.974. \quad \textbf{Ans.}$$

Example 8. *Draw the equivalent circuit for CE and CC configurations subject to the restriction that $R_L = 0$. Show that the input impedances of the two circuits are identical.*

Solution :

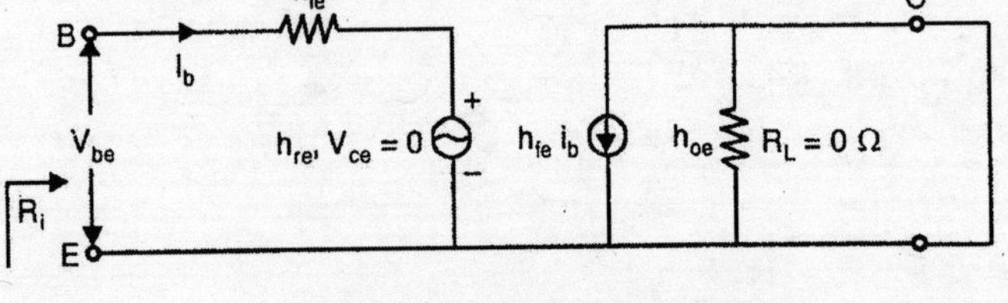

Fig. P (5.5a). *CE hybrid model.*

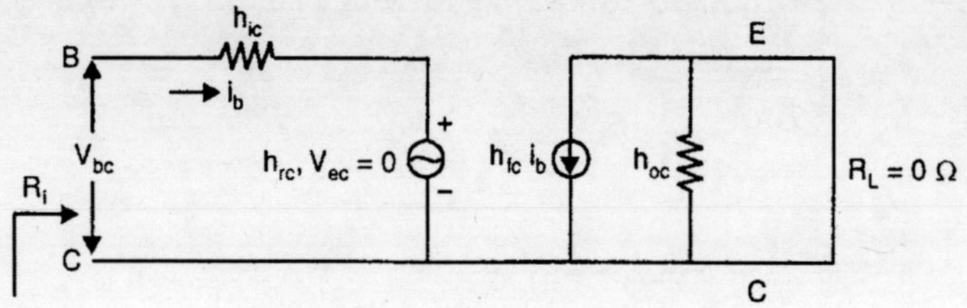

Fig. P (5.5*b*). *CC hybrid model.*

Input resistance in CE

$$R_i = \left.\frac{V_{be}}{i_b}\right|_{V_{ce}=0}$$

By definition $h_{ie} = \dfrac{V_{be}}{i_b}$

Hence, $R_i = h_{ie}$

Similarly for CC configuration

$$R_i' = \left.\frac{V_{bc}}{i_b}\right|_{V_{ec}=0} = h_{ic}$$

But we know that $h_{ie} = h_{ic}$

Hence, $R_i = R_{i'}$

Example 9. *For any single-transistor amplifier prove that,*

$$R_i = \frac{h_i}{1 - h_r A_V}.$$

Solution : For single stage amplifier, input resistance

$$R_i = h_i + h_r A_I R_L$$

and $$A_V = \frac{A_I R_L}{R_i}$$

$$\therefore \quad R_L = \frac{A_V R_i}{A_I}$$

Substituting R_L in the expression of R_i we get

$$R_i = h_i + \frac{h_r A_I A_V R_i}{A_I}$$

$$= h_i + h_r A_V R_i$$

$$\therefore \quad R_i = \frac{h_i}{1 - h_r A_V}$$ **Hence proved.**

Example 10. *For the transistor amplifier shown in Fig. P (5.6).*

$$h_{ie} = 1.5\ \Omega$$
$$h_{fe} = 100$$
$$h_{re} = 3 \times 10^{-4}$$
$$h_{oe} = 25\ \mu\ mho\ s = \frac{1}{40\ k}.$$

Calculate A_i, R_i, A_V, γ_o.

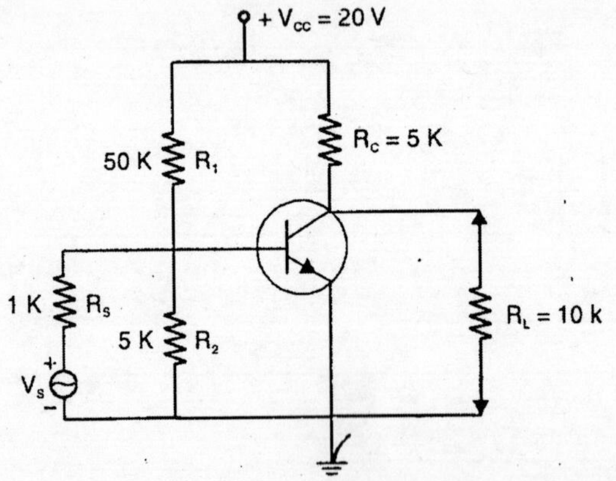

Fig. P (5.6).

Solution : The a.c. equivalent is as follows :

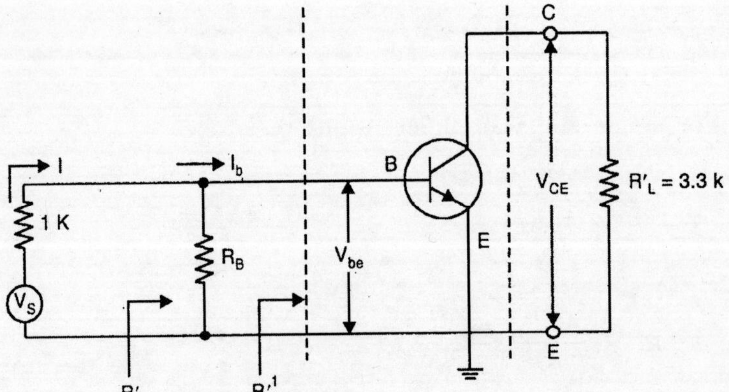

Fig. P (5.7). *AC equivalent of Fig. P(5.6).*

$$R_B = R_1 \parallel R_2 = \frac{R_1 R_2}{R_1 + R_2} = 4.5\ \text{k}$$

$$R_{L'} = R_L \parallel R_C = \frac{R_L R_C}{R_L + R_C} = 3.3\ \text{k}$$

(i) A_i (current gain for transistor)

$$= \frac{I_c}{I_b} = \frac{-h_{fe}}{1 + h_{oe} R_{L'}} = \frac{-100}{1 + \frac{3.3}{40}} = -92.4$$

The overall current gain

$$A_I = \frac{I_c}{I} = \left(\frac{I_c}{I_b}\right)\left(\frac{I_b}{I}\right)$$

$$= A_i + \frac{I_b}{I}$$

But $$\frac{I_b}{I} = \frac{R_B}{R_B + R_{i'}}$$

$$R_{i'} = h_{ie} + h_{re} A_I R_{L'} = 1.4 \text{ k}$$

$$\therefore \quad \frac{I_b}{I} = \frac{4.5}{4.5 + 1.4} = 0.76$$

Hence, $A_I = A_i \times 0.76 = -92.4 \times 0.76 = -70.2.$ **Ans.**

(ii) $R_1 = R_B \parallel R_{i'} = \dfrac{4.5 \times 1.4}{4.5 + 1.4} = 1.0 \text{ k}\Omega.$ **Ans.**

(iii) A_V (for transistor) $= \dfrac{V_{ce}}{V_{be}} = \dfrac{A_i R_{L'}}{R_{i'}}$

$$= \frac{-70.2 \times 3.3}{1.4} = -165.4$$

The overall voltage gain $A_{VS} = \dfrac{V_{ce}}{V_S} = \dfrac{V_{ce}}{V_{be}} \dfrac{V_{be}}{V_S} = A_V \cdot \dfrac{V_{be}}{V_S}$

But $$\frac{V_{be}}{V_S} = \frac{R_i}{R_S + R_i} = \frac{1}{1+1} = 0.5$$

$\therefore$ $A_{VS} = -165.4\ (0.4) = -82.7.$ **Ans.**

(iv) $$\gamma_o = h_{oe} - \frac{h_{re} h_{fe}}{h_{ie} + R_{S'}}$$

$$R_{S'} = \frac{R_S R_B}{R_S + R_B} = 818.18.$$

$\gamma_o = 1.2 \times 10^{-5}$ mho $= 12\ \mu$ mho. **Ans.**

Example 11. *A CE amplifier is driven by a voltage source of internal resistance R_S = 800 Ω. The load impedance is a resistance of 2 kΩ. The h-parameters are h_{ie} = 1.1 k,*

h_{fe} = 50, h_{oe} = 25 μ ohms, $h_{re} = 2.5 \times 10^{-4}$. Compute the current gain A_i, input resistance R_i, voltage gain A_v, output resistance R_o and output terminal resistance R_{o1}. Also calculate power gain A_p. Using (a) exact analysis (b) approximate analysis. Compute % error in A_I.

Solution : (*a*) Exact analysis

(*i*) Current gain

$$A_i = \frac{-h_{fe}}{1 + h_{oe} R_L} = \frac{-50}{1 + (25 \times 10^{-6} \times 2000)} = -47.6. \quad \textbf{Ans.}$$

(*ii*) Input resistance

$$R_i = h_{ie} + h_{re} A_i R_L$$
$$= 1100 - (2.5 \times 10^{-4}) \times 47.6 \times 2000$$
$$= 1076.2\ \Omega = 1.0762\ \text{k}\Omega. \quad \textbf{Ans.}$$

(*iii*) Voltage gain

$$A_V = \frac{A_i R_L}{R_i} = -88.46$$

Overall voltage gain (*i.e.*, including signal source) $A_{VS} = \dfrac{A_V R_i}{R_i + R_S}$

$$= \frac{-88.46 \times 1076.2}{800 + 1076.2} = -50.75. \quad \textbf{Ans.}$$

(*iv*) Output admittance

$$\gamma_0 = h_{oe} - \frac{h_{re} h_{fe}}{h_{ie} + R_S}$$

$$= 25 \times 10^{-6} - \frac{2.5 \times 10^{-4} \times 50}{1100 + 800} = 18.42\ \mu\ \text{ohms}$$

$$\therefore \quad R_0 = \frac{1}{\gamma_0} = 54.28\ \text{k}\Omega. \quad \textbf{Ans.}$$

(*v*) Power gain $A_p = A_i \times A_v = (-47.6)(-88.46) = 4210.6.$ **Ans.**

(*vi*) Output terminal resistance $R_{ot} = R_o \parallel R_L = \dfrac{54.28 \times 2}{54.28 + 2} = 1.92$ K. **Ans.**

(*b*) Approximate analysis

Assume $h_{re} = h_{oe} = 0$

(*i*) $A_i = \dfrac{-h_{fe}}{1 + h_{oe} R_L} \approx -h_{fe} \approx -50.$ **Ans.**

[∵ h_{oe} since $R_L = 25 \times 10^{-6} \times 2000 < 0.1$].

(*ii*) $R_i = h_{ie} + h_{re} A_i R_L \approx h_{ie} \approx 1100\ \Omega.$ **Ans.**

[∵ Since $h_{re} A_i R_L < 0.1\ h_{ie}$].

(*iii*) $A_V = \dfrac{A_i\, R_L}{R_i} \approx \dfrac{-h_{fe}\, R_L}{h_{ie}} \approx -\dfrac{50 \times 2000}{1100} \approx -90.91$

$A_{VS} = \dfrac{A_V\, R_i}{R_S + R_i} \approx \dfrac{A_V \times h_{ie}}{R_S + h_{ie}} \approx 52.62.$ **Ans.**

(*iv*) $\gamma_0 = 18.42\ \mu$ ohms ≈ 0 ohms

$\therefore\ R_0 = \dfrac{1}{\gamma_0} = 54.28$ k (large value). **Ans.**

(*v*) $A_P = A_i \times A_V = (-50)\ (-90.91) = 4545.5.$ **Ans.**

(*vi*) $R_{ob} = R_0 \parallel R_L \approx 20$ kΩ (*i.e.*, only R_L neglecting R_0).

$$\%\ \text{error} = \left(\frac{\text{Exact} - \text{approximate}}{\text{Exact value}}\right) \times 100\%$$

$\%$ error in $A_I = \dfrac{47.6 - (-50)}{47.6} \times 100 = 5.04\%$. **Ans.**

Example 12. *For Fig. P (5.8), calculate Z_i, A_i, A_v and Z_0.*

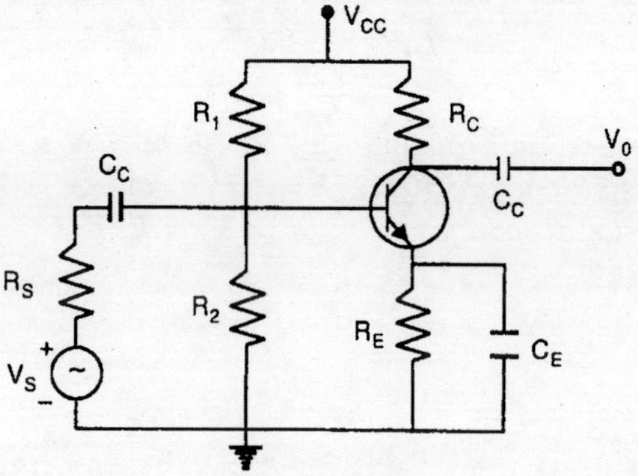

Fig. P (5.8).

Solution : By considering the basic r_c model we get the form as shown in Fig. P (5.9).

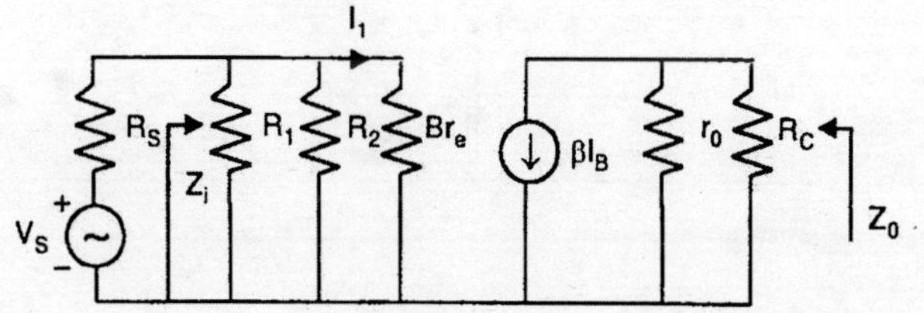

Fig. P (5.9).

Thus we get

$$Z_i = R_1 \parallel R_2 \parallel \beta r_e \textbf{ Ans.}$$

and $$Z_0 = R_C \parallel R_L \textbf{ Ans.}$$

And we obtain the A_V as follows:

$$A_V = -\frac{V_0}{V_i} = \frac{-(R_C \| r_0)}{r_e}$$

As we know from the above that

$$I_1 = \frac{V_i}{\beta r_e}$$

and $$V_0 = -\beta I_b (R_C \| r_0)$$

$$V_0 = -\beta\left(\frac{V_i}{\beta r_e}\right)(R_C \| r_0)$$

$$V_0 = -\frac{V_i}{r_e}(R_C \| r_0)$$

$$A_V = \frac{V_0}{V_i} = -\frac{(R_C \| r_0)}{r_e} \textbf{ Ans.}$$

$$A_1 = -A_V \frac{Z_i}{r_0} \Rightarrow \left(\frac{R_C \| r_0}{r_e}\right)\left(\frac{R_1 \| R_2 \| \beta r_e}{r_0}\right) \textbf{ Ans.}$$

Example 13. *A source with an internal resistance R_s = 600 Ω drives a non-ideal amplifier whose input resistance is 400 Ω. The source voltage is 1 mV. The amplifier feeds a load of 5 kΩ and the load voltage is found to be 100 mV. Determine*

(i) Voltage gain
(ii) Current gain } *of the amplifier*
(iii) Power gain

Solution : We can draw the circuit for the given information.

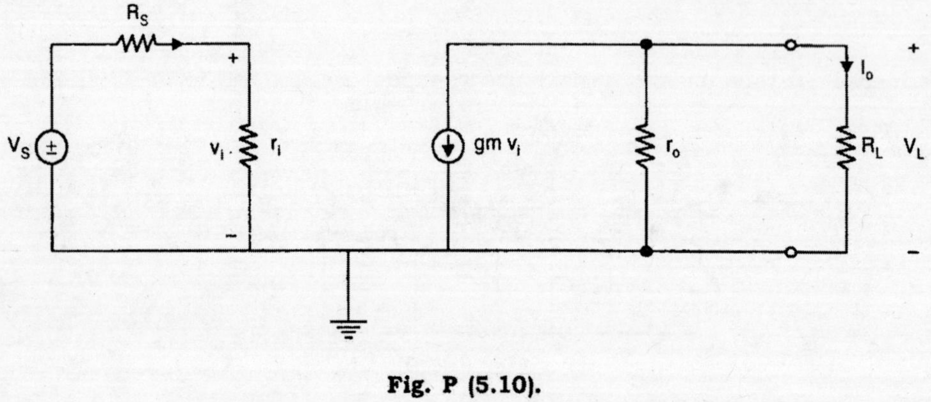

Fig. P (5.10).

From Fig. P (5.10)

$$\text{Input voltage, } V_i = V_s \times \frac{r_i}{(R_S + r_i)}$$

$$= 1 \text{ mV} \times \frac{400\ \Omega}{(400 + 600)\ \Omega}$$

$$= 0.4 \text{ mV}$$

$$\text{Input current, } I_i = \frac{V_S}{R_S + r_i}$$

$$= \frac{1 \text{ mV}}{(400 + 600)\ \Omega} = 1\ \mu\text{A}$$

$$\text{Output current, } I_o = \frac{V_o}{R_L}$$

$$= \frac{100 \text{ mV}}{5 \text{ k}\Omega} = 20\ \mu\text{A} \qquad (\because\ V_L = 100 \text{ mV},\ R_L = 5 \text{ k}\Omega)$$

$$\text{Voltage gain, } A_V = \frac{V_o}{V_i} = \frac{V_L}{V_i} = \frac{100 \text{ mV}}{0.5 \text{ mV}} = 250. \quad \textbf{Ans.}$$

$$\text{Current gain, } A_I = \frac{I_o}{I_i} = \frac{I_L}{I_i} = \frac{20\ \mu\text{A}}{1\ \mu\text{A}} = 20. \quad \textbf{Ans.}$$

$$\text{Power gain, } P = A_v \times A_i$$

$$= 250 \times 20 = 5000. \quad \textbf{Ans.}$$

? SELF-TEST (FEEDBACK)

1. What does you mean by feedback. What do you understand by negative and positive feedback ? Why is negative feedback applied in the high gain amplifier ?
2. Distinguish current feedback and voltage feedback with appropriate circuit block diagram.
3. Drive the formula for negative feedback amplifier gain in terms of A_f : A and B.
4. Explain, how is bandwidth increased by Negative feedback.
5. Draw the circuit of a transister amplifier with negative voltage feedback circuit. How to explained it ? Explain how in Negative voltage feedback to the input.
6. What is an oscillator ? How does it differ from an amplifier ? What are the essential parts of an oscillator circuit ? Explain the function of each part.
7. Explain why :

   (*i*) A Negative feedback is always employed in high gain amplifiers.

   (*ii*) Emitter-follower circuit is also called common-collector amplifier circuit.

(iii) A common emitter circuit without by pass capacitor is called a negative current feedback circuit.

8. State the merits and demerits of negative feedback in amplifiers.
9. Derive an expression to illustrate that the voltage gain in an amplifier circuit with negative feedback is some what stable even if the β of transistor changes due to its ageing or due to its replacement.
10. What do you mean by Miller's theorem ?
11. State Miller's dual theorem.
12. List the fire main advantages of Negative feedback amplifier.
13. Explain why the input impedance and output impedance of a transconductance amplifier should be high.
14. Explain different basic feedback topologies.
15. What do you mean by loop gain ?
16. How feedback effects stability ? Explain in detail.

SOLVED EXAMPLES

Example 14. *Calculate the gain of a negative-feedback amplifier with an internal gain A = 125 and feedback factor* $\beta = \frac{1}{10}$.

Solution : Given that A = 125

$$\beta = \frac{1}{10}$$

We know that gain of negative feedback is given by

$$A_f = \frac{A}{1 + A\beta}$$

$$= \frac{125}{13.5} = 9.259. \textbf{ Ans.}$$

Example 15. *An amplifier with negative feedback has a voltage gain of 100. It is found that without feedback, an input signal of 50 mV is required to produce a given output, whereas with feedback, the input signal must be 0.6 for the same output. Calculate the value of A and B.*

Solution : Given that $A_f = 100$.

The input voltage required to produce the same output voltage as for the amplifier without feedback is 0.6 V. Thus the output will be.

$$V_o = A_f V_i = 100 \times 0.6 = 60$$

According to question

If no feedback is employed, the required input to produce 60 V output is 50 mV.

Hence the internal gain of the amplifier is $A = \frac{V_o}{V_i} = \frac{60}{50 \times 10^{-3}} = 1200$

Now, from $$A_f = \frac{A}{1 + A\beta}$$

$$1_{00} = \frac{1200}{1 + 1200 \, . \, \beta} \Rightarrow 1 = \frac{12}{1 + 1200\beta} \Rightarrow 1200\ \beta = 11$$

$$\beta = \frac{11}{1200} \text{ Ans.}$$

Example 16. *To an amplifier of 60 dB gain, a feedback of β = 0.005 is applied. What would be the change in overall gain of the feedback amplifier if the internal amplifier is subjected to a gain reduction of 12%.*

Solution : Given : $A = 60\ dB = 1000$ $\quad \{\because$ gain in dB $= 20 \log_{10} \left(\frac{V_2}{V_1}\right)$

$\beta = 0.005$ $\quad 60 = 20 \log_{10} (A)$

or $A = 10^3 = 1000$

$$\frac{dA}{A} = -12\% = -0.12, \quad \frac{dA_f}{A_f} = ?$$

We know that $$\frac{dA_f}{A_f} = \frac{1}{1 + A\beta} \frac{dA}{A}$$

or $$\frac{dA_f}{A_f} = \frac{1}{1 + 1000 \times 0.005} \times (-0.12)$$

$$\frac{dA_f}{A_f} = -0.02 \text{ or } -2\%. \textbf{ Ans.}$$

Therefore, the overall gain of the feedback amplifier will be reduced by 2%.

Example 17. *An amplifier has gain A = 60 dB and output impedance Z_o = 12.6 kΩ. It is required to modify its output impedance to 500 Ω by applying negative feedback determine.*

(i) The value of feedback factor

(ii) The percentage change in overall gain for 20% change in the gain of the basic amplifier.

Solution : Given gain in dB = 60.

We know that gain in $$dB = 20 \log_{10} \left(\frac{V_2}{V_1}\right) = 20 \log_{10} (A)$$

$$60 = 20 \log_{10} (A)$$

So, $$A = 10^3 = 1000$$

$$Z_o = 12.6 \text{ k}\Omega$$

$$Z_{of} = 500 \ \Omega$$

We know that, $$Z_{of} = \frac{Z_o}{1 + A\beta}$$

$$500 = \frac{12600}{1 + 1000\beta}$$

$$5 + 5000 \ \beta = 126$$

$$\beta = \frac{121}{5000}$$

$$\beta = 0.0242$$

Now, $$\frac{dA_f}{A_f} = \frac{1}{1 + A\beta}\frac{dA}{A} = \frac{1}{1 + 1000 \times 0.0242} = 0.083\%. \textbf{ Ans.}$$

Example 18. *An amplifier without feedback gives a fundamental output of 36 V with 7% second harmonic distortion when the input is 0.028 V.*

(a) If 1.2% of the output is feedback into the input in a negative voltage series feedback circuit, what is the output voltage ?

(b) If the fundamental output is maintained at 36 V but the second-harmonic distortion is reduced to 1 percent, what is the input voltage.

Solution : Given that $V_o = 36$ V

$$V_i = 0.028 \text{ V}$$

$$A = \frac{V_o}{V_i} = \frac{36}{0.028} = 1285$$

$$\beta = \frac{V_f}{V_o} = \frac{1.2}{100} = 0.012$$

(*a*) Gain with the feedback is given by,

$$A_f = \frac{A}{1 + A\beta} = \frac{1285}{1 + 1285 \times 0.012} = 78.2$$

Also $$A_f = \frac{V_o}{V_s}$$

$$V_o = A_f V_s = 78.2 \times 0.028 = 2.19 \text{ V. } \textbf{Ans.}$$

(*b*) If the output is maintained constant at 36 V then the distortion generated by the device is unchanged. The reduction of the total distortion is caused by feedback.

$$D_f = \frac{D}{1 + A\beta}$$

or $$1 + A\beta = \frac{D}{D_f} = \frac{7}{1} \qquad \left\{\because \begin{array}{l} D_f = 1\% \\ D = 7\% \end{array}\right\}$$

$$A\beta = 7 - 1 = 6$$

Now, from equation, $$A_f = \frac{A}{1 + A\beta} = \frac{1285}{7} = 183.57$$

$$V_s = \frac{V_o}{A_f} = \frac{36}{183.57} = 0.196. \textbf{ Ans.}$$

Example 19. *An amplifier with open loop voltage gain of 1000 ± 10% is available. However we desire to build an amplifier whose gain does not vary by more than 0.1%. Find the required feedback ratio β and the corresponding closed loop voltage gain.*

Solution : Given, we know that

$$\frac{dA_f}{A_f} = \frac{dA}{A} \cdot \frac{1}{1 + A\beta}$$

$$\therefore \qquad 0.1 = \frac{10}{1 + A\beta}$$

$$1 + A\beta = 100$$

Given, $$A = 1000$$

$$\beta = \frac{100 - 1}{1000} = 0.099$$

So, closed loop gain $$A_f = \frac{A}{1 + A\beta} = \frac{1000}{100} = 10. \textbf{ Ans.}$$

Example 20. *An amplifier has open loop voltage gain of 1000 and delivers 10 watts output with 10% second harmonic distortion when the input is 10 mV. If 40 dB of negative feedback is applied, what will be the distortion ? How much input voltage should be applied to 10 watts output power ?*

Solution : The term 40 dB of feedback means $20 \log_{10} (1 + A\beta) = 40$ dB.

or $$\log_{10} (1 + A\beta) = \frac{40}{20}$$

or $$1 + A\beta = (10)^2$$

or $$1 + A\beta = 100 \qquad \ldots(i)$$

$$\therefore \qquad A_f = \frac{A}{1 + A\beta} = \frac{1000}{100}$$

$$= 10 \text{ \{Given } A = 1000, \text{and } 1 + A\beta = 100 \text{ from equation } (i)\}$$

∴ New value of second harmonic distortion

$$D_f = \frac{D_o}{1 + A\beta} = \frac{10}{100} = 0.1\%$$

Now value of input required = 10 *m*V × 100 = 1 V $\{\because V_s = V_i (1 + A\beta)\}$. **Ans.**

Example 21. *An amplifier has an input of 10 mV and a gain of 200 without feedback. The distortion produced at the output of the amplifier is 10%. It is desired to reduce the distortion to 1% by using negative feedback. Calculate the gain, input voltage and output voltage with feedback.*

Solution : Given :

$$D = 10\% = 0.1$$
$$D_f = 1\% = 0.01$$
$$A = 200$$
$$V_s = 10 \text{ mV}$$

We know that,
$$D_f = \frac{D}{1 + A\beta}$$

$$0.01 = \frac{0.1}{1 + 200 \,.\, \beta}$$

$$\beta = 0.045 \text{ or } 4.5\%$$

Gain with feedback,
$$A_f = \frac{A}{1 + A\beta} = \frac{200}{1 + 0.045 \times 200} = 20$$

Now output voltage,
$$V_o = A_f V_s = 20 \times 10 \text{ mV} = 0.2 \text{ V}$$

Now input voltage,
$$V_{in} = 0.01 \text{ V} + (-\,0.045 \times 0.2) = 0.001 \text{ V. } \textbf{Ans.}$$

Example 22. *An amplifier has an input impedance of 1 kΩ and output impedance of 10 kΩ and a voltage of 10,000. If a negative feedback of β = 0.02 is applied to it, determine the input and output impedance of the amplifier with feedback.*

Solution : Given :

$$A = 10{,}000$$
$$\beta = 0.02$$
$$Z_i = 1 \text{ k}\Omega$$
$$Z_o = 10 \text{ k}\Omega$$
$$Z_{if} = ?$$
$$Z_{of} = ?$$

Consider a voltage series feedback case in which feedback input impedance increases, while output impedance decreases and given by the relation.

$$Z_{if} = Z_i\,(1 + A\beta)$$

or
$$Z_{if} = 1 \text{ k}\Omega\,(1 + 10000 \times 0.02) = 1 \text{ k}\Omega\,(201) = 201 \text{ k}\Omega. \textbf{ Ans.}$$

and
$$Z_{of} = \frac{Z_o}{1 + A\beta}$$

or
$$Z_{of} = \frac{10 \text{ k}\Omega}{(1 + 10{,}000 \times 0.02)} = \frac{10}{201} = 0.4975 \text{ k}\Omega. \textbf{ Ans.}$$

Example 23. *An RC coupled amplifier has a mid-frequency gain of 400 and a frequency response from 200 Hz to 40 kHz. Determine the gain and frequency response when negative feedback with feedback ratio of 0.01 is introduced in the amplifier circuit.*

Solution : Given :

$$A = 400,$$

$$\beta = 0.01$$

We know that gain with feedback,

$$A_f = \frac{A}{1+\beta A}$$

$$= 400$$

Lower cut-off frequency with feedback

$$f_{LF} = \frac{f_L}{1+A\beta}$$

$$= \frac{200 \text{ Hz}}{1+400 \times 0.01} = 40 \text{ Hz.}$$

Upper cut-off frequency with feedback

$$f_{HF} = f_H (1 + A\beta)$$

$$= 40 \text{ kHz } (1 + 400 \times 0.01)$$

$$= 40 \times 5 = 200 \text{ kHz.}$$ **Ans.**

Example 24. *A negative feedback amplifier is shown in Fig. P (5.11) if the gain of the amplifier without feedback is 4000, find :*

(i) Feedback fraction or feedback ratio.

(ii) Overall voltage gain with feedback.

(iii) Output voltage if input voltage is 2 mV.

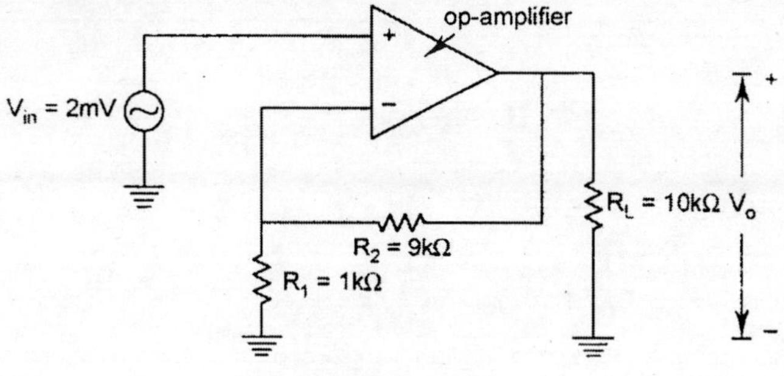

Fig. P (5.11)

Solution : Given

$$A = 4000$$

$$R_1 = 1 \text{ k}\Omega$$

$$R_2 = 9 \text{ k}\Omega$$

(i) Feedback fraction, $\beta = \frac{R_1}{R_1 + R_2} = \frac{1}{1+9} = 0.1.$ **Ans.**

(ii) Overall voltage gain with feedback $A_f = \dfrac{A}{1+A\beta}$

$$= \frac{4000}{1+4000\times 0.1} = \frac{4000}{1+400}$$

$$= 9.975. \textbf{ Ans.}$$

(iii) We know that, $A_f = \dfrac{V_o}{V_s}$ (here V_s = 2 mV)

$$V_o = A_f . V_s = 9.975 \times 2 \text{ mV} = 19.95 \text{ mV}. \textbf{ Ans.}$$

Example 25. *An amplifier has a gain of 54.8 dB without feedback. Find the change in gain if $\frac{1}{50}$ of the output voltage is feedback at the input. Determine the percentage reduction in harmonic distortion in the output due to feedback.*

Solution : Given : A in dB = 54.8

We know that gain in dB = 20 $\log_{10}$ (A)

$$\frac{54.8}{20} = \log_{10} (A)$$

or $$A = 10^{2.74}$$

or $$A = 549.5 \cong 550$$

Feedback factor, $$\beta = \frac{1}{50} = 0.02$$

Now, gain with feedback,

$$A_f = \frac{A}{1+A\beta} = \frac{550}{1+550\times 0.02}$$

Distortion with feedback,

$$D_f = \frac{D}{1+A\beta} = \frac{D}{1+550\times 0.02}$$

Percentage change in distortion

$$\frac{D_f}{D}\times 100 = \frac{1}{12}\times 100$$

$$= 8.33\%. \textbf{ Ans.}$$

Example 26. *(i) Find $\dfrac{V_f'}{V_o}$ for the network shown in Fig. P (5.12).*

(ii) Sketch the circuit of a phase shift FET oscillator using this feedback network.

(iii) Find the minimum gain required for the oscillation.

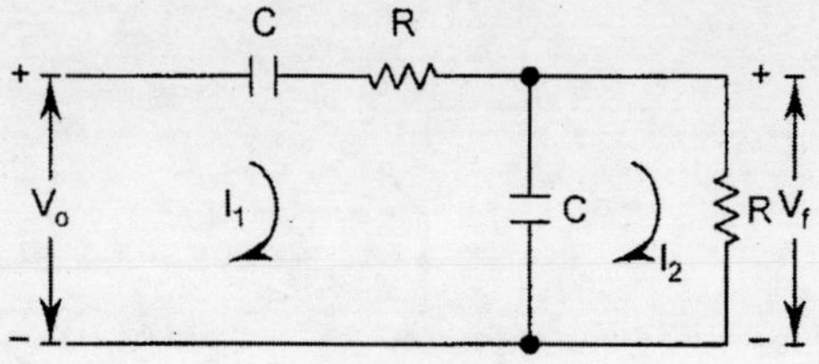

Fig. P (5.12)

Solution : (*i*) Applying KVL in the loop (1) and (2), we get

$$V_o = I_1\left(R + \frac{2}{j\omega C}\right) - I_2\frac{1}{j\omega C} \quad \text{...(A)}$$

$$0 = I_2\left(R + \frac{1}{j\omega C}\right) - I_1\frac{1}{j\omega C} \quad \text{...(B)}$$

From equations (A) and (B) eliminating current I_1

$$V_o = \frac{I_2\left(R + \frac{1}{j\omega C}\right)}{\frac{1}{j\omega C}}\left(R + \frac{2}{j\omega C}\right) - I_2\frac{1}{j\omega C}$$

$$\frac{V_o}{I_2} = j\omega C\left(R + \frac{1}{j\omega C}\right)\left(R + \frac{2}{j\omega C}\right) - \frac{1}{j\omega C}$$

or
$$\frac{V_o}{\frac{V_f'}{R}} = \frac{2 + 2Rj\omega C + Rj\omega C - R^2\omega^2C^2 - 1}{j\omega C} \quad (\because V_f = I_2R.)$$

or
$$\frac{V_f'}{V_o} = \frac{j\omega CR}{1 + 3j\omega CR - R^2\omega^2C^2}$$

or
$$\frac{V_f'}{V_o} = \frac{1}{3 + \frac{1}{j\omega CR} - \frac{R\omega C}{j}}$$

or
$$\frac{V_f'}{V_o} = \frac{1}{3 + j\left(R\omega C - \frac{1}{R\omega C}\right)} \quad \textbf{Ans.}$$

(*ii*) Circuit of phase-shift FET oscillator using the feedback network is shown in Fig. P (5.13).

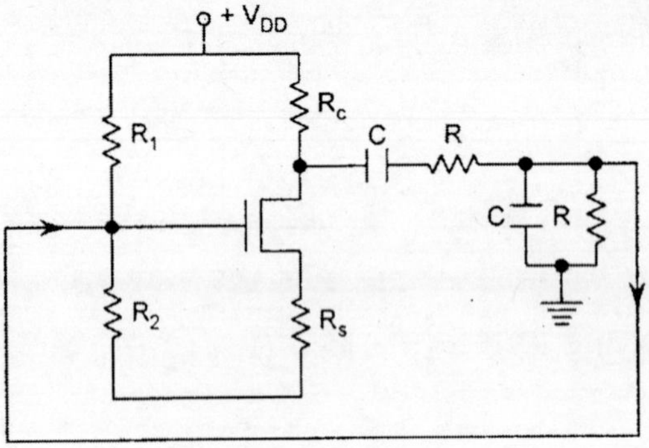

Fig. P (5.13)

(*iii*) We know that

$$\beta = \frac{V_f'}{V_o}$$

the frequency at which β is purely real. In other word imaginary part must be zero.

When in the expression $\frac{V_f'}{V_o}$ imaginary part equal to zero, then $\frac{V_f'}{V_o}$ becomes equal to $\frac{1}{3}$

i.e.,

$$\beta = \frac{1}{3}$$

Also

$$A\beta \geq 1$$

$$A \geq \frac{1}{\beta}$$

$$A \geq 3$$

$$A_{min} = 3 \textbf{ Ans.}$$

Example 27. *Determine the stability of an amplifier, given the loop gain function.*

$$T(f) = \frac{\beta(100)}{\left(1 + j\frac{f}{10^5}\right)^3}$$

Determine the stability of the amplifier for β = *0.20 and* β = *0.02.*

Solution : The loop gain can be written in terms of its magnitude and phase, as

$$T(f) = \frac{\beta(100)}{\left[\sqrt{1 + \left(\frac{f}{10^5}\right)^2}\right]^3} \angle -3\tan^{-1}\left(\frac{f}{10^5}\right)$$

The frequency f_{180} at which the phase becomes – 180 degree is

$$-3\tan^{-1}\left(\frac{f_{180}}{10^5}\right) = -180°, \text{ we get,}$$

$$f_{180} = 1.73 \times 10^5 \text{ Hz}$$

The magnitude of the loop gain at this frequency, for β = 0.20, is then

$$|T(f_{180})| = \frac{(0.20)(100)}{8} = 2.5$$

For β = 0.02, the magnitude is

$$|T(f_{180})| = \frac{(0.020)(100)}{8} = 0.25$$

NOTE : The loop gain magnitude at the frequency at which the phase is – 180 degree is 2.5 when β = 0.20 and 0.25 when β 0.02. The system is therefore unstable for β = 0.20 and stable for β = 0.02, since at β = 0.20 gain is 2.5 which is greater than 1. Hence, the system is unstable for β = 0.20.

Example 28. *Determine the required feedback transfer function* β *which gives a phase margin of 45 degrees.*

Given that :

$$T(f) = \frac{\beta(100)}{\left(1 + J\frac{f}{10^3}\right)\left(1 + J\frac{f}{5 \times 10^4}\right)\left(1 + J\frac{f}{10^6}\right)}$$

Solution : A phase margin of 45 degree implies that the phase of the loop gain is – 135 degree at the frequency at which the magnitude of the loop gain is unity. The phase of the loop gain is

$$\phi = -\left[\tan^{-1}\left(\frac{f}{10^3}\right) + \tan^{-1}\left(\frac{f}{5 \times 10^4}\right) + \tan^{-1}\left(\frac{f}{10^6}\right)\right]$$

Since the three poles are far apart the frequency at which the phase is – 135 degree is approximately equal to the frequency of the second pole. So in this example, $f_{135} \cong 5 \times 10^4$ Hz, so we have that

$$\phi = -\left[\tan^{-1}\left(\frac{5 \times 10^4}{10^3}\right) + \tan^{-1}\left(\frac{5 \times 10^4}{5 \times 10^4}\right) + \tan^{-1}\left(\frac{5 \times 10^4}{10^6}\right)\right]$$

or

$$\phi = -[88.9° + 45° + 2.86°] \cong -135°$$

Since we want the loop gain magnitude to be unity at this frequency, we have

$$|T(f)| = 1$$

$$= \frac{\beta(100)}{\sqrt{1 + \left(\frac{5 \times 10^4}{10^3}\right)^2}\sqrt{1 + \left(\frac{5 \times 10^4}{5 \times 10^4}\right)^2}\sqrt{1 + \left(\frac{5 \times 10^4}{10^6}\right)^2}}$$

or

$$1 \cong \frac{\beta(100)}{(50)(1.41)(1)}$$

? SELF-TEST (OSCILLATORS)

1. What do you understand by damped and undamped electrical oscillations ?

2. What is the condition of oscillation ?

3. Draw and explain tuned collector oscillator.

4. Draw the circuit and explain the operation of Colpitt's oscillator.

5. Discuss in detail Hartley oscillator.

6. Draw and discuss in detail the circuit of an R-C phase shift oscillator.

7. Draw the circuit of a Wein bridge oscillator and explained its working. Why is .negative feedback employed in this oscillator circuit in addition to the usual feedback positive feedback.

8. Explained the properties of quartz crystal which are responsible for its use in oscillator.

9. Explain why :

(*i*) Three R-C section are used R-C phase shift oscillator.

(*ii*) Negative feedback is provided in Wein bridge oscillation.

(*iii*) In Wien bridge oscillator, gangs are employed.

(*iv*) At low frequency (1 Hz to 100 kHz) application, we employ R-C oscillators and not L-C oscillator.

10. What are the factors which affect the frequency stability of an oscillator ?

11. What is Bark hausen criterion for the frequency stability of an oscillator ?

12. Explain how L-C tank circuits is used to generate oscillations in an electronic oscillator.

SOLVED EXAMPLES

Example 29. *An oscillatory circuit has L = 0.01 H and C = 10 pF, find the frequency of oscillations.*

Solution : Frequency of oscillation

$$f = \frac{1}{2\pi\sqrt{LC}} = \frac{1}{2\pi \times \sqrt{0.01 \times 10 \times 10^{-12}}}$$

$$= \frac{10^6}{2\pi\sqrt{0.1}} = 550 \text{ kHz. } \textbf{Ans.}$$

Example 30. *A tuned collector oscillator has a fixed inductance of 150 μH and has to be tunable over the frequency band of 500 kHz to 1500 kHz. Find the range of variable capacitor to be used.*

Solution : Resonant frequency is given by

$$f = \frac{1}{2\pi\sqrt{LC}}$$

$$C = \frac{1}{4\pi^2 f^2 L}$$

When, $f = 500$ kHz

$$C = \frac{1}{4\pi^2 \times \left(500 \times 10^3\right)^2 \times 150 \times 10^{-6}} = \frac{1}{4\pi^2 \times 250000 \times 100}$$

$= 1015$ pF

When $f = 500$ kHz

$$C = \frac{1}{4\pi^2 \times \left(500 \times 10^3\right)^2 \times 150 \times 10^{-6}} = 113 \text{ pF}$$

Hence, capacitor range required in (113 – 1015) pF **Ans.**

Example 31. *In a transistor colpitt's oscillator,*

(i) Operating frequency

(ii) Feedback fraction

(iii) Minimum gain to sustain oscillations

(iv) Emitter resistance if $R_C = 2.5\ \Omega$

Solution : (*i*) Given $L = 100\ \mu H$

$C_1 = 0.001\ \mu H$

$C_2 = 0.01\ \mu F$

Operating frequency, $f = \dfrac{1}{2\pi\sqrt{LC}}$

where $C = \dfrac{C_1 C_2}{C_1 C_2} = \dfrac{0.001 . 10^{-6} \times 0.01 \times 10^{-6}}{0.001 . 10^{-6} + 0.01 \times 10^{-6}}$

$$= \frac{0.001 \times 0.01 \times 10^{-6}}{0.011} = 0.0000909 \times 10^{-6}$$

Now, $f = \dfrac{1}{2\pi\sqrt{100 \times 10^{-6} \times 0.0000909 \times 10^{-6}}}$

$= 528$ kHz. **Ans.**

(*ii*) We know that feedback fraction (β) is given by

$$\beta = \frac{C_1}{C_2} = \frac{0.001 \times 10^{-6}}{0.01 \times 10^{-6}} = 0.1$$

(*iii*) Minimum gain to sustain oscillations,

$$A\beta \geq 1$$

$$A_{min}\ \beta = 1$$

$$A_{min} = \frac{1}{\beta} = \frac{1}{0.1} = 10.\ \textbf{Ans.}$$

(*iv*) Since,
$$A \cong \frac{R_C}{R_E}$$

or
$$R_E \cong \frac{R_C}{A} \cong \frac{2.5}{10}\ k\Omega = 0.25\ \Omega.\ \textbf{Ans.}$$

Example 32. *In a transistorised Hartley oscillator, if L_1 = 0.1 mH, L_2 = 10 mH and mutual inductance (M) between the two coils = 20 μH, calculate the value of capacitor C_1, of oscillatory circuit to obtain the frequency of 4110 kH_2.*

Solution : We know that in Hartley oscillator, the *tqo* coil L_1 and L_2 are connected in series and coupled magnetically

So, total inductance of the coils

$$L = L_1 + C_2 + 2M = 100\ \mu H = 150\ \mu H$$

Frequency of oscillations,
$$f = \frac{1}{2\pi\sqrt{LC}}$$

$$4110 \times 10^5 = \frac{1}{2\pi\sqrt{150 \times 10^{-6} \times C}}$$

$$C = \frac{1}{(2\pi)^2 \times 150 \times 10^{-6} \times (4110 \times 10^3)^2} = 10\ pF.\ \textbf{Ans.}$$

Example 33. *A resistance of 10 kΩ is connected in series with a capacitor. If an alternating frequency of 1 kHz is applied across the network, find the value of C for a phase shift of 60° (i.e., R_C phase shift oscillator).*

Solution : We know
$$\tan\phi = \frac{\alpha_C}{R} = \frac{1}{\omega CR}$$

$$\tan 60° = \frac{1}{2\pi \times 1 \times 10^3 \times C \times 10 \times 10^3}$$

$$1.732 = \frac{1}{2\pi \times 10^3 \times 10 \times 10^3 \times C}$$

$$C = 0.009\ \mu F.\ \textbf{Ans.}$$

Example 34. *In a Wein bridge oscillator given that R_1 = R_2 = 200 kΩ and C_1 = C_2 = 250 pF. Determine the frequency of oscillations.*

Solution : Frequency of oscillations in Wein bridge oscillator is given by

$$f = \frac{1}{2\pi\sqrt{R_1R_2C_1C_2}}$$

When $R_1 = R_2 = R$

and $C_1 = C_2 = C$, then

$$f = \frac{1}{2\pi RC}$$

or $$f = \frac{1}{2\pi \times 200 \times 10^3 \times 250 \times 10^{-12}}$$

or $f = 3177.9$ Hz. **Ans.**

Example 35. *Consider the two-section RC network shown in Fig. P (5.14), find the $\frac{V'_f}{V_o}$ function and verify that it is not possible to obtain 180° phase shif t with a finite attenuation.*

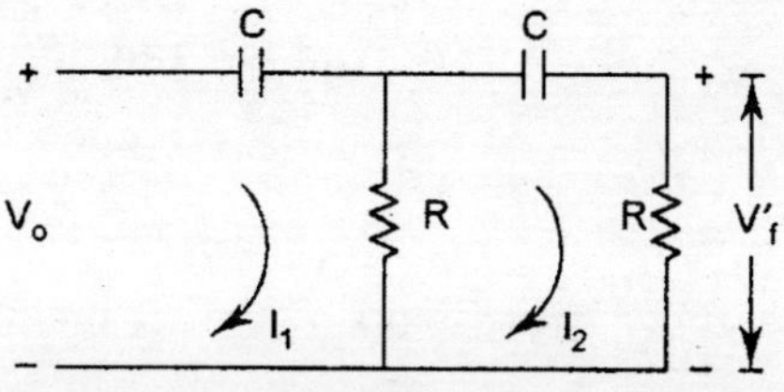

Fig. P (5.14)

Solution : Applying kVL in loop (1) and (2), we get

$$V_o = I_1\left(R + \frac{1}{j\omega C}\right) - I_2R \qquad ...(i)$$

$$0 = \left(R + \frac{1}{j\omega C}\right) I_2 - I_1R \qquad ...(ii)$$

From equations (*i*) and (*ii*)

$$V_o = \left(\frac{1}{R}\right)\left(2R + \frac{1}{J\omega C}\right) I_2 \left(R + \frac{1}{J\omega C}\right) - I_2R$$

$$V_o = \frac{I_2}{R}\left[\left(2R + \frac{1}{J\omega C}\right)\left(R + \frac{1}{J\omega C}\right) - R\right]$$

or $$V'_f = I_2R$$

$$I_2 = V'_f \; IR$$

$$V_o = \frac{V'_f}{R.R}\left[\left(2R + \frac{1}{J\omega C}\right)\left(R + \frac{1}{J\omega C}\right) - R\right]$$

or $$\frac{V_o}{V'_f} = \frac{J^3R\omega C + 1 - R^2\omega^2C^2}{-R^2\omega^2C^2}$$

or $$\frac{V'_f}{V_o} = \frac{\omega^2C^2R^2}{\omega^2C^2R^2 - 1 - 3RJ\omega C}$$

Put $$\frac{X_C}{R} = \frac{1}{\omega CR} = \alpha$$

$$\frac{V'_f}{V_o} = \frac{1}{1 - \alpha^2 - J3\alpha}$$

In order to have a phase shift of 180°, $\tan\theta = \tan 1800 = 0 - \frac{3\alpha}{1-\alpha^2}$. Hence either $\alpha = 0$ or $\alpha = 180°$.

If $\alpha = 0$, then from the above expression $\frac{V'_f}{V_o} = 1$ it means phase shift is 0° not 180°.

Note that $\alpha = 0$ requires either $R = \infty$ or $C = \infty$ which means that output is connected directly to the input. On the other hand if $\alpha = \infty$, then $\frac{V'_f}{V_o} = 0$ and this woulde require an amplifier with infinite gain. Note that $\alpha = \infty$ means either $C = 0$ or $R = 0$ so the attenuation is infinite.

Example 36. *A Colpitts oscillator has a coil with an inductance of 50 μH and is tuned by a capacitor of 300 pF across the output. Find the frequency of oscillation and the minimum gain for maintaining oscillation.*

Solution : For Colpitts oscillator the frequency is given as

$$f = \frac{1}{2\pi\sqrt{LC_{eq}}} \; ; \text{ where } C_{eq} = \frac{C_1C_2}{C_1 + C_2}$$

$$= \frac{300 \times 100}{300 + 100} = 75 \text{ pF}$$

$$L = 50 \; \mu H$$

so, $$f = \frac{1}{2\pi\sqrt{50 \times 10^{-6} \times 75 \times 10^{-6}}}$$

or $$f \approx 2.6 \text{ MHz. } \textbf{Ans.}$$

For maintaining oscillation

$$A_{loop} \geq 1$$

or $$A_{open\ loop} \times \beta = 1 \qquad \{A_{open\ loop} = \text{open loop gain or gain}\}$$

or $$A_{open\ loop} \times \frac{C_2}{C_1} = 1 \qquad \beta = \frac{C_2}{C_1}$$

or $$A_{open\ loop} = \frac{C_1}{C_2} = \frac{3}{1} = 3. \ \textbf{Ans.}$$

Example 37. *Prove that in a crystal the ratio of frequencies in series and parallel resonance is given by* $1 + \frac{1}{2}\frac{C}{C'}$.

Solution : The equivalent circuit of a crystal oscillator is given by

From Fig. P (5.15) series resonant frequency is given by

$$f_s = \frac{1}{2\pi\sqrt{LC}}$$

and parallel resonant frequency is given as

$$f_p = \frac{1}{2\pi\sqrt{LC_{eq}}} \ ;$$

where $C_{eq} = \dfrac{C \cdot C'}{C + C'}$

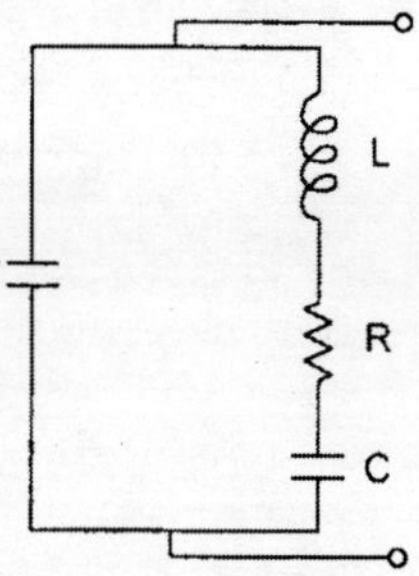

Fig. P (5.15). *Equivalent circuit of a crystal oscillator.*

f_s and f_p may be written as

$$f_s^2 = \frac{1}{(2\pi)^2 \, LC}$$

$$f_p^2 = \frac{1}{(2\pi)^2 \cdot LC_{eq}}$$

According to given condition,

$$\frac{f_p^2}{f_s^2} = \left(\frac{1}{(2\pi)^2 \, LC_{eq}}\right)\left(\frac{(2\pi)^2 \, LC}{1}\right)$$

or $$\frac{f_p^2}{f_s^2} = \frac{LC}{L.\dfrac{C.C'}{C+C'}}$$

or $$\frac{f_p^2}{f_s^2} = +\frac{C}{C'}$$

or $$\frac{f_p}{f_s} = \sqrt{1+\frac{C}{C'}}$$

or $$\frac{f_p}{f_s} = \left(1+\frac{C}{C'}\right)^2$$

or $$\frac{f_p}{f_s} = +\frac{1}{2}\frac{C}{C'}$$ (Neglecting the higher terms)

Hence proved.

□□□

UNIT—III

(Chapter 6)

At the end of this unit you will able to learn about the :

- ➢ Differential amplifier
- ➢ IC op-amps
- ➢ Analyse various types of op-amp configurations, such as inverting, non-inverting, unity gain amplifiers.
- ➢ Scale changer
- ➢ Differentiator
- ➢ Integrator

6

OPERATIONAL AMPLIFIERS

6.1. INTRODUCTION

Analog systems (both linear and non-linear) can be constructed with the **op-amp** or **differential amplifier** as the *basic building block*. In their early stages, these op-amps were constructed from discrete components (vacuum tubes and then transistors and resistors) and their cost was prohibitively high. In 1960's, with the development of IC's, set the stage for a first IC op-amp (μA702) developed by Fairchild semiconductors in 1963. In 1965, Fairchild introduced the μA709, regarded as *first generation op-amp*, which was the improved version of μA702. Fairchild, in 1968, came with the most popular IC op-amp μA741, also regarded as the *second generation op-amp*. One of the reasons for the popularity of op-amp is its *versatility* and, in fact, the IC op-amp has characteristics that closely approach assumed ideal op-amp.

6.2. DIFFERENTIAL AMPLIFIER

Let us consider the emitter-biased circuit. Fig. 6.1 shows two identical emitter-biased circuits in that transistor Q_1 has the same characteristics as transistor Q_2, $R_{C_1} = R_{C_2}$, $R_{E_1} = R_{E_2}$, and the magnitude of V_{CC} is equal to the magnitude of $-V_{EE}$. One point is noted that supply voltage V_{CC} and $-V_{EE}$ are measured with respect to ground.

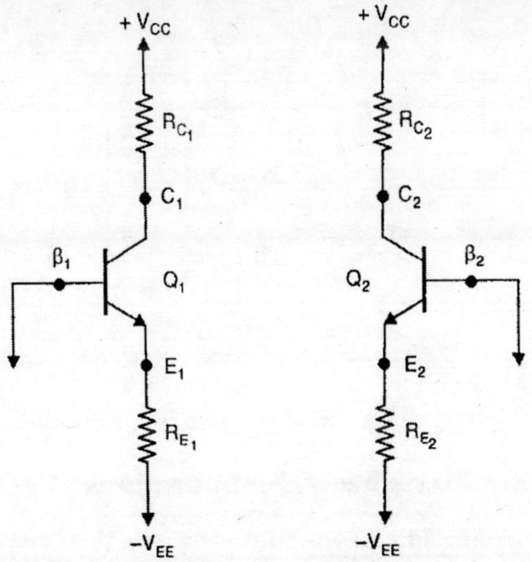

Fig. 6.1. *Two identical emitter-biased circuits.*

Now to obtain a single circuit such as shown in Fig. 6.2, we should reconnect these two circuits as follows :

1. Reconnect the emitter E_1 of transistor Q_1 to the emitter E_2 of transistor Q_2. Note that this reconnection places R_{E_1} in parallel with R_{E_2}.
2. Reconnect + V_{CC} supply voltages of the two circuits because the voltages are of the same polarity and amplitude. Similarly, reconnect $-V_{EE}$ supply voltages.
3. Show the input signal V_1 applied to the base B_1 of transistor Q_1 and V_2 applied to the base B_2 of transistor Q_2.
4. Label the voltage between the collectors C_1 and C_2 as V_0. Note that V_0 is the output voltage.

The circuit of Fig. 6.2 is referred as a differential amplifier. Note here one point that in the circuit of Fig. 6.2 $R_{C_1} = R_{C_2}$. In addition, the parallel combination of R_{E_1} and R_{E_2} is denoted by R_E.

The difference amplifier of Fig. 6.2 amplifies the difference between two input signals V_1 and V_2.

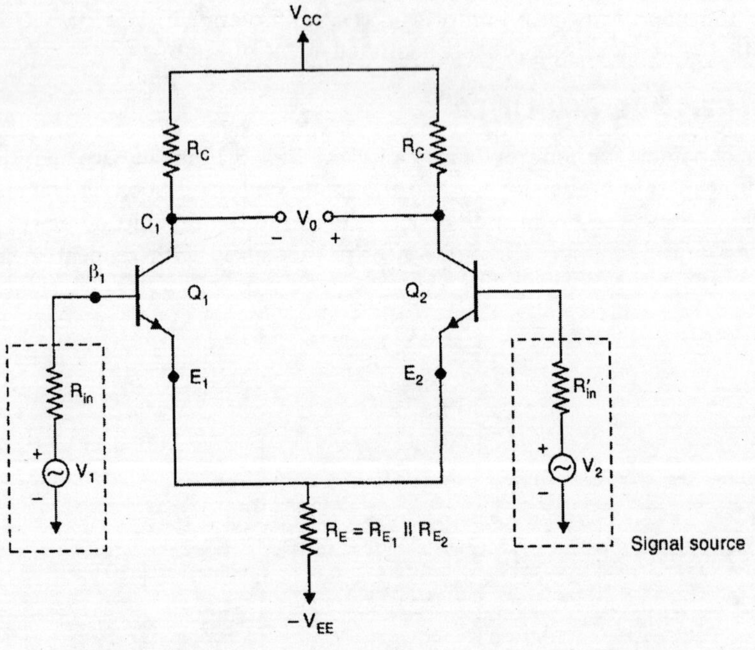

Fig. 6.2. *Dual input, balanced output differential amplifier.*

6.2.1. Differential Amplifier Circuit Configurations

The four differential amplifier configurations are the following :

1. Dual-input, balanced-output differential amplifier.
2. Dual-input, unbalanced-output differential amplifier.

3. Single-input, balanced-output differential amplifier.
4. Single-input, unbalanced-output differential amplifier.

The configurations are defined by the number of input signals used and the way the output voltage is measured. If we use two input signals, the configuration is said to be dual-input. If we use one input signal, then it is single input configuration. On the other hand, if the output voltage is measured between two collectors, it is referred to as a balanced output. However, if the output is measured at one of the collectors with respect to ground, the configuration is called unbalanced output.

6.3. CONCEPT OF OP-AMP

An operational-amplifier (op-amp), which is the basic building block for many analog system itself, consists of a very high gain, direct-coupled amplifier with feed-back, having high input impedance, a low output impedance and acting as differential amplifier.

The op-amp is basically a multistage amplifiers (since it consists many amplifier stages, directly coupled to previous one), so it can be represented as a block diagram as shown in Fig. 6.3.

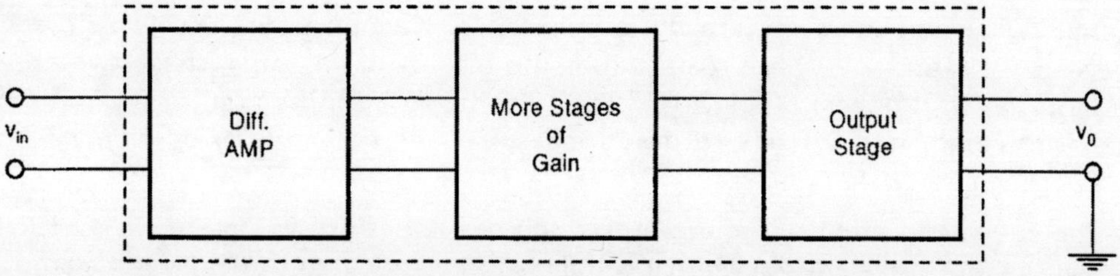

Fig. 6.3. *Block diagram of operational amplifier.*

The first block of the Fig. 6.3 is the *input stage,* consisting of ***D****ual* ***I****nput* ***B****alanced* ***O****utput* (DIBO) differential amplifier, providing most of the voltage gain and also establishes the input resistance of the op-amp.

The second block, consists of many stages of *direct coupled (DC)* ***D****ual* ***I****nput* ***U****nbalanced* ***O****utput* (DIUO, single ended) differential amplifiers, which provides additional gain.

The last block consists of two stages, one for DC level shifting (usually emitter follower) and the second is Class B power amplifier output stage. The well designed output stage provides low output resistance and also increases the output voltage swing (peak-to-peak variation) and raising the current supplying capability.

Fig. 6.4 (*a*), shows the schematic symbol of an op-amp and Fig. 6.4 (*b*), the equivalent circuit of op-amp.

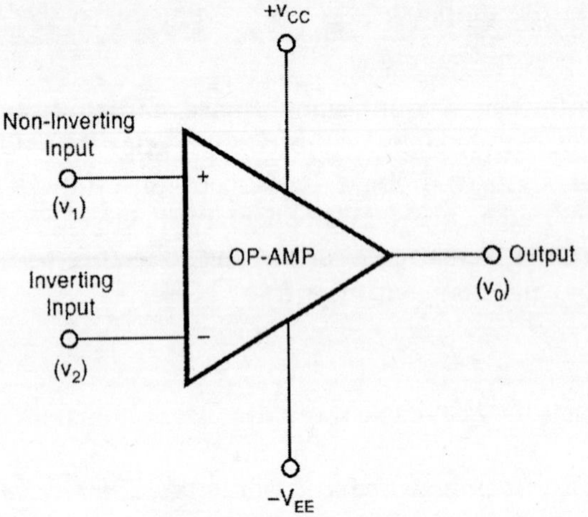

(a) Schematic symbol of op-amp

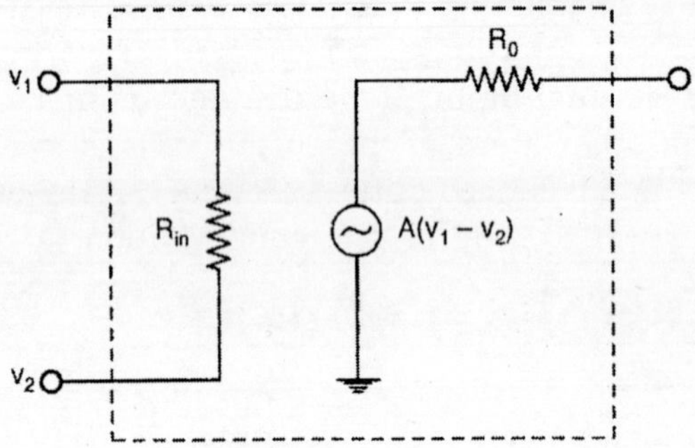

(b) Equivalent circuit of op-amp

Fig. 6.4.

From the signal point of view, the op-amp has three terminals, two input terminals and one output terminal. The two input terminals are **non-inverting** and **inverting** terminals. The amplifiers require dc supply for biasing purpose. The two dc sources, named as $+V_{CC}$ and $-V_{EE}$, are required for the biasing of op-amps, these two are equal in magnitude with opposite polarities. The reference grounding point in op-amp circuit is just the common point of two dc supplies, *i.e., no terminals of op-amp package is physically connected to ground.*

The equivalent circuit of an op-amp, includes important values, A, R_0 and R_{in}. The $A(v_1 - v_2)$ is an equivalent Thevenin's voltage source and R_0 is the Thevenin's equivalent resistance, looking back into the output terminals of an op-amp. The equivalent circuit is useful in analysing the basic operating principle and the effects of feedback arrangement. The output voltage can be written as :

$$v_0 = A(v_1 - v_2) \quad \ldots (6.1)$$

$$= A\, v_{id}$$

where, A = Large signal voltage gain or open loop voltage gain

$v_{id} = (v_1 - v_2)$ = The difference input voltage

v_1 = Voltage at the non-inverting input terminal

and v_2 = Voltage at the inverting input terminal.

Note that v_1 and v_2 are taken with respect to ground.

Thus from the eqn. (6.1), we deduce that, the op-amp only amplies the difference of the input signals, applied at the two input terminals. Another important point to be kept in mind is that the ideal op-amp does not suppose to draw any input current on both terminals (*i.e.,* $i_1 = i_2 = 0$ Amp or $R_{in} \simeq \infty$).

In eqn. (6.1), if $v_1 > v_2$, the v_0 is positive and if condition is reversed, *i.e.,* $v_2 > v_1$, v_0 is a negative quantity, thus acting as non-inverting and inverting amplifiers, respectively. The last case, results in zero output voltage, *i.e.*, $v_0 = 0$, when $v_1 = v_2$ and this is known as *common mode input configuration.*

6.4. IDEAL VOLTAGE TRANSFER CURVE

Equation (6.1) is the basic op-amp equation, in which the output offset voltage is assumed to be zero. This equation is useful in studying the op-amp's characteristics and in analyzing different circuit configurations that employ feedback. The graphic representation of this equation is shown in Fig. 6.5, where the output voltage V_0 is plotted against input difference voltage V_{id}, keeping gain A constant.

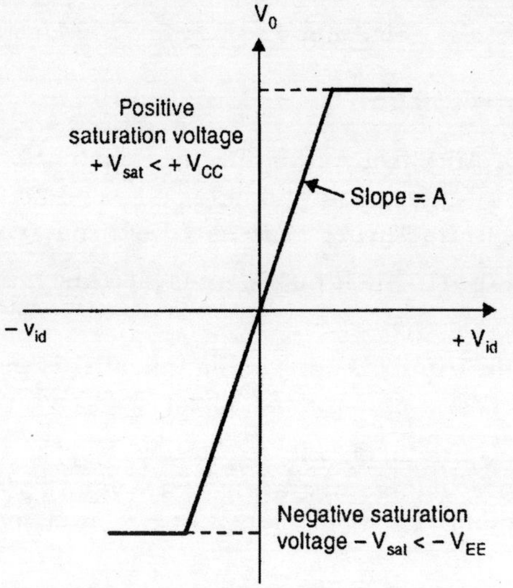

Fig. 6.5. *Ideal voltage transfer curve.*

Note one point that the output voltage cannot exceed the positive and negative saturation voltages. These saturation voltages are specified by an output voltage swing rating of the op-amp for given values of supply voltages. This means that the output voltage is directly proportional to the input difference voltage only until it reaches the saturation voltages and that thereafter output voltage remains constant as shown in Fig. 6.5.

The curve shown in Fig. 6.5 is called an ideal voltage transfer curve, ideal because output offset voltage is assumed to be zero. In normal op-amp use (with negative feedback), this voltage is near zero and is ignored for simplicity of calculation. Notice one point here that the curve is not drawn to scale. If drawn to scale, the curve would be almost vertical because of the very large values of A.

6.5. OP-AMP IC

An Operational Amplifier IC is a solid-state integrated circuit that uses external feedback to control its function. The op-amp without any external device is called 'open-loop' mode, referring actually to the so called ideal operational amplifier with infinite open-loop gain, input resistance, bandwidth and a zero output resistance. However, in practice no op-amp can meet these ideal characteristics. Since, the LM741/NE741/ μA741 op-amps are the most popular op-amps. Nowadays, the 741 is a frequency compensated device and widely used.

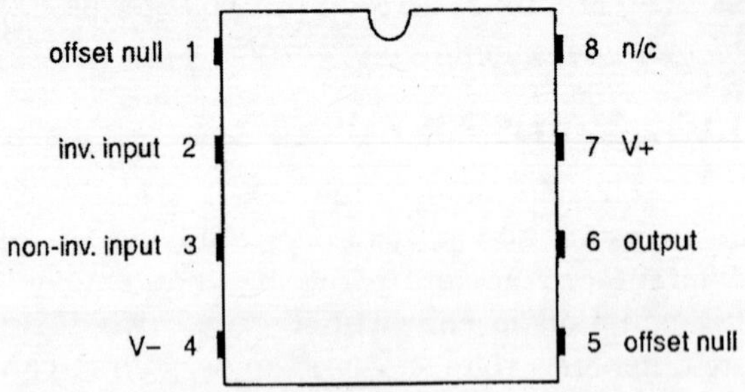

Fig. 6.6. *Pin Diagram of 741.*

Definition of 741 Pin Function

Pin 1 (Offset Null). An offset null-adjustment potentiometer is used to compensate for offset voltage. The null-offset potentiometer also compensate for irregularities in the operational amplifier manufacturing process which may cause an offset.

Pin 2 (Inverted Input). All input signals at this pin will be inverted at output pin 6.

Pin 3 (Non-inverted Input). All input signals at this pin will be processed normally without inversion. The result is same as pin 2.

Pin 4 (–V). The V-pin (also referred to as V_{ss}) is the negative supply voltage terminal. Supply negative voltage operating range for the 741 is –4.5 volts (minimum) to –18 volts (max.), and is specified for the operation between –5 and –15 V dc.

Pin 5 (Offset Null). See pin 1.

Pin 6 (Output). Output signal's polarity will be the opposite of the input's when this signal is applied to the op-amp's inverting input. For example, a sine wave at the inverting input will output a square wave in the case of an inverting comparator circuit.

Pin 7 (Posv). The V+ pin (also referred to as V_{CC}) is the positive voltage terminal of the 741 op-amp IC. Supply voltage operating range for the 741 is 4.5 volts (minimum) to 18 volts (maximum), and it is specified for operation between + 5 and + 15 V dc.

Pin 8 (N/C). The 'N/C' stands for 'Not Connected'. There is no other explanation. There is nothing connected to this pin, it is just there to make it a standard 8-pin package.

6.6. IDEAL OP-AMP

The ideal op-amp, are those op-amps, which have these characteristics, given below :

1. Infinite voltage gain A ($A = \infty$).
2. Infinite input resistance ($R_{in} = \infty$).
3. Zero output resistance ($R_0 = 0$).
4. Infinite bandwidth ($BW = \infty$).
5. Infinite common mode rejection ratio (CMRR = ∞).

6. Infinite slew rate (SR = ∞).

7. Zero output voltage where input difference voltage, v_{id}, is zero.

The properties mentioned above makes, the ideal op-amp, a perfect voltage amplifier and often referred to as a *voltage controlled voltage source (VCVS)*, as shown in Fig. 6.4 (*b*).

The practical op-amps can be made to behave nearly in unison with the ideal one by using the proper feedback circuitry.

CMRR (Common Mode Rejection Ratio) is defined as the ratio of *differential voltage gain* (A_d) to the *common mode voltage gain* (A_{CM}), *i.e.*,

$$\text{CMRR} = \frac{A_d}{A_{CM}} \quad \text{... (6.2)}$$

where A_{CM} is common mode voltage gain and is given by,

$$A_{CM} = \frac{V_{OCM}}{V_{CM}} \quad \text{... (6.3)}$$

V_{OCM} = Output common mode voltage

and V_{CM} = Input common mode voltage.

CMRR is often expressed in desibel's (dBs).

In practical case A_{CM} is very small (not exactly zero), and A_d is very large (as it is large signal gain, *i.e.*, *A*), therefore, CMRR is very large, but not exactly infinite, such as 90 dB for 741 IC op-amp.

A_{CM} is determined from the common mode configuration of op-amp, as shown in Fig. 6.7.

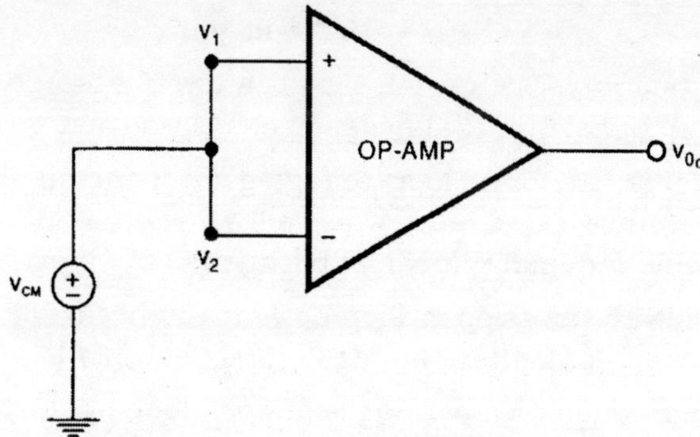

Fig. 6.7. *Common mode configuration.*

Slew Rate (SR) is the maximum rate of change of output voltage with respect to time, *i.e.*,

$$\text{SR} = \left.\frac{dv_0}{dt}\right|_{max} \quad \text{... (6.4)}$$

The unit of the Slew Rate is V/μs (volts per microsecond). It is the cause of non-linear distortion at the output.

6.7. OP-AMP APPLICATIONS

Initially the application areas of op-amp were analog computation and instrumentation, so named "op-amp", *i.e., the amplifier which performs many operations.* Originally, it was used to perform mathematical functions such as addition, multiplication by a constant, subtraction, differentiation, integration, etc., one now put to a variety of other uses, for *e.g.*, as comparator, pulse generator, wave generator, schimitt trigger, multivibrations, etc.

We can use the op-amps also as, logarithmic and antilogarithmic modules, linear differential eqn. solver and to solve other mathematical functions.

6.8. NEED OF FEEDBACK

As we know, op-amp amplifies only the difference of the two signals, applied at the non-inverting and inverting terminals, but the range of signals is very small due to the very large *open loop* (differential, large signal) voltage gain A_d (of the order of 10^5), because the maximum amplitude of the output signal cannot exceed to the maximum value of biasing voltage. Beyond this range op-amp is said to be in **saturation** with constant evelop of voltage signal. For example, if the maximum value of output voltage for an op-amp circuit is 20 V and the large signal voltage gain is 2×10^5, then the maximum value of signal, which can be amplified by op-amp is

$$v_{id} = \frac{v_0}{A_d}$$

or $$v_{id} = \frac{20}{2 \times 10^5}$$

or $$v_{id} = 10^{-4} \text{ volt}$$

which is a very small quantity. Such small signals are very susceptible to noise (unwanted and random electrical signals) and almost impossible to obtain in laboratory.

Besides, being large, the open loop voltage gain of the op-amp is not a constant, but varies with temperature, power supply as well as device to device. It makes the open loop op-amp unsuitable for many linear applications.

Also the bandwidth of most open loop op-amps is negligible small — almost zero. Therefore, the open loop configurations are impractical in AC applications.

Due to above mentioned disadvantages of open loop op-amps, these are generally not used in linear applications. We cannot only select, but also control the gain of the op-amp, simply by introducing the **feedback** (combining a portion of output signal with input signal either directly or indirectly) in the circuit. When the phase of input and feedback signals are same (*i.e.*, phase difference is 0°), the feedback is said to be *positive or regenerative feedback*. When the feedback signal is of opposite polarity (*i.e.*, out of phase by 180°) with respect to the input signal, the feedback is called *negative* or *degenerative feedback*. Since the discussion of "feedback" is beyond the scope of this book, so we will enlist only the advantages of negative feedback here.

In amplifiers, negative feedback have following characteristics :

- stabilizes the gain.
- increases bandwidth.
- changes input and output impedances.
- decreases the effect of distortion and noise.

} Advantages of Negative Feedback

- reduces the voltage gain (this is only disadvantage of negative feedback).

6.9. INVERTING AMPLIFIER

The *inverting amplifier* is most basic constant gain amplifier. It is also known as the *voltage-shunt feedback* amplifier. The basic circuit diagram is shown in Fig. 6.8. Here biasing power supplies are not shown for simplicity.

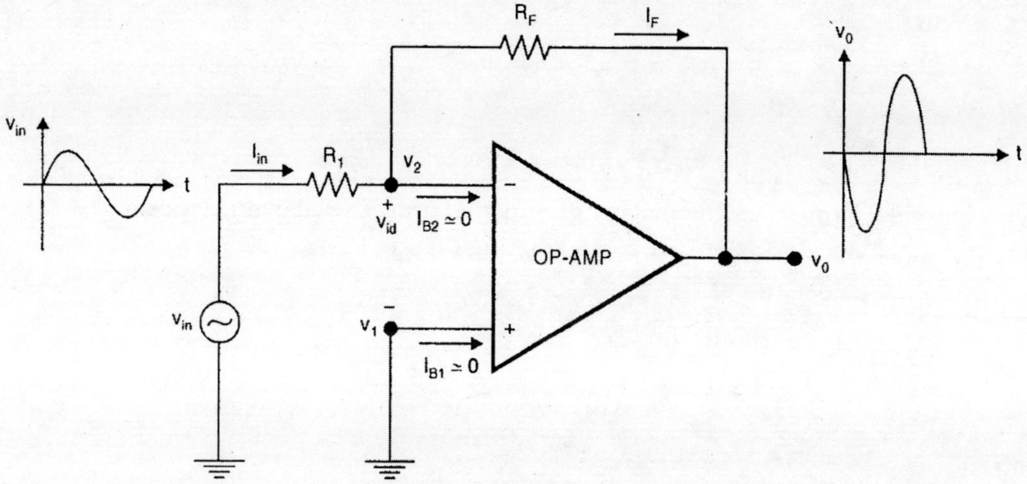

Fig. 6.8. *Inverting amplifier.*

An input voltage v_{in} drives the inverting input through resistor R_1. This results an inverting input v_{id}, which is amplified by the open-loop voltage gain to produce an inverted output voltage v_0. The output voltage is feedback to the input through the feedback resistor R_F. This arrangement forms a negative feedback, because the output is 180° out of phase with the input. Note that non-inverting terminal is grounded.

Concept of Virtual Ground

A point is any circuit is said to be grounded if the potential at that point is equal to the ground potential. If that point is connected to the ground with the help of any conducting (metallic) wire, then all the current from that point (node) flows towards the ground, and the point is said to be *physically* or *mechanically grounded.* In case of *virtual ground,* only the condition for potential is satisfied, *i.e.,* potential at that point becomes same as ground potential, but there is no path for current from that point to ground. The concept of virtual ground is a widely used shortcut for analysing an inverting amplifier.

The concept of virtual ground is based on an ideal op-amp, *i.e.*, the voltage gain of an ideal op-amp is infinite and input impedance is infinite. We can deduce the

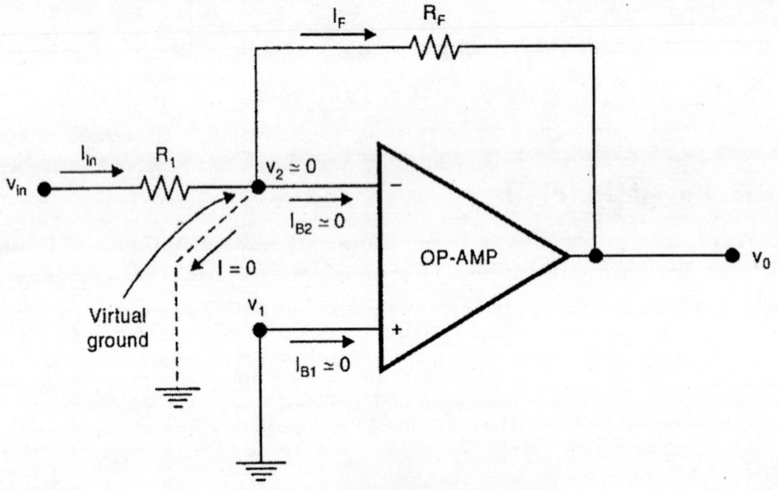

Fig. 6.9. *Concept of virtual ground.*

following ideal properties for inverting amplifier from the above discussion :

(1) since, $R_{in} \simeq \infty$, therefore, $I_{B2} = I_{B1} \simeq 0$,

i.e., $I_{in} \simeq I_F$.

(2) since, A_d is infinite (in ideal case),

$$\therefore \quad A_d = \frac{v_O}{v_{id}}$$

$$\Rightarrow \quad v_{id} = \frac{v_O}{A_d} \simeq \frac{v_O}{\infty} \simeq 0$$

$$\therefore \quad \boxed{v_{id} = (v_1 - v_2) \simeq 0 \text{ volt}} \quad \text{... (6.5)}$$

From Fig. 6.9 and eqn. (6.5), we see that,

$$v_2 \simeq 0 \text{ volt} = v_1$$

due to *virtual ground.*

Virtual ground is very unusual and just like half of a ground, because it is a short for voltage but an open for current.

Voltage Gain

Here, *voltage gain* means to the *closed loop* or *overall voltage gain* of inverting amplifier. It can be determined by two methods, either by feedback method or by virtual ground concept.

1. Virtual Ground Method (Ideal op-amp)

Consider the inverting amplifier configuration shown in Fig. 6.9. We have from virtual ground concept,

$$I_{in} \simeq I_F \qquad \ldots (6.6)$$

Here,
$$I_{in} = \frac{v_{in} - v_2}{R_1} \qquad \ldots (6.7)$$

$$\left\{ \text{as} \qquad \text{current} = \frac{\text{potential difference}}{\text{resistance}} \right\}$$

and
$$I_F = \frac{v_2 - v_0}{R_F} \qquad \ldots (6.8)$$

Therefore,
$$\frac{v_{in} - v_2}{R_1} = \frac{v_2 - v_0}{R_F}$$

but from eqn. (6.5),
$$v_2 \simeq 0$$

so,
$$\frac{v_{in}}{R_1} = -\frac{v_0}{R_F}$$

$$\Rightarrow \qquad \boxed{v_0 = -\frac{R_F}{R_1} v_{in}} \qquad \ldots (6.9)$$

or overall or closed loop voltage gain,

$$\boxed{A_0 = \frac{v_0}{v_{in}} = -\frac{R_F}{R_1}} \qquad \ldots (6.10)$$

2. Feedback Method (Practical op-amp)

In previous derivation, we assumed about ideal op-amp, but here we will consider the finite differential voltage gain A_{id}, which is very large (almost, but not exactly, infinite). Consider Fig. 6.10,

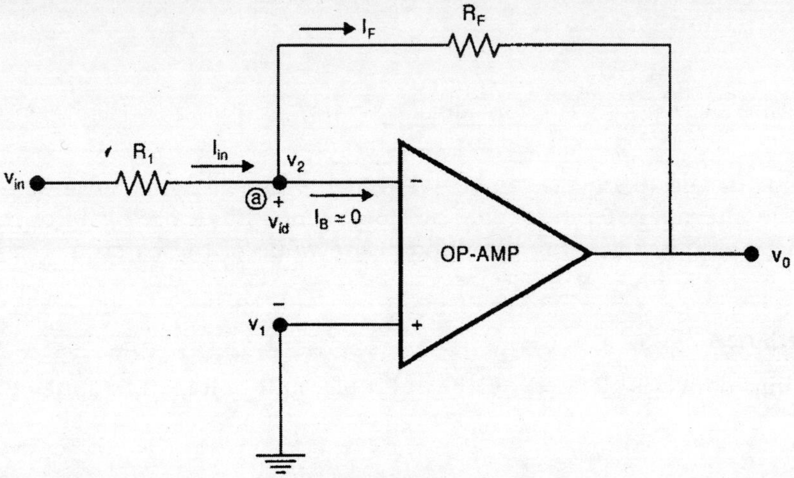

Fig. 6.10.

Applying KCL at node (*a*),

$$I_{in} = I_F + I_B$$

Since, R_{in} is very high (of the order of 2 MΩ), I_B is negligibly small, therefore,

$$I_{in} \simeq I_F \qquad ...(6.11)$$

Now from eqns. (6.7) and (6.8),

$$\frac{v_{in} - v_2}{R_1} = \frac{v_2 - v_0}{R_F} \qquad ...(6.12)$$

However,

$$v_1 - v_2 = \frac{v_0}{A_d}$$

Since,

$$v_1 = 0 \text{ V}$$

$$v_{id} = v_2 = -\frac{v_0}{A_d} \qquad ...(6.13)$$

substituting the value of v_2 from eqn. (6.13) into eqn. (6.12), we get,

$$\frac{v_{in} + \dfrac{v_0}{A_{id}}}{R_1} = \frac{-\dfrac{v_0}{A_d} - v_0}{R_F}$$

Solving,

$$A_0 = \frac{v_0}{v_{in}} = -\frac{A_d R_F}{R_1 + R_F + A_d R_1} \qquad ...(6.14)$$

since, internal differential gain A_{id} of the op-amp is very large (ideally infinite), therefore, $A_d R_1 >> R_1 + R_F$, hence eqn. (6.14) reduces to

$$\boxed{A_0 = \frac{v_0}{v_{in}} \simeq -\frac{R_F}{R_1}} \qquad ...(6.15)$$

which is same as eqn. (6.10) for ideal op-amp, or obtained using virtual ground concept.

Eqns. (6.10) and (6.14) also indicates that the **closed loop voltage gain** does not depend on the internal differential gain of the op-amp and desired gain or multiplier can be obtained by simply adjusting the appropriate value of the external resistors R_1 and R_F.

Input Impedance

Input impedance is defined as the ratio of input voltage to input current, *i.e.*,

$$R_i = \frac{v_{in}}{I_{in}} \qquad ...(6.16)$$

Now from eqn. (6.7), where $v_2 = 0$ V,

$$\boxed{R_i = R_1} \qquad ...(6.17)$$

6.10. NON-INVERTING AMPLIFIER

The **non-inverting amplifier** is another basic op-amp circuit uses negative voltage-series feedback to stabilize the overall voltage gain. It is also known as **voltage series feedback amplifier.** The circuit configuration of non-inverting op-amp is shown in Fig. 6.11.

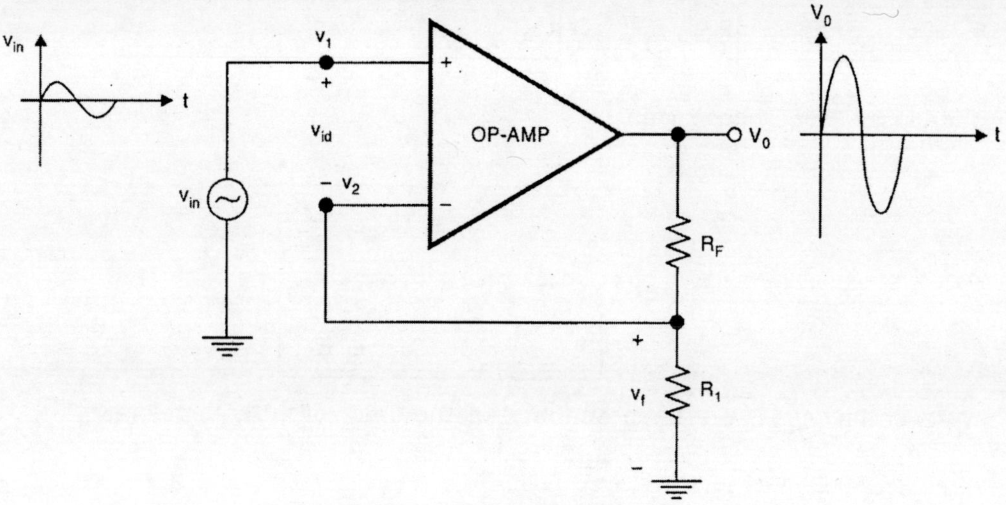

Fig. 6.11. *Non-inverting amplifier.*

Here, an input voltage source is connected at the non-inverting terminal of the op-amp. This input voltage is amplified to produce the **in phase** output voltage, as shown. Part of v_0, *i.e.*, v_f, is feedback to the input through a voltage divider network at the **inverting terminal.** Note here, that feedback voltage v_f is almost equal to the input voltage v_{in}, as $v_1 - v_2 \simeq 0$ V for op-amp. It can be easily verified by applying KVL at the input section. Since, the feedback voltage v_f, opposes the input voltage v_{in}, feedback is known as negative feedback. Almost zero voltage difference between the two terminals, *i.e.*, $v_1 - v_2 \simeq 0$, is due to almost infinite input resistance and almost infinite differential voltage gain of the operational amplifiers and known here as the **virtual short.** Because of the virtual short, the inverting input voltage follows the non-inverting input voltage. If the non-inverting voltage changes, the inverting input voltage immediately changes to the same value. This follow-to-leader action is called **Bootstrapping.** In other words, the inverting input is bootstrapped to the non-inverting input.

Voltage Gain

Consider the circuit of Fig. 6.12. If we apply KVL at the input side, we get,

$$v_{in} = v_{id} + v_f \quad \text{...(6.18)}$$

where,

$$v_{id} = v_1 - v_2$$

$$v_1 = v_{in}$$

and

$$v_f = v_z = \frac{R_1}{R_1 + R_F} v_0 \quad \text{(by voltage divider rule)}$$

However, $$v_0 = A_d(v_1 - v_2) \quad ...(6.19)$$
$$= A_d\, v_{id}$$

Therefore, $$v_0 = A_d\left(v_{in} - \frac{R_1 v_0}{R_1 + R_F}\right)$$

or $$v_0\left[1 + \frac{A_d\, R_1}{(R_1 + R_F)}\right] = A_d\, v_{in}$$

or the closed loop voltage gain is

$$A_0 = \frac{v_0}{v_{in}} = \frac{A_d}{1 + \frac{A_d\, R_1}{(R_1 + R_F)}}$$

$$\boxed{A_0 = \frac{A_d(R_1 + R_F)}{R_1 + R_F + A_d\, R_1}} \quad ...(6.20)$$

Generally, A_d is very large quantity (of the order of 10^5), therefore,

$$A_d R_1 >> (R_1 + R_F)$$

Thus, $$A_0 \simeq \frac{(R_1 + R_F)}{R_1}$$

or $$\boxed{A_0 = \left(1 + \frac{R_F}{R_1}\right) = \frac{v_0}{v_{in}}} \quad ...(6.21)$$

The same result, as in eqn. (6.21), can be obtained by virtual short concept as follows :

Rewriting eqn. (6.18) here, $$v_{in} = v_{id} + v_f$$

as $$v_{id} = v_1 - v_2 \simeq 0 \text{ V, we have,}$$
$$v_{in} \simeq v_f$$

but $$v_f = \frac{R_1}{R_1 + R_F} \cdot v_0$$

Substituting this value, we get,

$$v_{in} \simeq \frac{R_1}{R_1 + R_F} \cdot v_0$$

or closed loop voltage gain, $$A_0 = \frac{v_0}{v_{in}}$$
$$= \frac{R_1 + R_F}{R_1}$$

or $$\boxed{A_0 = 1 + \frac{R_F}{R_1}} \quad ...(6.22)$$

Input Impedance

Input impedance for this type of circuit is given as,

$$\boxed{R_i = R_{in}(1 + A_d\beta)} \quad \text{...(6.23)}$$

Where β is feedback circuit gain and given by :

$$\boxed{\beta = \frac{R_1}{R_1 + R_F}} \quad \text{...(6.24)}$$

6.11. OP-AMP APPLICATIONS

6.11.1. Unity Gain Amplifier

From the analysis of non-inverting amplifier, with feedback, we can conclude that non-inverting amplifier is **perfect voltage amplifier,** because it has stable and controllable voltage gain, very high input resistance, very low output resistance and large bandwidth.

Consider, here, the gain formula for non-inverting amplifier from eqn. (6.22), *i.e.*,

$$A_0 = 1 + \frac{R_F}{R_1}$$

From this eqn., we see that minimum constant gain that can be obtained using non-inverting configuration is unity, which can be obtained, simply by making $R_F = 0$ and $R_1 = \infty$ in Fig. 6.8, *i.e.*,

For $$R_F = 0 \text{ and } R_1 = \infty$$

$$\boxed{A_0 = 1} \quad \text{...(6.25)}$$

The circuit diagram of unity gain amplifier configuration is drawn in Fig. 6.12.

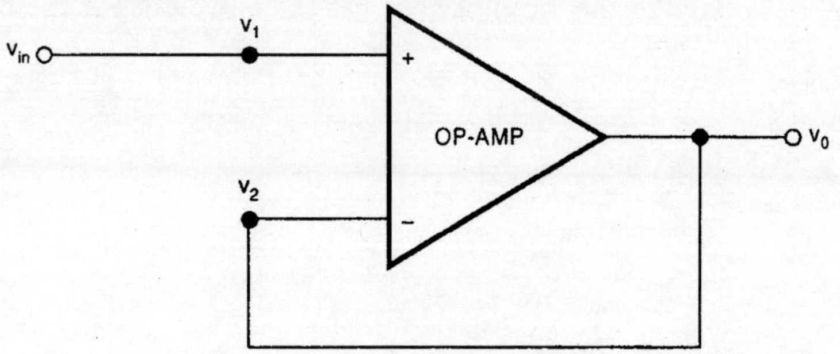

Fig. 6.12. *Unity gain amplifier.*

From eqn. (6.25) and Fig. 6.12, it reveals that in this circuit, output voltage is equal to input voltage and both are in phase, therefore, this configuration is often called as **op-amp voltage follower amplifier.** Here all voltage is feedback to the inverting input, therefore, negative feedback is maximum and the circuit is very close to ideal. The main application of this configuration is as a **non-inverting buffer.** When it is placed between two networks, it removes the loading on the first network, as it has very high input impedance and very low output impedance, just like a **emitter follower,** but better than that.

6.11.2. Sign Changer or Inverter

This configuration is an inverting amplifier but gain of the circuit is made unity

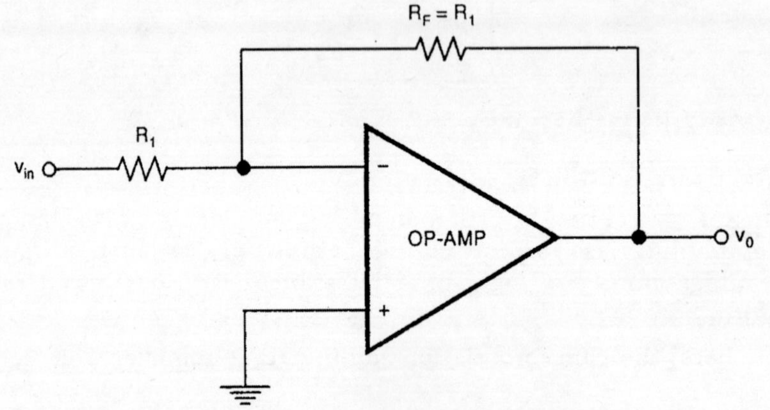

Fig. 6.13. *Inverter or sign changer.*

by selecting precisely equal resistors R_1 and $R_F = R_1$. This selection makes gain unity which produces output voltage, equal in magnitude but 180° out of phase with respect to input voltage, *i.e.*, the closed loop gain of inverter circuit is,

$$\boxed{A_0 = -1} \quad \text{...(6.26)}$$

6.11.3. Scale Changer or Constant Gain Multiplier

Both the configurations of op-amp, *i.e.*, inverting and non-inverting with feedback, can be used as scale changer or constant gain multiplier. In this configuration, when we apply an input voltage to the amplifier, we get scaled voltage at the output. For this purpose, we select the values of R_1 and R_F such that the ratio of these two, *i.e.*, $\frac{R_F}{R_1}$, is an integer constant, say, 'n'.

When we take,

$$\frac{R_F}{R_1} = n$$

We have, for inverting amplifier,

$$A_0 = -\frac{R_F}{R_1} = -n$$

i.e.,

$$\boxed{v_0 = -n\, v_{in}} \quad \text{...(6.27)}$$

and for non-inverting amplifier,

$$A_0 = 1 + \frac{R_F}{R_1} = (n + 1)$$

i.e.,
$$\boxed{v_0 = (n + 1)\, v_{in}} \quad \text{...(6.28)}$$

In this application, R_1 and R_F are selected as precision resistors.

6.11.4. Adders

Whenever, we need to combine two or more analog signals into a single output, we use **summing amplifier or adder.**

Inverting Configuration

The inverting configuration of a summing amplifier or adder is shown in Fig. 6.14. Only three input voltages are shown here for simplicity, but we can have as many inputs as needed for the application.

In Fig. 6.14, three input voltages v_a, v_b and v_c are connected at the inverting terminal through the resistors R_1, R_2 and R_3 respectively. The operation of circuit can be verified by the mathematical expression of v_0. The expression of v_0 can be obtained from Kirchhoff's current law, written at inverting node and applying virtual ground concept, we get,

$$i_a + i_b + i_c \simeq i_F \quad \text{...(6.29)}$$

as $i_B \simeq 0$ Amp and $v_1 = v_2 \simeq 0$ V.

Therefore,

$$\frac{v_a}{R_1} + \frac{v_b}{R_2} + \frac{v_c}{R_3} = -\frac{v_0}{R_F}$$

or
$$v_0 = -\left(\frac{R_F}{R_1} v_a + \frac{R_F}{R_2} v_b + \frac{R_F}{R_3} v_c\right) \quad \text{...(6.30)}$$

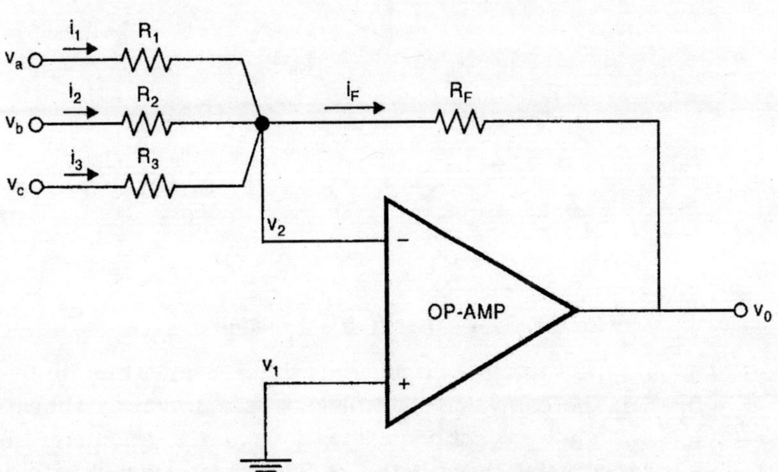

Fig. 6.14. *Summing amplifier (adder).*

This expression can also be verified using superposition theorem.

If we take, $R_1 = R_2 = R_3 = R$, then eqn. (6.30) reduces to,

$$v_0 = -\frac{R_F}{R}[v_a + v_b + v_c] \quad \text{...(6.31)}$$

Note that, in eqn. (6.31) output voltage is proportional to the sum of input voltage.

Obviously, when gain of circuit is 1, *i.e.*, $R = R_F$, the output voltage v_0 is equal to the **negative** sum of input voltages, *i.e.*,

$$\boxed{v_0 = -(v_a + v_b + v_c)} \quad \text{...(6.32)}$$

Non-Inverting Configuration

The non-inverting configuration can also be used as the **summing amplifier** if the input voltages to be added and corresponding resistors are connected at the non-inverting input terminals and selecting appropriate value of R_F and R_1 as shown in Fig. 6.15. Here, all the resistors connected with input voltages are taken equal, *i.e.*, $R_a = R_b = R_c = R$.

Since, virtual ground concept is not applicable in non-inverting configuration due to the feedback at the inverting terminal, the expression for output voltage v_0 can be obtained using **superposition theorem,** *i.e.*, by adding the output voltages, produced by the individual voltage when others are grounded.

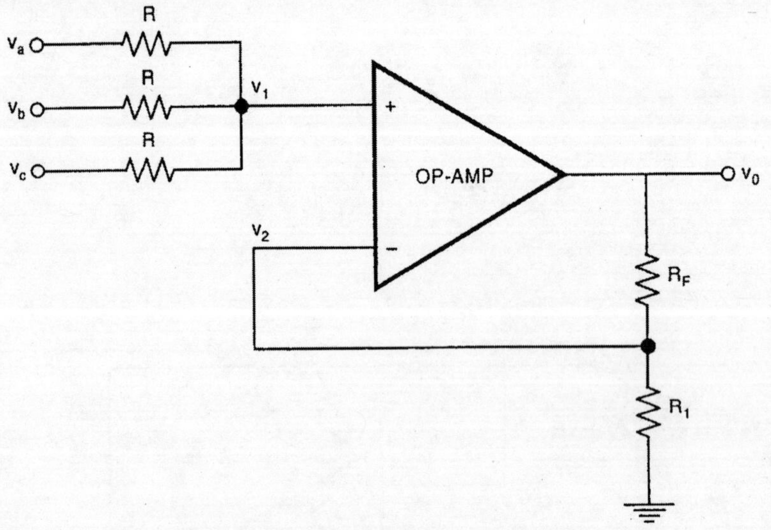

Fig. 6.15. *Non-inverting summer.*

First, consider only the voltage v_a, applied at the non-inverting input terminal and making $v_b = v_c = 0$, then the equivalent circuit can be redrawn as shown in Fig. 6.16. Here, the resistances which were connected with v_b and v_c and now in parallel and making the voltage divider network with the resistance connected with v_a.

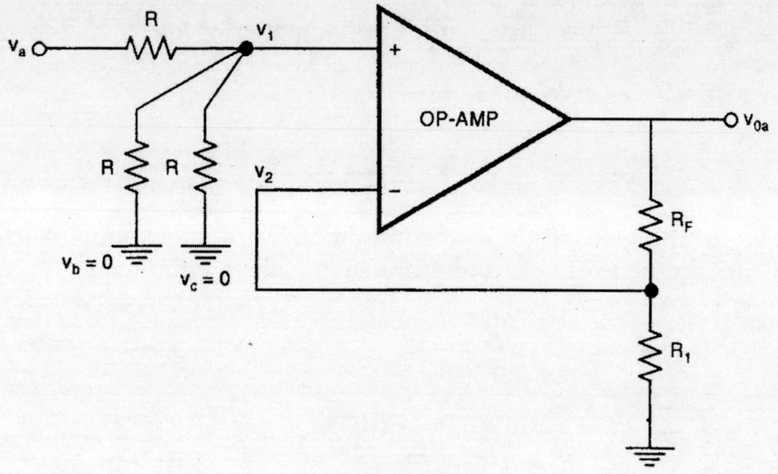

Fig. 6.16.

From Fig. 6.16, the voltage at non-inverting terminal is,

$$v_1 = v_a \cdot \frac{(R \| R)}{R + (R \| R)}$$

$$\therefore \quad v_1 = \frac{R/2}{R + R/2} \cdot v_a \qquad \text{...(6.33)}$$

Now, from eqn. (6.22),

$$v_{0_a} = \left(1 + \frac{R_F}{R_1}\right)\left(\frac{R/2}{R + R/2}\right) v_a \qquad \text{...(6.33}a\text{)}$$

Similarly, we can obtain the expression for the output voltage, v_{0_b} and v_{0_c} for the inputs v_b and v_c respectively, *i.e.*,

$$v_{0_b} = \left(1 + \frac{R_F}{R_1}\right)\left(\frac{R/2}{R + R/2}\right) v_b \qquad \text{...(6.34)}$$

and

$$v_{0_c} = \left(1 + \frac{R_F}{R_1}\right)\left(\frac{R/2}{R + R/2}\right) v_c \qquad \text{...(6.35)}$$

By superposition principle, we can write,

$$v_0 = v_{0_a} + v_{0_b} + v_{0_c}$$

$$= \left(1 + \frac{R_F}{R_1}\right)\left(\frac{R/2}{R + R/2}\right)(v_a + v_b + v_c)$$

$$v_0 = \left(1 + \frac{R_F}{R_1}\right)\left(\frac{1}{3}\right)(v_a + v_b + v_c) \qquad \text{...(6.36)}$$

if the gain, *i.e.*, $\left(1 + \frac{R_F}{R_1}\right)$ is made unity, the amplifier of Fig. 6.15 can be used as an **averaging circuit,** *i.e.*, eqn. (6.36) reduces to,

$$v_0 = \left(\frac{v_a + v_b + v_c}{3}\right) \quad ...(6.37)$$

and if the value of the gain of non-inverting amplifier is made equal to the number of inputs, then the circuit works as the summer or adder circuit, *i.e.*,

if, $$\left(1 + \frac{R_F}{R_1}\right) = n$$

$$= 3 \text{ (Here)}$$

We get, $$\boxed{v_0 = v_a + v_b + v_c} \quad ...(6.38)$$

Note that no sign change or phase reversal occurs between inputs and output.

6.11.5. Subtractor Circuit

The subtractor circuit can be implemented by using op-amp differential configuration, *i.e.*, by applying the voltages, to be subtracted, at the both (inverting and non-inverting) terminals simultaneously. Such a simple configuration is shown in Fig. 6.17.

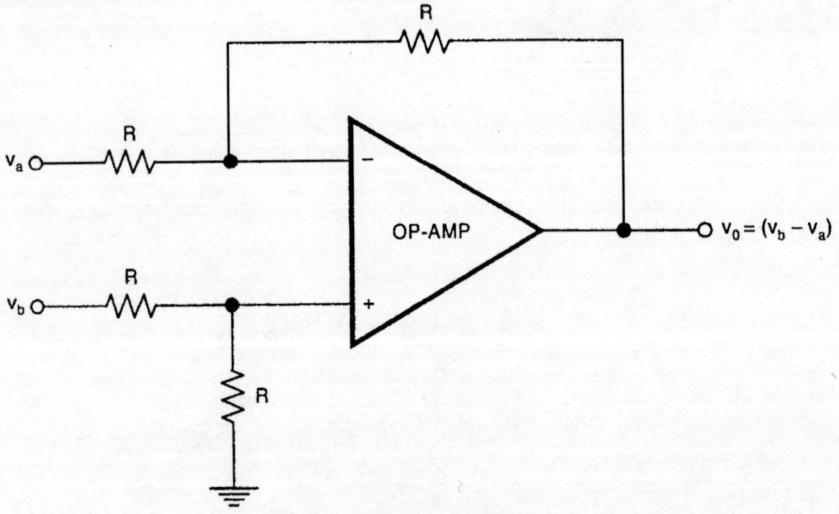

Fig. 6.17. *Subtractor.*

It can be shown easily by superposition principle, for the circuit of Fig. 6.17, that

$$\boxed{v_0 = v_b - v_a} \quad ...(6.39)$$

6.11.6. Integrator

If we place a capacitor C in the feedback path and a resistor R_1 in the input path, the circuit realizes the mathematical operation of integration and hence known as

integrator circuit. This is also known as **Miller integrator,** after its inventor and shown in Fig. 6.18. The integrator circuit is a **non-linear** application of op-amp.

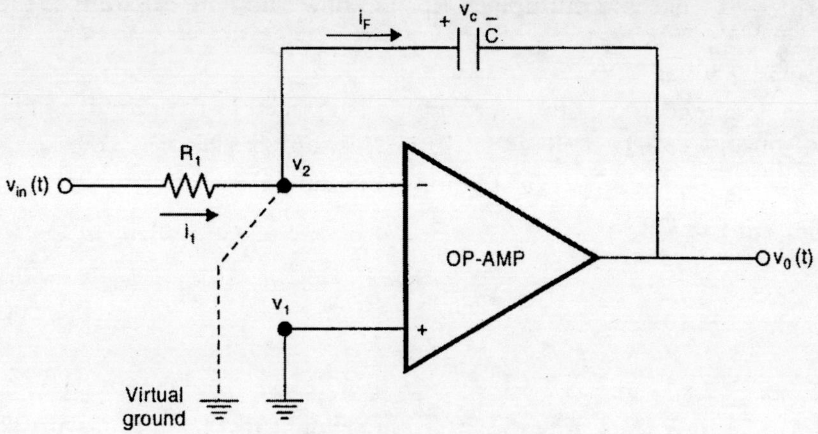

Fig. 6.18. *Integrator circuit.*

The function of integration using the circuit of Fig. 6.18 can be proved mathematically. Suppose that a time varying voltage $v_{in}(t)$ is applied at the input of circuit. Since, circuit is inverting type, we can use the concept of virtual ground to find the expression of output voltage. We can write,

$$i_1 \approx i_F$$

That is, voltage $v_{in}(t)$ appears in effect across R_1 and thus a time varying current i_1 flows through the capacitor C, causing charge to accumulate on C. If we assume that the input is applied just at the time instant $t = 0$, then at an arbitrary time t current i_1 will have deposited a charge equal to $\int_0^t i_1\, dt$ on capacitor C. If the initial voltage on C (*i.e.*, at $t = 0$, before applying the input) is $v_c(0)$, then voltage across C can be given as,

$$v_C(t) = v_C(0) + \frac{1}{C}\int_0^t i_1\, dt \qquad \text{...(6.40)}$$

but,

$$v_0(t) = -v_C(t)$$

and

$$i_1 = \frac{v_{in}(t)}{R_1} \quad \text{(From the concept of virtual ground)}$$

We have from eqn. (6.40),

$$v_0(t) = -\frac{1}{R_1 C}\int_0^t v_{in}(t)\, dt - v_C(0) \qquad \text{...(6.41)}$$

Here, $v_C(0)$ may be any constant value. If $v_C(0) = 0$ V, then,

$$\boxed{v_0(t) = -\frac{1}{R_1 C}\int_0^t v_{in}(t)\, dt} \qquad \text{...(6.42)}$$

It is revealed from eqn. (6.42) that output is proportional to the integration of the input, where $-\frac{1}{R_1C}$ is scale multiplier. R_1C is known as *time constant* of the *RC* circuit, and hence of the integrator.

Suppose a voltage of constant magnitude is applied to the circuit of Fig. 6.18, then the value of output can be calculated from eqn. (6.43) as,

$$\because \quad v_{in}(t) = K \text{ (constant)} \qquad \text{...(6.43)}$$

$\therefore$ From eqn. (6.43),

$$v_0 = -\frac{1}{R_1C}\int_0^t k\, dt$$

$$v_0 = -\frac{K}{R_1C}\cdot t \qquad \text{...(6.44)}$$

Therefore, output voltage is directly proportional to the time *t*, with scaling factor of $-\frac{K}{R_1C}$, *i.e.*, v_0 linearly varies with time. Such an output is shown in Fig. 6.19 for a constant input voltage, $K = 1$ and $R_1C = 1$ unit time. The output voltage is a ramp of slope $-\frac{K}{R_1C}$ which is 1 v/s here. It is an ideal case that if an input of constant magnitude (*DC*) is applied for infinite duration, output will tend to infinity, but such

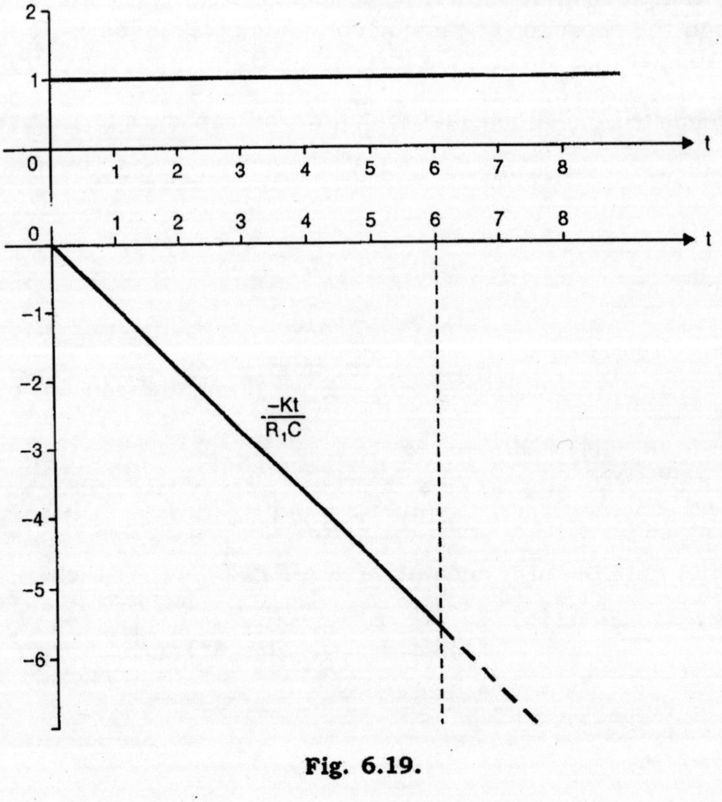

Fig. 6.19.

a case not possible practically. As described earlier too, the voltage can vary linearly only upto a certain permissible value, after that op-amp goes into saturation and a voltage of constant magnitude is appeared at the output terminal. It should be noted here that the product of R_1C controlles the slope of the output, *i.e.*, lower the product R_1C, steeper will be the ramp voltage.

For a sinusoidal input, output can be calculated as,

$$\because \quad v_{in}(t) = V_m \sin(\omega t)$$

$$\therefore \quad v_0(t) = -\frac{1}{R_1C}\int_0^t V_m \sin(\omega t)$$

$$= -\frac{V_m}{R_1C}\left[\frac{-\cos(\omega t)}{\omega}\right]_0^t$$

$$v_0(t) = \frac{V_m}{R_1C.\omega}(\cos\omega t - 1)$$

We clearly see here that output is 90° phase shifted to the input and integrated version of input.

If more than one inputs are applied at the input of integrator, Fig. 6.20, the output can be calculated by the superposition principle and given by eqn. (6.45), as,

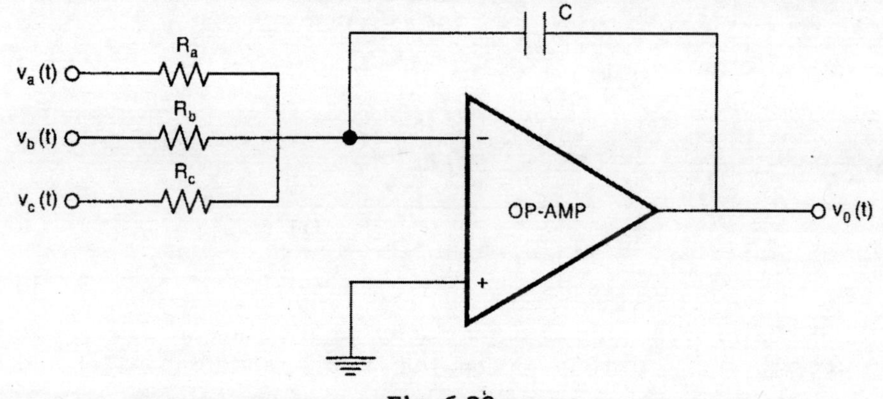

Fig. 6.20.

$$v_0(t) = -\left[\frac{1}{R_aC}\int v_a(t)dt + \frac{1}{R_bC}\int v_b(t)dt + \frac{1}{R_cC}\int v_c(t)dt\right] \quad ...(6.45)$$

The *main application* of integrator circuit is in *analog computers* to solve the *linear differential eqn.s with constant coefficients.* This circuit is also used in *Cathode Ray Oscilloscopes* (CROs) as a sweep generator and known there as *Miller sweep generator.*

6.11.7. Differentiator

If the position of resistor and capacitor is interchanged in the circuit of Fig. 6.18 circuit performs another interesting mathematical function, *i.e., differentiation* and known as *differentiator circuit.* This is also a non-linear amplifier and shown in Fig. 6.21.

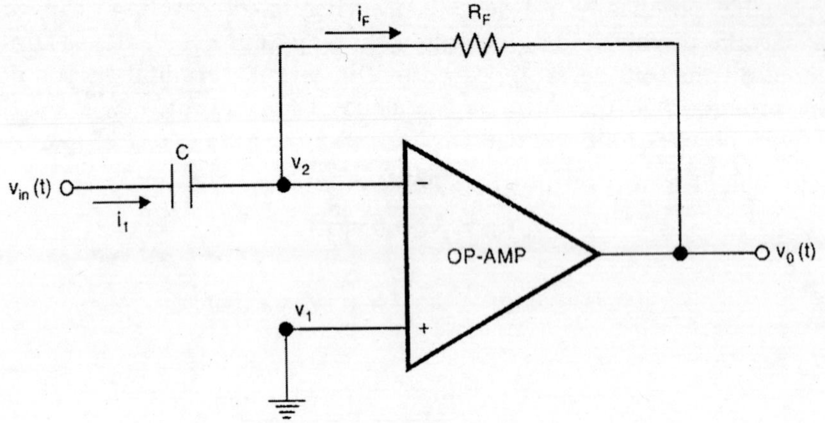

Fig. 6.21.

Using the concept of virtual ground we can write,

$$i_1 \simeq i_F$$

where, i_1 = Current through the capacitor

$$= C \frac{d}{dt} v_{in}(t)$$

and

$$i_F = -\frac{v_0(t)}{R_F}$$

Now, substituting these values we have,

$$-\frac{v_0(t)}{R_F} = C \frac{d}{dt} v_{in}(t)$$

i.e.,

$$\boxed{v_0 = -R_F C \frac{d\, v_{in}(t)}{dt}} \quad ...(6.46)$$

where, R_FC is the *time constant of differentiator* circuit and $-R_FC$ is known as *scale factor*. If a periodic square wave is applied, then to differentiate it R_FC should be much less than the full cycle time, *i.e.*, $R_FC << T$.

The application of differentiator circuit is in analog computers and to generate the short duration (edge) pulses for flip-flops (memory elements).

? SELF-TEST (OP-AMP)

1. Define op-amp.
2. What are the minimum number of terminals required in single op-amp ?
3. Which differential amplifier configurations are used in op-amp ?
4. Can op-amp amplify dc voltage ?

5. Which type of coupling is used between the various stages of op-amp and why ?

6. What are the main characteristics of an op-amp ?

7. Define CMRR and its significance.

8. Define Slew Rate.

9. Whether op-amp amplifies a large signal, small signal or both. Give the satisfactory conclusion.

10. What are the different application areas of op-amp ?

11. State the ideal characteristics of an operational amplifier and also the important mathematical operations of an OP-AMP can performs.

SOLVED EXAMPLES

Example 1. *For an op-amp, CMRR = 10^5 and differential gain $A_d = 10^5$. Determine the common mode gain A_{CM} of the op-amp.*

Solution : Given for an op-amp :

$$\text{CMRR} = 10^5$$

and
$$A_d = 10^5$$

Since, we know that,

$$\text{CMRR} = \frac{A_d}{A_{CM}}$$

$$\therefore \quad A_{CM} = \frac{A_d}{CMRR} = \frac{10^5}{10^5}$$

$$A_{CM} = 1. \quad \textbf{Ans.}$$

Example 2. *The output voltage of an op-amp circuit changes by 20 V in 4 μs. What is the slew rate ?*

Solution : Given :

Change in voltage, $\Delta V = 20$ V

Change in time, $\Delta t = 4$ μs

from the definition, **slew rate**

$$SR = \left.\frac{\Delta V}{\Delta t}\right|_{\max} = \frac{20 \text{ V}}{4 \text{ μs}}$$

$$SR = 5 \text{ V/μs}. \quad \textbf{Ans.}$$

Example 3. *The input voltage to an op-amp is a large voltage step. The output is an exponential waveform that changes 0.75 V in 50 ns. What is the* **slew rate** *of the op-amp ?*

Solution : Given :

$$\Delta V = 0.75 \text{ V}$$
$$\Delta t = 50 \text{ ns}$$

∵ From definition :

$$\text{Slew rate } (SR) = \left.\frac{\Delta V}{\Delta t}\right|_{max}$$

$$= \frac{0.75 \text{ V}}{50 \text{ ns}} = \frac{0.75 \text{ V}}{50 \times 10^{-3} \mu\text{s}}$$

$$\therefore \quad SR = 15 \text{ V}/\mu\text{s}. \quad \textbf{Ans.}$$

? SELF-TEST (INVERTING)

1. Why an open-loop op-amp is not suitable for linear applications ?
2. What is feedback ?
3. What are the advantages of negative feedback ?
4. Which type of feedback is suitable for oscillators and non-linear applications ?
5. What is virtual ground ?
6. What are the properties of virtual ground ?
7. Which type of feedback is used in inverting amplifier configurations ?
8. What will happen to closed loop gain A_0 if load resistor R_L is connected at the output terminal and why ?
9. Suppose, a voltage source of 2 V is connected at the non-inverting terminal. Is the concept of virtual ground is still valid ?
10. What is the formula for the gain of inverting amplifier ?

SOLVED EXAMPLES

Example 4. *Consider the circuit of Fig. 6.6. Determine the value of closed loop voltage gain and input impedance if* $R_1 = 1\ k\Omega$ *and* $R_F = 4.7\ k\Omega$.

Solution : Given for the inverting amplifier,

$$R_1 = 1 \text{ k}\Omega$$

and $$R_F = 4.7 \text{ k}\Omega$$

We know that,

closed loop voltage gain, $$A_0 = \frac{v_0}{v_{in}} = -\frac{R_F}{R_1}$$

$$= -\frac{4.7\text{ k}\Omega}{1\text{ k}\Omega}$$

$$A_0 = -4.7. \quad \textbf{Ans.}$$

and input impedance, $R_i = R_1$

$\Rightarrow$ $R_i = 1\text{ k}\Omega.$ **Ans.**

Example 5. *Consider the circuit in Fig. 6.6. Determine the values for R_1 and R_2 such that the closed loop gain is – 20 and the input impedance is 5 kΩ.*

Solution : Given for inverting amplifier configuration :

$$A_0 = -20$$

$$R_i = 5\text{ k}\Omega$$

$$R_1 = ?$$

$$R_F = ?$$

We know that, $R_i = R_1$

$\therefore$ $R_1 = 5\text{ k}\Omega.$ **Ans.**

Also, closed loop voltage gain for inverting amplifier is given as :

$$A_0 = -\frac{R_F}{R_1}$$

Substituting the values, $-20 = -\dfrac{R_F}{5\text{ k}\Omega}$

$\Rightarrow$ $R_F = 20 \times 5\text{ k}\Omega$

$\therefore$ $R_F = 100\text{ k}\Omega.$ **Ans.**

Example 6. *What output voltage results for an input of 1.25 V in the circuit of Fig. P (6.1).*

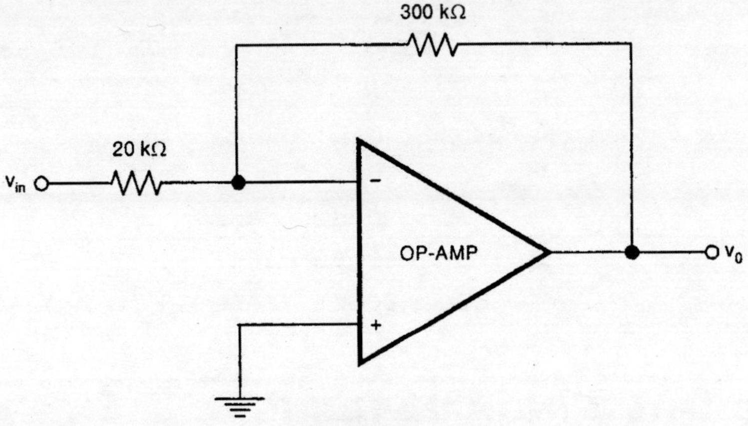

Fig. P (6.1).

Solution : Given in Fig. P (6.1),

$$R_1 = 20\text{ k}\Omega$$

$$R_F = 300\text{ k}\Omega$$

and $v_{in} = 1.25$ V

$v_0 = ?$

We know that,

$$A_0 = \frac{v_0}{v_{in}}$$

$$\therefore \quad v_0 = A_0 v_{in}$$

$$= -\frac{R_F}{R_1} v_{in}$$

$$\therefore \quad v_0 = -\frac{300}{20} \times 1.25$$

$$v_0 = -18.75 \text{ volt.} \quad \textbf{Ans.}$$

NOTE : This output voltage can be achieved only when the maximum voltage swing is greater than or equal to ± 19 V.

Example 7. *Determine the values of R_1 and R_F for an inverting amplifier configuration for which gain is – 4 V/V and the total resistance used is 100 kΩ.*

Solution : Given for an inverting amplifier circuit,

$$A_0 = -4 \text{ V/V}$$

$$R_1 + R_F = 100 \text{ k}\Omega \quad \text{...(1)}$$

$$R_1 = ?$$

$$R_F = ?$$

We have, $A_0 = -\frac{R_F}{R_1}$

$$\therefore \quad -4 = -\frac{R_F}{R_1}$$

$$\therefore \quad R_F = 4R_1 \quad \text{...(2)}$$

Substituting the value of R_F from eqn. (2) into eqn. (1).

$$R_1 + 4R_1 = 100$$

$$\therefore \quad R_1 = 20 \text{ k}\Omega. \quad \textbf{Ans.}$$

and $R_F = 4 \times 20$ kΩ

$$R_F = 80 \text{ k}\Omega. \quad \textbf{Ans.}$$

? SELF-TEST (NON-INVERTING)

1. Which type of feedback is used in non-inverting amplifier configuration ?

2. At which terminal, a part of output is feedback ?

3. What is bootstrapping ? How is it comes into non-inverting amplifier ?

4. What is virtual short ?
5. What is the significance of virtual short in non-inverting configuration ?
6. What is the basic difference between inverting and non-inverting amplifiers ?
7. Sketch the properties of an ideal OP-AMP.
8. With a neat circuit diagram, explain the operation of an adder using an ideal OP-AMP.
9. Draw the circuit of an OP-AMP voltage follower and find an expression for $\frac{v_0}{v_i}$.
10. Explain the following terms in connection with an OP-AMP :
    (*i*) Input offset voltage.
    (*ii*) Bias current.
    (*iii*) Slew rate.
11. What will happen if R_F is made zero in non-inverting amplifier ?
12. What is the effect of feedback at the input impedance of non-inverting amplifier ?
13. Give the formula for voltage gain of non-inverting amplifier.
14. What is the phase relationship between the input, output and feedback voltages in non-inverting amplifier configuration ?

SOLVED EXAMPLES

Example 8. *Consider the circuit of Fig. 6.12. What output voltage results in the circuit for an input voltage of – 0.25 V, if* R_1 *= 15 kΩ and* R_F *= 450 kΩ ?*

Solution : Given for a non-inverting amplifier :

$$R_1 = 15 \text{ k}\Omega$$

$$R_F = 450 \text{ k}\Omega$$

and

$$v_{in} = -0.25 \text{ V}$$

$$v_0 = ?$$

We have,

$$v_0 = A_0 \, v_{in}$$

where,

$$A_0 = 1 + \frac{R_F}{R_1}$$

so,

$$v_0 = \left(1 + \frac{R_F}{R_1}\right) v_{in}$$

Substituting the values, we get,

$$v_0 = \left(1 + \frac{450}{15}\right) (-0.25)$$

$$v_0 = 7.75 \text{ volts.}$$ **Ans.**

Example 9. *Design a circuit based on the topology of the non-inverting amplifier to obtain a gain of + 1.5 V/V, using only 10 kΩ resistors.*

Solution : Given for a non-inverting amplifier,

Desired closed loop gain, $A_0 = +1.5$ V/V

Available value of resistors, $R = 10$ kΩ

$R_1 = ?$

$R_F = ?$

We have,
$$A_0 = \left(1 + \frac{R_F}{R_1}\right)$$

$$1.5 = 1 + \frac{R_F}{R_1}$$

$$\therefore \quad \frac{R_F}{R_1} = 0.5$$

This relation of R_1 and R_F can be implemented in two ways, these are described below.

(*i*) $R_F = 0.5\ R_1$

If $R_1 = R = 10$ kΩ. **Ans.**

Then, $R_F = 0.5\ R_1$

$R_F = 5$ kΩ. **Ans.**

i.e., $R_F = R \parallel R$

The circuit diagram for this configuration is shown in Fig. P (6.2).

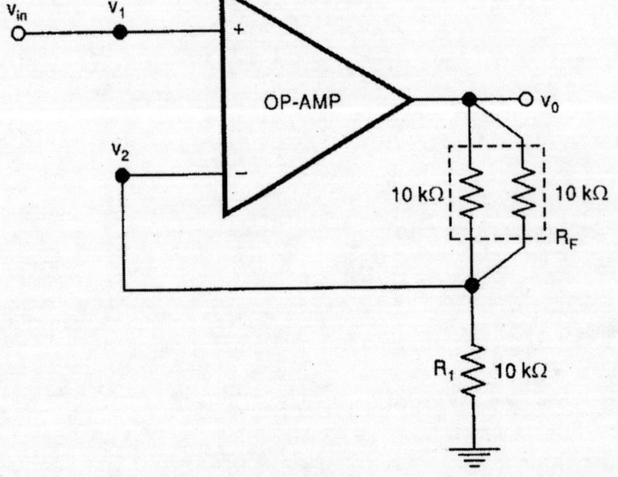

Fig. P (6.2).

(*ii*) $R_1 = 2\,R_F$

Now, If $R_F = 10\text{ k}\Omega.$ **Ans.**

Then, $R_1 = 2\,R_F$

$= 2\,R = R + R$

$\Rightarrow$ $R_1 = 20\text{ k}\Omega.$ **Ans.**

The circuit diagram for this configuration is shown in Fig. P (6.3).

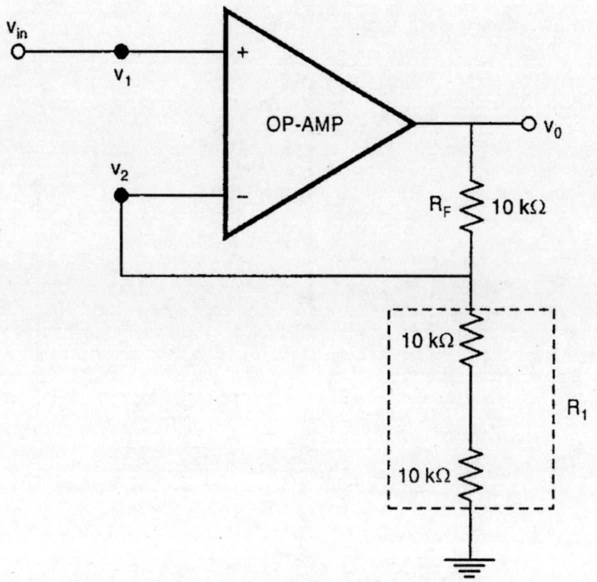

Fig. P (6.3).

Example 10. *Design a non-inverting amplifier circuit that is capable of providing a voltage gain of 10. Assume an ideal op-amp. (Resistor should not exceed 30 kΩ)*

Solution : Desired closed loop gain of non-inverting amplifier,

$$A_0 = 10$$

$$R_1 = ?$$

$$R_F = ?$$

We know that, $$A_0 = \left(1 + \frac{R_F}{R_1}\right)$$

$\therefore$ $$10 = 1 + \frac{R_F}{R_1}$$

or $$\frac{R_F}{R_1} = 9$$

or $$R_F = 9\,R_1$$

Now, according to the given condition, maximum value of resistor should not exceed 30 kΩ. According to above relation between R_F and R_1, we can select 27 kΩ resistor for R_F, *i.e.*,

$$R_F = 27 \text{ k}\Omega. \quad \textbf{Ans.}$$

and

$$R_1 = \frac{27}{9}$$

or

$$R_1 = 3 \text{ k}\Omega. \quad \textbf{Ans.}$$

Circuit diagram is shown in Fig. P (6.4).

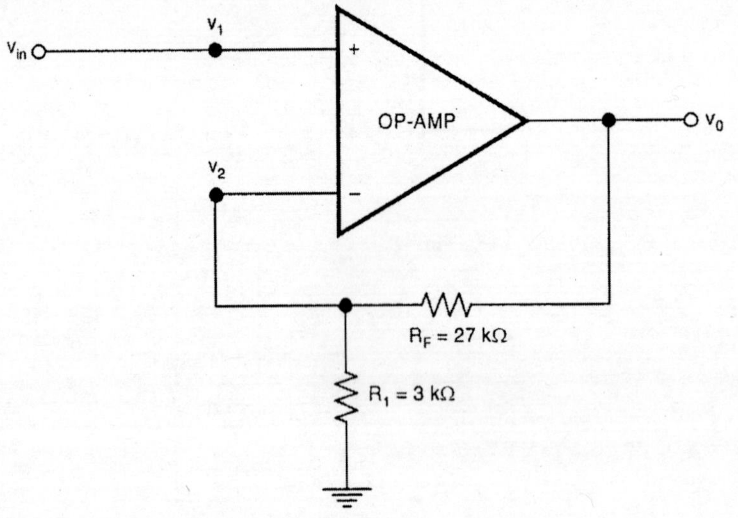

Fig. P (6.4).

SELF-TEST

1. List any two linear and non-linear applications of op-amp with the expression for output voltage ?
2. What are different applications of non-inverting amplifier configuration ?
3. What is main property of unity gain amplifier and how is it advantageous in coupling circuits ?
4. What is the significance of time constant *RC*, (product) of resistor and capacitor, in integrator and differentiator ?
5. What are the main application areas of integrators ?
6. What factors limit the maximum range of output voltage in op-amp configurations ?
7. What are the ideal characteristics of an ideal operational amplifier derive the expression for non-inverting amplifier ?

8. List six characteristics of the ideal OP-AMP. Draw the schematic diagram of an ideal non-inverting OP-AMP with voltage series feedback. Derive the expression for the voltage gain.
9. Draw the schematic diagram of an OP-AMP connected as
   (*i*) An inverter
   (*ii*) A scale changer
   (*iii*) A phase shifter
   (*iv*) An adder.
10. Draw the circuit of an OP-AMP integrator and explain its operation. What measure(s) are taken against dc-saturation of an OP-AMP ?
11. Draw the circuit diagram of an inverting integrator and determine its output equation.
12. With a neat sketch diagram, explain the operation of an adder using ideal OP-AMP.

SOLVED EXAMPLES

Example 11. *Calculate the output voltage of the circuit, shown in Fig. P (6.5).*

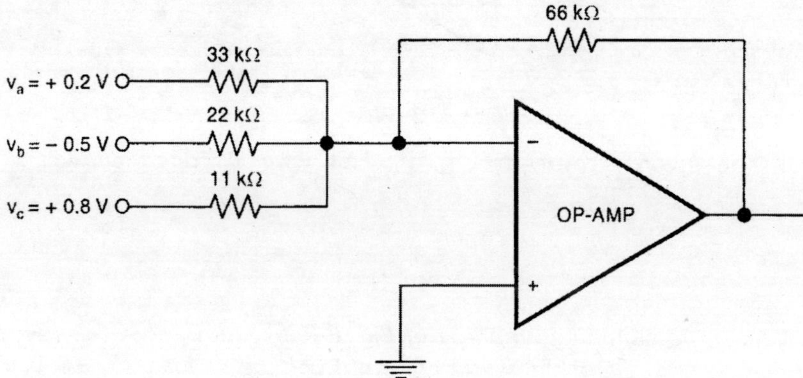

Fig. P (6.5).

Solution : Given in Fig. P (6.5), that,

$$v_a = +0.2 \text{ V}, \quad R_a = 33 \text{ k}\Omega$$
$$v_b = -0.5 \text{ V}, \quad R_b = 22 \text{ k}\Omega$$
$$v_c = +0.8 \text{ V}, \quad R_c = 11 \text{ k}\Omega$$
$$R_F = 66 \text{ k}\Omega$$
$$v_0 = ?$$

Using superposition principle, we can write that,

$$v_0 = -\frac{R_F}{R_a}v_a - \frac{R_F}{R_b}v_b - \frac{R_F}{R_c}v_c$$

$$= -\frac{66\text{ k}\Omega}{33\text{ k}\Omega} \times (0.2\text{ V}) - \frac{66\text{ k}\Omega}{22\text{ k}\Omega} \times (-0.5\text{ V}) - \frac{66\text{ k}\Omega}{11\text{ k}\Omega}(0.8\text{ V})$$

$$= -0.4\text{ V} + 1.5\text{ V} - 4.8\text{ V}$$

or $v_0 = -3.7$ V. **Ans.**

Example 12. *Calculate the output voltage for the circuit shown in Fig. P (6.6), for v_{in} = 5 sin (1000 t).*

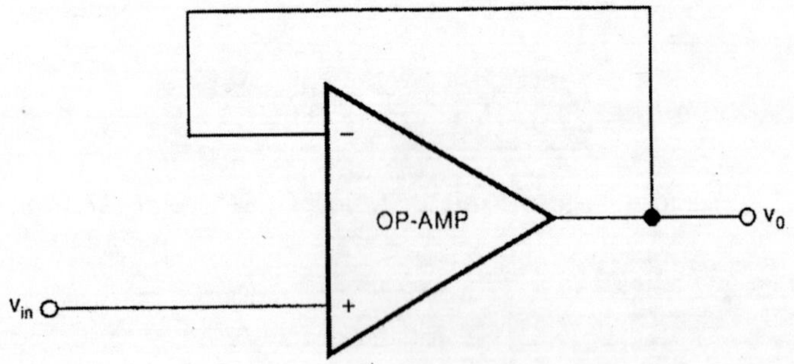

Fig. P (6.6).

Solution : The given circuit in Fig. P (6.6) is a *voltage follower circuit,* for which,

$$v_0 = v_{in}$$

Therefore, $v_0 = 5\sin(1000\,t)$. **Ans.**

Example 13 : *If the summing amplifier of Fig. P (6.5) has R_F = 10 kΩ, R_a = 10 kΩ, R_b = 2.5 kΩ and R_c = 4 kΩ, find the value of output voltage v_0 for v_a = + 6 V, v_b = – 3 V and v_c = – 0.75 V.*

Solution : Given for the summing amplifier in Fig. P (6.5).

$v_a = +6$ V, $R_a = 10$ kΩ

$v_b = -3$ V, $R_b = 2.5$ kΩ

$v_c = -0.75$ V, $R_c = 4$ kΩ

and $R_F = 10$ kΩ

$v_0 = ?$

We can write for the circuit in Fig. P (6.5),

$$v_0 = -\frac{R_F}{R_a}v_a - \frac{R_F}{R_b}v_b - \frac{R_F}{R_c}v_c$$

Substituting the values, we get,

$$v_0 = -\frac{10\ \text{k}\Omega}{10\ \text{k}\Omega} \times 6\ \text{V} - \frac{10\ \text{k}\Omega}{2.5\ \text{k}\Omega} \times (-3\ \text{V}) - \frac{10\ \text{k}\Omega}{4.0\ \text{k}\Omega} \times (-0.75\ \text{V})$$

$$= (-6 + 12 + 1.875)\ \text{V}$$

or $\quad v_0 = 7.875$ volts. **Ans.**

Example 14. *For the differential amplifier circuit, shown in Fig. P (6.7), calculate the closed loop differential gain and output voltage. Assume the op-amp is ideal.*

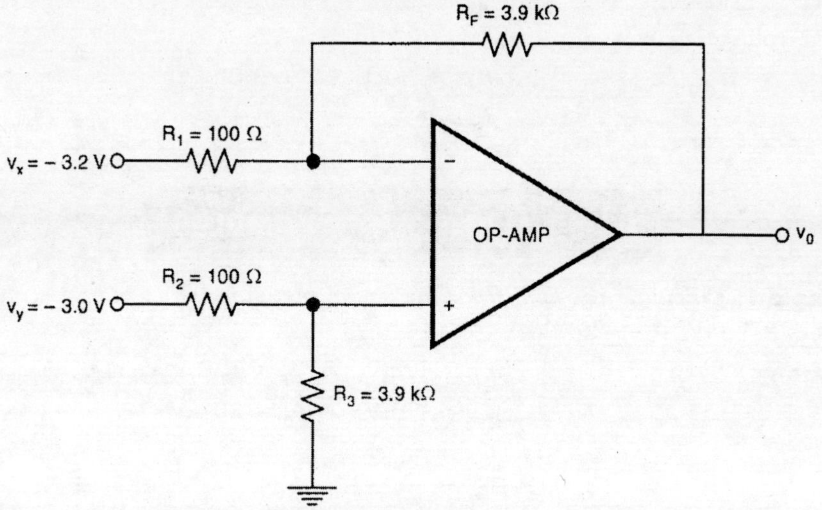

Fig. P (6.7).

Solution : Given for the differential amplifier circuit,

$$R_1 = R_2 = 100\ \Omega$$
$$R_F = R_3 = 3.9\ \text{k}\Omega$$
$$v_x = -3.2\ \text{V}$$
$$v_y = -3.0\ \text{V}$$
$$v_0 = ?$$
$$A_{0d} = ?$$

To find v_0, we use the superposition principle. Output due to v_x only, *i.e.*,

$$v_{0x} = -\frac{R_F}{R_1} v_x \qquad \text{...(1)}$$

and output due to v_y only, *i.e.*, $\quad v_{0y} = \frac{R_3}{(R_2 + R_3)}\left(1 + \frac{R_F}{R_1}\right) v_y \qquad \text{...(2)}$

Now, if $R_1 = R_2$ and $R_F = R_3$, overall output can be written as,

$$v_0 = v_{0x} + v_{0y}$$

$$= -\frac{R_F}{R_1}v_x + \frac{R_F}{R_1}v_y$$

or $$v_0 = -\frac{R_F}{R_1}(v_x - v_y)$$

$\Rightarrow$ Closed loop differential gain is,

$$A_{0d} = \frac{v_0}{(v_x - v_y)} = -\frac{R_F}{R_1} = -\frac{3.9\text{ k}\Omega}{100\ \Omega}$$

$$A_{0d} = -39. \quad \textbf{Ans.}$$

and output voltage v_0 is given as,

$$v_0 = -39\,(-3.2 + 3.0)$$

$$v_0 = 7.8\text{ Volt.} \quad \textbf{Ans.}$$

MISCELLANEOUS EXAMPLES

Example 1. *Calculate the CMRR for the circuit measurements of v_{id} = 1 mV, v_0 = 120 mV and v_{CM} = 1 mV, v_0 = 20 μV.*

Solution : Given : v_{id} = 1 mV, v_0 = 120 mV

and v_{CM} = 1 mV, v_0 = 20 μV

$\because$ $$CMRR = \frac{A_d}{A_{CM}}$$

where, $$A_d = \frac{v_0}{v_{id}} = \frac{120\text{ mV}}{1\text{ mV}}$$

$$A_d = 120$$

and $$A_{CM} = \frac{v_0}{v_{CM}} = \frac{20\ \mu\text{V}}{1\text{ mV}}$$

$$A_{CM} = 20 \times 10^{-3}$$

$\therefore$ $$CMRR = \frac{120}{20 \times 10^{-3}}$$

$$CMRR = 6 \times 10^3. \quad \textbf{Ans.}$$

Example 2. *Determine the closed loop gain of the circuit shown in Fig. P (6.8), and what range of output would you expect for the input of 30 mV ?*

Solution : Given for the inverting op-amp circuit in Fig. P (6.8),

$$R_1 = 10\text{ k}\Omega$$

$$R_F = 1000\text{ k}\Omega$$

$$A_0 = ?$$

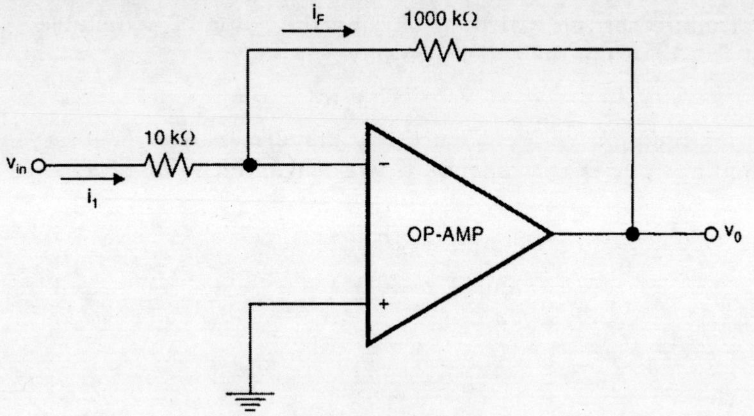

Fig. P (6.8).

We have from the concept of virtual ground,

$$\frac{v_{in}}{R_1} = -\frac{v_0}{R_F}$$

$$\Rightarrow \qquad \frac{v_0}{v_{in}} = A_0 = -\frac{R_F}{R_1}$$

i.e.,

$$A_0 = -\frac{1000\ \text{k}\Omega}{10\ \text{k}\Omega}$$

$$A_0 = 100\ \text{k}\Omega$$

and,

$$v_0 = A_0\, v_{in}$$

$$\Rightarrow \qquad v_0 = -100 \times 30\ \text{mV}$$

$$v_0 = -3\ \text{volts.}$$ **Ans.**

Example 3. *What is the range of the voltage-gain adjustment in the circuit of Fig. P (6.9).*

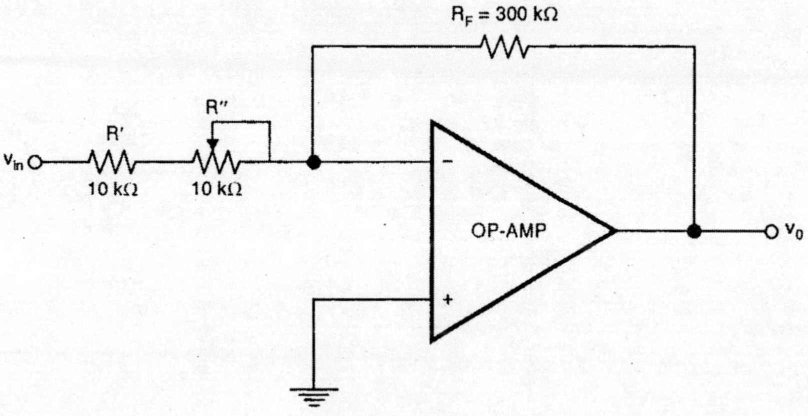

Fig. P (6.9).

Solution : It should be noted here that the value of input resistance, connected at inverting terminal can be varied by varying the value of adjustable resistance R''. Since R' and R'', both one in series, we have

$$R_1 = R' + R''.$$

R_1 has the minimum value, when the adjustable resistor R'' is set at $R'' = 0$ kΩ and the maximum value is corresponds to the maximum value of R'', *i.e.*, $R'' = 10$ kΩ, therefore,

$$R_1\,(\text{min}) = 10 \text{ k}\Omega$$

and

$$R_1\,(\text{max}) = 20 \text{ k}\Omega.$$

For inverting op-amp amplifier,

$$A_0 = -\frac{R_F}{R_1}$$

Now, closed loop voltage gain corresponding to R_1 (min) is,

$$A_0 = -\frac{300 \text{ k}\Omega}{10 \text{ k}\Omega}$$

or

$$A_0 = -30$$

and closed loop voltage gain corresponding to R_1 (max) is given as,

$$A_0 = -\frac{300 \text{ k}\Omega}{20 \text{ k}\Omega}$$

$$A_0 = -15.$$

Thus, the closed loop gain of the circuit can be adjusted at any value between – 15 to – 30 with the help of variable resistor R''.

Example 4. *What range of output voltage is developed by the circuit of Fig. P (6.10).*

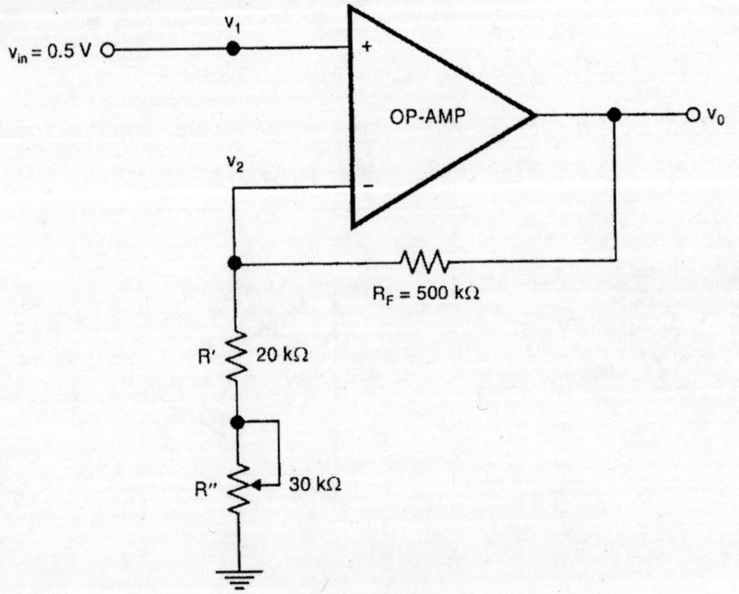

Fig. P (6.10).

Solution : Given for the non-inverting op-amp circuit of Fig. P (6.10),

$$R_F = 500\ \text{k}\Omega$$

$$R' = 20\ \text{k}\Omega$$

$$R'' = 0 - 30\ \text{k}\Omega \text{ (continuous variable)}$$

$$v_{in} = 0.5 \text{ volts}$$

$$v_0 = ?$$

In the circuit, R' and R'' makes a series circuit, whose effective (equivalent) resistance R_1 can be varied by adjusting the nobe of variable resistor. To find the range of output voltage, we consider the minimum and maximum gain of the circuit, which actually depends on the value of resistor R''.

Therefore, for $R'' = 0$, $R_1\,(\text{min}) = 20\ \text{k}\Omega$

and for $R'' = 30\ \text{k}\Omega$, $R_1\,(\text{max}) = 50\ \text{k}\Omega$

Since, for non-inverting configuration, we have,

$$A_0 = \frac{v_0}{v_{in}} = \left(1 + \frac{R_F}{R_1}\right)$$

$$\Rightarrow \qquad v_0 = v_{in}\left(1 + \frac{R_F}{R_1}\right)$$

$\therefore$ output voltage corresponding to R_1 (min),

$$v_0 = 0.5\left(1 + \frac{500\ \text{k}\Omega}{20\ \text{k}\Omega}\right)$$

$$v_0 = 13 \text{ volts}$$

and output voltage corresponding to R_1 (max),

$$v_0 = 0.5\left(1 + \frac{500\ \text{k}\Omega}{50\ \text{k}\Omega}\right)$$

$$v_0 = 5.5 \text{ volts}$$

Therefore, output voltage can be continuously varied between 5.5 Volts to 13 Volts by adjusting the variable resistance 30 kΩ to 0 kΩ, respectively.

Example 5. *The adjustable resistor of Fig. P (6.11) can be varied from 0 to 100 kΩ. Calculate the minimum and maximum closed-loop voltage gain. Assume ideal op-amp.*

Solution : Given for the inverting amplifier configuration,

$$R_1 = 2\ \text{k}\Omega$$

$$R' = 2\ \text{k}\Omega$$

$$R'' = 0 - 100\ \text{k}\Omega \text{ (continuous variable)}$$

$$A_0\,(\text{min}) = ?$$

$$A_0\,(\text{max}) = ?$$

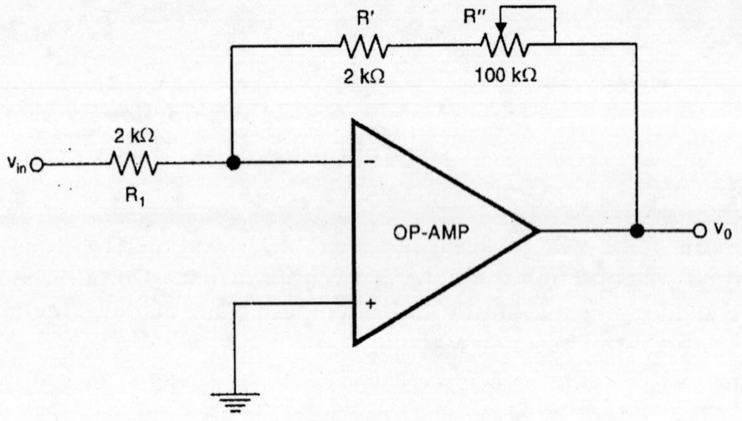

Fig. P (6.11).

It reveals from the Fig. P (6.11), that resistors R' and R'' makes the series circuit and give the effective value of feedback resistor R_F. Therefore,

$$R_F\,(\text{min})\big|_{R''\,=\,0\ \text{k}\Omega} = 2\ \text{k}\Omega$$

and
$$R_F\,(\text{max})\big|_{R''\,=\,100\ \text{k}\Omega} = 102\ \text{k}\Omega$$

As we have, closed-loop gain of inverting amplifier,

$$A_0 = -\frac{R_F}{R_1}$$

which implies that closed-loop voltage gain is directly proportional to the feedback resistor R_F. Hence,

$$A_0\,(\text{min}) = -\frac{R_F\,(\text{min})}{R_1}$$

$$= -\frac{2\ \text{k}\Omega}{2\ \text{k}\Omega}$$

$$A_0\,(\text{min}) = -1.$$ **Ans.**

and
$$A_0\,(\text{max}) = -\frac{R_F\,(\text{max})}{R_1}$$

$$= -\frac{102\ \text{k}\Omega}{2\ \text{k}\Omega}$$

$$A_0\,(\text{max}) = -51.$$ **Ans.**

Therefore, closed-loop voltage gain varies from – 1 to – 51 by adjusting resistor R'' in the range of 0 to 100 kΩ.

Example 6. *Calculate the maximum and minimum closed-loop voltage gain for the non-inverting amplifier shown in Fig. P (6.12). Assume ideal op-amp.*

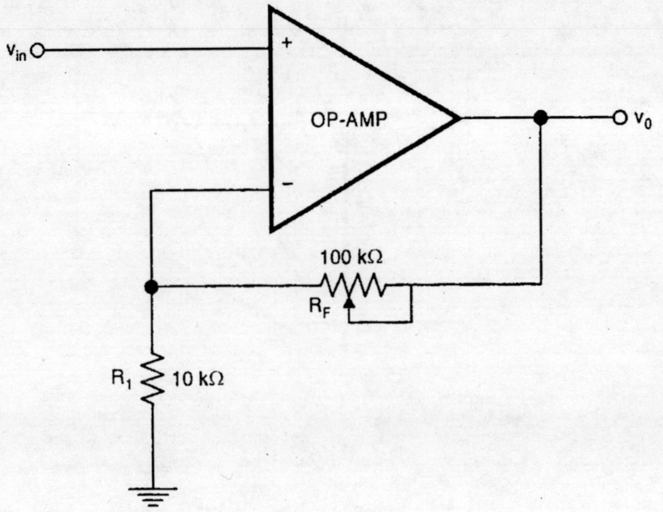

Fig. P (6.12).

Solution : Given for the non-inverting amplifier,

$$R_F = 100 \text{ k}\Omega \text{ (adjustable)}$$

$$R_1 = 10 \text{ k}\Omega \text{ (fixed)}$$

$$A_0 = ?$$

As we know, for non-inverting amplifier,

$$A_0 = \left(1 + \frac{R_F}{R_1}\right)$$

which implies that,

$$A_0\,(\text{min}) = \left.\left(1 + \frac{R_F}{R_1}\right)\right|_{R_F\,=\,0\text{ k}\Omega}$$

$$A_0\,(\text{min}) = 1. \quad \textbf{Ans.}$$

and

$$A_0\,(\text{max}) = \left.\left(1 + \frac{R_F}{R_1}\right)\right|_{R_F\,=\,100\text{ k}\Omega}$$

$$= \left(1 + \frac{100 \text{ k}\Omega}{10 \text{ k}\Omega}\right)$$

$$A_0\,(\text{max}) = 11. \quad \textbf{Ans.}$$

Thus, the closed-loop voltage gain of non-inverting amplifier can be adjusted to any value between + 1 and + 11, by varying the value of R_F between 0 – 100 kΩ .

Example 7. *In Fig. P (6.13), what is the closed-loop voltage gain for each switch position ? Assume ideal op-amp.*

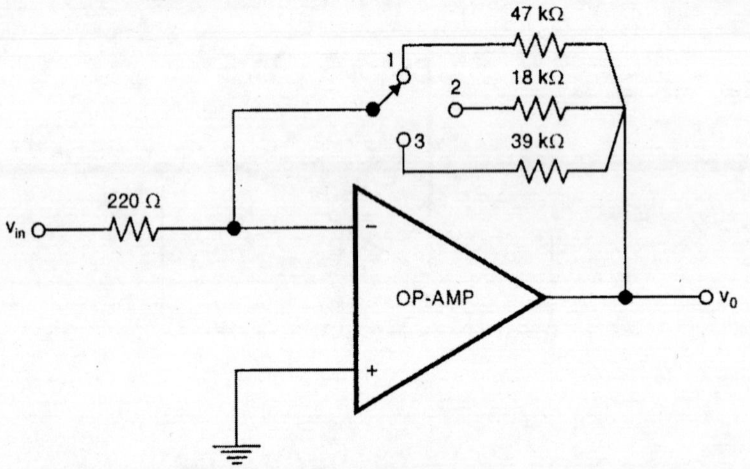

Fig. P (6.13).

Solution : As we know that the closed-loop voltage gain for an inverting amplifier is given as,

$$A_0 = -\frac{R_F}{R_1}$$

and it reveals from the circuit that for each switch position the value of feedback resistor is changed.

(*a*) **For switch position-1 :**

$$R_1 = 220\ \Omega$$

and $R_F = 47\ \text{k}\Omega$

hence, $A_0 = -\dfrac{47\ \text{k}\Omega}{220\ \Omega}$

or $A_0 = -213.64.$ **Ans.**

(*b*) **For switch position-2 :**

$$R_1 = 220\ \Omega$$

and $R_F = 18\ \text{k}\Omega$

hence, $A_0 = -\dfrac{18\ \text{k}\Omega}{220\ \Omega}$

or $A_0 = -81.82.$ **Ans.**

(*c*) **For switch position-3 :**

$$R_1 = 220\ \Omega$$

and $R_F = 39\ \text{k}\Omega$

hence, $$A_0 = -\frac{39\ \text{k}\Omega}{220\ \Omega}$$

or $$A_0 = -177.27. \quad \textbf{Ans.}$$

Example 8. *What is the closed-loop voltage gain for each switch position of Fig. P (6.14)? Assume ideal op-amp is used in circuit. (Please see Fig. P (6.14))*

Solution : Given for the non-inverting amplifier circuit in Fig. P (6.14),

$$R_F = 120\ \text{k}\Omega$$

$$R_1' = 6\ \text{k}\Omega$$

$$R_1'' = 3\ \text{k}\Omega$$

op-amp is ideal.

$$A_0 = ?$$

We know that, $$A_0 = 1 + \frac{R_F}{R_1}$$

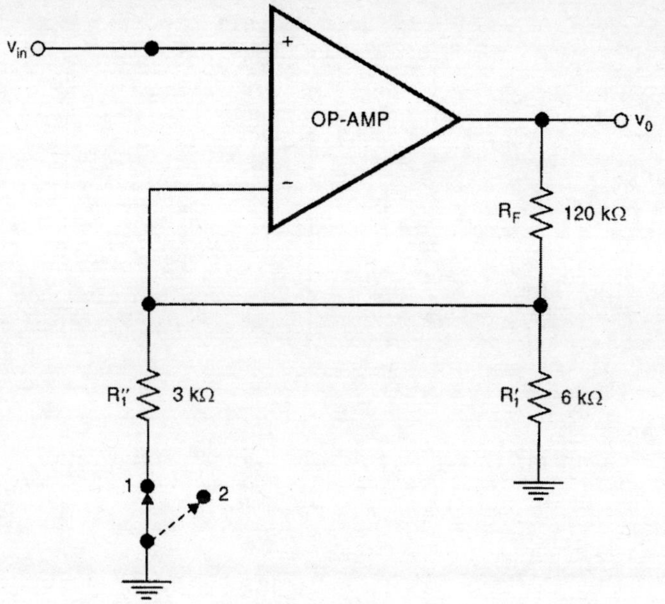

Fig. P (6.14).

(*a*) **For switch position-2**

Since no current flows through resistor R_1'' due to open circuit. Therefore, only R_F and R_1' contributes in the voltage divider network. Therefore,

$$R_1 = R_1' = 6\ \text{k}\Omega$$

$$\therefore \quad A_0 = 1 + \frac{120\ \text{k}\Omega}{6\ \text{k}\Omega}$$

$$A_0 = 21. \quad \textbf{Ans.}$$

(*b*) **For switch position-1**

Here, as the switch takes its position at 1, the path is completed through R''_1 and hence R''_1 is in parallel now. Therefore,

$$R_1 = R'_1 \parallel R''_2$$

$$= 6 \parallel 3 = \frac{6 \times 3}{6+3}$$

$$\therefore \quad R_1 = 2 \text{ k}\Omega$$

and closed loop gain,

$$A_0 = \left(1 + \frac{120 \text{ k}\Omega}{2 \text{ k}\Omega}\right)$$

$$\therefore \quad A_0 = 61. \quad \textbf{Ans.}$$

Example 9. *Determine output voltage v_o, for the circuit shown in Fig.* P (6.15).

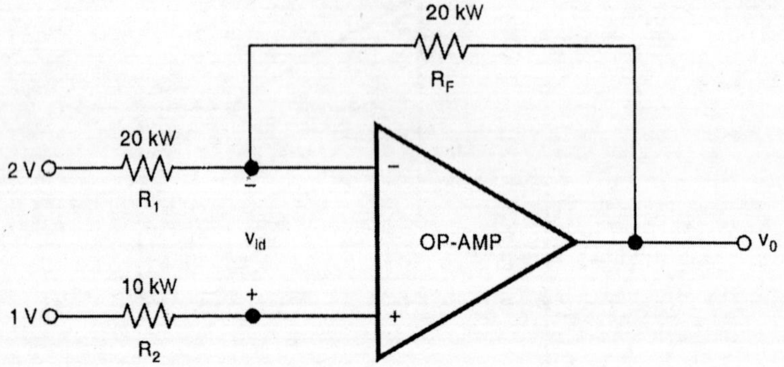

Fig. P (6.15).

Solution : Given for the differential amplifier configuration,

$$R_F = 20 \text{ k}\Omega$$
$$R_1 = 20 \text{ k}\Omega$$
$$R_2 = 10 \text{ k}\Omega$$
$$v_{in_1} = 2 \text{ V}$$
$$v_{in_2} = 1 \text{ V}$$
$$v_0 = ?$$

Since, the op-amp is ideal, current through resistor R_2 is zero and hence all the voltage (v_{in_2}) is applied at the non-inverting terminal.

By using superposition principle, we can write,

$$v_0 = -\frac{R_F}{R_1} v_{in_1} + \left(1 + \frac{R_F}{R_1}\right) v_{in_2}$$

$$= -\frac{20\ \text{k}\Omega}{20\ \text{k}\Omega} \times 2 + \left(1 + \frac{20\ \text{k}\Omega}{20\ \text{k}\Omega}\right) \times 1$$

$$= -2 + 2$$

$$v_0 = 0. \quad \textbf{Ans.}$$

Example 10. *For the circuit in Fig. P (6.16), calculate the voltage, produced at output terminal, i.e., v_0. (Please see Fig. P (6.16))*

Solution : Given for the closed-loop differential amplifier configuration,

$$R_F = 60\ \text{k}\Omega$$

$$R_1 = 20\ \text{k}\Omega$$

$$v_{in_1} = 2\ \text{V}$$

$$v_{in_2} = 0.1\ \text{V}$$

$$v_0 = ?$$

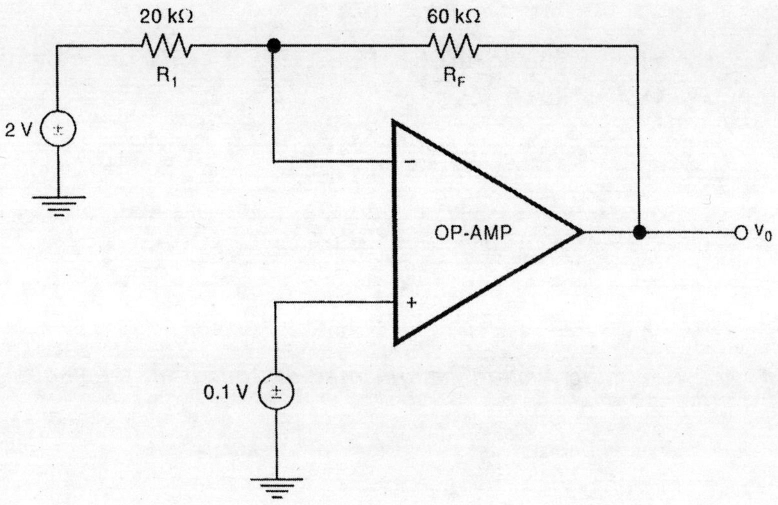

Fig. P (6.16).

By using superposition principle,

$$v_0 = -\frac{R_F}{R_1} v_{in_1} + \left(1 + \frac{R_F}{R_1}\right) v_{in_2}$$

Substituting the values, $$v_0 = -\frac{60\ \text{k}\Omega}{20\ \text{k}\Omega} \times 2 + \left(1 + \frac{60\ \text{k}\Omega}{20\ \text{k}\Omega}\right) \times 0.1$$

$$= -6 + 0.4$$

or $$v_0 = -5.6 \text{ volts.} \quad \textbf{Ans.}$$

Example 11. *Calculate the output voltage, v_0 for the circuit shown in Fig. P (6.17). Assume, op-amp is ideal.*

Given : *$R_1 = 10\ \text{k}\Omega$, $R_2 = 20\ \text{k}\Omega$, $R_3 = 10\ \text{k}\Omega$ and $R_F = 20\ \text{k}\Omega$.*

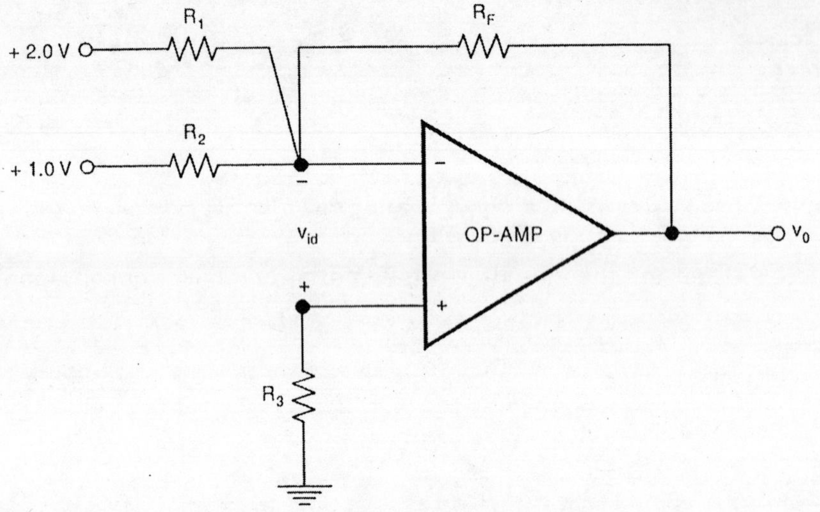

Fig. P (6.17).

Solution : The circuit shown in Fig. P (6.17), performs the summing amplifier, therefore, output voltage is given by,

$$v_0 = -\frac{R_F}{R_1} \times v_{in_1} - \frac{R_F}{R_2} \times v_{in_2}$$

$$v_0 = -\frac{20\text{ k}\Omega}{10\text{ k}\Omega} \times 2 - \frac{20\text{ k}\Omega}{20\text{ k}\Omega} \times 1$$

$$v_0 = -4 - 1$$

or $$v_0 = -5 \text{ volt.}$$ **Ans.**

Example 12. *How much voltage is appeared at the output terminal in Fig. P (6.18)? Assume, ideal op-amp.*

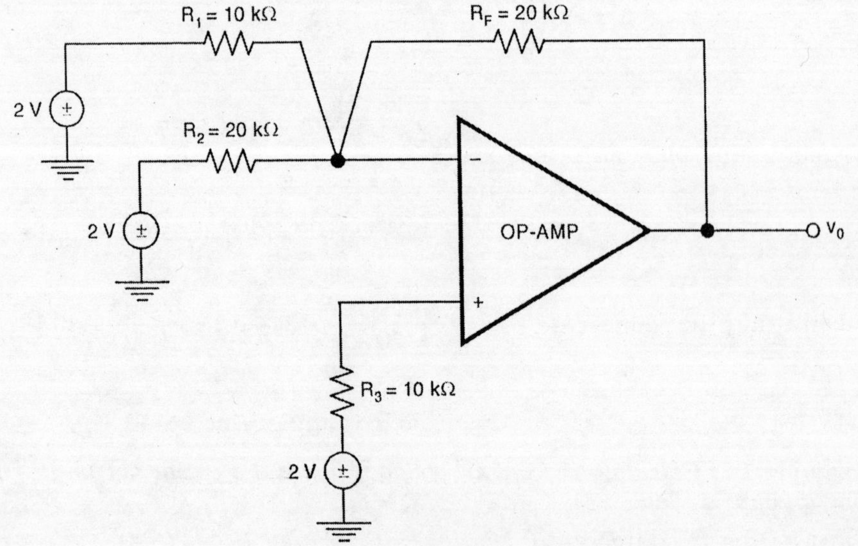

Fig. P (6.18).

Solution : For the differential amplifier configuration, we can write by superposition principle,

$$v_0 = -\frac{R_F}{R_1} v_{in_1} - \frac{R_F}{R_2} v_{in_2} + 1 + \frac{R_F}{(R_1 \| R_2)} v_{in_3}$$

Substituting the values from circuit diagram,

$$v_0 = -\frac{20 \text{ k}\Omega}{10 \text{ k}\Omega} \times 2 - \frac{20 \text{ k}\Omega}{20 \text{ k}\Omega} \times 2 + 1 + \frac{20 \text{ k}\Omega}{(10||20) \text{ k}\Omega} \times 2$$

$$= -4 - 2 + 8$$

$\therefore \quad v_0 = 2$ volt. **Ans.**

Example 13. *What is the output voltage, v_o in Fig. P (6.19) ? Assume, op-amp is ideal.*

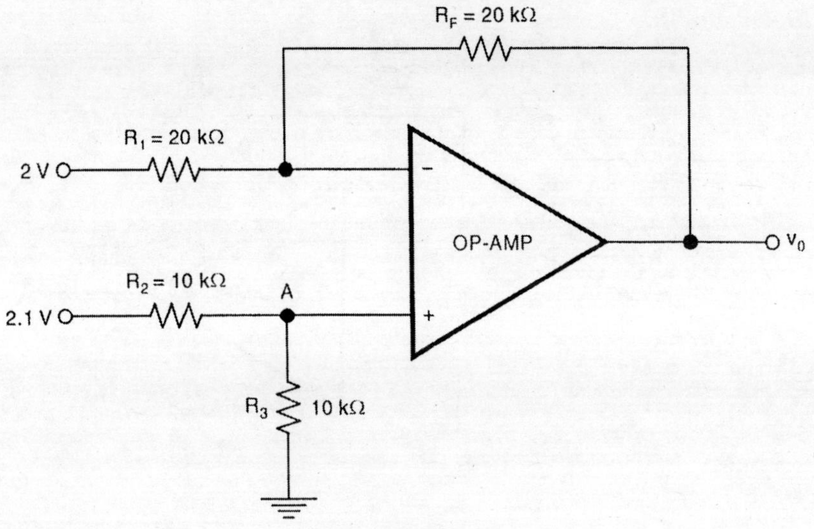

Fig. P (6.19).

Solution : Given for the differential amplifier configuration,

$$R_1 = 20 \text{ k}\Omega$$

$$R_2 = R_3 = 10 \text{ k}\Omega$$

$$R_F = 20 \text{ k}\Omega$$

$$v_{in_1} = 2 \text{ volt}$$

$$v_{in_2} = 2.1 \text{ volt}$$

The input voltage at the non-inverting terminal,

$$V_A = \frac{10 \times 2.1}{20} \quad \text{(Voltage divider rule)}$$

$$= 1.05 \text{ volt.}$$

Now, by using superposition principle, we get,

$$v_0 = -\frac{R_F}{R_1} v_{in_1} + \left(1+\frac{R_F}{R_1}\right)\left(\frac{R_3}{R_2+R_3}\right) v_{in_2}$$

Substituting the values,

$$v_0 = -\frac{20\ \text{k}\Omega}{20\ \text{k}\Omega} \times 2 + \left(1+\frac{20\ \text{k}\Omega}{20\ \text{k}\Omega}\right) 1.05$$

$$= -2 + 2\ (1.05)$$

$$\therefore \quad v_0 = 0.1 \text{ volt.} \quad \textbf{Ans.}$$

Example 14. *Sketch the circuit of summer using OP-AMP to get* $V_0 = -v_1 + 2v_2 - 3v_3$.

Solution : We know that following is the circuit for the inverting summing amplifier shown in Fig., P (6.20).

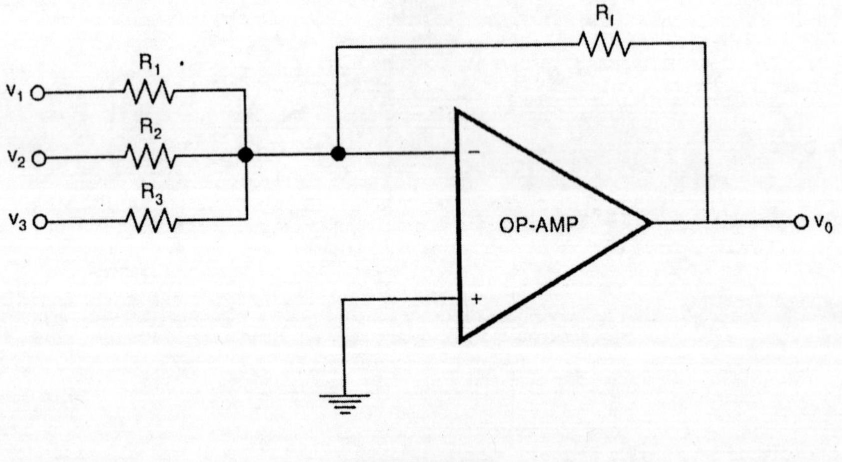

Fig. P (6.20).

Here the output :

$$v_0 = \frac{-R_f}{R_1} v_1 - \frac{R_f}{R_2} v_2 - \frac{R_f}{R_3} v_3$$

or

$$v_0 = -R_f\left(\frac{v_1}{R_1} + \frac{v_2}{R_2} + \frac{v_3}{R_3}\right)$$

But in equation we need v_2 positive so polarity inversion is required for v_2. So, following circuit shown in Fig. P (6.21) can work for the required equation.

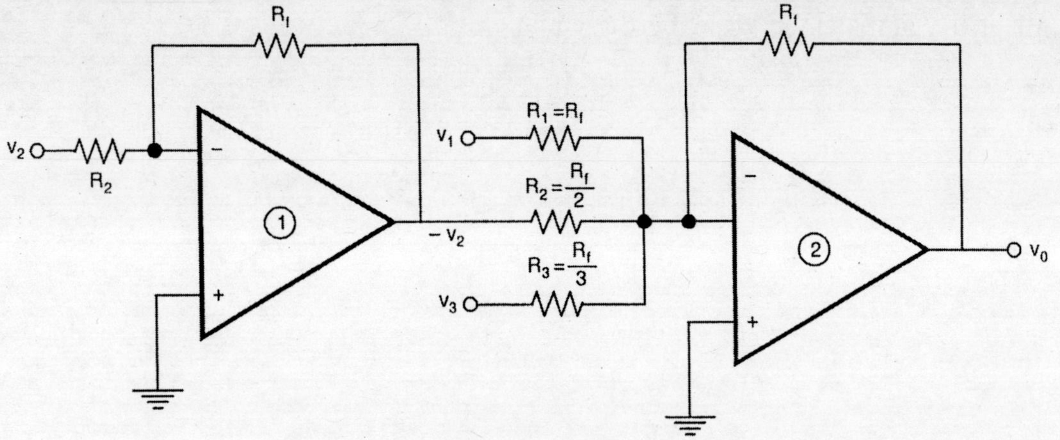

Fig. P (6.21).

Here in Fig. P (6.21) OP-AMP first works as an inverting buffer so the output of first OP-AMP will be $-v_2$. The second OP-AMP works as inverting summing amplifier. It's output will be,

$$v_0 = -R_f\left[\frac{v_1}{R_1}+\frac{(-v_2)}{R_2}+\frac{v_3}{R_3}\right]$$

by setting $R_1 = R_f;$

$$R_2 = \frac{R_f}{2};$$

$$R_3 = \frac{R_f}{3};$$

We get, $v_0 = -v_1 + 2v_2 - 3v_3$. **Ans.**

Example 15. *The variable resistance varies from 0 to 100 K ohm, find out the maximum loop voltage gain as shown in Fig. P (6.22).*

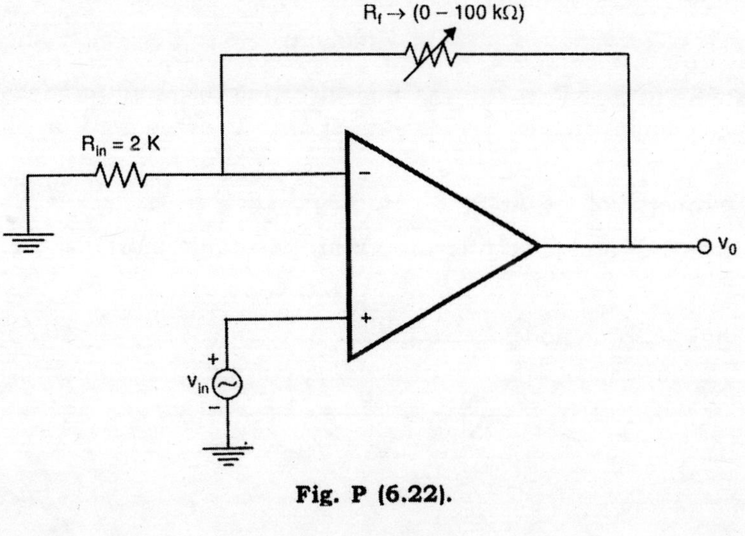

Fig. P (6.22).

Solution : The given circuit is a non-inverting amplifier using OP-AMP. Here the output of the amplifier is given by the equation.

$$v_0 = \left(1 + \frac{R_f}{R_{in}}\right) v_{in}$$

and voltage gain $(A_v) = \dfrac{v_0}{v_{in}} = 1 + \dfrac{R_f}{R_{in}}$

When $R_f = 0$

$(A_v)_{min} = 1$

When $R_f = 100$ K

$(A_v)_{max} = 1 + \dfrac{100}{2} = 51.$ **Ans.**

Example 16. *For the circuit shown in Fig. P (6.23), find* $\dfrac{v_0}{v_{in}}$.

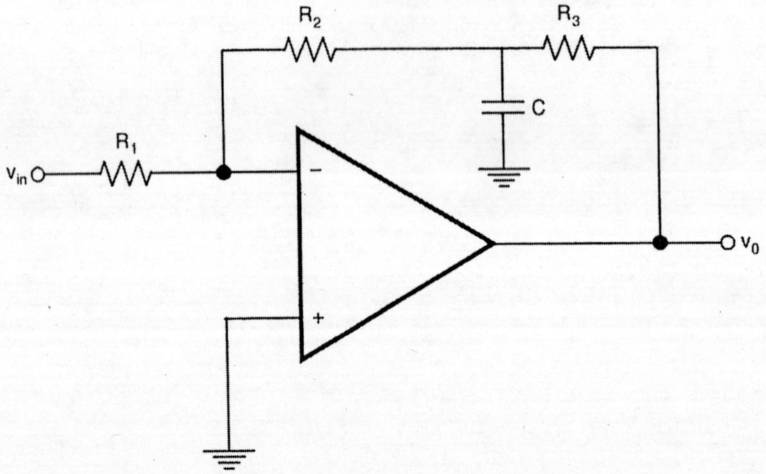

Fig. P (6.23).

Solution : As in Fig. P (6.23), the non-inverting terminal at the zero potential so, the potential at point A in Fig. P (6.24) as shown below is zero (apply the concept of virtual ground).

i.e., from the property of OP-AMP.

Its non-inverting and inverting inputs are virtually short.

$$v_A = v_N = 0 \text{ volt}$$

On applying KCL at node A, we get,

$$\frac{v_{in} - v_A}{R_1} = \frac{v_A - v_B}{R_2}$$

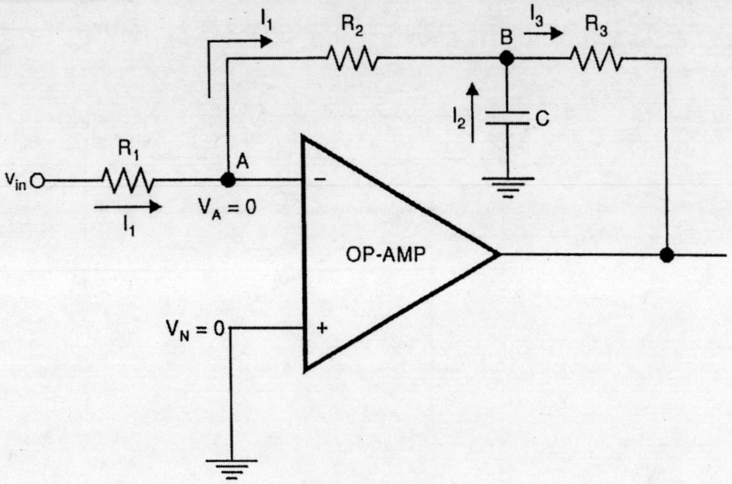

Fig. P (6.24).

as $$v_A = 0$$

so $$\frac{v_{in}}{R_1} = -\frac{v_B}{R_2}$$

or $$v_B = -\frac{R_2}{R_1}\cdot v_{in} \quad ...(1)$$

Again applying KCL at node B, we get,

$$I_1 + I_2 + I_3 = 0$$

$$\frac{0 - v_B}{R_2} + \frac{0 - v_B}{X_C} + \frac{v_0 - v_B}{R_3} = 0 \qquad \left\{\text{where } X_C = \frac{1}{j\omega c}\right\}$$

or $$-\left(\frac{v_B}{R_2} + \frac{v_B}{X_C} + \frac{v_B}{R_3}\right) = -\frac{v_0}{R_3}$$

or $$v_B\left(\frac{1}{R_2} + \frac{1}{R_3} + \frac{1}{X_C}\right) = \frac{v_0}{R_3} \quad ...(2)$$

From eqns. (1) and (2), we have,

$$\left(-v_{in}\frac{R_2}{R_1}\right)\left(\frac{1}{R_2} + \frac{1}{R_3} + \frac{1}{X_C}\right) = \frac{v_0}{R_3}$$

or $$\frac{v_0}{v_{in}} = -\frac{R_2 R_3}{R_1}\left[\frac{R_2 + R_3 + R_2 R_3 / X_C}{R_2 R_3}\right]$$

or $$\frac{v_0}{v_{in}} = -\frac{1}{R_1}[R_2 + R_3 + R_2 R_3/X_C].$$ **Ans.**

Example 17. *For an integrator shown in Fig. P (6.25), find v_0.*

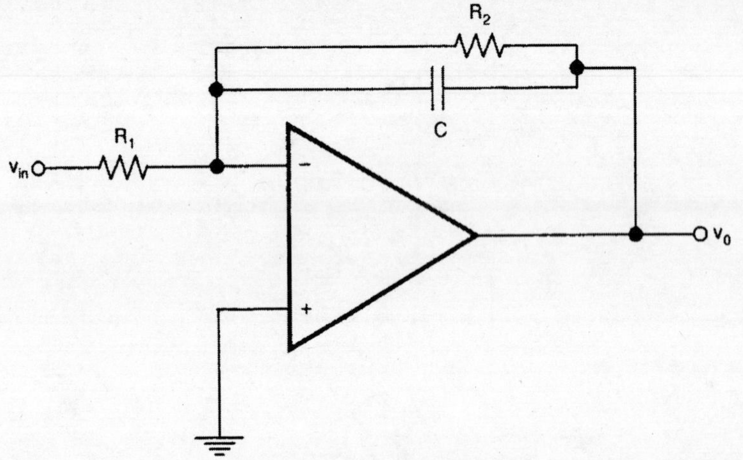

Fig. P (6.25).

Solution : For easy to understand, circuit is redrawn and different current and voltages are assumed in Fig. P (6.26).

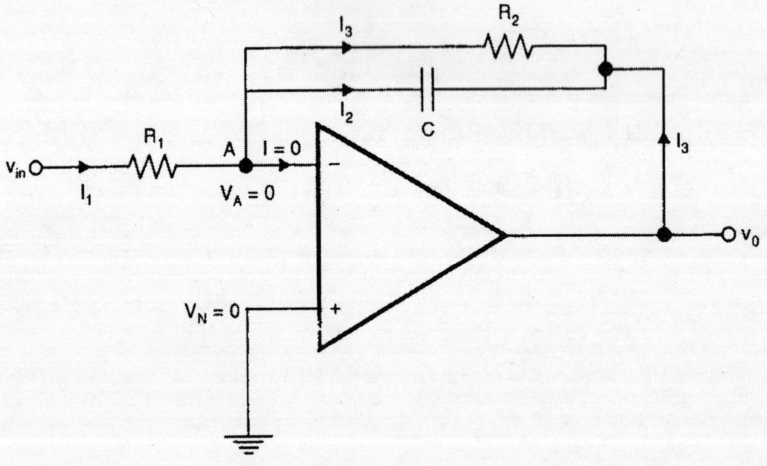

Fig. P (6.26).

From Fig. P (6.26), we see that v_A = 0 volt due to virtual short with non-inverting terminal, as current into inverting terminal is zero.

On applying KCL at node A, we get,

$$I_1 = I_2 + I_3$$

$$\frac{v_{in} - v_A}{R_1} = \frac{v_A - v_0}{X_C} + \frac{v_A - v_0}{R_2} \quad ...(1) \quad \left\{ \begin{aligned} \text{where } X_c &= \frac{1}{JwC} \\ \text{or } X_c &= \frac{1}{J2\pi fC} \end{aligned} \right\}$$

But $v_A = 0$, so eqn. (1) reduces

$$\frac{v_{in}}{R_1} = \frac{-v_0}{X_C} - \frac{v_0}{R_2}$$

or
$$v_0 = v_{in}\left[\frac{1}{R_1\left(\frac{1}{X_C}+\frac{1}{R_2}\right)}\right]$$

or
$$v_0 = v_{in}\left[\frac{R_2}{R_1(1+J(2\pi f R_2C))}\right].$$ **Ans.**

Example 18. *The two input terminals of an OP-AMP are connected to voltage signals of strength 745 μV and 740 μV respectively. The gain of the OP-AMP in differential mode is 5 × 10⁵ and its CMRR is 80 dB. Calculate the output voltage and percentage error due to common mode.*

Solution : We know that the CMRR in dB is given by the relation,

$$\text{CMRR (dB)} = 20 \log_{10}\left(\frac{A_d}{A_{CM}}\right)$$

where
A_d = differential mode gain
A_{CM} = common mode gain

given
CMRR = 80 dB
$A_d = 5 \times 10^5$
$v_1 = 745\ \mu\text{V}$
$v_2 = 740\ \mu\text{V}$

Now,
$$80 = 20 \log_{10}\left(\frac{5\times10^5}{A_{CM}}\right)$$

or
$$\log_{10}\left(\frac{5\times10^5}{A_{CM}}\right) = 4$$

or
$$\frac{5\times10^5}{A_{CM}} = 10^4$$

or
$$A_{CM} = \frac{5\times10^5}{10^4} = 50$$

so, output in differential mode gain

$$v_{0(d)} = A_d(v_1 - v_2)$$

or
$$v_{0(d)} = 5 \times 10^5 (745 \times 10^{-6} - 740 \times 10^{-6})$$

or
$$v_{0(d)} = 2.5 \text{ volts}$$

output due to common mode gain

$$v_{0\,(CM)} = A_{CM}\left(\frac{v_1 + v_2}{2}\right)$$

or $$v_{0\,(CM)} = 50\left(\frac{745\times10^{-6} + 740\times10^{-6}}{2}\right)$$

or $$v_{0\,(CM)} = 25 \times 1485 \times 10^{-6}$$

or $$v_{0\,(CM)} = .037125 \text{ volt.}$$

Now, $$\%\ \text{error} = \frac{v_{0(CM)}}{v_{0(d)}} \times 100$$

or $$\%\ \text{error} = \frac{.037125}{2.5} \times 100 = 1.48\%. \quad \textbf{Ans.}$$

Example 19. *Compute the input impedance, voltage gain and power gain of the amplifier circuit shown in Fig. P (6.27). Assume the OP-AMP is ideal.*

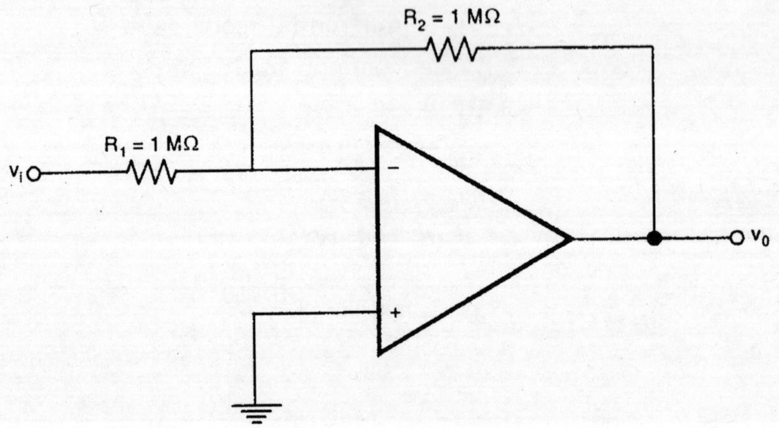

Fig. P (6.27).

Solution : As the OP-AMP is ideal its inverting and non-inverting inputs will be short circuited. So, node A will be at virtual short.

So, input impedance of the circuit will be,

$$R_{in} = R_1 = 1\ \text{M}\Omega. \quad \textbf{Ans.}$$

Since, this is inverting amplifier hence voltage gain,

$$A_{vf} = \frac{-R_2}{R_1} = \frac{-1\ \text{M}\Omega}{1\ \text{M}\Omega} = -1. \quad \textbf{Ans.}$$

Example 20. *Calculate the output voltage for the following circuits.*

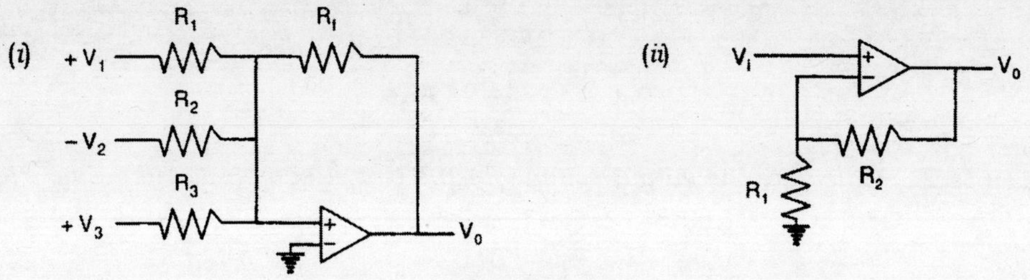

Fig. P (6.28).

Solution : (*i*)

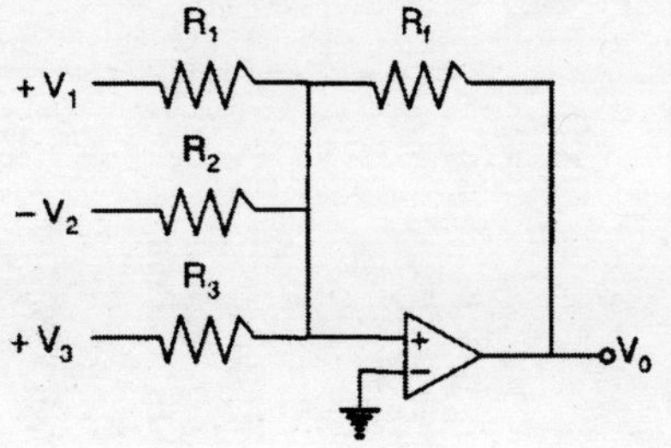

Fig. P (6.29).

Applying KCL at junction we get,

$$-\frac{V_2}{R_2} + \frac{V_1}{R_1} + \frac{V_3}{R_3} = \frac{V_0}{R_f}$$

or

$$V_0 = R_f\left[\frac{V_1}{R_1} + \frac{V_3}{R_3} - \frac{V_2}{R_2}\right]$$

(*ii*)

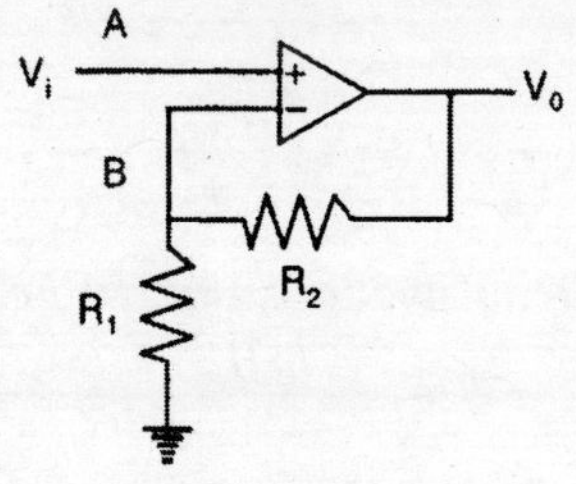

Fig. P (6.30).

Negative terminal of op-amp have voltage $\approx V_i$

So we get,

$$\frac{V_i}{R_1} = \frac{V_0 - V_i}{R_2}$$

$$V_i\left(\frac{1}{R_1} + \frac{1}{R_2}\right) = \frac{V_0}{R_2}$$

$$\frac{V_0}{V_i} = R_2\left(\frac{1}{R_1} + \frac{1}{R_2}\right)$$

or $$V_0 = V_i R_2\left(\frac{1}{R_1} + \frac{1}{R_2}\right) \quad \textbf{Ans.}$$

Example 21. *Calculate the output voltage for the circuit shown below in Fig. P(6.31).*

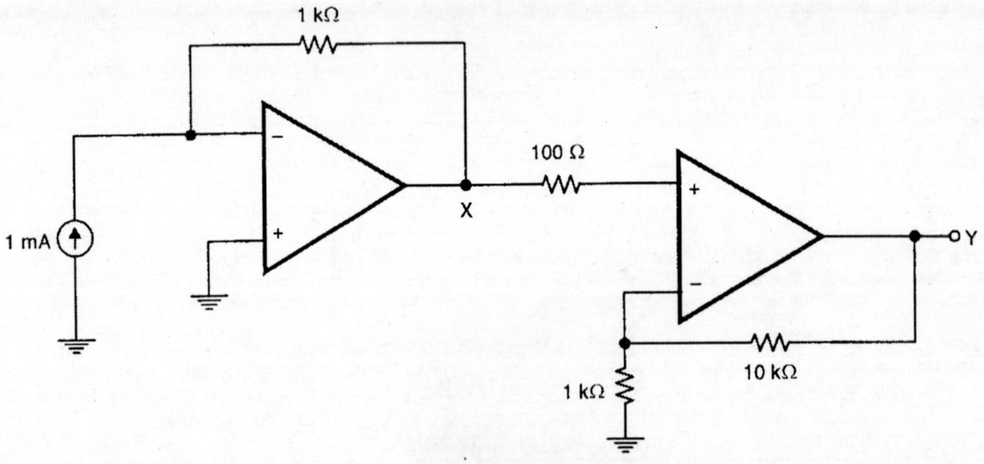

Fig. P (6.31).

Solution : The input current for first op-amp = 1 mA

Now the output voltage at first op-amp stage

$$= -i_1 R_F = -1 \text{ mA} \times 1 \text{ k}\Omega$$
$$= -1 \text{ volt}$$

Here this – ve value of one voltage is applied to + ve terminal of the second stage of op-amp *i.e.*,

$$V_{g_1} = -1\left(1+\frac{10}{1}\right) = -11 \text{ volts.} \quad \textbf{Ans.}$$

Example 22. *The adder circuit shown below in Fig. P (6.32). Calculate output voltage v_0. ($R_{S_1} = R_{S_2} = R_{S_3} = 1$ kΩ and $R_f = 10$ kΩ).*

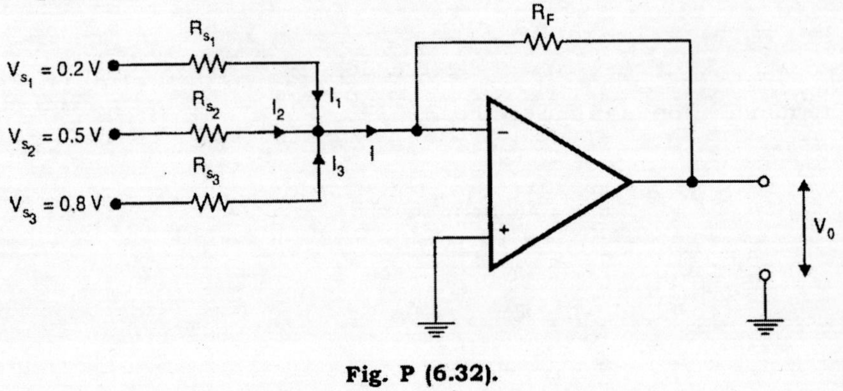

Fig. P (6.32).

Solution : For this circuit shown in Fig. P (6.32), we have

$$R_{S_1} = R_{S_2} = R_{S_3} = 1 \text{ k}\Omega$$

$$R_f = 10 \text{ k}\Omega$$

Now the current is above circuit is

$$I = I_1 + I_2 + I_3$$

$$\Rightarrow \quad I = \frac{V_{S_1}}{R_{S_1}} + \frac{V_{S_2}}{R_{S_2}} + \frac{V_{S_3}}{R_{S_3}}$$

Also $\quad I = -I_f \quad$ and $\quad V_0 = -I_f R_f$

$$\Rightarrow \quad V_0 = -R_F\left(\frac{V_{S_1}}{R_{S_1}} + \frac{V_{S_2}}{R_{S_2}} + \frac{V_{S_3}}{R_{S_3}}\right)$$

$$\therefore \quad V_0 = \frac{-R_F}{R_{S_1}}[V_{S_1} + V_{S_2} + V_{S_3}] \qquad \{\because R_{S_1} = R_{S_2} = R_{S_3}\}$$

Using the values given, we get

$$V_0 = -\frac{10 \text{ k}\Omega}{1 \text{ k}\Omega}[0.2 + 0.5 + 0.8]$$

$$\therefore \quad V_0 = 1.5 \text{ V} \quad \textbf{Ans.}$$

But the supply voltage of 10 V is used, so the op-amp will reach is saturation. Therefore, output voltage is – 10 volt.

Example 23. *An op-amp set up consists of Z_S which has R is parallel with capacitor 'C' and $Z_f = R_f$. The input voltage $V_i = kt$. Assume $A_i = \infty$, find V_0.*

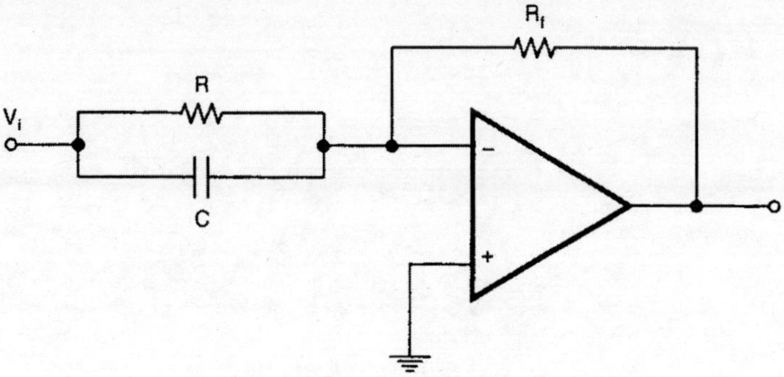

Fig. P (6.33)

Solution : Here we have

$$V_i = kt, \qquad A_i = \text{gain} = \infty$$

$$Z_S = r \times \frac{1}{S_C} = \frac{r}{1 + rC_S}$$

$$\Rightarrow \qquad Z_S = \frac{1}{R + \frac{1}{S_C}}$$

Now,
$$V_0 = -\frac{R_f}{Z_S} \cdot V_i$$

$$\therefore \qquad V_0 = -\frac{R_F}{1 + RC_S} \cdot kt. \qquad \textbf{Ans.}$$

Example 24. *If an input of $V_i = kt$ is applied to as integrator. Calculate the output voltage.*

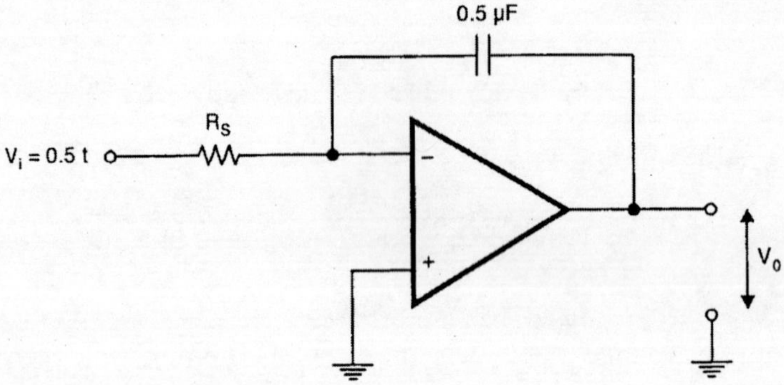

Fig. P (6.34).

Solution : We know that for as integrator

$$V_0 = -\frac{1}{RSCP}\int V_i.dt$$

$$= -\frac{1}{1 \times 10^6 \times 0.5 \times 10^{-6}}\int 0.5\, t\, dt$$

$$= -\frac{1.0}{0.5}\left(0.5\frac{t^2}{2}\right)$$

$$\therefore \qquad V_0 = -0.5\, t^2. \qquad \textbf{Ans.}$$

Example 25. *Find out the current through R_L resistor for the circuit shown in Fig. P (6.35).*

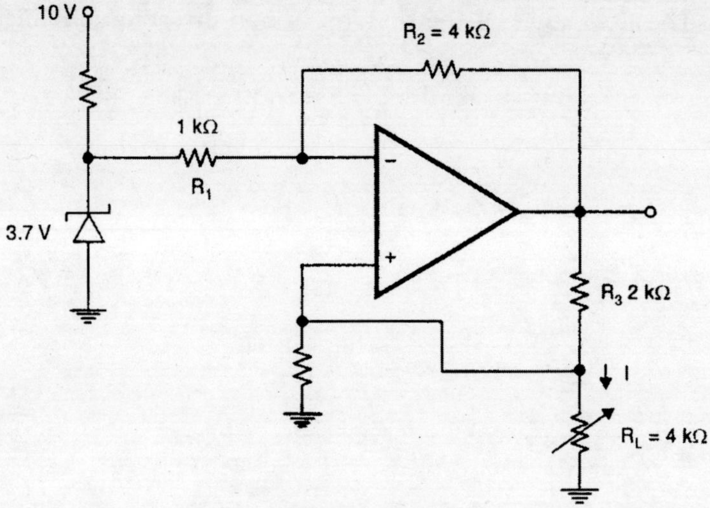

Fig. P (6.35)

Solution : Here from the above Fig. P (6.35), we have

$$\text{Ratio of } \frac{R_2}{R_1} = \frac{R_4}{R_3} = 4$$

$$\text{and Current through } R_L = -\frac{V_i}{R_3} \qquad \{\because V_i = 3.7 \text{ V}\}$$

$$I_L = -\frac{3.7}{2 \text{ k}} = -1.85 \text{ mA.} \qquad \textbf{Ans.}$$

Example 26. *Find the output voltage waveform at the output for the circuit shown below in Fig. P (6.36).*

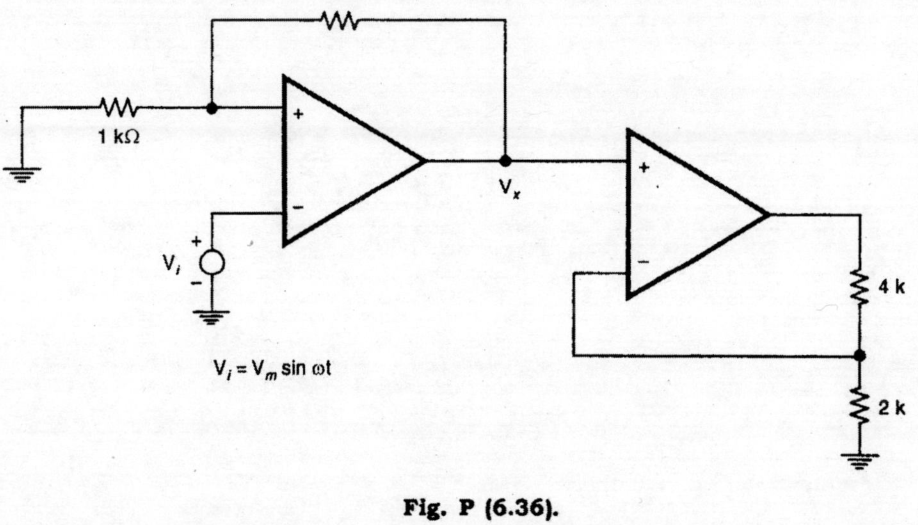

Fig. P (6.36).

Solution : The first stage of the op-amp is non-inverting amplifier stage. Here,

$$V_x = V_i\left(1+\frac{R_f}{R_S}\right) = V_i\left(1+\frac{1}{1}\right)$$

$\Rightarrow$ $$V_x = 2V_i$$

$\therefore$ $$V_x = 2\ V_m \sin \omega t$$

Now the second stage is as $\frac{V}{I}$ *i.e.,*

$$I_2 = \frac{V_x}{R_S} = \frac{2\ V_m \sin \omega t}{2\ \text{k}\Omega}$$

$$I_2 = V_m \sin \omega t \text{ mA.}$$ **Ans.**

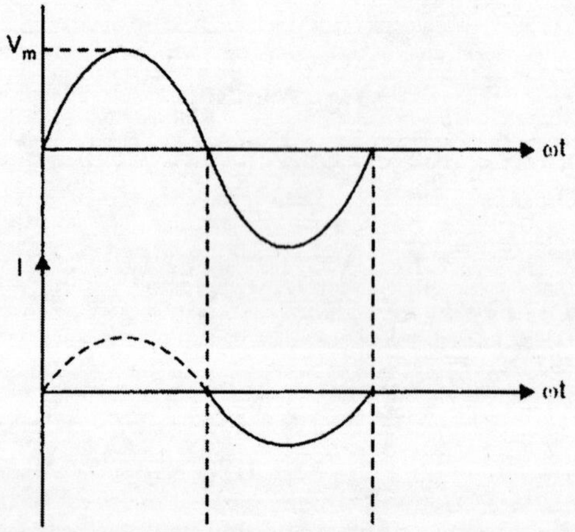

Fig. P (6.37).

□□□

UNIT—IV

(Chapters 7, 8 and 9)

At the end of this unit you will able to learn about the :

- ➢ JFET and its characteristics
- ➢ MOSFET and its characteristics
- ➢ SCR
- ➢ TRIAC
- ➢ UJT
- ➢ Applications of various power electronics devices.

7

JUNCTION FIELD EFFECT TRANSISTOR

7.1. INTRODUCTION

In the previous unit, we have studied about Bipolar Junction (BJT). We discussed several configurations and operating regions alongwith concerned biasing techniques. The forthgoing unit describes another class of transistors, the field-effect transistors abbreviated as FET. The FET is also a three terminal device. It has got numerous applications in electronic circuits like BJT. There are several similarities and dissimilarities between FET and BJT. Constructional features, operating mechanism and performance characteristics differ greatly from those of BJTs.

The main difference between BJT and FET is that former has got two junctions (J_E and J_C) but later has no any junction. Because of this advantage, several qualities have been enhanced in the transistors which were lacking in BJT. Another interesting difference between the BJT and FET is that a bipolar junction transistor is a current-current device. That is why, the output characteristics of this device are controlled by the base current and not by the base voltage. While the FET is a voltage-controlled device. That is why, the ouput characteristics of FET are controlled by input voltage and not by the input current. In this unit, we will study about the FET and BJT comparatively and will be able to know why and how FET replaced the BJT in most specific applications.

7.2. CLASSIFICATION OF FIELD-EFFECT DEVICES

Depending upon the constructional features and operating principles, FETs broadly classified in two main types field-effect transistors.

1. Junction field-effect transistor (JFET)
2. Metal oxide semiconductor field-effect transistor (MOSFET) or insulated gate field-effect transistor (IGFET).

Both of these type further divided into *n*-channel and *p*-channel. For better understanding about the classification tree form is shown in Fig. 7.1.

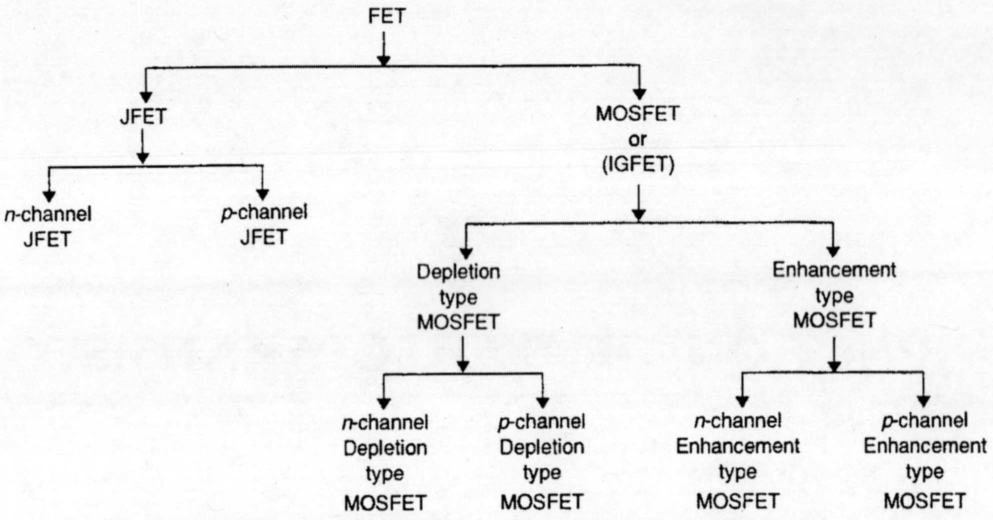

Fig. 7.1. *Different types of field effect transistors.*

7.3. COMPARISON BETWEEN BJT AND FET

The basic differences between BJT and FET is that the former is a current-controlled device and later is a voltage-controlled device. In the case of BJT the prefix 'Bi' comes out from the fact that a BJT have both types of charge carriers *i.e.*, electrons and holes, but in the of a FET, either electrons or holes constitute the drain current, so named as unipolar transistor. Depending upon the type of charge carriers, FETs have been classified as *n*-channel or *p*-channel FET, discussed later or in other words we can say that the operation of FET depends upon the flow of majority carries only. There are a large number of advantages of FET over BJT. But, the main disadvantage of FET is its relatively small gain bandwidth-product in comparison with that which can be obtained with a conventional bipolar transistor.

Fig. 7.2 illustrates the difference between BJT and FET.

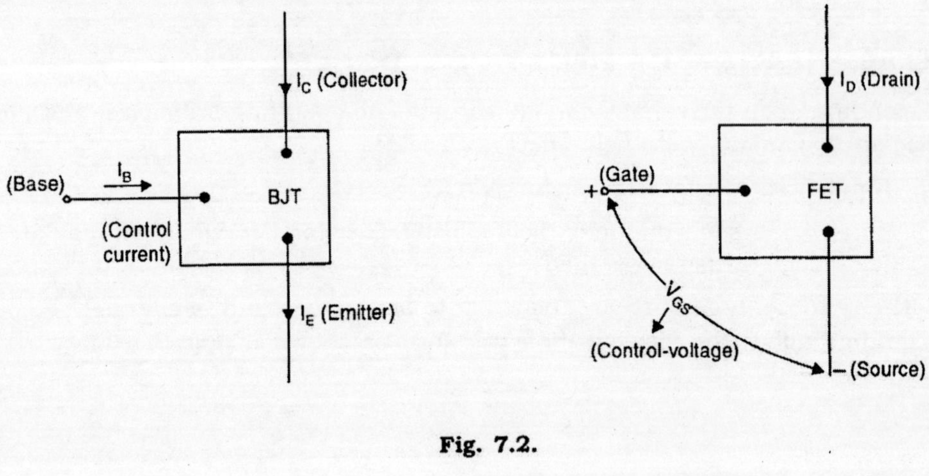

Fig. 7.2.

Following are the few differences between BJT and FET which must be known to the students before discussing FET's operation in detail.

FET	***BJT***
1. FET is an unipolar semiconductor device because its operation depends upon the flow of majority carriers *i.e.*, either holes or electrons as the case may be.	1. BJT is a bipolar semiconductor device because the current constituting elements are both majority carriers as well as minority carriers in this case.
2. The input impedance of FET is much more larger (ranging in Megaohms) than BJT. The reason behind this is that the input terminal *i.e.*, gate to source of FET is reverse biased and reverse bias offers ideally infinite resistance.	2. The input impedance of BJT is very less in comparison to FET.
3. FET is a voltage controlled device.	3. BJT is a current controlled device.
4. FET is less noisy. Because there are no junctions.	4. Much noisy than FET.
5. Higher frequency response.	5. Frequency variation affects the performance.
6. Good thermal stability because of absence of minority carriers.	6. Temperature dependent, thermal runaway may cause.
7. Costlier than BJT.	7. Relatively cheaper.
8. Small sized.	8. Comparatively bigger.
9. In FET, relationship between input and output quantities is nonlinear due to square term in shockley's equations $I_D = I_{DSS}\left(1 - \frac{V_{GS}}{V_P}\right)^2$	9. The BJT is an almost linear device or we can say that BJT works linearly in active region as an amplifier.
10. No offset voltage; so it works better as a switch or chopper.	10. There is always an offset voltage before switching.
11. Small gain bandwith product.	11. Greater than FET.

7.4. STRUCTURE AND OPERATION OF JFET OR FET

The junction field-effect transistors (JFET's) can be divided depending upon their structure into the following two categories :

(*i*) *n*-channel JFET

(*ii*) *p*-channel JFET

The basic construction of an *n*-channel JFET is as shown in Fig. 7.3(*a*). It consists of an *n*-type semiconductor substrate (here substrate is nothing but a very lightly doped bar which is used as channel) with two *p*-type heavily doped regions diffused on opposite sides of its middle part. The *p*-type regions form two *pn*-junctions. The space between the junctions (*i.e.*, *n*-type region) is called a channel. Both the *p*-type regions are connected internally and a single wire is taken out in the form of a terminal called the gate (*G*). The electrical connections are made to both ends of the *n*-type semiconductor and are taken out in the form of two terminals called drain (*D*) and source (*S*).

Thus the typical structure of a junction field-effect transistor consists of four basic elements :

(*a*) **Source.** Referring to Fig. 7.3 given below the source *S* is the terminal through which the majority carriers enter the bar. Conventional current entering the bar at *S* is designated by I_S.

(*b*) **Drain.** The drain *D* is the terminal through which the majority carriers leave the bar. Conventional current entering at the bar at *D* is designated by I_D. The drain to source voltage V_{DS} is positive when drain is more positive than source.

(*c*) **Gate.** On both sides of the *n*-type bar, heavily droped *p* regions of acceptor impurities have been formed by alloying or by diffusion, or by any other means. These impurity regions are called the gate *G* and between the gate and source a reverse bias voltage $-V_{GS}$ is applied. The conventional current entering the bar at *G* is designated by I_G.

(*d*) **Channel.** The region of *n*-type material between two gate regions is the channel through which the majority carriers move from source to drain.

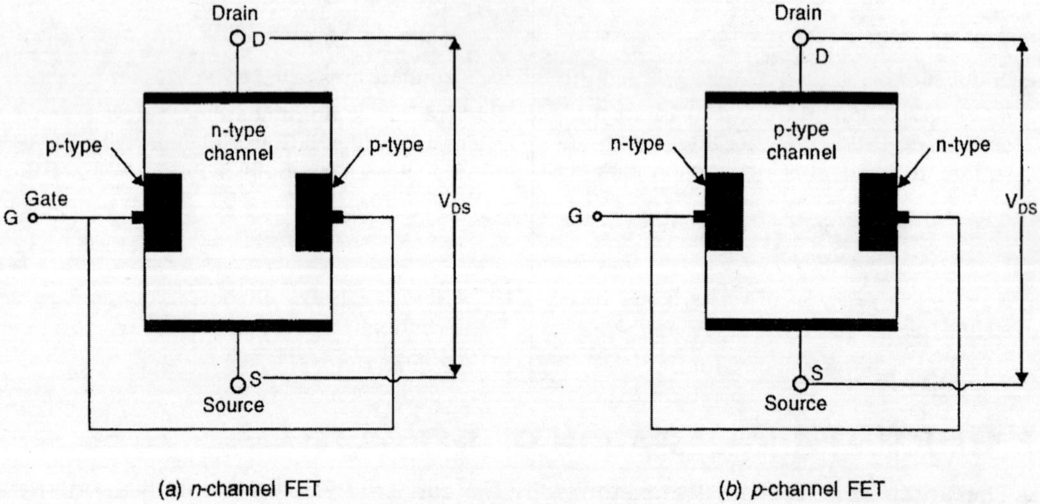

Fig. 7.3. *JFET's type.*

Whenever a voltage is applied across the drain and source terminals, a current flows through the *n*-channel. The current consists of only one type of carriers (*i.e.*, electrons) therefore the FET (field-effect transistor) is called a unipolar device. The symbols for JFET's for both n-channel and *p*-channel devices are given in Fig. 7.4.

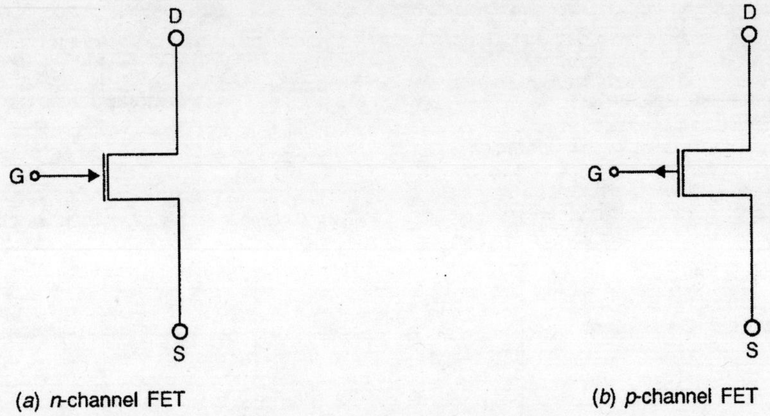

Fig. 7.4. *Symbols for FET's.*

A *p*-channel JFET is shown in Fig. 7.4(*b*). Its construction is similar to that of *n*-channel JFET, except that it consists of *p*-channel and *n*-type junctions. The current carriers in *p*-channel JFET are the holes. The operation of *p*-channel JFET is similar to *n*-channel except that all voltage and currents are reversed.

Analogies are seldom perfect and at times can be misleading, but the water analogy as shown in Fig. 7.5 does provide a sense for the JFET control at the gate terminal and the source of water pressure can be likened to the applied voltage (*i.e.*, V_{DS}) from drain to source that will establish a flow of water (electrons) from the source. The "gate" through an applied potential which controls the flow of water (*i.e.*, here charge) to the "Drain".

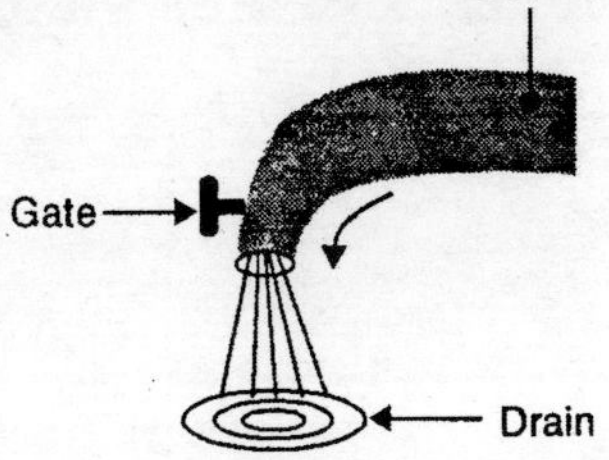

Fig. 7.5.

7.4.1. Operation of JFET

Consider an *n*-channel JFET as shown in Fig. 7.6(*a*) here the *p*-type gate and *n*-type channel constitutes *pn*-junction. This PN-junction is always reverse biased in JFET operation.

When the circuit with very small reverse voltage is applied across the gate (V_G) as shown in Fig. 7.6(*a*). Due to this reverse bias the width of the depletion layer is increased over the *pn*-junction. **Since, the doping of the *p*-type region is more than n-type, therefore depletion layer penetrates, *n*-type region deeply, so that the width of the *n*-type channel is decreased.** Now, the electrons coming from the source have to pass through narrow width *n*-channel. Because of reduced channel-width, the resistance of the path of electron is increased and hence the drain current (I_D) is decreased.

As discussed earlier that *p*-region of a *n*-channel JFET is heavily doped, as compared to the *n*-channel. Incidently, when there is no applied voltage between the drain and the source, the depletion region is symmetrical around the whole junction. The conductivity of depletion region is zero because there are no mobile charge carriers in this region. Hence the effective width of *n*-channel is reduced which is better illustrated

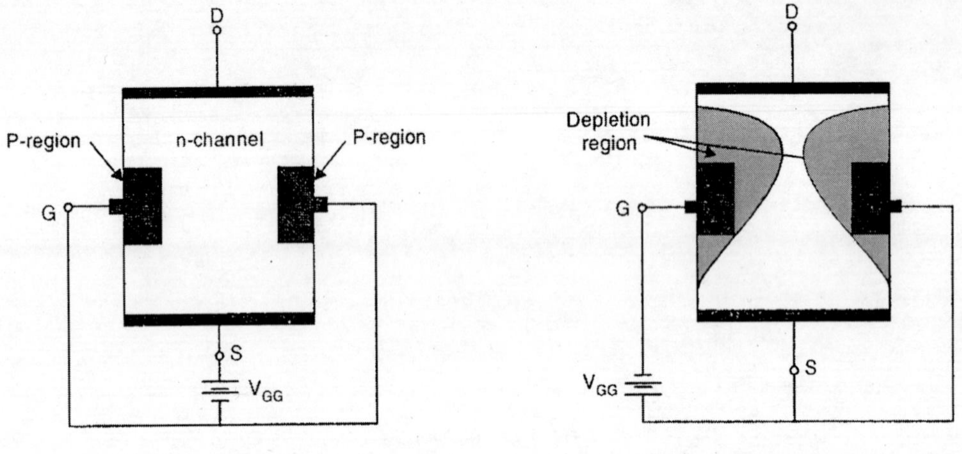

(*a*) *n*-channel FET (*b*) depletion regions in FET

Fig. 7.6. *Basic operation of FET.*

in Fig. 7.7. It further reduces with the increased reverse-bias voltage applied across the gate and source terminals of the JFET.

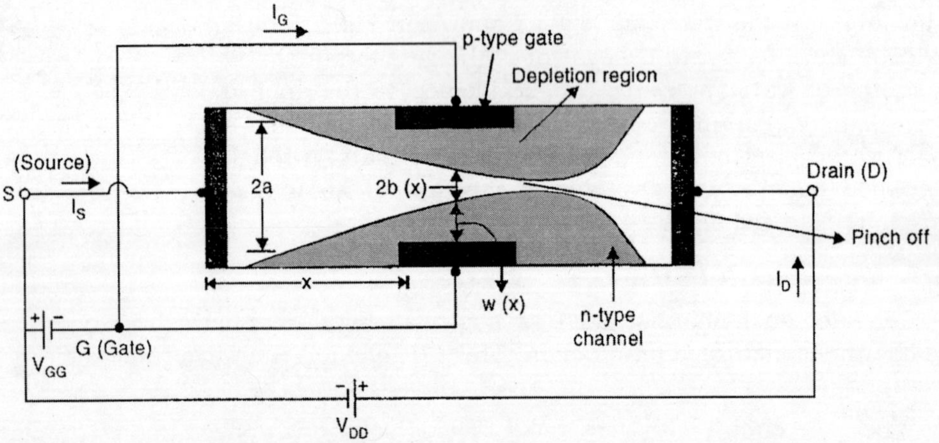

Fig. 7.7. *Detailed structure of an n-channel FET.*

From Fig. 7.7 we see that :

$2a$ = width of channel.

$2b\ (x)$ = effective width of the channel after the formation of depletion region.

$w\ (x)$ = is the width of depletion region.

Both the effective width $b\ (x)$ and width of depletion region $w\ (x)$ are the function of x, it means both vary with varying the reverse voltage applied across gate terminal and the source terminal. The reverse-bias across the gate source junction of a JFET may also be achieved by applying a voltage across the drain and source terminal as shown in Fig. 7.8. It may be noted that drain (*D*) is connected to the positive terminal of the d.c. supply (V_{DD}) and source (*S*) is connected to the negative terminal.

7.4.2. Static Characteristics of FET

Before the study of the characteristics in detail, we should memorize Table 7.1 for JFET.

Table 7.1

S. No.	*Particular*	*n-channel FET*	*p-channel FET*
1.	V_{DS}	+ ve	– ve
2.	V_{GS}	– ve	+ ve
3.	I_D	*D* to S	*S* to *D*
4.	I_G	incoming	outgoing

BJT to FET terminals analogy

BJT	*Collector (C)*	*Base (B)*	*Emitter (E)*	I_C	V_{CE}
FET	Drain (*D*)	Gate (*G*)	Source (*S*)	I_D	V_{DS}

There are two types of FET characteristics.

1. Drain characteristics
2. Transfer characteristics

7.4.2(a). Drain Characteristics

It is the relationship between V_{DS} and I_D at constant V_{GS} *i.e.*, we assume any fixed value for V_{GS}. Now, depending upon the values of V_{DS}, following cases arise.

Case 1. When $V_{GS} = 0$ and $V_{DS} = 0$. In this condition the drain current is zero.

Case 2. When $V_{GS} \leq 0$ and $V_{DS} = 0$. Under this condition *pn*-junction being reverse biased and increases the width of the depletion layer. Now, if V_{GS} is increased more negatively then the situation of pinch off occurs. At this instant, $I_{D_{max}}$ also decreases because channel becomes narrower due to reverse bias.

Case 3. When $V_{GS} = 0$ and $V_{DS} > 0$, *i.e.*, a positive voltage V_{DS} has been applied across the channel. The gate and source are connected directly *i.e.*, $V_{GS} = 0$. In this situation $I_D = I_S$ (*i.e.*, $V_{DS} = V_{DD}$). Under this condition **the flow of charge is relatively uninhabited and limited only by the resistance of the *n*-channel between drain and source.**

It is important to note that depletion region is wider near the drain side as compared to source side because when current flows in silicon bar, there occurs a voltage drop across the channel along whole length due to this reverse bias voltage across the drain and the gate is comparatively much larger than the reverse bias voltage across the gate and source. This can be easily understood by analysing Fig. 7.8(*b*) given as follows :

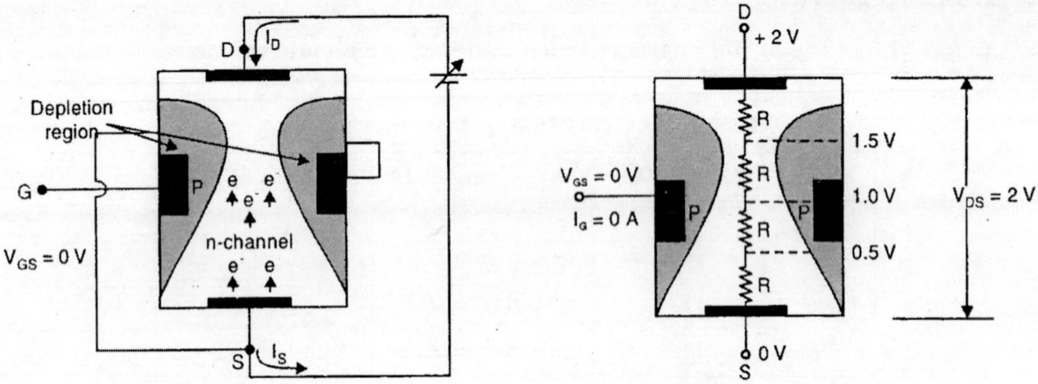

(a) *N*-channel FET

(*b*) Diagram showing voltage drops in succession through channel

Fig. 7.8.

Fig. 7.8(*b*) shows that the reverse bias voltage is directly proportional to the depletion region. Thus, we can say that the width of the depletion layer is maximum at drain terminal side where there is maximum reverse bias voltage.

It is to be noted here that I_G = 0 is the important characteristic of the JFET. The drain current I_D depends upon following factors :

1. Number of majority carriers
2. Length of the channel
3. Cross-sectional area of the channel
4. V_{DS}

Mathematically,

$$\boxed{I_D = \frac{V_{DS}}{R} = \frac{V_{DS}A}{\rho l}} \qquad \text{...(7.1)}$$

where,

V_{DS} = Voltage between drain and source

R = Resistance of the channel

A = Area of the channel

ρ = Resistivity of the material used

l = Length of the channel.

In order to explain, the typical shape of drain characteristics, let us select the curve with V_{GS} = 0 volt as shown in Fig. 7.9. The curve may be subdivided into the following regions.

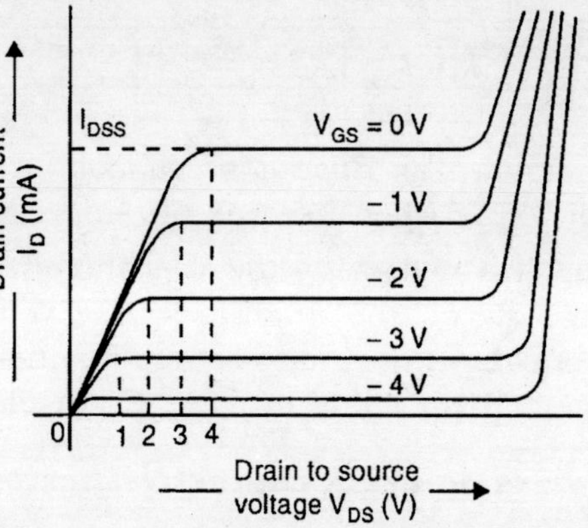

Fig. 7.9. *Drain characteristics of JFET.*

1. **Ohmic region.** This region is shown as a curve OA in Fig. 7.10. In this region, the drain current increases linearly with the increase in drain-to-source voltage, obeying Ohm's law. The linear increase in drain current is due to the fact that *n*-type semiconductor bar acts like a simple resistor.

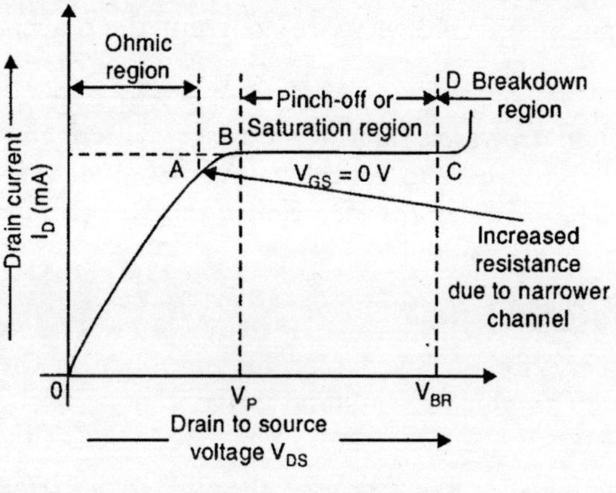

Fig. 7.10. *Drain characteristics with V_{GS} = 0 volt.*

2. **Curve AB.** In this region, the drain current increase at the reverse square law rate with the increase in drain-to-source voltage. It means that drain current increases slowly as compared to that in ohmic region. It is because of the fact, that with the increase in drain-to-source voltage, the drain current increases. This in turn increases the reverse bias voltage across the gate-source junction. As a result of this, the depletion region grows in size, thereby reducing the effective width of channel. At the drain-to-source voltage, corresponding to point B, the channel width is reduced to a minimum value and is known as pinch off. The drain-to-source voltage, at which the channel pinch-off occurs, is known as *pinch-off voltage (V_p).*

3. **Pinch off region.** This region is shown by the curve BC. It is also called saturation region or constant current region. In this region, the drain current remains constant at its maximum value (*i.e.*, I_{DSS}). The drain, current in the pinch off region, depends upon the gate-to-source voltage and is given by the relation,

$$I_D = I_{DSS}\left(1 - \frac{V_{GS}}{V_P}\right)^2$$

The above relation is known as *Shockly's equation.* The pinch off region is the normal operating region of JFET, when used as an amplifier.

4. **Breakdown region.** This region is shown by the curve *CD*. In this region, the drain current increases rapidly as the drain-to-source voltage is also increased. It happens because of the breakdown of gate-to-source junction due to avalanche effect. The drain-to-source voltage corresponding to point *C* is called breakdown voltage.

Important points regarding drain characteristics of n-channel FET

1. Drain current is maximum when V_{GS} = 0 *i.e.*, V_D = I_{DSS} (saturation current)
2. In FET, generally avalanche breakdown occurs (due to excessive amount of electric field)
3. V_P is a particular value of V_{GS} with either positive or negative sign depending upon the case of the FET.
4. From the characteristics, it is clear that there is a region from end of ohmic portion up to pinch off voltage where current decreases due to increase in mobility.

There are also called transconductance curves, which gives us the relationship between drain current (I_D) and gate to source voltage (V_{GS}) for a constant value of drain-to-source voltage (V_{DS}). The transfer characteristics may be obtained by using the circuit arrangement as shown in Fig. 7.11.

7.4.2(b). Tansfer Characteristics

As mentioned earlier, the FET works as an amplifier in the region beyond pinch off (or current saturation region or constant current pentrode). It may be noted that curve is part of parabola.

Following equation governs the transfer characteristics of FET.

$$I_D = I_{DSS}\left(1 - \frac{V_{GS}}{V_P}\right)^2 \quad \text{...(7.2)}$$

where, I_D = Drain current

I_{DSS} = Saturated (max) value of drain current

V_{GS} = Gate to source voltage (*i.e.*, control variable)

V_P = Pinch-off voltage.

Here in eqn. (7.2), V_{GS} is control variable and I_{DSS} is the saturation current (having constant value). *The linear relationship does not exist between the output and input quantities of a JFET. The above relation is known as* **the Shockley's equation.**

The approximate parabolic equation given in eqn. (7.2) is a best fit for transfer characteristics obtained for FET by experimental approach *i.e.*, diffusion process.

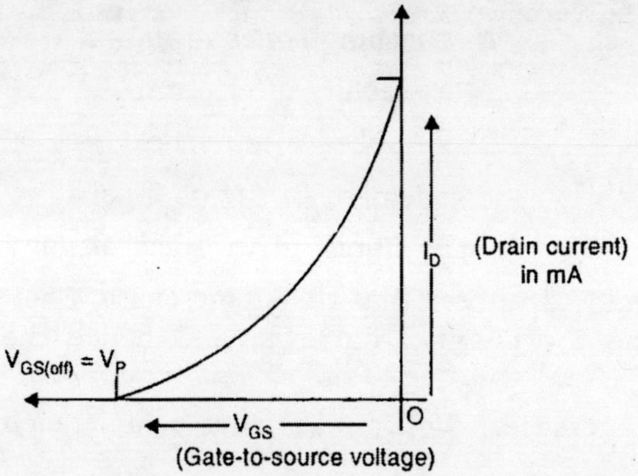

Fig. 7.11. *Transfer characteristics of n-channel.*

7.5. PINCH OFF VOLTAGE

As we have already discussed that if the magnitude of reverse bias voltage is applied to gate is increased sufficiently, drain current I_D becomes zero. This situation of JFET operation is known as *pinch-off.*

To understand the concept of pinch-off voltage is very necessary. It can be better understand by Fig. 7.12.

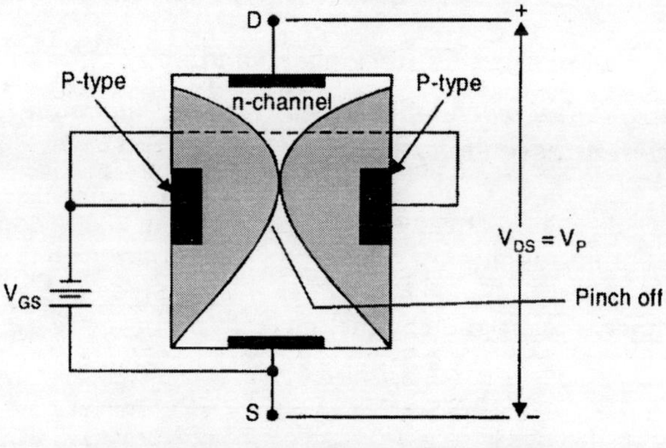

Fig. 7.12. *The situation of pinch-off.*

Above discussion reveals that when gate to source voltage (V_{GS}) approaches to pinch off voltage (*i.e.*, V_P), drain current (I_D) becomes zero, but in fact it is not so. This can be easily understood by the mathematical relation of drain current :

$$I_D = A\, q\, N_D\, \mu_n\, E$$

or $$\boxed{I_D = 2b(x)\, w\, q\, N_D\, \mu_n\, E} \quad ...(7.3) \quad \left\{ \begin{array}{l} \text{Area } A = 2b(x).w, \text{ From Fig. 7.7} \\ V_{DS} = E \times L \end{array} \right\}$$

where, I_D = Drain current

$2b(x)$ = Effective channel width

w = Depletion region width

N_D = Doping concentration

μ_n = Mobility of electron

E = Applied electric field

L = Length of the channel

q = Change of the electron (or hole)

Thus, from relation (7.3) we see that as V_{DS} increases, E and I_D increases, whereas $b(x)$ decreases because the channel narrows, and hence the current density $J = \frac{I_D}{2b(x).w}$ $\left\{\therefore J = \frac{I}{A}\right\}$ increases. We now see that complete pinch-off (b = 0) cannot takes place because, if it did, J would become infinite, which is a physically impossible condition. It is found experimentally that the mobility is a function of electric field (as discussed in unit 2) intensity and remains constant only for $E < 10^4$ V/cm in n-type silicon. For moderate fields, 10^3 to 10^4 V/cm, the mobility is approximately inversely proportional to the square root of the applied field $\left(i.e., \mu_n \propto \frac{1}{\sqrt{E}}\right)$. For still higher fields, such as encountered at pinch-off μ_n is inversely proportional to the applied electric field $\left(i.e., \mu_n \propto \frac{1}{E}\right)$. In this region the drift velocity of the electrons ($N_d = \mu_n E$) remains constant, and ohm's law is no longer valid.

Finally we concluded that pinch-off voltage (V_p) is the minimum drain-to-source voltage at which the drain current essentially becomes constant.

NOTE 1 : As we have discussed already that the P-region is highly doped as compared to the n-channel, it means number of acceptor ions in a p-type is very much higher than the Donar-ions (*i.e.*, $N_A >> N_D$). Under this situation the space-change width $w(x)$ at a distance x along the channel in Fig. 7.7 :

$$W(x) = a - b(x)$$

or

$$W(x) = \left\{\frac{2E}{qN_D}(V_0 - V_{(x)})\right\}^{1/2} \quad ...(7.4)$$

where

$\in$ = Dielectric constant of the channel material

q = Magnitude of electronic charge

N_D = Donar concentration

V_0 = Junction contact potential at x.

$V(x)$ = Applied potential across space-charge region at x and is negative number for an applied reverse bias.

If the drain current is zero, $b(x)$ and $V(x)$ are independent of x and $b(x) = b$. If in eqn. (7.4) we substitute $b(x) = b = 0$ and solve for V, on the assumption that $|V_0| << |V|$, we obtain pinch off voltage V_p.

NOTE 2 : Relation between the pinch-off voltage (V_p), doping concentration (N_D), channel width a is given by the equation :

$$|V_p| = \frac{q\,N_D}{2\,\epsilon}a^2 \qquad ...(7.5)$$

where
V_p = Pinch-off voltage
N_D = Doping concentration
a = Channel width

$$\epsilon = \begin{cases} 12\epsilon_0 \text{ for Si} \\ 16\epsilon_0 \text{ for Ge} \end{cases} \quad \text{and } \epsilon_0 = 8.86 \times 10^{-12} \frac{\text{Newton meter}^2}{\text{Coloumb}^2}$$

NOTE 3 : Relation between pinch-off voltage (V_p) and gate-to-source voltage is given by equation :

$$V_{GS} = \left(1 - \frac{b}{a}\right)^2 V_P \qquad ...(7.6)$$

where,
V_{GS} = Gate-to-source voltage
V_P = Pinch of volage
a = Channel width
b = Effective channel width.

7.6. SMALL SIGNAL MODEL FOR JFET OR FET

The linear small signal model for JFET is shown in Fig. 7.13.

This small signal model is valid both for JFET and MOSFET (which is discussed in the next chapter) is used to relate small changes in FET current and voltages about the quicescent operating point. In this model, the gate-to-source junction is represented by an open circuit and no current is drawn by the input terminal of the field-effect transistor. It is because of the fact, that the input resistance (*i.e.*, the resistance of gate-source junction) is very large. For all practical purpose, its input resistance can be considered to be so large that the input of FET can be considered as an open circuit.

It will be interesting to know that although the gate-source junction appears as an open circuit, yet the gate-to-source voltage affects the value of drain current. It is indicated by a voltage-controlled current source ($g_m \cdot V_{GS}$) whose value is proportional to the gate-to-source voltage.

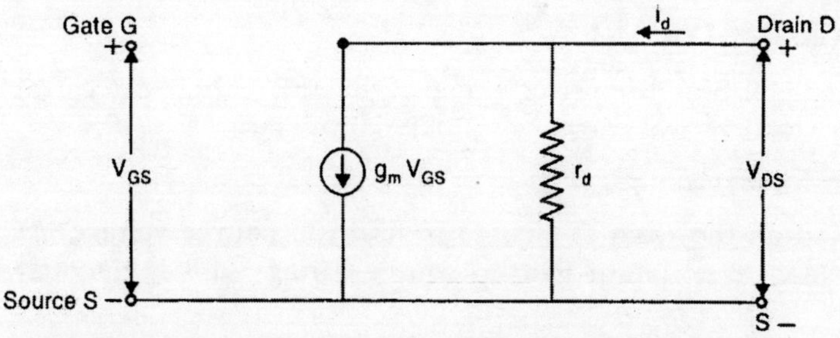

Fig. 7.13. *The low frequency small-signal model for JFET.*

The FET drain resistance (also called FET output resistance) is represented by the resistance (γ_d). Typical values of drain resistance are from 100 kΩ to 1 MΩ for JFET's and 1 kΩ to 50 kΩ for MOSFET's.

Now, we shall discuss about the terms involved and their significance in regards of above small-signals model :

We can express the drain current i_d as a function of the gate voltage V_{GS} and drain voltage V_{DS} by

$$i_D = f\,(V_{GS},\ V_{DS})$$

The transconductance g_m and drain resistance r_d is given

As we know that,

$$i_D = f\,(V_{GS},\ V_{DS}) \qquad \ldots(7.7a)$$

Now, if both gate and drain voltages are varied then the change in drain current is given by :

$$\Delta i_D = \left.\frac{\partial i_D}{\partial V_{GS}}\right|_{V_{DS} = \text{Const.}} \Delta V_{GS} + \left.\frac{\partial i_D}{\partial V_{DS}}\right|_{V_{GS} = \text{Const.}} \Delta V_{DS} \qquad \ldots(7.7b)$$

From the model, (using Taylor's series)

$$i_d = g_m\ V_{GS} + \frac{V_{DS}}{r_d} \qquad \ldots(7.7c)$$

where

$$g_m = \left.\frac{\partial i_D}{\partial V_{GS}}\right|_{V_{DS} = \text{Const.}} \approx \left.\frac{\Delta i_D}{\Delta V_{GS}}\right|_{V_{DS} = \text{Const.}} = \left.\frac{i_d}{V_{GS}}\right|_{V_{DS} = \text{Const.}} \qquad \ldots(7.7d)$$

is the mutual conductance or transconductance. It is often referred to as common-source forward transadmittance.

The second parameter r_d in eqn. (7.7c) is the drain resistance or output resistance and is defined by :

$$r_d = \left.\frac{\partial V_{DS}}{\partial i_D}\right|_{V_{GS} = \text{Const.}} \approx \left.\frac{\Delta V_{DS}}{\Delta i_D}\right|_{V_{GS} = \text{Const.}}$$

i.e., r_d is the inverse ratio of change in drain voltage to change in drain current at constant gate to source voltage.

7.7. JFET PARAMETERS

In order to determine the performance of the JFET, there are certain characteristics described below which are known as JFET parameters :

(1) **AC drain resistance, (r_d)**

It is defined as the ratio of change in drain to source voltage (ΔV_{DS}) to change in drain current (ΔI_D) at constant gate to source voltage and is denoted by r_d.

Mathematically,

$$r_d = \left.\frac{\Delta V_{DS}}{\Delta I_D}\right|_{V_{GS} = \text{Const.}}$$

It is also known as *dynamic drain resistance.*

(2) **Transconductance**

Besides change in V_{DS}, V_{GS} can also control over the change in drain-current. The parameter transconductance of the JFET corresponds to the relationship between change in gate to source voltage and change in drain current. It may be defined as the ratio of change in drain current to the change in gate to source voltage at constant drain to source voltage.

Mathematically,

$$g_m = \left.\frac{\Delta I_D}{\Delta V_{GS}}\right|_{V_{DS} = \text{Const.}}$$

It is also known as *forward transadmittance.* Generally, it is expressed in mA/volt or micro-seiemens.

When I_D, the drain current gets saturated *i.e.*, attains the value of I_{DSS}, then g_m is denoted by g_{mo}.

Relation between g_m and g_{mo}

As per the Shockley's equation,

$$I_D = I_{DSS}\left[1 - \frac{V_{GS}}{V_P}\right]^2$$

Differentiating both sides of Shockley's equation, we get,

$$\frac{dI_D}{dV_{GS}} = 2I_{DSS}\left(1 - \frac{V_{GS}}{V_P}\right)\left(-\frac{1}{V_P}\right)$$

or
$$\frac{dI_D}{dV_{GS}} = 2I_{DSS}\left(1 - \frac{V_{GS}}{V_P}\right)\left(-\frac{1}{V_P}\right)$$

or
$$g_m = \frac{-2I_{DSS}}{V_P}\left(1 - \frac{V_{GS}}{V_P}\right)$$

Substituting $V_{GS} = 0$ in above expression then

$$\boxed{g_{mo} = \frac{-2I_{DSS}}{V_P}} \qquad (\therefore\ I_D = V_{DSS} \text{ when } V_{GS} = 0)$$

Now, using eqn. (7.7*b*) and (7.7*c*), we get

$$\boxed{g_m = g_{mo}\left(1 - \frac{V_{GS}}{V_P}\right)} \qquad \text{...(7.8)}$$

(3) **Amplification-factor**

As in V_{GS} and V_{DS} (ΔV_{GS}) can change the drain-current I_D. Now, we can have a relationship between change in gate to source voltage and change in drain to source voltage (ΔV_{GS}) for incorporating same change in drain current.

Thus, amplification factor is defined as the ratio of change in drain to source voltage to the change in gate to source voltage at constant drain-current. It is denoted by μ.

Mathematically,

Amplification factor,

$$\mu = \left.\frac{\Delta V_{DS}}{\Delta V_{GS}}\right|_{I_D}$$

Relation between amplification factor, transconductance and drain-resistance

We have amplification factor,

$$\mu = \frac{\Delta V_{DS}}{\Delta V_{GS}}$$

$$= \frac{\Delta V_{DS}}{\Delta I_D} \times \frac{\Delta I_D}{\Delta V_{GS}}$$

$$\boxed{\mu = r_d \times g_m} \qquad \ldots(7.9)$$

Thus, Amplification factor = ac drain resistance × transconductance.

(4) **DC drain resistance**

It is also called as the static or ohmic resistance of the channel and is defined as *ratio of drain to source voltage and drain-current* I_D. *It is represented by* R_{DS}.

$$\boxed{R_{DS} = \frac{V_{DS}}{I_D}}$$

NOTE 1 : Drain resistance, r_d is not to be confused with the drain to source resistance, (R_{DS}). Because r_d is a dynamic or ac resistance whereas R_{DS} is the dc resistance of the channel when the depletion regions are absent.

NOTE 2 : Important relationships.

JFET		*BJT*
$I_D = I_{DSS}\left(1 - \frac{V_{GS}}{V_P}\right)^2$	⇔	$I_C = \beta I_B$
$I_D = I_S$	⇔	$I_C \cong I_E$
$I_G \cong 0$ Amp	⇔	$V_{BE} \cong 0.7$ V.

7.8. BIASING OF JFET

As JFETs have only majority carriers responsible for the current flow, thermal run away does not occur in FETs unlike BJTs. In order to obtain sharp dc operating point (Q-point) for the FET to work as an amplifier following biasing schemes are employed.

1. **Fixed bias circuit**

DC bias of a FET device needs setting of gate to source voltage V_{GS} to give desired drain-current I_D. For a JFET drain current is limited by I_{DSS}. Since the FET has such a high input impedance which do not allow to flow the gate current and the dc voltage of the gate set by a voltage divider or a fixed battery is not affected or loaded by the FET.

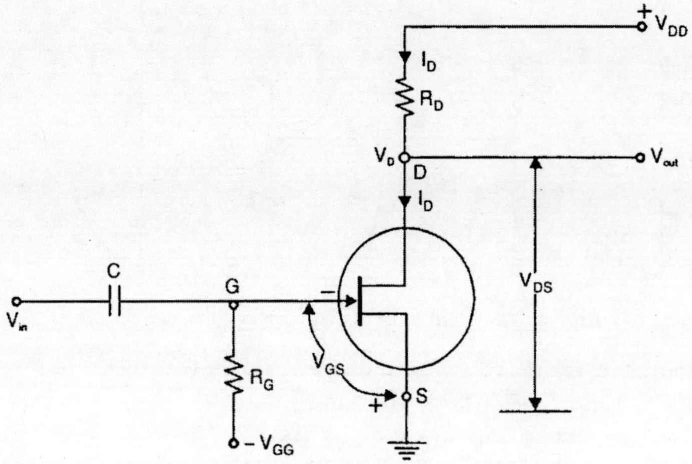

Fig. 7.14. *Fixed biasing circuit for JFET.*

Referring to above Fig. 7.14 fixed bias circuit is obtained using a battery V_{GG}. This battery ensures that the gate is always negative with respect to source and no current flows through resistor R_G and gate terminal *i.e.*, $I_G = 0$. The battery provides a voltage V_{GS} to bias the n-channel JFET, but no resulting current is drawn from the battery V_{GG}. Resistor R_G is included to allow any ac signal applied through capacitor C to develop across R_G. While any ac signal will develop across R_G, the dc voltage drop across R_G is equal to $I_G R_G$ *i.e.*, 0 volt.

The gate to source voltage V_{GS} is then

$$\boxed{V_{GS} = V_G - V_S = -V_{GG} - 0 = -V_{GG}}$$

The drain-source current I_D is then fixed by the gate-source voltage. This current then causes a voltage drop across the drain resistor R_D and is given as :

$$V_{DD} = I_D \cdot R_D + V_{DS}$$

or

$$\boxed{I_D = \frac{V_{DD} - V_{DS}}{R_D}}$$

2. **Self-bias circuit**

This is the most common method for biasing a JFET. Self bias circuit for *n*-channel FET is as shown in Fig. 7.15.

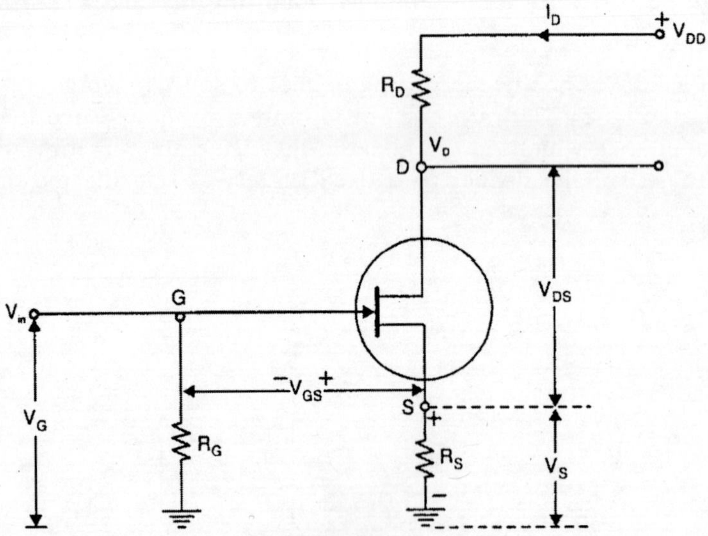

Fig. 7.15. *Self bias circuit for n-channel JFET.*

Referring to above Fig. 7.15 no gate current flows through the reverse-biased gate-source, the gate current $I_G = 0$ and, therefore,

$$V_G = I_G R_G = 0 \text{ V}$$

With a drain current I_D the voltage at the S is

$$V_S = I_D R_S$$

The gate-source voltage is then

$$V_{GS} = V_G - V_S = 0 - I_D R_S = -I_D R_S$$

Now,
$$V_{DD} = I_D R_D + V_{DS} + I_D R_S$$

$$\boxed{V_{DS} = V_{DD} - I_D (R_D + R_S)}$$

So, voltage drop across resistance R_S provides the biasing voltage V_{GS} and no external source is required for biasing and this is the reason that it is called self-biasing.

3. **Potential divider biasing**

Another class of biasing network is the potential divider biasing. This biasing is most widely useful in practical circuit and shown in Fig. 7.16.

The resistors R_{G_1} and R_{G_2}, form a potential divider across the drain supply V_{DD}. The voltage V_2 across R_{G_2} provides the necessary bias. The additional gate resistor R_{G_1} from gate to supply voltage facilitates in larger adjustment of the dc bias point.

The gate is reverse biased so that $I_G = 0$ and gate voltage

$$V_G = V_2 = \frac{V_{DD}}{R_{G_1} + R_{G_2}} \times R_{G_2}$$

and $$V_{GS} = V_G - V_S$$

or $$\boxed{V_{GS} = V_G - I_D R_S}$$

The circuit is so designed that $I_D R_S$ is larger than V_G (or V_2) so that V_{GS} is negative. This provides exact bias voltage. The operating point can be determined as :

$$\boxed{I_D = \frac{V_2 - V_{GS}}{R_S}} \quad \text{and} \quad \boxed{V_{DS} = V_{DD} - I_D (R_D + R_S)}$$

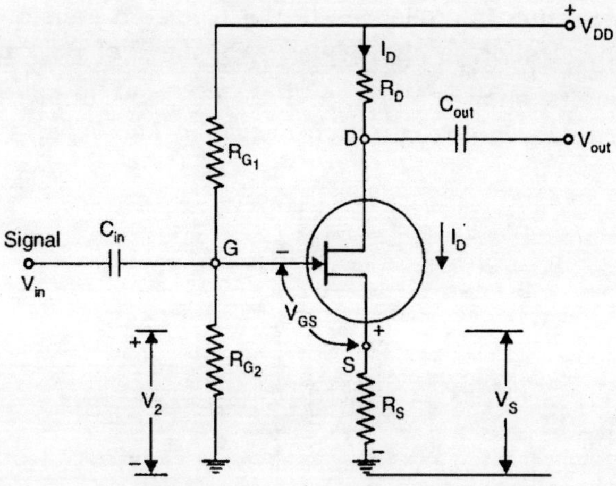

Fig. 7.16. *Potential divider bias circuit for n-channel JFET.*

7.9. FET AS AN AMPLIFIER

As in case of BJT as an amplifier, we have three configurations like common emitter (CE), common collector (CC) and common base (CB), similarly in case of FET's we have got following three configurations

(*i*) Common drain (CD)

(*ii*) Common source (CS)

(*iii*) Common gate (CG)

The following diagrams depicts the picture of (CS) and (CD) configurations of the JFET as an amplifier.

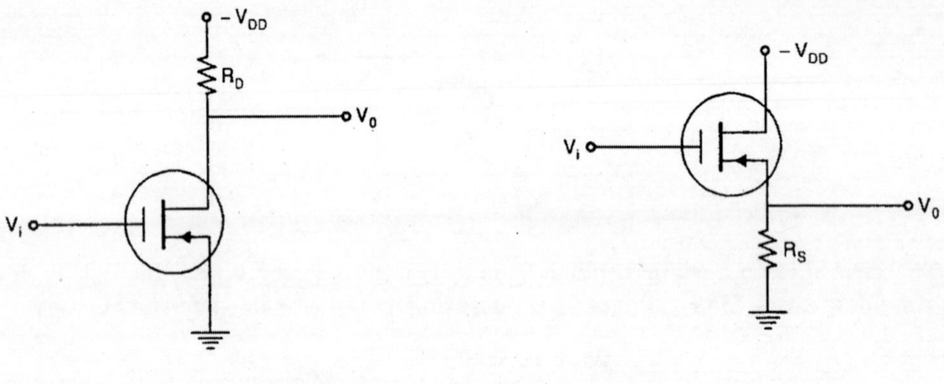

(a) Common Source Configuration (CS) (b) Common Drain Configuration (CD)

Fig. 7.17. *CS and CD configurations of FET as an amplifier.*

The CS configuration is analogous to the bipolar transistor CE amplifier and the common drain configuration is analogous to CC stage of BJT amplifier.

Both the circuits shown above in Fig. 7.17 are analysed simultaneously by considering the generalized configuration shown in Fig. 7.18.

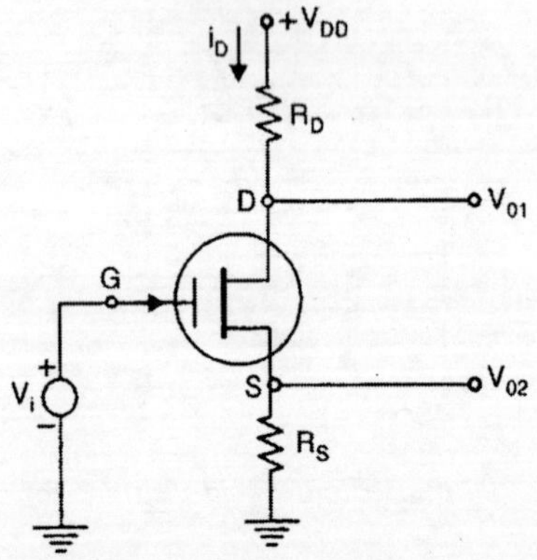

Fig. 7.18. *Generalised FET amplifier configuration.*

From the above configuration if output is taken from the drain (D) *i.e.*, V_{01} it is called common-source (CS) amplifier, however if the output is taken from the source (S) *i.e.*, V_{02}, it is called common-drain (CD) or the source follower (analogous to the *emitter* follower for a bipolar junction transistor).

7.9.1. Computation of A_v, R_i, R_o of Single FET Amplifiers

In order to calculate the different parameter like voltage gain (A_v), input impedance (R_i) and output impedance (R_o) of different configuration, small signal equivalent circuit of the above generalized amplifier configuration may be used so for easy calculation the small signal equivalent circuit of the above generalised amplifier configuration is shown in Fig. 7.19.

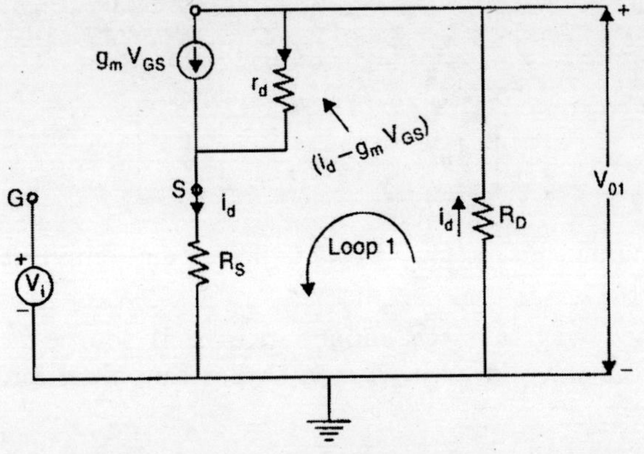

Fig. 7.19. *The small signal equivalent circuit for FET amplifier.*

Applying KVL to the loop 1, we have

$$i_d R_d + (i_d - g_m V_{GS}) r_d + i_d R_S = 0 \qquad ...(7.10a)$$

Also, $$V_{GS} = V_i - i_d R_S \qquad ...(7.10b)$$

Combining eqns. (7.10*a*) and (7.10*b*) and using $\mu = r_d g_m$

We have,

$$i_d = \frac{\mu V_i}{r_d + R_d + (\mu+1) R_S}$$

Case 1. The CS amplifier with an unbypassed source resistance.

From Fig. 7.19 $V_{01} = - i_d R_d$, then

We have, $$V_{01} = \frac{-\mu V_i R_d}{r_d + R_d + (\mu+1) R_S} \qquad ...(7.10c)$$

From eqn. (7.10*c*), we obtain the Thevenin's equivalent circuit of Fig. 7.20 given under here :

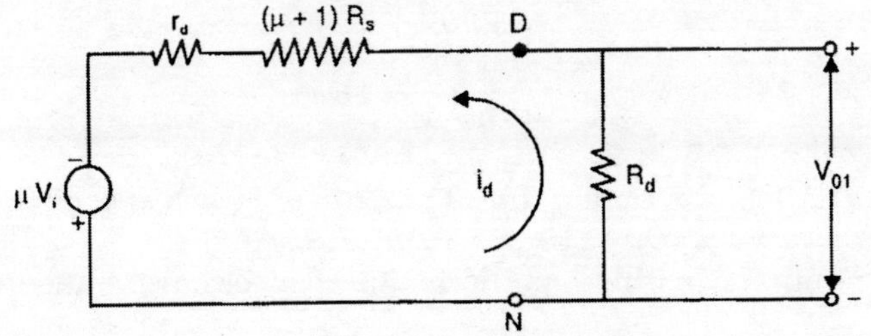

Fig. 7.20. *Equivalent circuit for the generalized FET amplifier when output is taken from the drain.*

Voltage gain (A_v)

$$A_V = \frac{V_{01}}{V_i}$$

or
$$A_V = \frac{-\mu V_i R_d}{r_d + R_d + (\mu+1) R_S} \cdot \frac{1}{V_i}$$

or
$$\boxed{A_V = \frac{-\mu R_d}{r_d + R_d + (\mu+1) R_S}} \qquad \text{...(7.10}d\text{)}$$

NOTE : Here the minus sign in eqn. (7.10*d*) indicates that output is 180° out of phase with the input.

If R_S is bypassed with a large capacitance or if source is grounded, the above relation in eqn. (7.10*d*) is valid with $R_S = 0$. Under this situation voltage gain becomes

$$A_V = \frac{-\mu R_d}{r_d + R_d} \qquad \{\because R_S = 0\}$$

or
$$\boxed{A_V = -g_m \cdot R_d'} \qquad \text{where } \left\{ \begin{array}{l} \mu = g_m r_d \text{ and} \\ R_d' = R_d \parallel r_d \end{array} \right\}$$

Output resistance (R_0)

Referring from the Fig. 7.20

$$\boxed{R_0 = (\mu + 1)\, R_S + r_d}$$

Case 2. Common drain (*CD*) or source follower with drain resistance.

From Fig. 7.18, we have

$$V_{02} = i_d R_S$$

$$V_{02} = \frac{\mu V_i R_S}{r_d + R_d + (\mu+1) R_S} \qquad \{\text{Using the equation for } i_d\}$$

or
$$V_{02} = \frac{\left[\dfrac{\mu V_i}{\mu+1}\right] . R_S}{\dfrac{(r_d + R_d)}{\mu+1} + R_S} \qquad \text{...(7.10}e\text{)}$$

From eqn. (7.10*e*), we obtain the Thevenin's equivalent circuit as in Fig. 7.21 working into the source. In this equation, the open circuit voltage is $\frac{\mu V_i}{(\mu+1)}$, and the output resistance is $R_0 = \left(\frac{r_d + R_d}{\mu+1}\right)$.

The voltage gain is $A_0 = \frac{V_{02}}{V_i}$. *There is no phase shift between input and output.*

If $R_d = 0$ and if $(\mu + 1)\ R_S > r_d$, then

$$A_V \cong \frac{\mu}{\mu+1} \approx 1 \text{ for } \mu > 1$$

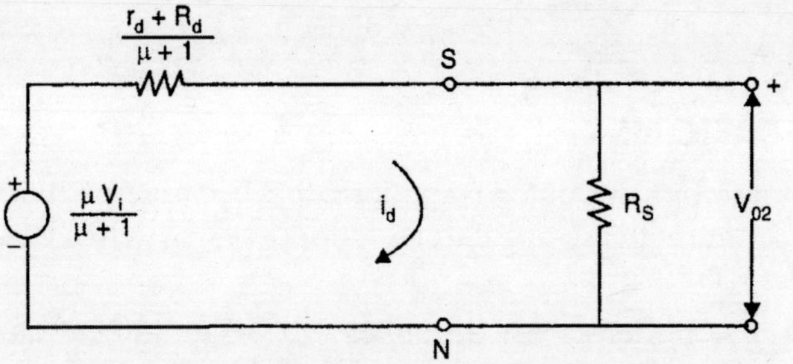

Fig. 7.21. *Equivalent circuit for the generalized FET amplifier when the output is taken from the source.*

Hence for source follower with $R_d = 0$ and $\mu >> 1$

$$R_0 = \frac{r_d + R_d}{\mu + 1}$$

or $$R_0 = \frac{r_d}{\mu} \qquad \{R_d = 0,\ \mu >> 1\}$$

or $$R_0 = \frac{r_d}{g_m r_d}$$

or $$\boxed{R_0 = \frac{1}{g_m}}$$

Unity voltage gain means that the output (at the source) follows the input (at the gate). Hence, the CD configuration is called a *source follower* (analogous to the emitter follower for a BJT amplifier).

Comparison of n-channel and *p*-channel devices

As we have discussed earlier that the construction of *p*-channel is exactly same as that of the *n*-channel devices. However there are some differences between the *n*-channel and *p*-channel, are given in Table 7.2.

Table 7.2

S.No.	***Particular***	***n-channel***	***p-channel***
1.	Nature of charge carriers	In an *n*-channel JFET the current carriers or charge carriers are electrons.	Here, charge carriers are holes.
2.	Mobility of the charge carriers	Mobility of electrons in *n*-channel is higher because their masses are smaller than holes.	The mobility of holes in *p*-channel devices is lower than that of electrons in *n*-channel devices.

3.	Noise	They are less noisy.	They are more noisy than *n*-channel.
4.	Transconductance	Larger.	Smaller.
5.	Manufacturing process	Its manufacturing process is very difficult.	Its manufacturing process is comparatively easy.

7.10. FET APPLICATIONS

JFET's have numerous industrial applications because of its inherent properties. They can be used as amplifiers, oscillators, choppers, switches etc. Following are the few applications of FET :

1. ***As a buffer.*** A buffer is an amplifier whose amplification factor is one. The purpose of inserting a buffer between stages of amplifying system to provide isolation of one stage from its successive stage. For this very purpose the device applied as a buffer should have very large input impedance so that it draws negligibly small current from the stage attached to its input. At the same time, the ouput impedance of the buffer should be as minimum as possible because the buffer has to operate very large loads at its output terminals. Here from heavy loads mean that the loads which draw more currents because of their low resistances. The FET has input impedance in the range from several mega ohms to hundred ohms and low output impedance (ranging in few hundred ohms). Thus, FET can work as a buffer between two stages to avoid loading effects.

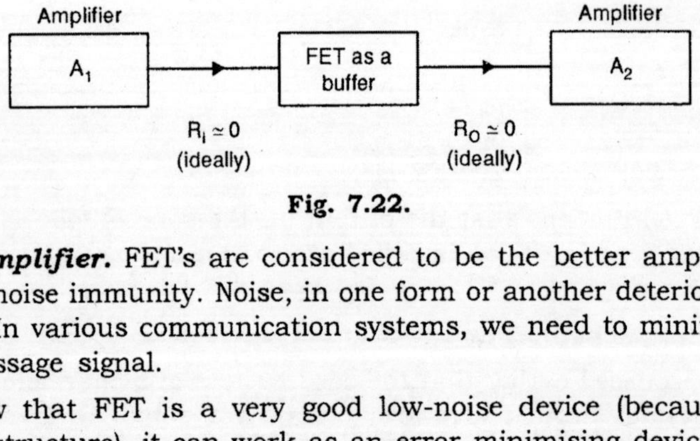

Fig. 7.22.

2. ***As an amplifier.*** FET's are considered to be the better amplifiers than BJT's because of good noise immunity. Noise, in one form or another deteriorates the original message signal. In various communication systems, we need to minimise the effect of noise on the message signal.

As we know that FET is a very good low-noise device (because of absence of junctions in its structure), it can work as an error minimising device when employed before the point end of receivers and other electronic equipments.

3. ***As a phase shift oscillator.*** The high input impedance of FETs make them valuable for designing phase shift oscillators where they avoid loading effects.

4. ***FET as a voltage variable resistor.*** As we know from the characteristics of the FET that it works most effectively in the constant-current portion. But if it is operated on the region prior to pinch-off (*i.e.*, when V_{DS} is very small) it will behave as a voltage variable resistor (*VVR*).

It is due to the fact that in this region drain to source resistance R_{DS} can be controlled by varying the bias voltage V_{GS}. In such applications the FET is also referred to as a voltage-variable resistor or voltage dependent resistor.

The following mathematical relationship describes the control of voltage variations over the resistance of the device.

$$r_d = \frac{r_0}{1-\left(\frac{V_{GS}}{V_P}\right)^2}$$

where r_0 is the resistance with V_{GS} = 0 V and r_d is the resistance at a particular level of V_{GS}.

The drain to source conductance

$$g_d = \frac{I_D}{V_{DS}}$$ can be expressed as,

$$g_d = g_{do}\left[1-\left(\frac{V_{GS}}{V_P}\right)^{\frac{1}{2}}\right] \qquad \left\{g_{do} = \frac{-2I_{DSS}}{V_P}\right\}$$

where g_{do} is the value of drain conductance when the bias voltage V_{GS} is zero. The variation of her r_d with V_{GS} can be closely approximated by the empirical expression :

$$r_d = \frac{r_0}{1-kV_{GS}}$$

where r_d = Drain resistance at zero gate bias

k = Constant whose value depends upon the type of the FET used.

V_{GS} = Gate-to-source voltage.

FET as a voltage variable resistor can be used in Automatic Gate Control (AGC) circuit of a multistage amplifier.

? SELF-TEST (JFET)

1. Draw the symbol and characteristics of an *n*-channel JFET and mark the linear region, saturation region and breakdown region.
2. What do you mean by drain and transfer characteristics in reference to JFET ?
3. What is meant by depletion region in JFET ? Explain with suitable diagrams. What are the basic difference between BJT and JFET ?
4. What are the relative merits and demerits of a FET amplifier over a transistor amplifier ?
5. With the help of a neat sketch explain the working of a FET amplifier.
6. Define the parameters of JFET.
7. Define pinch off voltage of JFET.

8. Explain with the help of circuit diagram how an FET used as a VVR ?

9. List different methods of biasing used in FET.

10. How does a *p*-channel JFET differ from an *n*-channel JFET ?

11. Why FETs have better thermal stability ?

12. Why does a FET acts as an excellent buffer amplifier ?

SOLVED EXAMPLES

Example 1. *Fig. P (7.1) show the transfer characteristics of a JFET, write the equation for drain current.*

Solution : Referring to the transfer characteristic in Fig. P (7.1), we have

$$I_{DSS} = 14 \text{ mA}$$
$$V_{GS(off)} = -6 \text{ V}$$

From the relation :

$$I_D = I_{DSS}\left(1 - \frac{V_{GS}}{V_{GS(off)}}\right)^2$$

$$I_D = 14\left(1 + \frac{V_{GS}}{6}\right)^2 \text{ mA. } \textbf{Ans.}$$

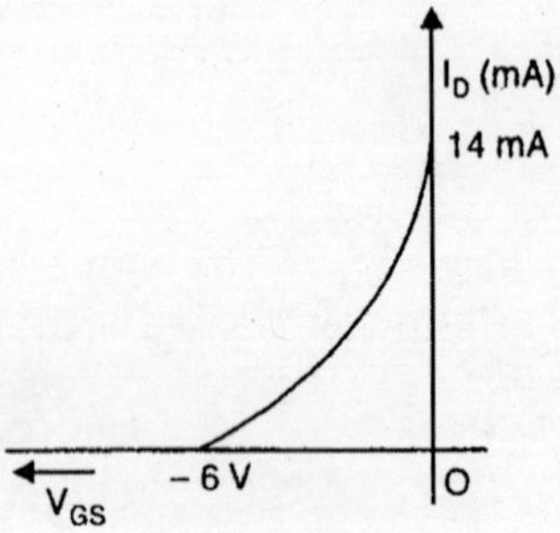

Fig. P (7.1).

Example 2. *A JFET has the following parameters :*
$I_{DSS} = 30$ *mA* ; $V_{GS(off)} = -8$ *V* ; $V_{GS} = -5$ *V. Find the value of drain current.*

Solution : We know that,

$$I_D = I_{DSS}\left(1 - \frac{V_{GS}}{V_{GS(off)}}\right)^2$$

$$= 30\left[1 - \left(\frac{-5}{-8}\right)\right]^2$$

$$= 30\,(1 - .625)^2$$

$$= 30 \times .140625$$

$$= 4.218 \text{ mA. } \textbf{Ans.}$$

Example 3. *In a self-bias n-channel JFET, the operating point is to be set at* $I_D = 1.5$ *mA and* $V_{DS} = 10$ *V. The JFET parameters are* $I_{DSS} = 5$ *mA and* $V_P = -2$ *V. Find the value of* R_S *and* R_D*. Given that* V_{DD} *20 V.*

Solution : Fig. P (7.2) shows the circuit arrangement.

$$I_D = I_{DSS}\left(1 - \frac{V_{GS}}{V_P}\right)^2$$

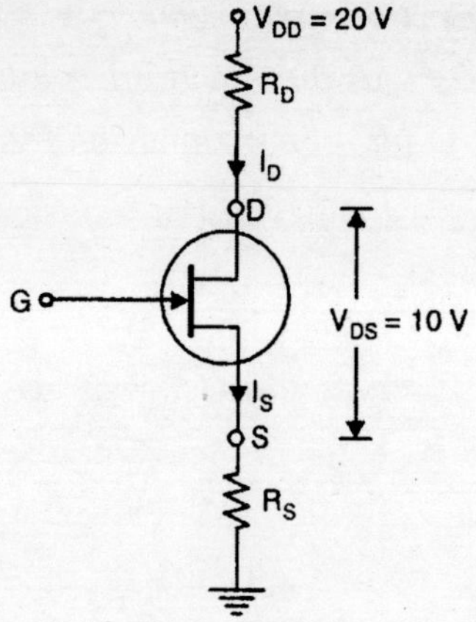

Fig. P (7.2).

or $$1.5 = 5\left(1 + \frac{V_{GS}}{2}\right)^2$$

or $$\left(\frac{V_{GS}}{2} + 1\right) = \sqrt{\frac{1.5}{5}}$$

or $$\frac{V_{GS}}{2} = .55 - 1$$

or $$V_{GS} = -0.9 \text{ V}$$

Now, $$V_{GS} = V_G - V_S$$

$$V_{GS} = 0 - V_S \qquad \{V_G = 0 \text{ V}\}$$

or $$V_S = -V_{GS} = -(-.9) = 0.9 \text{ V}$$

∴ $$R_S = \frac{V_S}{I_D} = \frac{0.9 \text{ V}}{1.5 \text{ mA}} = 0.6 \text{ k}\Omega. \quad \textbf{Ans.}$$

Now, applying KVL at the output section, we have

$$V_{DD} = I_D R_D + V_{DS} + I_D R_S$$

$$20 = 1.5 \times R_D + 10 + 1.5 \times .6$$

$$20 = 1.5\, R_D + 10 + .9$$

or $$1.5\, R_D = 20 - 10.9$$

$$R_D = \frac{9.1}{1.5} \cong 6 \text{ k}\Omega. \quad \textbf{Ans.}$$

Example 4. *In the JFET circuit shown in Fig. P (7.3), find (i)* V_{DS} *(ii)* V_{GS}.

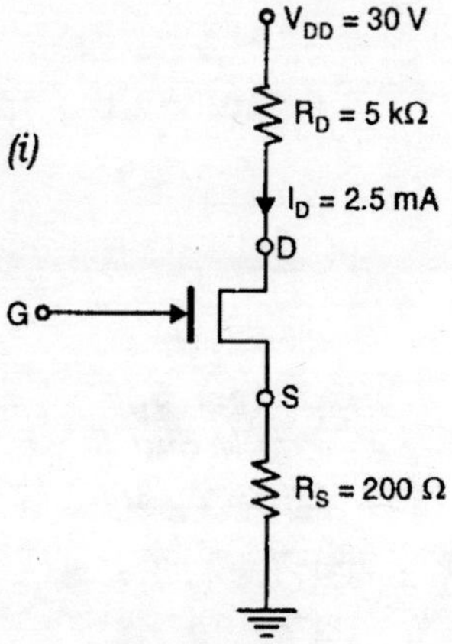

Fig. P (7.3).

Solution : (*i*) Applying KVL, we have

$$V_{DD} = I_D R_D + V_{DS} + I_D R_S$$

or $$V_{DS} = V_{DD} - I_D (R_D + R_S)$$

or $$V_{DS} = 30 - 2.5 \text{ mA } (5000 + 200)$$

or $$V_{DS} = 30 - (2.5 \times 10^{-3} \times 5200)$$

or $$V_{DS} = 30 - 13 = 17 \text{ V}. \quad \textbf{Ans.}$$

(*ii*) $$V_{GS} = -I_D R_S$$

or $$V_{DS} = -2.5 \times 10^{-3} \times 200$$

or $$V_{GS} = -0.5 \text{ V}. \quad \textbf{Ans.}$$

Example 5. *For an n-channel silicon FET with $a = 3 \times 10^{-4}$ cm and $N_D = 10^{15}$ electrons/cm^3, find*

(a) the pinch off voltage and

(b) the channel half-width for $V_{GS} = 1/2\ V_P$ and $I_D = 0$.

Given that, $N_D = 10^{15}$ *electrons/cm^3*

or $N_D = 10^{21}$ *electrons/m^3*

$\epsilon = 12\epsilon_0$

Solution : (*a*) As we know that for Si, the relative dielectric constant is 12, so $\epsilon = 12\epsilon_0$

As per the relation

$$|V_P| = \frac{q\,N_D}{2\epsilon}a^2 \qquad ...(1)$$

Substituting the values of q, ϵ, N_D and a in MKS units in above expression (1)

$$|V_P| = \frac{1.60 \times 10^{-19} \times 10^{21} \times (3 \times 10^{-6})^2}{2 \times 12 \times (36\pi \times 10^9)^{-1}} = 6.8 \text{ V.} \quad \textbf{Ans.}$$

(*b*) We have for V_{GS},

$$V_{GS} = \left(1 - \frac{b}{a}\right)^2 V_P \qquad ...(2)$$

Solving eqn. (2) for *b*, we have

$$b = a\left[1 - \left(\frac{V_{GS}}{V_P}\right)^{\frac{1}{2}}\right]$$

Substituting the approximate values given,

$$b = (3 \times 10^4)\left[1 - \left(\frac{1}{2}\right)^{\frac{1}{2}}\right] = 0.87 \times 10^{-4} \text{ cm.} \quad \textbf{Ans.}$$

Example 6. *The amplifier given in Fig. P (7.4) utilizes an n-channel FET for which $V_P = -2.0$ V and $I_{DSS} = 1.645$ mA. It is desired to bias the circuit at $I_D = 0.8$ mA, using $V_{DD} = 24$ V. Assume $r_d >> R_d$. Find*

(*a*) V_{GS}

(*b*) g_m

(*c*) R_S

(*d*) R_d, *such that the voltage gain is at least 20 dB, with R_S bypassed with a very large capacitance C_S.*

Solution : (*a*) Fig. P (7.4) is an arrangement of self-bias circuit. Using following relationship :

$$I_D = I_{DSS}\left(1 - \frac{V_{GS}}{V_P}\right)^2$$

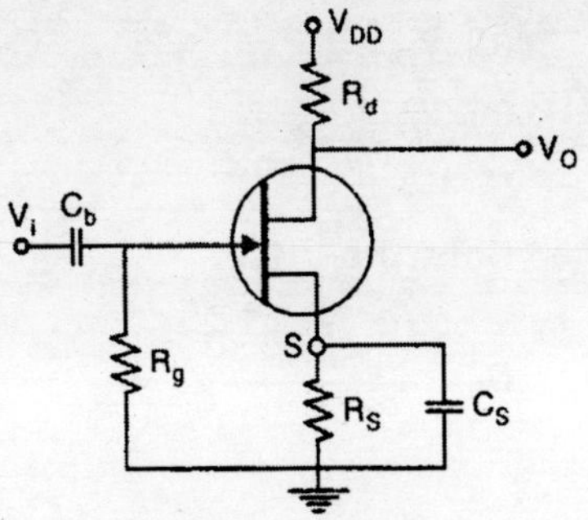

Fig. P (7.4).

we have,

$$0.8 = 1.65\left(1 + \frac{V_{GS}}{2.0}\right)^2$$

Solving above expression

$$V_{GS} = -0.62 \text{ V. } \textbf{Ans.}$$

(*b*) Also,

$$g_{mo} = \frac{-2I_{DSS}}{V_P}$$

Now,

$$g_{mo} = \frac{2(1.65)}{2} = 1.65 \text{ mA/V}$$

$$g_m = g_{mo}\left(1 - \frac{V_{GS}}{V_P}\right)$$

Substituting, required values in above expression

$$g_m = (1.65)\left(1 - \frac{0.62}{2.0}\right)$$

$$= 1.14 \text{ mA/V. } \textbf{Ans.}$$

(*c*)

$$R_S = \frac{-V_G}{I_D}$$

$$= \frac{0.62}{0.8} = 0.77 \text{ k} = 770\ \Omega. \textbf{ Ans.}$$

(*d*) It is given that gain is 20 dB

Now,

$$A_{dB} = 20 \log_{10} (A)$$

$$20 = 20 \log_{10} (A)$$

$\therefore$

$$A = 10$$

It is given that,

$$r_d >> R_d$$

$\therefore$

$$|A_V| = g_m R_d \geq 10$$

or

$$R_d \geq 10/1.14 = 8.76 \text{ k}\Omega. \textbf{ Ans.}$$

Example 7. *Determine the following for the network given in Fig. P (7.5) (where Q indicates quiescent condition)*

(*a*) V_{GS}

(*b*) I_{DQ}

(*c*) V_{DS}

(*d*) V_D

(*e*) V_G

(*f*) V_S.

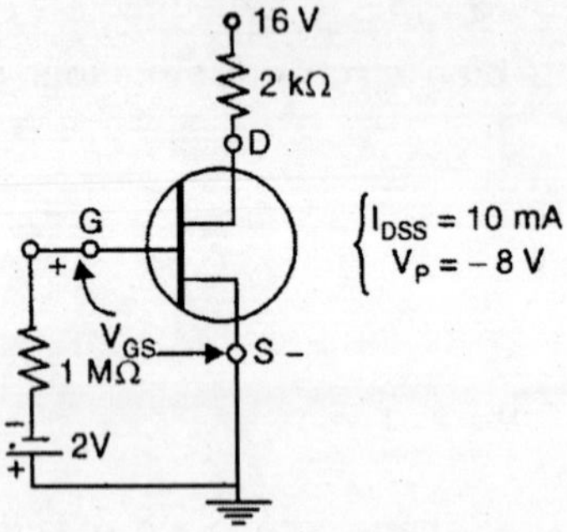

Fig. P (7.5).

Solution : (*a*) $V_{GS} = -V_{GG} = -2\text{V}$. **Ans.**

(*b*) $$I_{DQ} = I_{DSS}\left(1 - \frac{V_{GS}}{V_P}\right)^2$$

$$= 10 \text{ mA}\left(1 - \frac{-2\text{V}}{-8\text{V}}\right)^2$$

$= 5.625$ mA. **Ans.**

(*c*) $V_{DS} = V_{DD} - I_D R_D$

$= 16 - 5.625 \times 10^{-3} \times 2 \times 10^3$

$= 16 - 11.25 = 4.75$ V. **Ans.**

(*d*) $V_D = V_{DS} = 4.75$ V. **Ans.**

(*e*) $V_G = V_{GS} = -2$ V. **Ans.**

(*f*) $V_S = 0$ V. **Ans.**

Example 8. *When a reverse voltage of 10 V is applied between gate and source of JFET the gate current is 0.001 μA. Determine resistance between gate and source.*

Solution : Gate to source voltage, $V_{GS} = 10.0$ V

Gate current $I_G = 0.001\ \mu\text{A} = 1 \times 10^{-9}$ A

Now, gate to source resistance,

$$R_{GS} = \frac{V_{GS}}{I_G} = \frac{10}{1 \times 10^{-9}} = 10000 \text{ M}\Omega. \quad \textbf{Ans.}$$

Example 9. *When a drain-source voltage is changed by 1.5 volts, the change in drain current is of 120 μA, the gate-source voltage remaining unchanged. Determine the ac drain resistance of the JFET.*

Solution : Change in drain-source voltage,

$$\Delta V_{DS} = 1.5 \text{ V}$$

Change in drain current,

$$\Delta I_D = 120 \times 10^{-6} \text{ A}$$

∴ AC drain resistance of the JFET,

$$r_d = \frac{\Delta V_{DS}}{\Delta I_D} = \frac{1.5}{120 \times 10^{-6}} = 12.5 \text{ k}\Omega. \quad \textbf{Ans.}$$

Example 10. *In a JFET the drain current changes from 1.2 mA to 1.5 mA when the gate to source voltage is varied from – 4.25 V to – 4.10 V, keeping the drain source voltage constant. Determine the transconductance for the given JFET.*

Solution : Change in drain currents,

$$\Delta I_D = I_{D_2} - I_{D_1}$$
$$= 1.5 - 1.2 = 0.3 \text{ mA}$$

Change in gate to source voltage,

$$\Delta V_{GS} = V_{GS_2} - V_{GS_1}$$
$$= -4.10 - (-4.25)$$
$$= 0.15 \text{ V}$$

∴ Transconductance, $$g_m = \frac{\Delta I_D}{\Delta V_{GS}} = \frac{0.3 \text{ mA}}{0.15 \text{ V}}$$
$$= 2 \text{ mA/V. } \textbf{Ans.}$$

Example 11. *An FET follows the following relation,*

$$I_D = I_{DSS}\left[1 - \frac{V_{GS}}{V_P}\right]^2$$

I_{DSS} = 8.4 mA, V_P = – 3 V, what is the value of I_D for V_{GS} = – 1.5 V ? Find g_m at this point.

Solution : Drain source saturation current,

$$I_{DSS} = 8.4 \text{ mA}$$

Pinch off voltage, $V_P = -3$ V

Gate-source voltage, $V_{GS} = -1.5$ V

Drain current $$I_D = I_{DSS}\left(1 - \frac{V_{GS}}{V_P}\right)^2$$
$$= 8.4\left(1 - \frac{-1.5}{-3}\right)^2$$
$$= 2.1 \text{ mA. } \textbf{Ans.}$$

Transconductance, $$g_{mo} = \frac{-2\, I_{DSS}}{V_P}$$
$$= \frac{-2 \times 8.4}{-3} = 5.6 \text{ mA/V}$$

Transconductance, $$g_m = g_{mo}\left(1 - \frac{V_{GS}}{V_P}\right)$$
$$= 5.6\left(1 - \frac{-1.5}{-3}\right)$$
$$= 2.8 \text{ mA/V. } \textbf{Ans.}$$

Example 12. *For a p-channel silicon FET a = 2 × 10^{-4} cm, ρ = 10 Ω. cm, find*

(*a*) *The pinch off voltage* (V_p) *for Si*

(*b*) *The pinch off voltage* (V_p) *for Ge*

Given that for Si, $\epsilon = 12\epsilon_0$

For Ge, $\epsilon = 16\epsilon_0$

where $\epsilon_0 = 8.86 \times 10^{-12}\ c^2/Nm^2$

$\mu_p = 500\ cm^2/V\text{-sec for Si}$

$\mu_p = 1800\ cm^2/V\text{-sec for Ge.}$

Solution : We know that $$V_p = \frac{q N_A a^2}{2\epsilon} \qquad ...(1)$$

Also, we know that $$\sigma = n\mu q$$

or $$\sigma = N_A \mu q$$

or $$\frac{1}{\rho} = N_A \mu_q$$

$$N_A = \frac{1}{\rho\mu q} \qquad ...(2)$$

Now, from eqns. (1) and (2)

$$V_P = \frac{q}{2\epsilon}\frac{1}{\rho\mu q}a^2 = \frac{a^2}{2\mu\epsilon\rho}$$

or $$V_P = \frac{a^2}{2\mu\epsilon\rho}$$

For *Si*, $$V_P = \frac{a^2}{2\mu_p(12\epsilon_0)\rho} \qquad \{\because \epsilon = 12\epsilon_0 \text{ for Si}\}$$ **Ans.**

For *Ge*, $$V_P = \frac{a^2}{2\mu_p(16\epsilon_0)\rho} \qquad \{\because \epsilon = 16\epsilon_0 \text{ for Ge}\}$$ **Ans.**

Example 13. *Show that for small values of* V_{GS} *compared with* V_P, *the drain current is given approximately by* $I_D \cong I_{DSS} + g_{mo} V_{GS}$.

Solution : We know that

$$I_D = I_{DSS}\left(1 - \frac{V_{GS}}{V_P}\right)^2$$

$$= I_{DSS}\left[1 + \left(\frac{V_{GS}}{V_P}\right)^2 - \frac{2V_{GS}}{V_P}\right]$$

$$= I_{DSS} + I_{DSS}\left\{\left(\frac{V_{GS}}{V_P}\right)^2 - 2\frac{V_{GS}}{V_P}\right\}$$

$$= I_{DSS} + V_{GS}\left\{I_{DSS} \cdot \frac{V_{GS}}{V_P{}^2} - \frac{2I_{DSS}}{V_P}\right\}$$

$\therefore$ Provided that $V_{GS} << V_P$

$$\therefore \quad \frac{V_{GS}}{V_P{}^2} \simeq 0$$

$$I_D \cong I_{DSS} + V_{GS}\left(\frac{-2I_{DSS}}{V_P}\right)$$

or $$I_D \cong I_{DSS} + V_{GS} \cdot g_{mo.}$$ **Hence proved** $\left\{\because g_{mo} = \frac{2I_{DSS}}{V_P}\right\}$

Example 14. *For the FET given in Fig. P (7.6). Determine*

(*i*) g_m , r_d

(*ii*) Z_i

(*iii*) Z_O

(*iv*) A_V

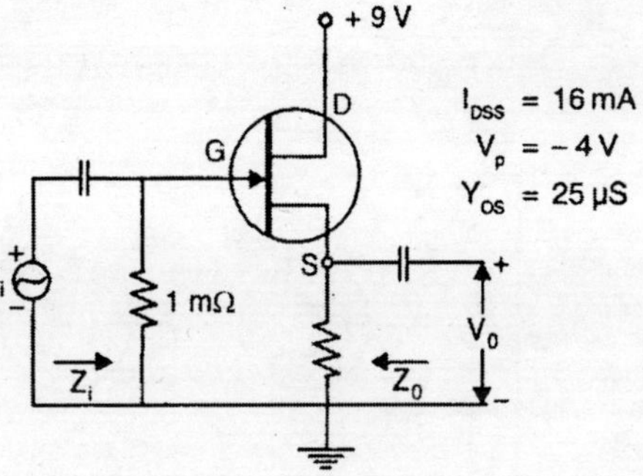

Fig. P (7.6).

Solution : (*i*) $$g_m = g_{mo}\left(1 - \frac{V_{GS}}{V_P}\right)$$

$$= \frac{2I_{DSS}}{|V_P|}\left(1 - \frac{V_{GS}}{V_P}\right)$$

$$= \frac{2 \times 16}{4}\left(1 - \frac{2.86}{4}\right) = 2.28 \text{ mS or mA/V}$$ **Ans.**

Now, $$r_d = \frac{1}{Y_{OS}} = \frac{1}{25 \times 10^{-6}} = 40 \text{ k}\Omega$$ **Ans.**

(*ii*) $Z_i = 1 \text{ M}\Omega$ **Ans.**

(*iii*) Calculation for Z_0

$$Z_{0F}(\text{int}) = \frac{r_d + R_d}{\mu + 1}$$

int → *internal*

$$= \frac{40 \times 10^3}{1 + 91.2} = \frac{40}{92.2} \times 10^3$$

$$\left\{ \begin{aligned} \mu &= r_d \times g_m \\ &= 40 \times 10^3 \times 2.28 \times 10^{-3} \\ &= 40 \times 2.28 = 92.2 \end{aligned} \right\}$$

Now,
$$Z_0 = Z_{0F}\ (\text{int}) \parallel 2.2 \times 10^3$$

$$= \frac{40 \times 10^3}{92.2} \parallel 2.2 \times 10^3$$

$$= 362.52\ \Omega$$ **Ans.**

(*iv*)
$$A_v = \frac{\mu R_S}{r_d + R_d + (\mu + 1) R_S}$$

$$= \frac{92.2 \times 2.2 \times 10^{-3}}{40 \times 10^3 + (92.2)\, 2.2 \times 10^3}$$

$$\left\{ \begin{aligned} R_d &= 10 \\ R_S &= 2.2 \\ \mu &= 92.2 \end{aligned} \right.$$

$$= 0.83$$ **Ans.**

Example 15. *If two FETs are connected in parallel as shown in Fig. P (7.7), find the effective g_m and r_d.*

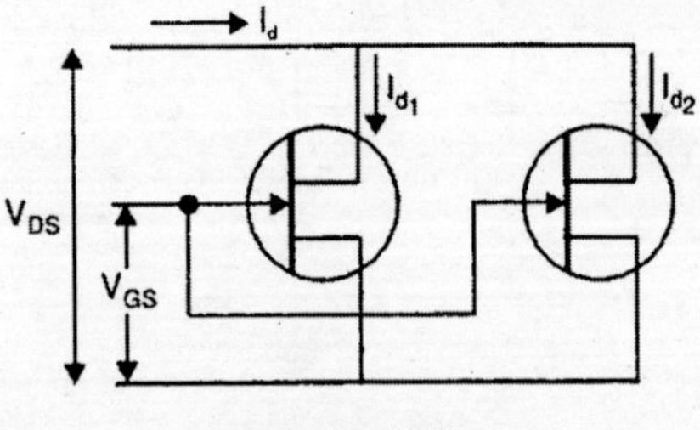

Fig. P (7.7).

Solution : As we know that

$$I_d = g_m V_{GS} + \frac{1}{r_d} V_{DS}$$

Consider that $V_{DS} = 0$

then $I_d = g_m V_{GS}$ ...(1)

We know that $I_d = I_{d_1} + I_{d_2}$

$$g_m = g_{m_1} \times V_{GS} + g_{m_2} V_{GS} \quad \text{...(2)}$$

or
$$g_m = (g_{m_1} + g_{m_2}) V_{GS}$$

{on comparing eqn. (1) and (2)}

Thus, effective $g_m = g_{m_1} + g_{m_2}$ **Ans.**

Now, consider $V_{GS} = 0$

then $$I_d = \frac{V_{DS}}{r_d}$$

$$\therefore \quad I_d = I_{d_1} + I_{d_2}$$

$$= \frac{V_{DS}}{r_{d_1}} + \frac{V_{DS}}{r_{d_2}} = V_{DS}\left(\frac{1}{r_{d_1}} + \frac{1}{r_{d_2}}\right)$$

$$\therefore \quad \text{Effective } r_d = \frac{V_{DS}}{I_d} = \frac{1}{\frac{1}{r_{d_1}} + \frac{1}{r_{d_2}}} = \frac{r_{d_1} r_{d_2}}{r_{d_1} r_{d_2}} \quad \textbf{Ans.}$$

NOTE : Effective g_m when two FETs are connected in parallel will be equal to $g_{m_1} + g_{m_2}$ and effective r_d in parallel will be equal to $\frac{r_{d_1} r_{d_2}}{r_{d_1} + r_{d_2}}$.

Example 16. *(a) Calculate the voltage gain of the FET amplifier shown in Fig. P (7.8) assuming blocking capacitor C_b to be large. $g_m = 2$ mA/V and $r_d = 10$ K.*

(b) Repeat if $C_b = 2$ nF at frequency of input 1 kHz.

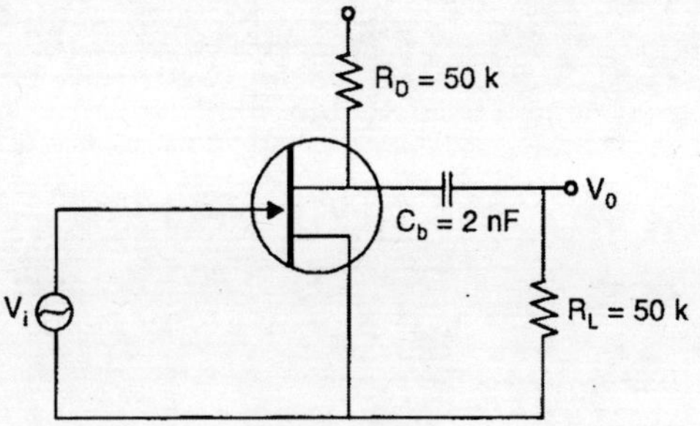

Fig. P (7.8).

Solution : (*a*) If C_b is large, it acts as a short circuit $\left(\therefore X_C = \frac{1}{2\pi f C_b}\right)$

so, $$\text{effective load} = r_d \parallel R_D \parallel R_L$$

$$= 10 \parallel 50 \parallel 50$$

$$= 7.14 \text{ k}$$

$\therefore$ Voltage gain $A_V = -g_m R = -2 \times 7.14 = -14.28$

(*b*) For small value of C_b, consider R_L separately

Ignoring R_L, $A_V = -g_m r_d \parallel R_D$

$$= -2 \times \frac{10 \times 50}{10 + 50} = -16.67$$

$$R_0 = \text{output impedance} = r_d \parallel R_D$$
$$= 8.33 \text{ k}$$

$$X_C = \frac{1}{\omega C} = \frac{1}{6.28 \times 1000 \times 2 \times 10^{-9}} = 75.5 \text{ k}\Omega$$

Now consider the loading effect of $R_L - jX_C$

$\therefore$
$$A_V = 16.67 \times \frac{R_L}{-jX_C + R_0 + R_L}$$

$$= 16.67 \frac{50}{-j79.5 + 8.33 + 50}$$

$$= 8.44 \angle 53.6°. \quad \textbf{Ans.}$$

Example 17. *The amplifier shown in Fig. P (7.9) uses a FET with* I_{DSS}, $= 3$ *mA ;* $V_{GS} = -0.47$ *V;* $V_P = -2.4$ *V,* $r_d >> R_D$. *Find the quiescent drain current, quiescent drain to source voltage, and small signal voltage gain* A_V.

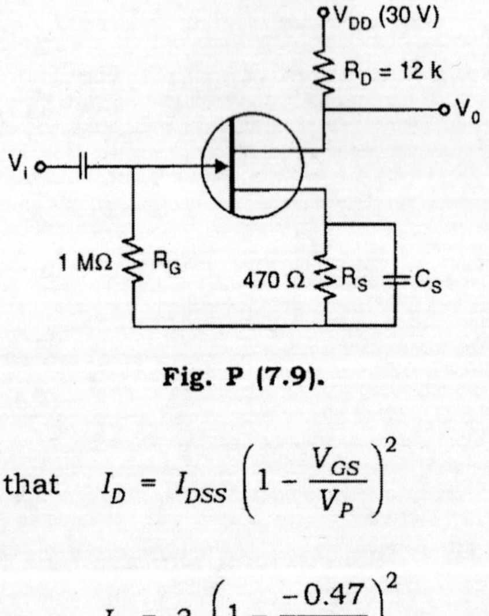

Fig. P (7.9).

Solution : We know that $I_D = I_{DSS}\left(1 - \frac{V_{GS}}{V_P}\right)^2$

$\therefore$
$$I_D = 3\left(1 - \frac{-0.47}{-2.4}\right)^2$$

$\therefore$
$$\frac{1}{3} I_D = (1 - 0.1958\ I_d)^2$$

$\therefore$
$$0.03835\ I_d^2 - 0.7249 + 1 = 0$$

$\therefore$
$$I_d = \frac{0.7249 \pm [0.5255 - 0.1534]^{1/2}}{2 \times 0.03834}$$
$$= 1.496 \text{ mA} \simeq 1.5 \text{ mA}$$

$\therefore$
$$V_D = V_{DD} - I_d R_D$$
$$= 30 - 1.5 \times 12 = 12 \text{ V}$$

$$V_S = I_d R_S = 0.47 \times 1.5 = 0.7 \text{ V}$$

∴ Quiescent point $\quad V_{DS} = 12 - 0.7 = 11.3 \text{ V}$

$$g_m = g_{mo}\left(1 - \frac{V_{GS}}{V_P}\right) = \frac{-2I_{DSS}}{V_P}\left(1 - \frac{V_{GS}}{V_P}\right)$$

$$= \frac{-2 \times 3}{-2.4}\left(1 - \frac{-0.7}{-2.4}\right) = 1.76 \text{ mA/V}$$

∴ $\quad A_V = -g_m R_D$

$$= -1.76 \times 12 = -21.18. \quad \textbf{Ans.}$$

Example 18. *The FET of given Fig. P (7.10) has I_{DSS} = 5.6 mA and V_P = – 4 V. Find the voltage gain A_V of the source follower circuit using the small signal ac model and g_m = 2 mA/V. Find the changes in V_0 due to a 10 V change in V_i.*

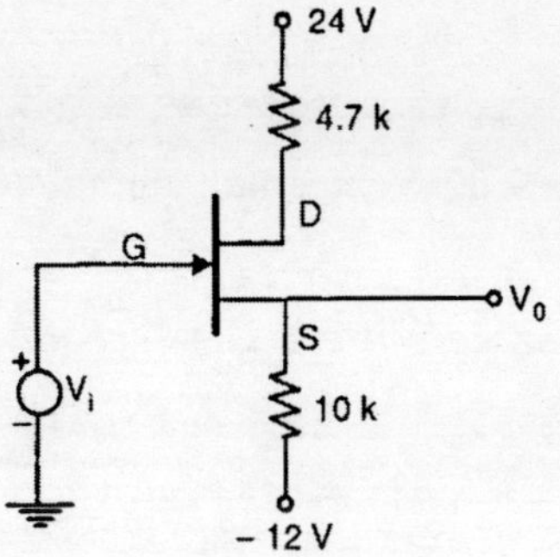

Fig. P (7.10).

Solution : Using the source follower formula,

$$A_V = \frac{g_m R_S}{1 + g_m R_S} = \frac{2 \times 10}{1 + 2 \times 10} = 0.952$$

∴ $\quad \Delta V_0 = A_v \, \Delta V_i = 9.52 \text{ V} \qquad \{\because V_i = 10 \text{ V}\}$ **Ans.**

Example 19. *Referring to source follower circuit of Fig. P (7.10), compare the result obtained by using the large signal model and V_i swings from 0 to 10 V.*

Solution : Consider $\quad V_i = 0$

∴ $$I_D = I_{DSS}\left(1 - \frac{V_{GS}}{V_P}\right)^2$$

$$= 5.6\left(1 - \frac{12 - 10\, I_D}{-4}\right)^2$$

On simplifying we get,

$$35\, I_D^2 - 113\, I_D + 89.6 = 0$$

∴ $\quad I_D = 1.4 \text{ mA or } 2.32 \text{ mA}$

Taking lower value, $\quad I_D = 1.4 \text{ mA}$

$$\therefore \quad V_0 = V_{SS} + I_D R_S$$
$$= -12 + 1.4 \times 10 = 2 \text{ V}$$

Now consider $V_i = 10$ V

$$\therefore \quad I_D = I_{DSS}\left(1 - \frac{V_{GS}}{V_P}\right)^2$$

$$= 5.6\left(1 - \frac{10 + 12 - 10\, I_D}{-4}\right)^2$$

Simplifying, we get

$$6.25\, I_D^2 - 32.678\, I_D + 42.225 = 0$$

$$\therefore \quad I_D = 2.34 \quad \text{or} \quad 2.78 \text{ mA}$$

Again taking lower value, $I_D = 2.34$ mA

$$\therefore \quad V_0 = -12 + 10 \times 2.34 = 11.4 \text{ V}$$

$$\therefore \quad \Delta V_0 = 11.4 - 2 = 9.4 \text{ V}$$

This compares well with 9.52 V of small signal model. **Ans.**

Example 20. *Find the quiescent values of V_{DS}, I_d and V_{GS} for the amplifier shown in Fig. P (7.11). The FET parameters are $I_{DSS} = -4$ mA, $V_P = 4$ V.*

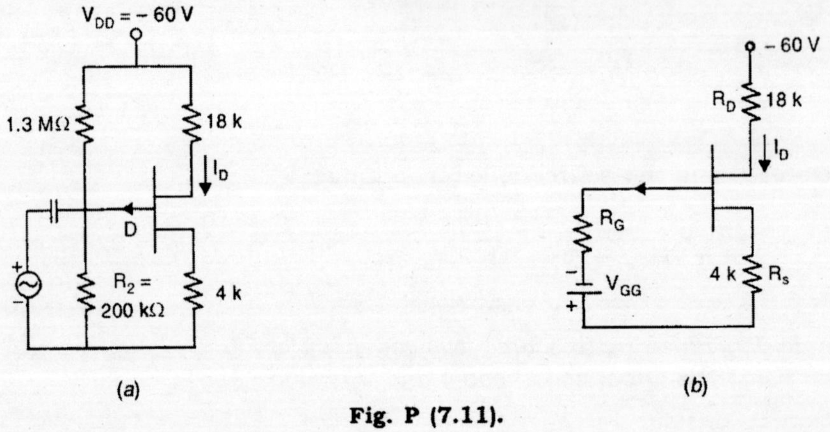

Fig. P (7.11).

Solution : Replace the biasing network of Fig. P (7.11 *a*) by its Thevenin equivalent as shown in Fig. P (7.11 *b*).

Now,
$$V_{GG} = -60 \times \frac{200}{200 + 1300} = -8 \text{ V}$$

$$R_G = \frac{200 \times 1300}{200 + 1300} = 173.3 \text{ k}$$

$$I_d = I_{DSS}\left(1 - \frac{V_{GS}}{V_P}\right)^2 \qquad \text{...(1)}$$

Since $V_{GS} = V_G - V_S$

or $V_{DS} = -8 - 4\,I_d$ ...(2)

Now, from eqns. (1) and (2), we have

$$I_d = -4\left(1 - \frac{-8 - 4I_d}{4}\right)^2$$

$\therefore$ $-0.25\,I_d = (3 + I_d)^2 = I_d^2 + 6I_d + 9$

$\therefore$ $I_d^2 + 6.25\,I_d + 9 = 0$

$\therefore$ $I_d = \dfrac{-6.25 \pm (6.25^2 - 36)^{1/2}}{2} = -2.25$ mA. **Ans.**

$\therefore$ $V_{GS} = -8 - 4 \times (-2.25) = 1$ V. **Ans.**

$\therefore$ $V_{DS} = V_{DD} - I_D\,(R_D + R_S)$

$= -60 + 2.25\,(18 + 4)$

$V_{DS} = -10.5$ V. **Ans.**

Example 21. *For the network shown in Fig. P (7.12), determine*

(*a*) I_{DQ} *and* V_{GSQ}

(*b*) V_{DS}.

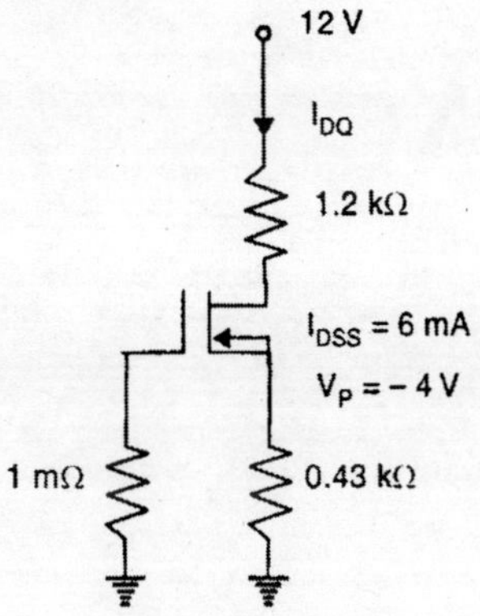

Fig. P (7.12).

Solution : Since $I_G \cong 0$ A

$V_G = I_G R_G = 0$ V

$V_{GSQ} = V_G - V_S = 0 - I_S R_S$

For FETs, $I_D = I_S$

So, $V_{GSQ} = -I_D R_s$

We know that

$$I_{DQ} = I_{DSS}\left(1 - \frac{V_{GS}}{V_P}\right)^2$$

$$= 6\text{ mA}\left(1 - \frac{(-I_D R_S)}{(-4)}\right)^2$$

$$I_D = 6\left(1 - \frac{I_D 0.43}{4}\right)^2$$

or $0.069\,I_D^2 - 2.2\,I_D + 6 = 0$

or $I_{DQ} = \dfrac{2.2 \pm 1.78}{0.138} = 28.84$ mA, 3.04 mA

Neglecting I_D = 28.84 mA (because for this value of I_{DQ}, voltage drop at R_D = 1.2 kΩ will be ($1.2 \times 10^3 \times 28.84 \times 10^{-3}$ = 34.60 V) which is impossible with a supply of 12 V)

So, I_{DQ} = 3.04 mA **Ans.**

and $V_{GSQ} = -I_D R_S = -(3.04)(0.43) = -1.307$ V.

(*b*) Apply KVL we get

$$V_{DS} - I_D R_D - V_{DS} - I_S R_S = 0$$

or $$V_{DS} = V_{DD} - I_D(R_D + R_S)$$

or $$V_{DS} = 12 - 3.04 \text{ mA } (1.2 + 0.43) \times 10^3 \ \Omega$$

or $$V_{DS} = 12 - 3.04 \times 10^{-3} \times 1.63 \times 10^3$$

or $$V_{DS} = 7.045 \text{ V}$$ **Ans.**

□□□

8

METAL OXIDE SEMICONDUCTOR FIELD EFFECT TRANSISTOR (MOSFET)

8.1. INTRODUCTION

In the preceding chapter we have discussed the drain and transfer characteristics and small signal properties of junction field-effect transistor (JFET). We now turn our attention to the metal oxide-semiconductor field effect transistor (MOSFET) or insulated-gate field-effect transistor (IGFET). The MOSFET transistor has become one of the most important devices used in the design and construction of integrated circuits for digital computers. Its thermal stability and other general characteristics make it extremely popular in computer circuit design. MOSFET can be broadly classified in depletion-type MOSFET and enhancement type MOSFET and each is further divided in *n*-channel and *p*-channel MOSFET. The characteristics and properties of depletion-type MOSFET is same as that of JFETs, both follow the *Shokley equation*. However, the properties of enhancement-type MOSFET and JFETs are quite different. The basic difference between the depletion and enhancement MOSFET is that in the depletion-type MOSFET the channel is already formed while in the enhancement-type MOSFET channel is not initially formed. Other main important difference between these two is that in the depletion gate is always reverse biased with the drain however, in the enhancement MOSFET gate is forward biased with drain that's why drain current increases with increasing the gate voltage *i.e.*, enhance the drain current, that's why called enhancement MOSFET.

One more interesting difference between the JFET and MOSFET is that in the MOSFET there is no direct electrical connection between the gate terminal and the channel of MOSFET which were present in the JFET, this is due to the fact that MOSFET transistor uses an additional insulator layer of SiO_2 which provides good electrical isolation, this is the region why the input impedance of MOSFET is very high as compare to the JFET.

8.2. DEPLETION-TYPE MOSFET

The basic construction of the *n*-channel depletion-type MOSFET is shown in Fig. 8.1. For the *n*-channel depletion type MOSFET substrate (which is very highly doped acts as a base) is *p*-type formed. It may be noted that substrate is the foundation upon which the device will be constructed. In some cases the substrate is internally connected to the source terminal. However, may discrete devices provide an additional terminal labelled *SS*, resulting in a four terminal device.

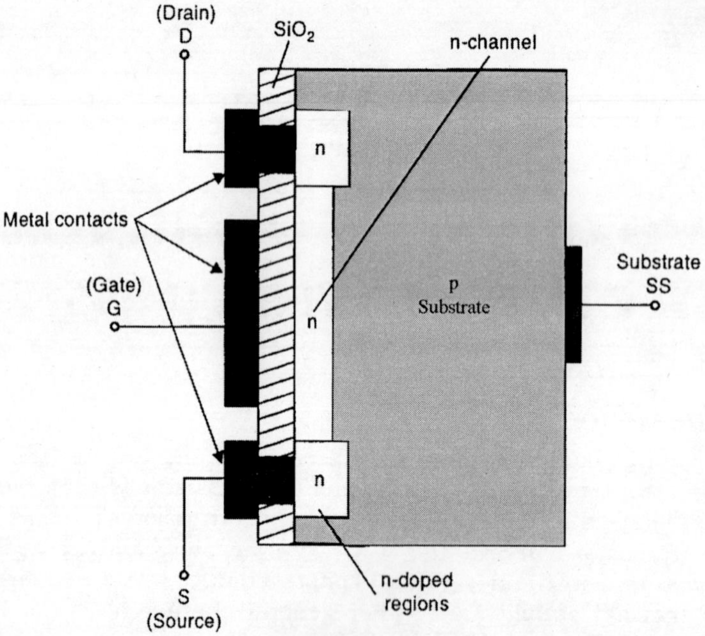

Fig. 8.1. *n-channel depletion-type MOSFET.*

From Fig. 8.1 it is obvious that source (*S*) and drain (*D*) terminals are connected through metallic contacts to *n*-doped regions linked by *n*-channel. The gate is also connected to a metal contact surface but remains insulated from the *n*-channel by a very thin silicon dioxide (SiO_2) layer. This SiO_2 layer acts as a dielectric that sets up opposing electric fields within the dielectric when exposed to an externally applied field.

Reason why SiO_2 layer is used as a dielectric. There are two main reason why SiO_2 layer is used as a dielectric :

(1) It provides better electrical isolation between gate and channel, or in other words we can say that there is no electrical connection between the gate terminal and the channel of a MOSFET.

(2) Due to better electrical isolation its input impedance becomes very large. Which is very good to overcome the loading effect, which is generally occurred in the device.

The very high input impedance continues to fully support the fact that the gate current (I_G) is essentially zero amperes for dc-biased configurations.

8.2.1. Basic Operation and Characteristics of n-channel Depletion-Type MOSFET

Fig. 8.2 shows *n*-channel depletion type MOSFET with a gate to source voltage (V_{GS}) is zero volt and applied voltage V_{DD}.

As soon as an voltage V_{DS} (or V_{DD}) is applied across the drain-to-source terminals. The result is an attraction for the positive potential at the drain by the free electrons of the *n*-channel and a current similar to that established through the channel of the JFET. From Fig. 8.2 it is obvious that the resulting current with V_{GS} = 0 V is I_{DSS} (*i.e.*, saturation current)

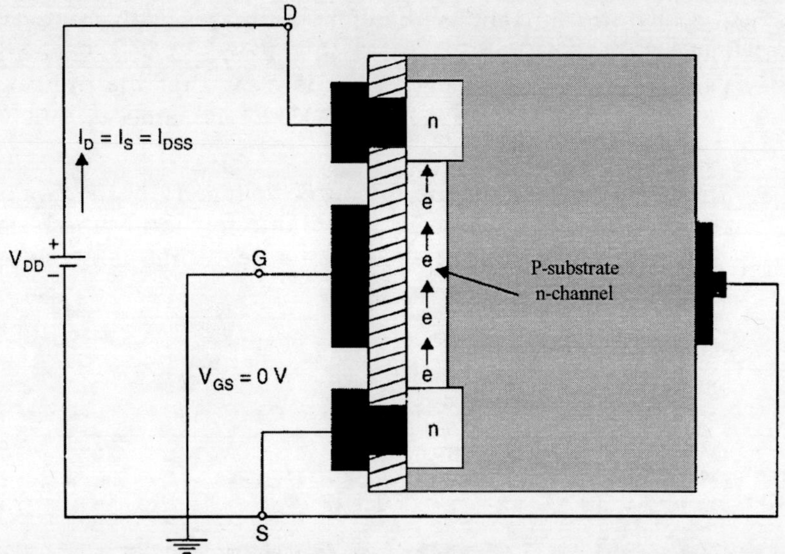

Fig. 8.2. *n-channel depletion type MOSFET with V_{GS} = 0 V and an applied voltage V_{DD}.*

Now, consider the situation when the drain-to-source (V_{DS}) voltage is held constant while the gate-to-source (V_{GS}), voltage is varied. As V_{GS} has been set a negative voltage such as – 1 V, the negative voltage at the gate will tend to pressure electrons towards the *p*-type substrate, this is due to the fact that dielectric creates the opposite sign on the left side while the same sign (*i.e.*, potential) on the right side.

Resulting this attract opposite charges and repel the like charges (*i.e.*, electrons here). Due to this result the *recombination* (as discussed earlier) phenomenon takes place. This can be better understand by Fig. 8.3. Depending on the magnitude of the negative bias established by V_{GS}, a level of recombination between electrons and holes will occur that will reduce the number of free electrons in the *n*-channel available for conduction.

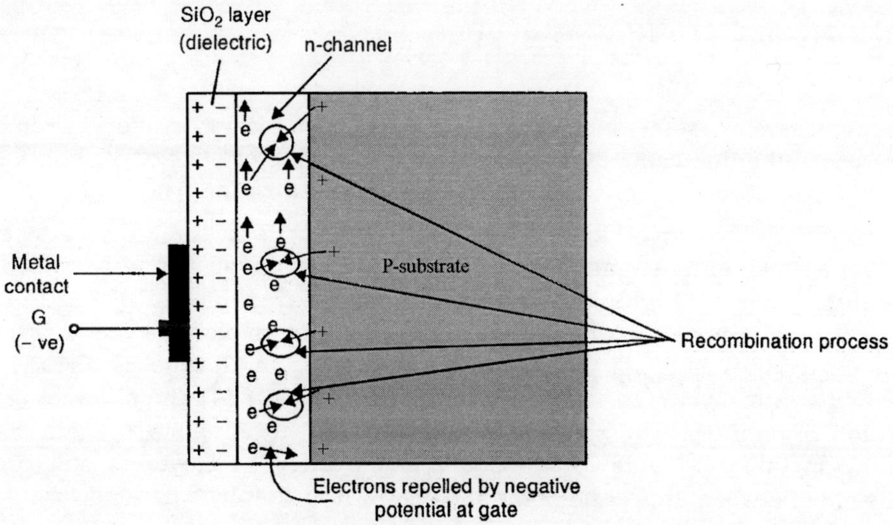

Fig. 8.3. *Reduction in free carriers in channel due to a negative potential at the gate terminal.*

Thus, the value of drain current is therefore decreases with increasing negative bias for V_{GS} as show in Fig. 8.4. For V_{GS} = – 1 V, – 2 V, – 3 V, – 4 V, and so on, to the pinch off level of – 8 V. From the Fig. 8.4 it is clear that the drain and transfer characteristics for an *n*-channel depletion type MOSFET is same as that were exactly described for the JFET.

However, for the positive value of gate-to-source voltage (V_{GS}) the positive gate will draw additional electrons from the *p*-type substrate due to the reverse leakage current and establish new carriers through the collisions resulting between accelerating particles.

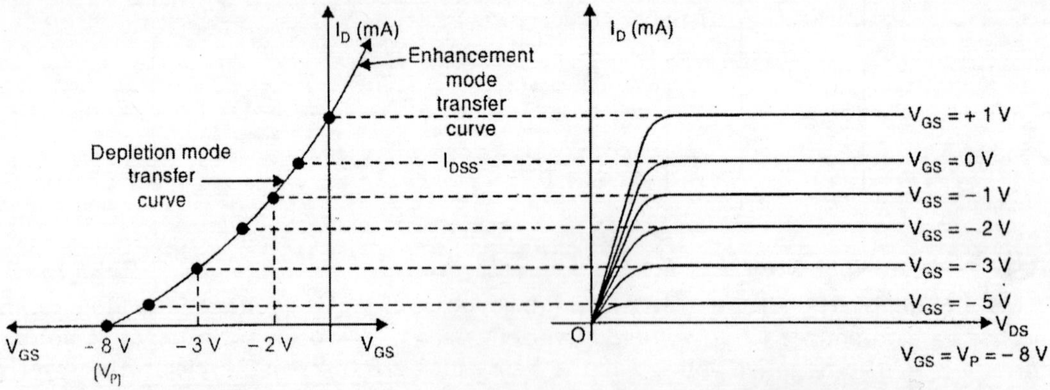

(*a*) Transfer characteristics for an n-channel depletion type MOSFET

(*b*) Drain characteristics for an n-channel depletion type MOSFET

Fig. 8.4.

From Fig. 8.4, it is clear that, the application of a positive gate-to-source voltages (V_{GS}) has enhanced the level of free electrons (charge carriers) in the channel compared to that encountered with V_{GS} = 0 V. For this reason the region of positive gate voltages on the drain or transfer characteristics is often referred to as the *enhancement region.*

8.3. ENHANCEMENT-TYPE MOSFET

8.3.1. Basic Construction

Although there are some similarities in construction and mode of operation between depletion-type and enhancement-type MOSFET, the characteristics of the enhancement-type MOSFET are quite different from anything discussed so far. The basic construction of *n*-channel enhancement type MOSFET is shown in Fig. 8.5.

The *n*-channel enhancement-type MOSFET consists of a highly doped *p*-type substrate into which two highly doped *n*-regions are diffused. From Fig. 8.5 it is obvious that the main difference between the construction of depletion-type MOSFET and enhancement-type MOSFET is that in the former case the channel is already formed, however in the latter case (*i.e.*, in enhancement type MOSFET) the channel is initially no formation of channel. The source (*S*) and drain (*D*) terminals are again connected through the metallic contacts to *n*-doped regions. The SiO_2 layer is still present to isolate the gate between drain and source, but now it is simply separated from a section of the *p*-type material.

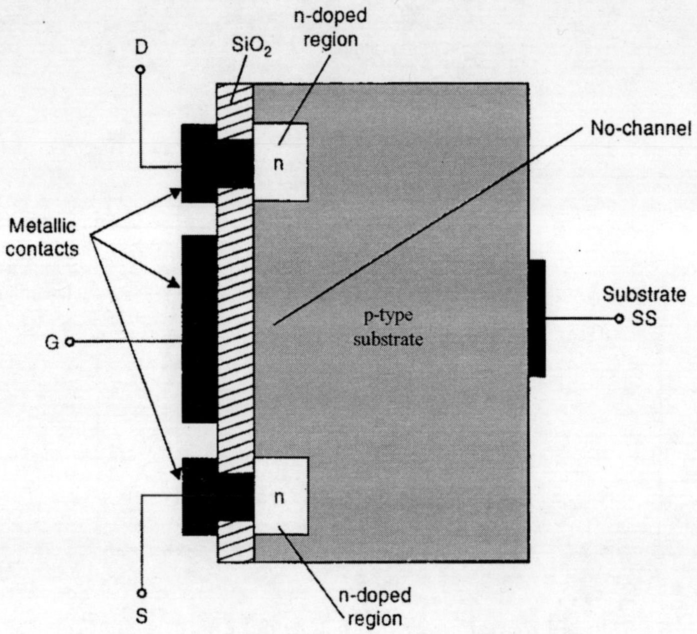

Fig. 8.5. *n-channel enhancement-type MOSFET.*

Finally we conclude that the construction of an enhancement-type MOSFET is quite similar to that of the depletion-type MOSFET except for the absence of a channel between the drain and source terminal.

8.3.2. Basic Operation and Characteristics of Enhancement-Type MOSFET

If V_{GS} is set at 0 V and a voltage applied the drain and source of the device of Fig. 8.5, the absence of an *n*-channel will result in a current (I_D) effectively zero amperes quite different from the depletion-type MOSFET and JFET where $I_D = I_{DSS}$ at $V_{GS} = 0$V.

When both $V_{GS} > 0$ V and $V_{DS} > 0$ V. When both V_{DS} and V_{GS} have been set at positive voltage greater than zero volts, establishing the drain and gate at a positive potential with respect to the source as shown in Fig. 8.6.

As the positive potential is applied between gate and source, due to the presence of SiO_2 layer which acts as a dielectric, is attracts the charge carrier (*i.e.*, electrons) from the substrate. The SiO_2 layer and its insulating qualities will prevent the negative carriers from being absorbed at the gate terminal.

As V_{GS} increases in magnitude the concentration of electrons-near the SiO_2 surface increases until eventually the induced *n*-type region can support a measurable flow between drain and source, resulting the formation of *inversion layer.* This inversion layer is formed when a certain gate-to-source voltage (V_{GS}) is applied. Thus, the minimum value of gate voltage at which inversion of semiconductor surface-take place is known as *threshold voltage* (V_T).

Since the channel is non-existent with $V_{GS} = 0$ V and "enhanced" by the application of a positive gate-to-source voltage, this type of MOSFET is called an enhancement-type MOSFET.

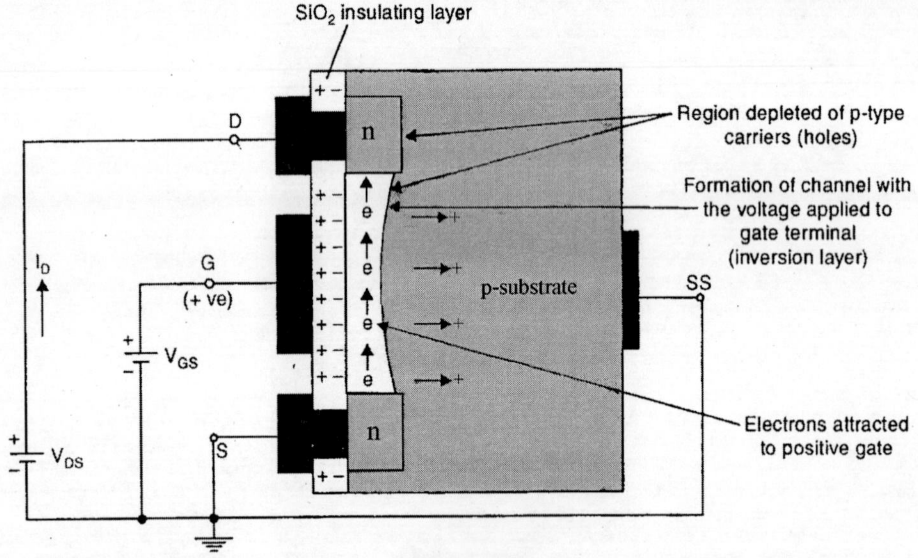

Fig. 8.6. *Channel formation in the n-channel enhancement-type MOSFET.*

As V_{GS} is increased beyond the threshold level the density of free carriers in the induced channel will increase, resulting in an increased level of drain current. However, if we hold V_{GS} constant and increase the level of V_{DS}, the drain current will eventually reach a saturation level as occurred for the JFET and depletion-type MOSFET. In short :

→ At $V_{DS} < V_{GS} - V_T$: An increase in V_{DS} causes the I_D to increase linearly and the MOSFET behaves as a resistance.

→ At $V_{DS} = V_{GS} - V_T$: This results in a slow increase of I_D with increase in V_{DS}.

→ At $V_{DS} > V_{GS} - V_T$: Further increase in V_{DS} produce no change in current and the current saturation occur.

The drain and transfer characteristics of *n*-channel enhancement-type MOSFET is shown in Fig. 8.7.

From Fig. 8.7 it is clear that for the values of V_{GS} less than the threshold level the drain current of an enhancement-type MOSFET is 0 mA.

For levels of $V_{GS} > V_T$ the drain current is related .to the applied gate-to-source voltage by the following non-linear relationship :

$$\boxed{I_D = K\,(V_{GS} - V_T)^2} \qquad \text{...(8.1)}$$

where V_{GS} = Applied gate-to-source voltage

V_T = Threshold voltage

I_D = Drain current

K = Constant that is a function of the construction of the device.

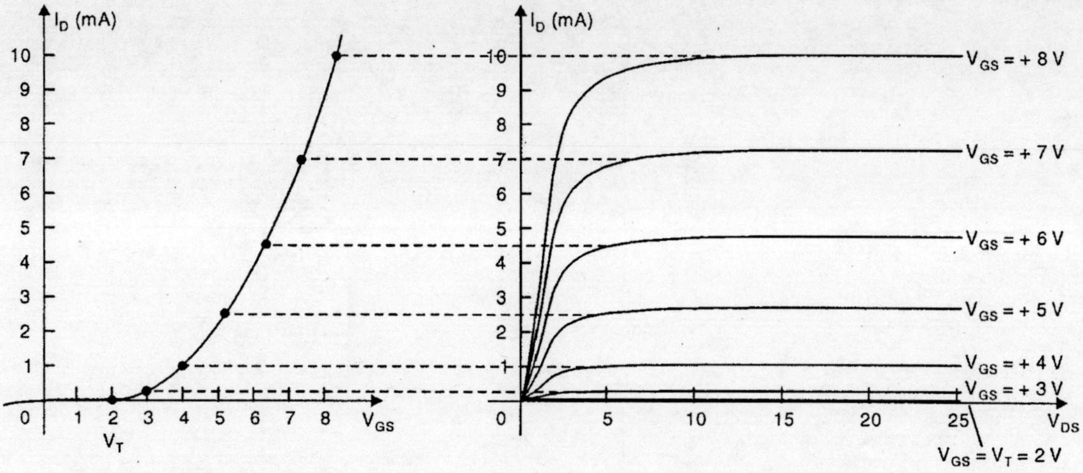

(*a*) Transfer characteristics (*b*) Drain characteristics

Fig. 8.7. *Transfer and drain characteristics of n-channel enhancement-type MOSFET.*

The value of K can be determined from the following equation (derived from eqn. 8.1) where $I_{D(on)}$ and $V_{GS(on)}$ are the values for each at a particular point on the characteristics of the device.

$$K = \frac{I_{D(on)}}{\left(V_{GS(on)} - V_T\right)^2} \qquad \text{...(8.2)}$$

It may be noted that for all the practical purpose the value of threshold voltage (V_T) should be as low as possible, because a low value of threshold voltage allows :

— the use of low power supply voltage

— smaller switching time due to the smaller voltage swing during switching.

— higher packing densities.

8.3.3. Methods of Reducing Threshold Voltage (V_T)

Methods which are generally used to lower the magnitude of V_T are given below :

(1) The silicon nitride approach makes use of a layer Si_3N_4 and SiO_2, whose dielectric constant is about the twice that of SiO_2 alone.

(2) Polycrystalline silicon doped with boron is used as the gate electrode instead of aluminium.

8.4. CIRCUIT SYMBOLS FOR FET AND MOSFET

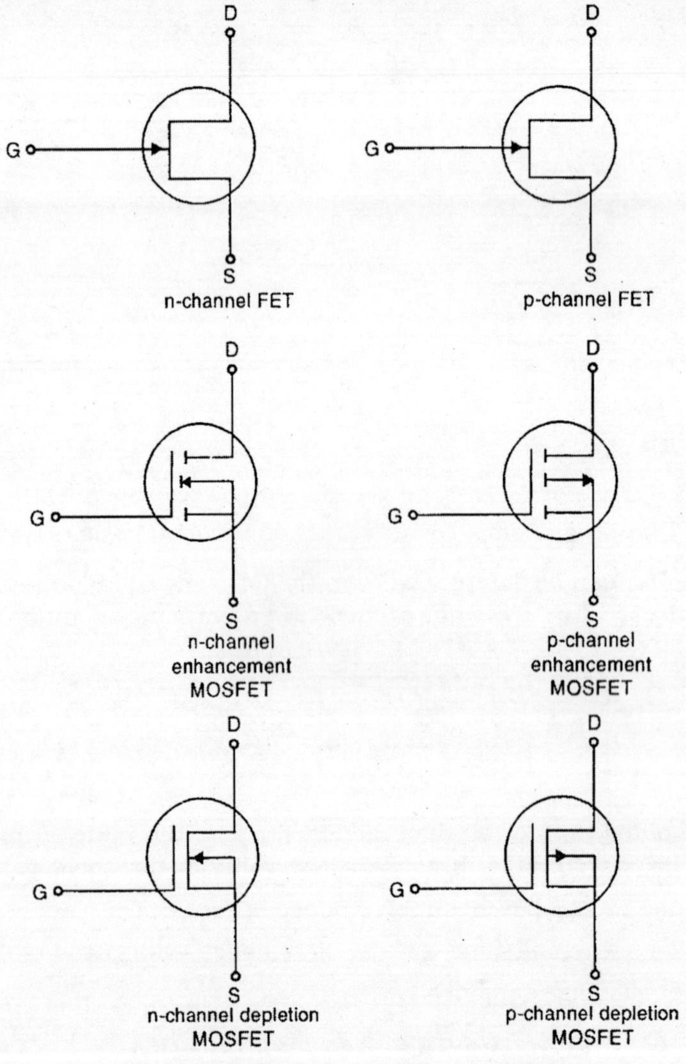

Fig. 8.8. *Symbols for JFET and MOSFET.*

8.5. COMPARISON OF JFET AND MOSFET

JFETs and MOSFETs are quite similar in their operating principles and in their electrical characteristics. However, they differ in some aspects, as discussed below :

1. JFETs can only be operated in the depletion mode whereas MOSFETs can be operated in either depletion or in enhancement mode. In a JFET, if the gate is forward biased, excess carrier junction occurs and the gate current is substantial. Thus channel conductance is enhanced to some degree due to excess carriers but the device is never operated with gate forward biased because gate current is undesirable.

2. MOSFETs have input impedance much higher than that of FET. This is due to negligibly small leakage current.

3. JFETs have characteristics curves more flatter than those of MOSFETs indicating a higher drain-resistance.

4. When JFET is operated with a reverse bias on the junction, the gate current I_G is larger than it would be in a comparable MOSFET. The current caused by minority carrier extraction across a reverse biased junction is greater, (per unit area) than the leakage current that is supported by the oxide layer in a MOSFET. Thus MOSFET devices are more useful in electrometer applications than the JFETs.

For the above reason and also because MOSFETS are somewhat easier to manufacture, they are more widely used than the JFETs.

? SELF-TEST (MOSFET)

1. Explain the basic construction of a *n*-channel depletion type MOSFET. Draw and explain its characteristics.
2. What is the significant difference between the construction of an enhancement-type MOSFET and a depletion-type MOSFET ?
3. What are the advantages of MOSFET over JFETs ?
4. What do you mean by threshold voltage V_T in the case of MOSFET ?
5. List the methods used to reduce the threshold voltage (V_T).
6. Explain the basic operation and characteristics of enhancement-type MOSFET.
7. Draw the symbol of enhancement and depletion type MOSFET.
8. What are the similarities and dissimilarities between the JFET and MOSFET.

SOLVED EXAMPLES

Example 1. *Sketch the transfer characteristics for an n-channel depletion-type MOSFET with I_{DSS} = 10 mA and V_P = – 4 V.*

Solution : At $V_{GS} = 0$ V, $I_D = I_{DSS} = 10$ mA

$$V_{GS} = V_P = -4 \text{ V}, \; I_D = 0 \text{ mA}$$

$$V_{GS} = \frac{V_P}{2} = \frac{-4}{2} = -2 \text{ V},$$

$$I_D = \frac{I_{DSS}}{4} = \frac{10}{4} = 2.5 \text{ mA}.$$

At $$I_D = \frac{I_{DSS}}{2}, \; V_{GS} = 0.3$$

$$V_P = 0.3\,(-4) = -1.2 \text{ V}.$$

All these values are shown in Fig. P (8.1).

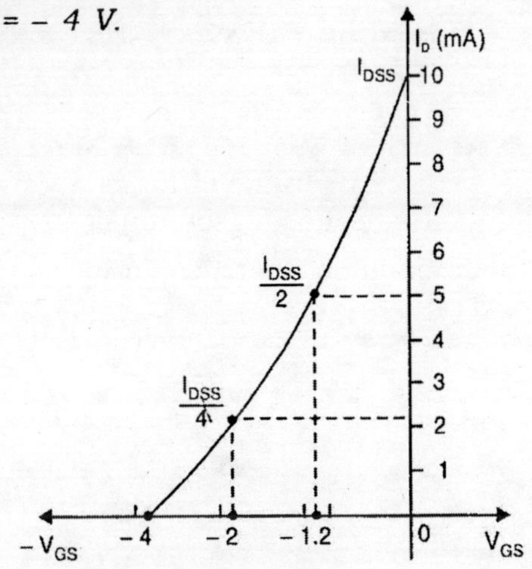

Fig. P (8.1).

Example 2. *Determine the V_{DS} for the network of Fig. P (8.2).*

Solution : From Fig. P (8.2),

$$V_{GS} = 0 \text{ V. Therefore, } I_P = I_{DSS} = 10 \text{ mA.}$$

so $$V_{DS} = V_{DD} - I_D R_D$$

or $$V_{DS} = 20 - 10 \times 10^{-3} \times 1.5 \times 10^9$$

or $$V_{DS} = 20 - 15$$

or $$V_{DS} = 5 \text{ V.} \quad \textbf{Ans.}$$

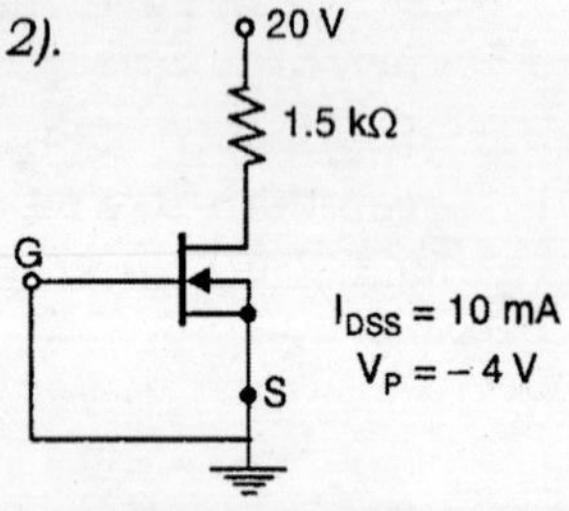

Fig. P (8.2).

Example 3. *An n-channel E-MOSFET has the following parameters*

$$I_{D\,(ON)} = 5 \text{ mA at } V_{GS} = 8 \text{ V and } V_{GST} = 4 \text{ V}$$

Determine the drain current when V_{GS} = 6 V.

Solution : As per the equation :

$$K = \frac{I_{D(ON)}}{\left(V_{GS(ON)} - V_{GST}\right)^2}$$

$$K = \frac{5 \text{ mA}}{(8 - 4)^2} = 0.3125 \text{ mA/V}^2$$

Now, drain current, $$I_D = k\,(V_{GS} - V_{ST})^2$$

$$= 0.3125\,(6 - 4)^2$$

$$= 1.25 \text{ mA.} \quad \textbf{Ans.}$$

Example 4. *The data sheet for a certain enhancement-type MOSFET reveals that $I_{D(on)}$ = 10 mA at V_{GS} = – 12 V and $V_{T(on)}$ = – 3 V. Is this device P-channel or N-channel? Find the value of I_D when V_{GS} = – 6 V.*

Solution : Given

$$I_{D(on)} = 10 \text{ mA}$$

$$V_{GS} = -6 \text{ V}$$

$$V_{T(on)} = -3 \text{ V}$$

Since the value of V_{GS} is negative for the enhancement-type MOSFET, this indicated that the device is P-channel.

We know that the drain current,

$$I_D = K\,[V_{GS} - V_{T(on)}]^2 \quad \text{...(1)}$$

and $$K = \frac{I_{D(on)}}{\left(V_{GS} - V_{T(on)}\right)^2}$$

or $$K = \frac{10 \times 10^{-3}}{\{-12 - (-3)\}^2}$$

or $$K = \frac{10}{81} = .12 \text{ mA/V.}$$

Substituting this value of K and $V_{GS} = -6$ V in eqn. (1), we have

$$I_D = 0.12 \times 10^{-3} \{-6 - (-3)\}^2$$

or $$I_D = 1.08 \text{ mA.} \quad \textbf{Ans.}$$

Example 5. *Find for the Network shown in Fig. P (8.3)*

(*i*) I_{DQ}

(*ii*) V_{DS}.

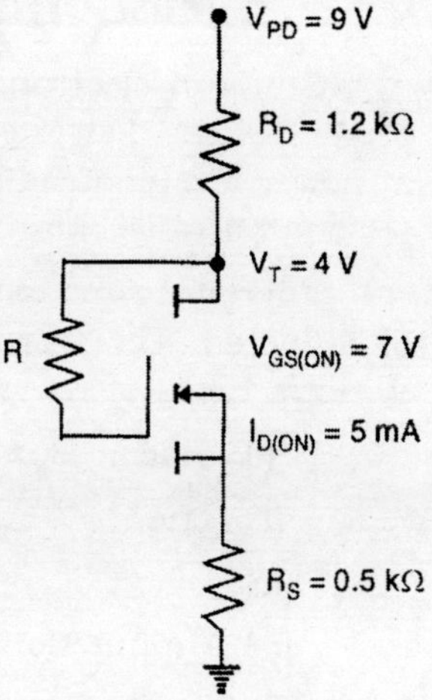

Fig. P (8.3)

Solution : (*i*) We know that,

$$K = \frac{I_{D(ON)}}{(V_{GS(ON)} - V_T)^2}$$

or $$K = \frac{5mA}{(7-4)^2} = 0.55 \times 10^{-3} \text{ A/V}^2$$

$$I_D = K(V_{GS} - V_T)^2$$

$$= 0.55 \times 10^{-3} (7 - 4)^2$$

$$= 0.55 \times 10^{-3} \times 9$$

$$= 4.95 \times 10^{-3} \text{ Amp} = 4.95 \text{ mA.}$$

(*ii*) $$V_{DS} = V_{DD} - I_D R_D - I_D R_S$$

or $$V_{DS} = 9 - I_D (R_D + R_S)$$

or $$V_{DS} = 9 - 4.95 \times 10^{-3}(1.2 + 0.5) \times 10^3$$

or $$V_{DS} = 9 - 4.95 \times 1.7$$

or $$V_{DS} = 9 - 8.415$$

or $$V_{DS} = 0.585 \text{ V} \quad \textbf{Ans.}$$

□□□

9

POWER ELECTRONIC DEVICES

9.1. INTRODUCTION

Study of electronic devices and circuits would be incomplete without the study of power devices. Power electronics basically deals with electronic devices that are operated at power level instead at signal level. They are used in aerospace, commercial, industry, residential, telecommunications and high voltage d.c. transmission.

Thyristors are solid state devices capable of handling large currents upto hundreds of amperes. The word THYRISTOR is a combination of THYRatron and transISTOR. Various types of thyristors are :

(*i*) Silicon Controlled Rectifier (SCR)

(*ii*) Triac

(*iii*) Diac

(*iv*) Unijunction Transistor (UJT)

9.2. SILICON CONTROLLED RECTIFIER (SCR)

General Diagram

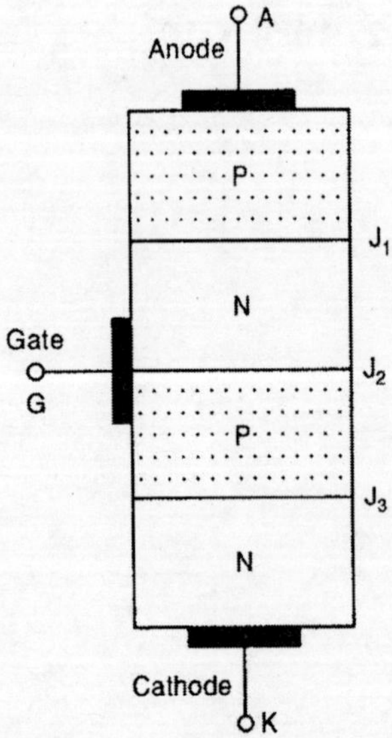

(*a*) Structure

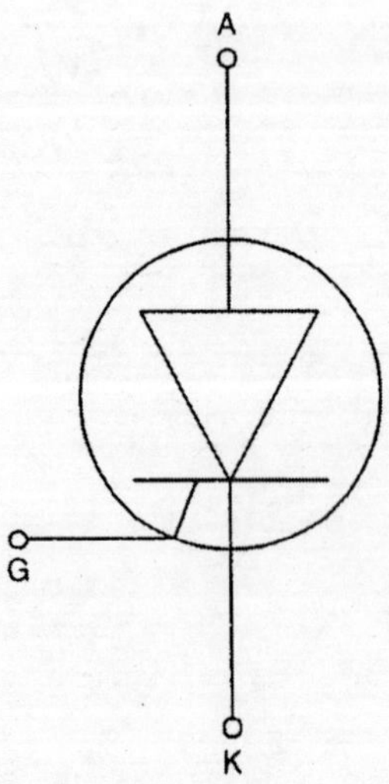

(*b*) Symbol

Fig. 9.1. *General Diagram of SCR.*

9.2.1. Construction

It is a four layer semiconductor device being alternate of *P*-type and *N*-type silicon. as shown in Fig. 9.1. Fig. 9.1 shows that it consists of three junctions namely J_1, J_2, J_3. J_1 and J_3 operates in forward direction whereas J_2 in reverse direction. It also consists of anode (*A*), cathode (*K*) and the gate (*G*). Gate controls the firing of SCR. It conducts only in one direction *i.e.,* from anode to cathode and hence it is a UNIDIRECTIONAL DEVICE. As SCR's are solid state device, therefore, they are compact and possess high reliability and have low loss.

9.2.2. Working

The operation of SCR is easy. Two ways to study are when the gate is open as shown in Fig. 9.2 (*a*), when the gate is positive w.r.t. cathode as shown in Fig. 9.2 (*b*). When the gate is open, no voltage is applied at the gate. Under this condition, J_1 and J_3 are forward biased, J_2 is reverse biased, no current flows through R_L, SCR is cut-off. But, their is a certain critical value of voltage called the BREAKDOWN VOLTAGE at which junction breakdown (J_2) occurs. The SCR now conducts heavily and is said to be ON state.

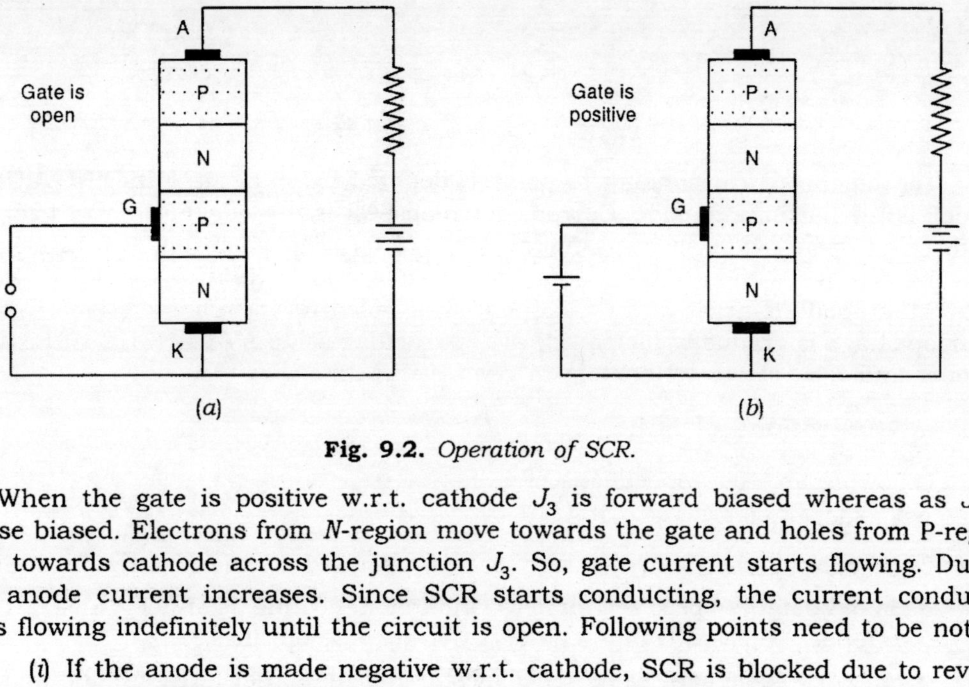

Fig. 9.2. *Operation of SCR.*

When the gate is positive w.r.t. cathode J_3 is forward biased whereas as J_2 is reverse biased. Electrons from *N*-region move towards the gate and holes from P-region move towards cathode across the junction J_3. So, gate current starts flowing. Due to this, anode current increases. Since SCR starts conducting, the current conductor keeps flowing indefinitely until the circuit is open. Following points need to be noted :

(*i*) If the anode is made negative w.r.t. cathode, SCR is blocked due to reverse biased junctions J_1 and J_3. This shows it is a unidirectional device.

(*ii*) Either SCR does not conduct or conducts heavily.

9.2.3. V-I Characteristics of SCR

The forward characteristics is obtained when anode is positive w.r.t. cathode and the curve *OA BC* above represents it. When the supply voltage is increased from zero, a point A is reached when SCR starts conducting. When the gate is open, the BREAKOVER VOLTAGE is defined as the minimum forward voltage at which SCR starts conducting heavily.

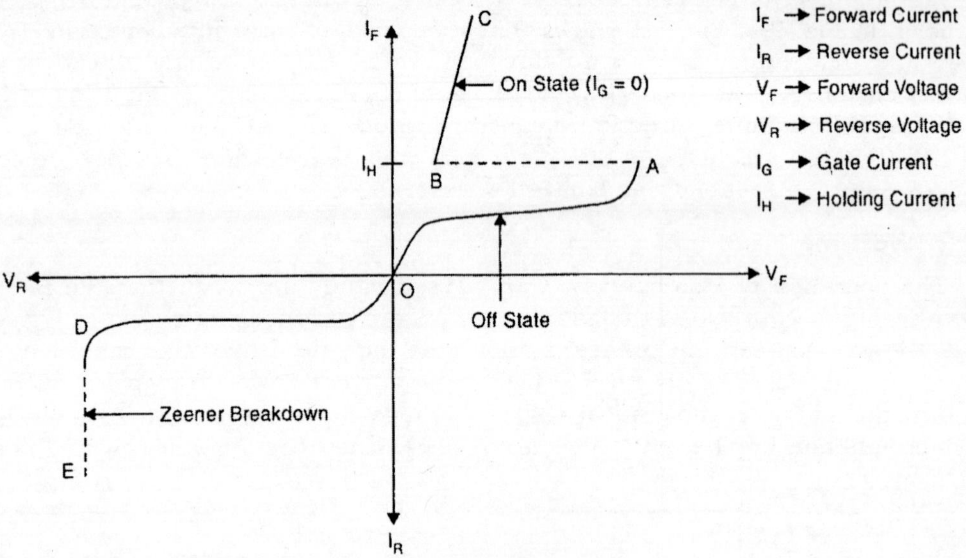

Fig. 9.3. *Characteristics of SCR.*

The current corresponding to point B denoted by I_H is the HOLDING CURRENT which is the maximum anode. Current at which SCR is turned OFF from ON condition, gate being open. There is a maximum value of current which on SCR is capable of passing through without destruction. This is known as FORWARD CURRENT RATING. When the anode is negative w.r.t. cathode it is called reverse characteristic. When the anode voltage is gradually increased, at a particular value AVALANCHE BREAKDOWN occurs and SCR starts conducting in reverse direction.

9.2.4. Applications of SCR

Following are the applications of SCR :

1. In phase controlled rectifiers.
2. In subway cars, trolley buses, battery-operated vehicles, battery charging etc.
3. In adjustable speed AC drivers, induction heating, stand by air-craft power supplies, ups for computers, HVDC transmission lines.
4. As AC voltage controllers for domestic and industrial heating, lighting control, speed control of 1 ϕ and 3 ϕ AC drivers and starting of induction motors.
5. As cyclo convertors in speed control of high power AC drivers, induction heating, static VAR generation, in air-crafts or ship-boards.

9.3. THE TRIAC

General Diagram

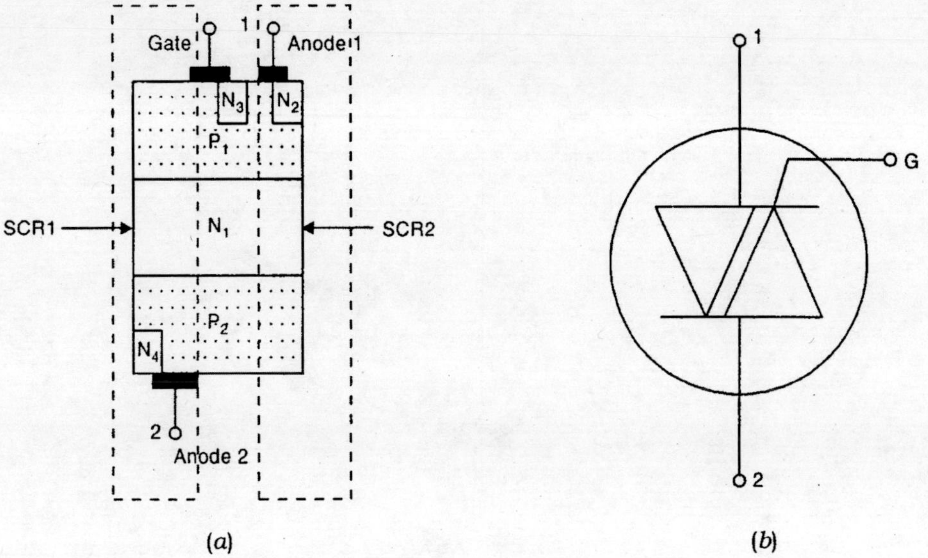

Fig. 9.4. *TRIAC.*

9.3.1. Construction

It is a four layer semiconductor device with three terminals : terminal 1 (Anode 1), terminal 2 (Anode 2) and gate terminal. Since, triac corresponds to both negative and positive voltages at the anode, hence the concept of cathode is dropped. A pulse of either polarity can switch the triac from OFF state to ON state. A triac may be considered as two SCRs connected in parallel but in opposite direction. The gate terminal makes ohmic contacts with both N and P materials.

Working : The circuit used for the purpose is shown in Fig. 9.5. Let the supply voltage is below breakover voltage of the triac. The operation is as follows :

(*i*) When switch S is open, there will be no gate current and triac is cut-off.

(*ii*) When switch S is closed, gate current starts flowing. Two situations are possible :

(*a*) When anode 1 is positive w.r.t. anode 2, the triac turns ON, the current flows from anode 1 to 2.

(*b*) When anode 1 is negative w.r.t. anode 2, the triac turns ON, the current flows from anode 2 to anode 1.

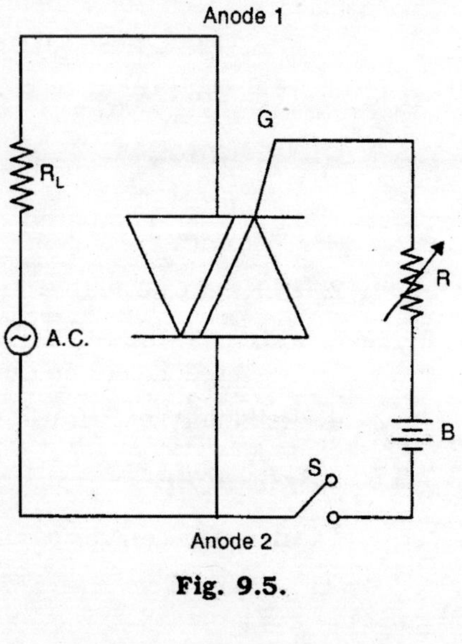

Fig. 9.5.

9.3.2. V-I Characteristics

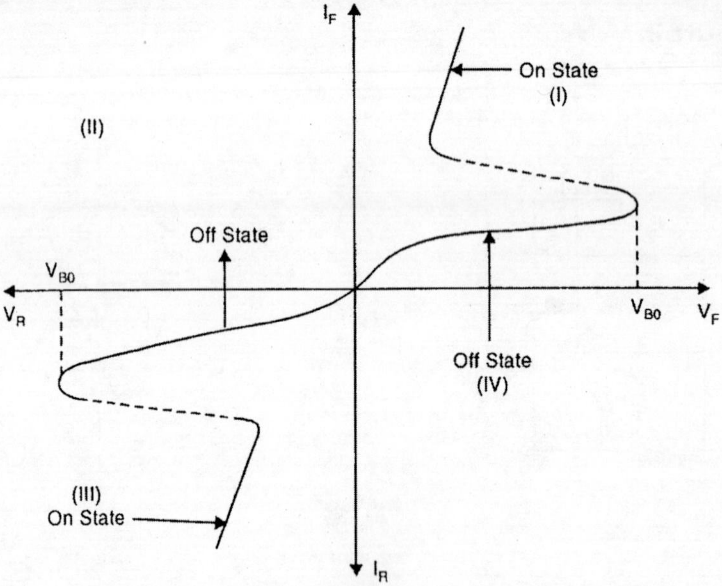

Fig. 9.6.

The triac is said to be positively biased when anode 1 is positive w.r.t. anode 2 (Quadrant I). The triac is said to be negative biased when anode 1 is negative w.r.t. anode 2 (Quadrant III). The triac can be operated either with positive or negative gate voltage. The gate voltage is positive in quadrant I and negative in quadrant II. The greater the gate current, the smaller the supply voltage at which triac is turned ON.

9.3.3. Advantages

(*i*) The TRIAC can be triggered with positive or negative polarity voltages.

(*ii*) The TRIAC needs a single heat sink of slightly larger size.

(*iii*) The TRIAC needs a single fuse protection, this simplifies construction.

(*iv*) This way works without a diode safely in either direction.

9.3.4. Disadvantages

1. Low $\left(\frac{dv}{dt}\right)$ compared to SCRs.
2. SCRs are available in larger ratings compared to TRIACs.
3. A trigger circuit in TRIAC needs careful consideration, as the TRIAC can be triggered in either direction.
4. Reliability of TRIACs is less than that of SCRs.

9.3.5. Applications

1. In speed control of universal motors.
2. In food mixers and portable drills.

3. Lamp dimming and heat control.
4. In zero voltage switched ac relay.
5. It is used as high power lampswitch.
6. To control ac power to a load.

9.4. UNIJUNCTION TRANSISTOR (UJT)

General Diagram

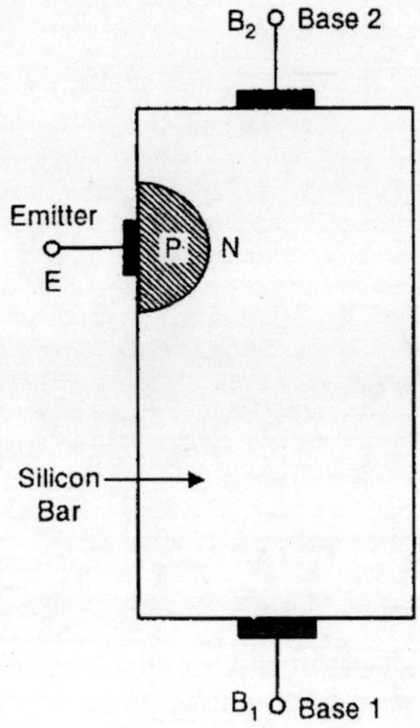

(a) Structure

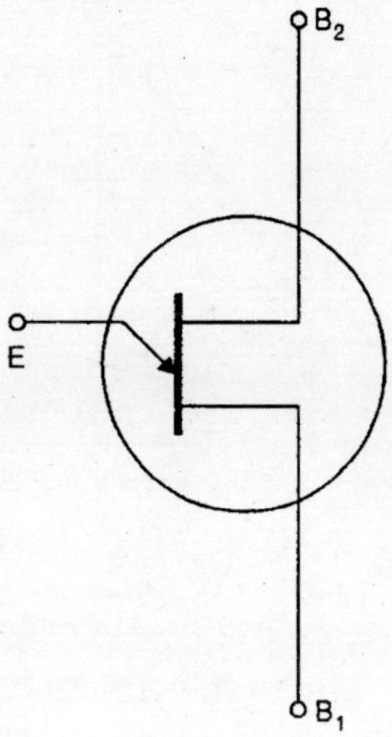

(b) Symbol

Fig. 9.7.

9.4.1. Construction

It is a three terminal semiconductor switching device. It consists of a lightly-doped N-type silicon bar with a heavily doped *P*-type material alloyed to its one-side for producing a single *P-N* junction. At the ends of silicon bar two connections are made for Base 1 (B_1) and Base 2 (B_2). The terminal from *P*-region is for the Emitter (*E*) connection. Following points are to be noted :

(*i*) It has only one *p-n* junction.

(*ii*) It has three leads and thus differs from an ordinary diode.

(*iii*) It is operated with emitter junction forward biased.

(*iv*) It has the ability to control a large ac power with a small signal.

9.4.2. Working

The UJT works in the following manner :

(*i*) When the emitter diode is reverse biased, only a small emitter current flows. Under this condition R_{B1} is at its normal high value. This is UJTs "OFF" state.

(*ii*) When the emitter diode becomes forward biased R_{B1} drops to a very low value so that the total resistance between E and B_1 becomes very low, allowing emitter current to flow readily. This is the 'ON' state.

9.4.3. V-I Characteristics

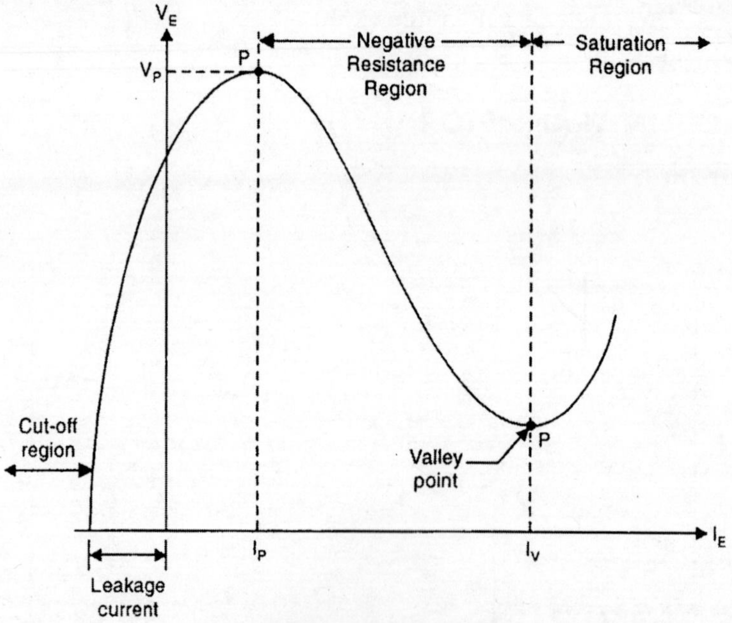

Fig. 9.8. *V-I characteristics.*

Fig. 9.8 shows the curve between emitter voltage V_E and emitter current I_E of a UJT at a given voltage V_{BB} between the bases. In the cut-off region, a small leakage current *V* starts flowing from B_2 to emitter. This is due to minority carries at reverse biased diode. Above a particular value of emitter voltage, emitter current starts to flow. The current increases until the peak current and peak voltage are reached at peak point *P*. V_P and I_P corresponding peak voltage and peak current. Now, holes are injected into *N*-region which increase the conductivity of the region between emitter and base B_1. The process continues until valley point *V* is reached. At valley point, the current is increased to such an extent that no further increase in conductivity can take place. Beyond valley point, the device behaves as a conventional forward-biased *p-n* junction diode.

9.4.4. Applications of UJT

UJT is used in a variety of circuit applications. Some of which are :

(*i*) Relaxation oscillator to generate pulses

(*ii*) Switching device

(*iii*) Switching generator or saw tooth generator

(*iv*) Timing and trigger circuit

(*v*) Phase control

(*vi*) Voltage or current regulated supplies

(*vii*) Over voltage detector.

(*viii*) UJT as relaxation oscillator.

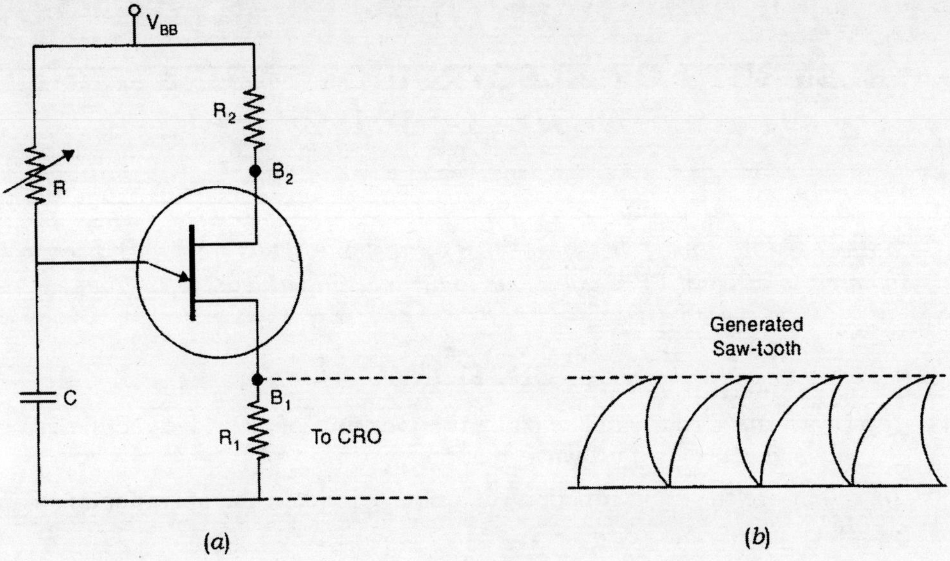

Fig. 9.9.

SELF-TEST (SCR, UJT, TRIAC)

1. What is SCR ?
2. Draw equivalent circuit of SCR.
3. Explain working and construction of SCR.
4. Draw *V-I* characteristics of SCR.
5. What is TRIAC ?
6. "TRIAC can be triggered by either a positive or a negative control pulse." Explain.
7. Explain working and construction of TRIAC.
8. How does the current rating of a TRIAC differ from that of an SCR ? Why ?
9. Describe the construction and working of UJT.
10. Draw the equivalent circuit of UJT and discuss its working from the circuit.
11. Draw and discuss static emitter characteristics of a UJT.

□□□

Roll No. ....

Total No. of Pages : 2

BT-1/D06 **8040**

ELEMENTS OF ELECTRONICS ENGINEERING

Paper—EL-101 E

Time : 3 Hours] *[Maximum Marks : 100*

Note : *In the question paper, there are* **EIGHT** *questions,* **TWO** *from each section. You are required to attempt* **FIVE** *questions in all selecting at least ONE from each section.*

SECTION A

1. (*a*) Draw the circuit and explain the working of a bridge-rectifier. Why it is preferred over a Full-wave rectifier ? **10**

(*b*) What are the main functions of a filter ? Explain the operation of a capacitive filter. **10**

2. (*a*) Explain with reference to the Zener-diode characteristics, Zener voltage, breakdown-current and dynamic-impedance. How does the dynamic impedance vary with Zener current ? **10**

(*b*) What is LED ? Give its principle of working, construction, circuit symbol, merits, demerits and applications. **10**

SECTION B

3. (*a*) Sketch typical *CB* output-characteristics curves for an NPN transistor. Label all variables. Explain the shapes of the curves qualitatively. **10**

(*b*) Draw the load line on the output characteristics of a common emitter amplifier. What is its significance ? Show how the transistor acts as an amplifier. **10**

4. (*a*) Draw a simple circuit of a common collector amplifier and explain its operation. Obtain an expression for its voltage gain. **10**

(*b*) What is feedback in amplifier ? Prove that negative feedback in amplifier increases the bandwidth and reduces the distortion. **10**

SECTION C

5. (*a*) With a neat diagram, explain the operation and transfer characteristics of differential amplifier. **10**

(*b*) Explain the basic principle of working of a UJT. Draw the static emitter characteristics and explain the shape of the curves. **10**

6. (*a*) Draw the *V-I* characteristics of SCR and explain its behaviour using two-transistor analogy and hence suggest various methods of triggering an SCR. **10**

(*b*) Explain the working of a JFET. Define the parameters of a JFET and develop its equivalent circuit. **10**

SECTION D

7. (*a*) Compare the characteristics of transistor amplifiers in the three possible configurations. **6**

(*b*) Sketch the CE transistor input and output characteristics and explain them. **7**

(*c*) Discuss the working of a Full-wave centre-tapped rectifier circuit. **7**

8. Write short notes on the following :

(*i*) op-amp as non-inverting amplifier.

(*ii*) Characteristics and construction of MOSFETs.

(*iii*) Diode as voltage multiplier.

(*iv*) A.C. load-line and operating point selection. **4 × 5 = 20**

□□□

Roll No. .... Total No. of Pages : 2

BT-2/M06 **8245**

ELEMENTS OF ELECTRONICS ENGINEERING (2003–04)
Paper—EL-101 E

Time : 3 Hours] *[Maximum Marks : 100*

Note : *Attempt any* ***FIVE*** *Questions.*

1. (*a*) Draw and explain *V-I* characteristics of P-N Junction diode. **10**

   (*b*) If we assume ideal diode in Fig. (1) and Fig. (2), then draw wave shape of output signal for input as shown in Fig. **10**

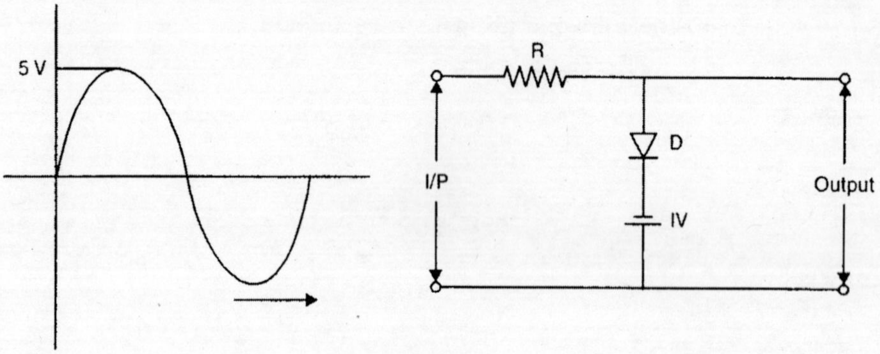

Fig. 1

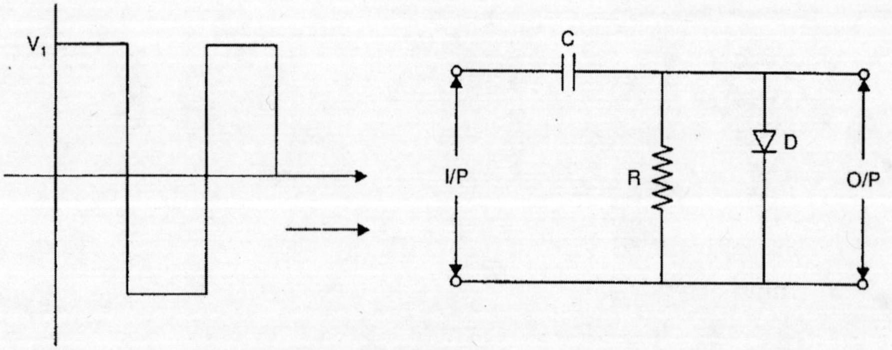

Fig. 2

2. (*a*) For what purpose we are using LED ? Explain and draw *V-I* characteristics of LED. **7**

   (*b*) How we can use Zener-diode as voltage regulator ? **7**

   (*c*) What is the use of using filter in rectifier ? Explain various types of filters. **6**

3. (*a*) Explain early effect of BJT. **10**

(*b*) Why we need biasing Ckt in BJT ? What are basic requirements of biasing Ckts ? **10**

4. Draw *h*-Parameters model of C.E. configuration transistor Amplifier. Also calculate current gain, voltage gain, Input and Output Impedance of C.E. amplifier in terms of *h*-Parameters. **20**

5. (*a*) What is difference between drift and diffusion current in semiconductor ? **10**

(*b*) Draw input and output *V-I* characteristics of transistor in C.E. configuration. **10**

6. (*a*) Explain Input offset current, Input offset voltage, output offset voltage, power supply rejection ratio and slew rate of op-amp. **10**

(*b*) How we can use op-amp as adder and scale changer ? Explain with diagram. **10**

7. (*a*) For the op-amp shown in Fig. (3). Find output voltage. **10**

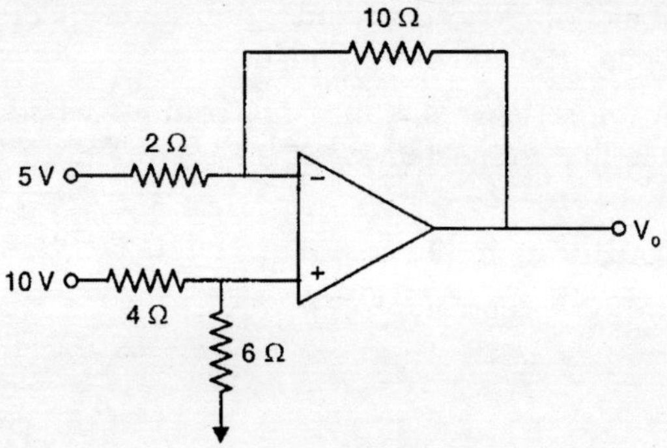

Fig. 3.

(*b*) Draw the structure of *n*-channel FET. Draw and explain *V-I* characteristics of *n*-channel FET. **10**

8. Write short notes on any of **Two** :

(*i*) Silicon controlled rectifier

(*ii*) Unijunction transistor

(*iii*) TRIAC. **10 × 2 = 20**

□□□

Roll No. ....   Total No. of Pages : 2

BT-1/D05   **8045**

ELEMENTS OF ELECTRONICS ENGINEERING (2004–05)

Paper—EL-101 E

Time : 3 Hours]   *[Maximum Marks : 75*

Note : *Answer* ***FIVE*** *Questions.*

1. (*a*) Explain conduction in extrinsic semiconductors. What is the effect of temperature on conductivity ? **5**

(*b*) What do you understand by d.c. and a.c. resistance of a crystal diode ? How will you determine them from *V-I* characteristics of a crystal diode ? **5**

(*c*) Describe the action of capacitor input π filter and chock input LC filter. **5**

OR

2. (*a*) Draw symbols of Schottky diode, Varactor diode, Photodiode and Light emitting diode. Explain briefly also. **7½**

(*b*) What value of series resistance is required when 10 W, 10 V, 1000 mA Zener-diodes are connected in series to obtain 20 V regulated output from a 35 V d.c. source ? **7½**

3. (*a*) What do you mean by ?

(*i*) Saturation region

(*ii*) Cut-off region

(*iii*) Active region **5**

(*b*) Draw the circuit diagram and output characteristics of a CE transistor amplifier and write down the equation of d.c. load line. **7**

(*c*) Explain why a transistor should be biased. **3**

OR

4. (*a*) Find *h*-parameters of the circuit shown : **5**

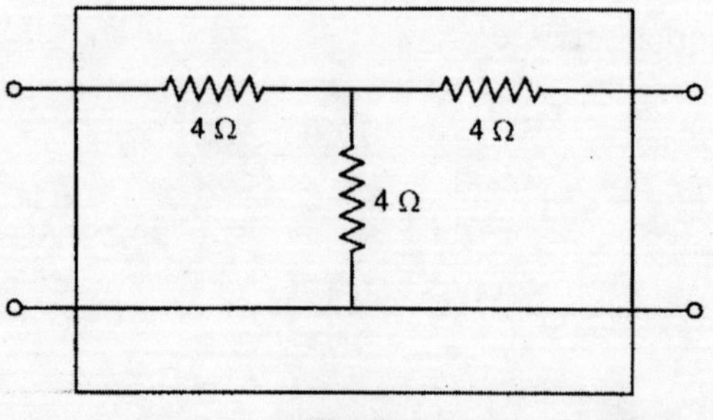

Fig 1.

(*b*) What do you understand by class A, class B, class C power amplifier ? **5**

(*c*) Write a short note on Wein bridge oscillator. **5**

5. (*a*) Explain the characteristics of ideal operational amplifier. **7½**

(*b*) Explain the operation of op-amp as differentiator and integrator. **7½**

6. (*a*) Explain the construction and working of MOSFET. **7½**

(*b*) How will you determine the drain characteristics of JFET ? What do they indicate ? **7½**

OR

7. (*a*) Draw the equivalent circuit of UJT and explain its working. **5**

(*b*) Sketch the *V-I* characteristics of a TRIAC and explain. **5**

(*c*) Explain the application of UJT as relaxation oscillator. **5**

8. Write short notes on any **Three :**

(*a*) op-amp as scale changer

(*b*) Concept of feedback in amplifiers

(*c*) Transfer characteristics of differential amplifier

(*d*) CMRR. **5 × 3 = 15**

□□□

Roll No. .... Total No. of Pages : 1

BT-2/M06 **8246**

ELEMENTS OF ELECTRONICS ENGINEERING (2004–05)
Paper—EL-101-E

Time : 3 Hours] *[Maximum Marks : 75*

Note : *Attempt any* ***FIVE*** *Questions.*

1. (*a*) What is difference between metal, insulator and semiconductor ? **5**

   (*b*) What is intrinsic semiconductor ? What is the effect of temperature on conductivity of semiconductor ? **5**

   (*c*) For what purpose we are using filter circuit in rectifier ? Also explain various type of filters. **5**

2. (*a*) Draw V-I characteristics of *p-n* junction diode. Also explain the factors on which reverse saturation current is dependent. **8**

   (*b*) For what purpose we are using Zener-diode ? Draw V-I characteristics of Zener-diode. What is difference between Zener and avalanche breakdown ? **7**

3. Draw *h-parameter* model of C.E. configuration transistor amplifier. Also calculate current gain, voltage gain, input and output impedance of C.E. amplifier in terms of *h-parameters.* **15**

4. For what purpose we are using Oscillator ? What type of feedback is being used in Oscillator ? Explain working of phase shift Oscillator. **15**

5. Find voltage gain of ideal and practical operational amplifier for inverting and non-inverting amplifier. **15**

6. (*a*) Explain input offset current, input offset voltage, output offset voltage, power supply rejection ratio and slew rate of op-amp. **7**

   (*b*) How we can obtain integration and differentiation of input voltage using op-amp. **8**

7. Draw the structure of *n*-channel enhancement and depletion MOSFET. Also draw and explain V-I characteristics of enhancement and depletion MOSFET. Draw symbol of *n*-channel MOSFET. **15**

8. Write short notes on any **Three :**

   (*i*) Silicon controlled rectifier

   (*ii*) Unifunction transistor

   (*iii*) Field-effect transistor

   (*iv*) TRIAC

   (*v*) Clipper and Clamper **5 × 3**

□□□

Roll No. .... Total No. of Pages : 2

BT-1/J05 8046

ELEMENTS OF ELECTRONICS ENGINEERING

Paper—EL-101 E

(w.e.f. 2004–05)

Time : 3 Hours] *[Maximum Marks : 75*

***Note :** Attempt any **FIVE** Questions.*

1. (*a*) Explain the mechanism of hole-current flow in a semiconductor. **4**

   (*b*) Derive the expression for efficiency for a full wave rectifier. **4**

   (*c*) For the Ckt given below, find maximum and minimum values of Zener-diode current. **7**

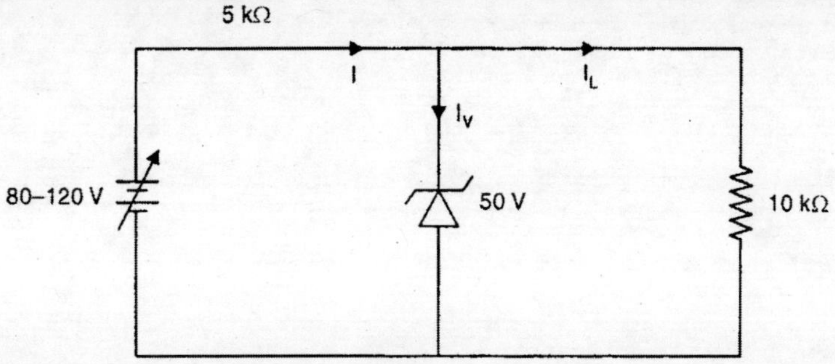

Fig. 1

OR

2. (*a*) Explain the operation of L-C filter, R-C filter and π filter. **6**

   (*b*) How diodes works as clipper and how a clipper Ckt. differ from clamper ? Explain. **6**

   (*c*) Explain the effect of temperature on leakage current. **3**

3. (*a*) Explain the terms β, I_{CBO}, I_{CEO}, input and output dynamic resistance. **5**

   (*b*) Draw ac load line of CE amplifier and explain. **5**

   (*c*) Draw the I/P and O/P characteristics of CB connection and explain. **5**

OR

4. (*a*) Explain performance of a linear circuit in h-parameters. **7½**

   (*b*). Explain the effects of negative current feedback on amplifiers. **7½**

5. (*a*) With a neat diagram, explain the transfer characteristics of differential amplifier. **7½**

(*b*) How op-amp can be used as integrator ? Explain with the help of neat diagram. **7½**

OR

6. (*a*) Explain working of pulse generator using an op-amp. **5**

(*b*) Explain operation of S/H switch circuit. **5**

(*c*) Explain characteristics of ideal operational amplifier. **5**

7. (*a*) Differentiate JFET with a bipolar transistor. **5**

(*b*) How TRIAC works as an a.c. switch ? **5**

(*c*) Discuss the application of a DIAC. **5**

8. Write short note on any **Three :**

(*a*) MOSFET

(*b*) UJT

(*c*) Hartly's oscillator

(*d*) SCR. **5 × 3 = 15**

□□□

Roll No. .... **Total No. of Pages : 2**

BT-2/M05 **8235**

ELEMENTS OF ELECTRONICS ENGINEERING (2004–05)

Paper—EL-101 E

Time : 3 Hours] *[Maximum Marks : 75*

***Note :** Attempt **FIVE** Questions.*

1. (*a*) What is meant by cut-in voltage ? Why is its value higher for silicon than for germanium ?

(*b*) Differentiate between biased and unbiased Clipper circuit.

(*c*) Explain why is it necessary to use a voltage regulator in a power supply.

(*d*) Derive the expression for ripple factor of full wave rectifier. **15**

OR

2. (*a*) Explain why a *p-n* junction possesses capacitance.

(*b*) What are the two types of reverse breakdowns in a *p-n* junction diode which occurs at lower voltages ?

(*c*) Explain why a series inductor filter cannot be used with a half wave rectifier.

(*d*) Determine V_o for the circuit shown below, V_z is zener voltage of the Zener-diode. **15**

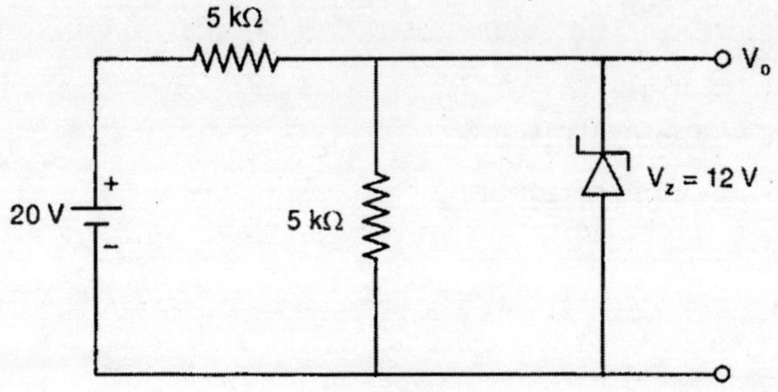

Fig. 1.

3. (*a*) Draw the ac equivalent circuit of CE transistor amplifier and derive the expressions for voltage gain ,current gain, input impedance and output impedance of the amplifier in terms of *h*-parameters.

(*b*) State the merits and demerits of negative feedback in amplifiers. Calculate the gain of a negative feedback amplifier with an internal gain of 100 and feedback factor is 1/10. **15**

OR

4. (*a*) Explain why a transistor action cannot be achieved by connecting two diodes back to back.

(*b*) What is the basic principle of producing oscillations ? Discuss the working of phase shift oscillator.

(*c*) Draw the input and output characteristics of common emitter (CE) configuration. Compare the parameters of CB and CE configuration. **15**

5. (*a*) Describe the equivalent circuit of an operational amplifier. What is the significance of virtual ground ? Discuss the working of OP-AMP as scale changer.

(*b*) Explain the construction and working of JFET and draw its characteristics. **15**

OR

6. (*a*) Explain the triggering process of an SCR.

(*b*) What do you mean by open-loop and closed-loop gain of an OP-AMP ?

(*c*) Derive the expression for voltage gain of non-inverting OP-AMP.

(*d*) Give the applications of UJT. **15**

7. Compare the following :

(*a*) Positive clamper and negative clamper

(*b*) BJT and JFET

(*c*) JFET and MOSFET

(*d*) Inverting and non-inverting amplifier. **15**

8. Write short note on the following :

(*a*) LED and its applications

(*b*) Diode as voltage multiplier

(*c*) Selection of operating point

(*d*) TRIAC. **15**

□□□

Roll No. .... **Total No. of Pages : 2**

BT-1/D06 **8041**

ELEMENTS OF ELECTRONICS ENGINEERING (2004–05)

Paper—EL-101 E

Time : 3 Hours] *[Maximum Marks : 75*

***Note :** Attempt any **FIVE** Questions. All questions carry equal marks.*

1. (*a*) What is the effect of temperature on *V-I* characteristics of semiconductor diode ? The forward current through Si diode is 10 mA at room temperature (27°C). The corresponding forward voltage is 0.75 V. Calculate the reverse saturation current.

(*b*) Discuss the working of bridge rectifier. Derive the expression for ripple factor and rectification efficiency. Compare its performance with centre tap full wave rectifier. **6 + 9**

2. (*a*) What is the difference between biased and unbiased clipper ?

(*b*) Plot the transfer characteristics for the following clipper circuit and give the response of the circuit to sinusoidal input.

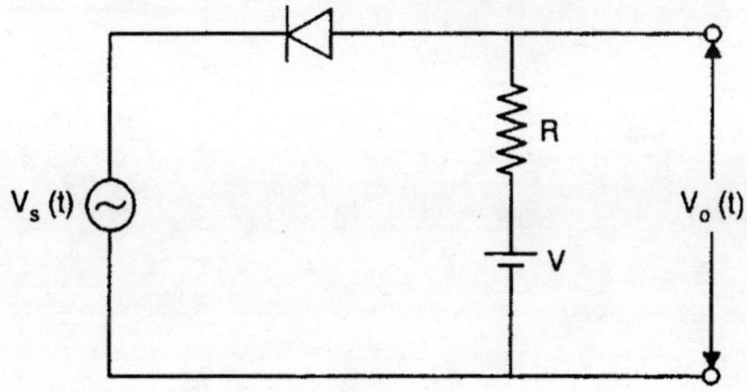

Fig. 1

(*c*) Draw a practical clamper circuit that clamps the negative extremities to zero. Give applications of clampers. **5 + 5 + 5**

3. (*a*) Explain why a transistor action cannot be achieved by connecting two diodes back to back ?

(*b*) Discuss the working of transistor amplifier in CE configuration. Write the equation for d.c. load line and plot it on the output characteristics. Define operating point.

(*c*) What is early effect and what are its effects on the transistor characteristics ?

4 + 6 + 5

4. (*a*) Plot the input and output characteristics for transistor in CE configuration and identify the various regions of operation. Why are CE output characteristics more sloping than CB characteristics ?

(*b*) What is the importance of h-parameters ? Draw the h-parameters equivalent circuit of transistor in CE configuration. Derive the expression for voltage gain and input impedance in terms of h-parameters.

(*c*) What are the merits and demerits of negative feedback in amplifiers.

5 + 6 + 4

5. (*a*) Give the electrical characteristics of ideal OP-Amp.

(*b*) Define the parameters : input offset voltage, CMRR, slew rate, bandwidth and PSRR. What are their values for IC 741 ? Give the procedure for measuring input offset voltage in IC 741. **5 + 10**

6. (*a*) What is the difference between open-loop and closed-loop configurations in op-amp ?

(*b*) Discuss the application of OP-Amp as non-inverting amplifier and differentiation. **5 + 10**

7. (*a*) Compare the performance of BJT and JFET.

(*b*) Explain the construction and working principle of MOSFET. Plot its drain current characteristics and transfer characteristics. Discuss the operation of MOSFET in depletion and enhancement mode. **5 + 10**

8. (*a*) Discuss the turn on and turn off characteristics of SCR.

(*b*) What is UJT ? Describe its working. Discuss the application of UJT as relaxation oscillator. **6 + 9**

□□□

Roll No. .... Total No. of Pages : 2

BT-2/J04 **8223**

ELEMENTS OF ELECTRONICS ENGINEERING
Paper—EL-101 E

Time : 3 Hours] *[Maximum Marks : 100*

Note : *Attempt **FIVE** Questions.*

1. (*a*) How is a *n*-type semiconductor prepared ? Explain. **5**

(*b*) Realise a positive clipper and a clamper circuits using diodes. Explain their working and list uses of both the circuits. **10**

(*c*) The junction capacitance of a diffused junction is 2 pF under zero bias. The built in potential is 0.5 V. Calculate the capacitance with reverse bias of 12 V. Assume doping profile to be that of an abrupt junction. **5**

OR

2. (*a*) Why are filters used ? Explain the working of a π type filter and give its uses. **6**

(*b*) Draw the circuit of a full wave rectifier. Explain its operation and obtain the expression for output voltage at any instant. **8**

(*c*) Explain the construction and characteristics of LEDs. List their uses. **6**

3. (*a*) Differentiate between a *p-n* junction diode and a Zener-diode on the basis of their construction and working. Give applications of Zener-diodes. **8**

(*b*) Explain the working of a UJT and draw its characteristics. **6**

(*c*) Give the applications of JFETs and MOSFETs. **6**

4. (*a*) Explain the working of an emitter-follower and draw its characteristics. **6**

(*b*) Determine g_m and r_0 for a *p*-channel JFET.

Given I_D = 0.6 mA, I_{DSS} = −1.2 mA, V_P = 2 V and $\lambda = -10^{-2}\ V^{-1}$. **8**

(*c*) How can a diode work as voltage multiplier ? Explain. **6**

5. (*a*) Explain the working of a transistor as an amplifier. Use common base configuration for the purpose. **6**

(*b*) Compare CE, CB and CC configurations. Which of them is best suited for power amplification and why ? Explain. **8**

(*c*) Draw the circuit of a Colpitt oscillator ? Give its applications. **6**

OR

6. (*a*) What is the need of feedback in amplifiers ? Explain. List advantages of negative feedback. **6**

(*b*) Explain the purpose of ac load line and explain its use by giving an example. **6**

(*c*) Draw the small signal model of a transistor and derive the expression for output resistance and power gain of the transistor. **8**

7. (*a*) Explain the following terms with reference to IC op-amps :

PSRR, Input offset current and unit gain BW. Also explain their significance. **10**

(*b*) Explain the working of an op-amp based scale changer and obtain the expression for the output voltage. **10**

OR

8. (*a*) How do you use an OP-Amp as an integrator ? Draw the circuit and obtain expression for the output voltage. **10**

(*b*) What is a TRIAC ? Where do you use it ? Explain its characteristics. **10**

□□□

INDEX

A

Acceptor, 23, 24

B

Breakdown Characteristic of a Zener Diode, 150
Breakdown Mechanism, 148
Basic Construction of a Junction Transistor, 182
Bark Hausen Criterion, 284
Basic Concept of Feedback, 266
Basic Operation and Characteristics of Enhancement, 437
Basic Operation and Characteristics of *n*-channel Depletion Type MOSFET, 434

C

Comparison between Electron Tubes and Solid State Devices, 2
Conductor, 15
Conduction in Intrinsic Semiconductors, 32
Covalent Bond, 9
Capacitor Input or π-Filter, 103
Choke Input or L-section Filter, 101
Clamping Circuits, 108
Clipping Circuits, 104
Comparison of Rectifier Circuits, 98
Comparison in Different Configurations, 193
Circuit Components, 299
Classification of Amplifiers, 245
Classification of Negative Feedback, 269
Colpitts Oscillator, 295
Common-Collector Configuration, 262
Common-Emitter Configuration, 264
Crystal Oscillator, 296, 281
Current Amplifier, 275, 276
Concept of op-amp, 333
Concept of Virtual Ground, 339
Classification of Field-Effect Devices, 393
Comparison between BJT and FET, 394
Computation of A_v, R_i, R_o of Single FET Amplifiers, 412
Comparison of JFET and MOSFET, 440

D

Donors, 23
Doping, 16
Damped Oscillations, 281
Depletion-Type MOSFET, 433

E

Electron-Hole Concept, 10
Electronic Components, 3
Energy Bands, 12
Energy Levels, 14
Extrinsic Semiconductors, 17
Electrical Equivalent Circuit of a Crystal, 296
Enhancement-type MOSFET, 436

F

Fermi Level, 21
Filters, 98
Full Wave Bridge Rectifier, 95
Full Wave Rectifier (FWR), 89
Feedback Factor, 268

Feedback Network, 277
Feedback Ratio, 268
Frequency Stability of Oscillator, 299
FET Applications, 416
FET as an Amplifier, 411

G

Group-IV Materials, 7
Graphical Analysis of CE Amplifier, 246

H

Hole Mechanism, 11
h-Parameter Model for Low Frequency, 251
Hartley and Colpitts Oscillator, 294

I

Important Parameters Regarding to Semiconductor Materials, 8
Insulators, 15
Intrinsic Semiconductors, 17
Ideal op-amp, 336
Inverting Amplifier, 339

J

JFET Parameters, 406

L

Limitation of RC Oscillators, 296
Loop Gain, 268

M

Majority and Minority Charge Carriers, 20
Meaning of Electronics, 1
Metal Oxide Semiconductor Field Effect Transistor (MOSFET), 433
Methods of Transistor Biasing, 203

N

N-type Semiconductors, 18
Negative Current Feedback, 269
Negative Feedback, 268
Negative Resistance Oscillator, 281
Negative Voltage Feedback, 269
Need of Feedback, 338
Non-Inverting Amplifier, 343

O

Operating Point of the Active Device, 299
Output Load, 300
Op-amp Applications, 338, 345

P

P-N Junction, 37
P-type Semiconductors, 19
Parallel Resonant Frequency, 299
Phase Shift Oscillator, 286
Piezoelectric Crystals, 296
Positive Feedback, 268
Principle of Feedback in Amplifiers, 275
Pinch Off Voltage, 403

R

Resistivity (ρ), 8
Rectification or Power Conversion Efficiency, 85
Rectifiers, 79
Relation between α and β, 189
Reason behind the Amplification Action of a Transistor, 193
Regions of a Transistor, 195
RC Oscillators, 286, 280
RC Phase Shift Oscillator, 286

S

Semiconductor, 15
Simplified Representation of Energy Level Diagram, 14
Static and Dynamic Resistance of a PN-Junction Diode, 47
Series Inductor Filter, 100
Shunt Capacitor Filter, 100
Stability Factor, 196
Sampling Network, 277
Series Resonant Frequency, 299
Signal Source, 277
Single Loop Feedback Amplifier, 277
Sinusoidal Oscillators, 280
Supply Voltage, 299
Small Signal Model for JFET or FET, 405
Structure and Operation of JFET or FET, 395

T

Temperature Dependency of Mobility, 28
The Mass-Action Law, 25
Transformer Utilization Factor (TUF), 86, 94
Transistor, 181
Transistor Biasing, 201
Transconductance Amplifier, 276
Transfer Ratio or Gain, 278
Transistor Amplifier, 283
Transistor Crystal Oscillator, 298
Transistor Oscillator, 284
Transresistance Amplifier, 276, 277
Tuned Circuit Oscillator, 280
Types of Feedback, 268
Types of Transistor Configuration, 185

U

Undamped Oscillations, 281, 282

V

V-I Characteristics of a PN-Junction Diode, 43
Various Solid State Devices and Applications, 3
Voltage Regulation, 85
Voltage Amplifier, 275, 276
Voltage Series Feedback, 270
Voltage Shunt Feedback, 272

Z

Zener Diode, 150
Zener Resistance, 151

□□□